HANDBOOK OF PROBABILITY
AND STATISTICS
WITH TABLES

OTHER McGRAW-HILL HANDBOOKS OF INTEREST

AMERICAN INSTITUTE OF PHYSICS · American Institute of Physics Handbook
AMERICAN SOCIETY OF MECHANICAL ENGINEERS · ASME Handbooks:

Engineering Tables	Metals Engineering—Processes
Metals Engineering—Design	Metals Properties

ARCHITECTURAL RECORD · Time-Saver Standards
BAUMEISTER AND MARKS · Standard Handbook for Mechanical Engineers
BERRY, BOLLAY, AND BEERS · Handbook of Meteorology
BLATZ · Radiation Hygiene Handbook
BRADY · Materials Handbook
BURINGTON · Handbook of Mathematical Tables and Formulas
CHOW · Handbook of Applied Hydrology
CONDON AND ODISHAW · Handbook of Physics
CONSIDINE · Process Instruments and Controls Handbook
CONSIDINE AND ROSS · Handbook of Applied Instrumentation
ETHERINGTON · Nuclear Engineering Handbook
FINK AND CARROLL · Standard Handbook for Electrical Engineers
FLÜGGE · Handbook of Engineering Mechanics
GRANT · Hackh's Chemical Dictionary
HAMSHER · Communication System Engineering Handbook
HARRIS AND CREDE · Shock and Vibration Handbook
HENNEY · Radio Engineering Handbook
HUNTER · Handbook of Semiconductor Electronics
HUSKEY AND KORN · Computer Handbook
IRESON · Reliability Handbook
JURAN · Quality Control Handbook
KAELBLE · Handbook of X-rays
KALLEN · Handbook of Instrumentation and Controls
KING AND BRATER · Handbook of Hydraulics
KLERER AND KORN · Digital Computer User's Handbook
KOELLE · Handbook of Astronautical Engineering
KORN AND KORN · Mathematical Handbook for Scientists and Engineers
LANDEE, DAVIS, AND ALBRECHT · Electronic Designer's Handbook
LANGE · Handbook of Chemistry
MACHOL · System Engineering Handbook
MANTELL · Engineering Materials Handbook
MARKUS · Electronics and Nucleonics Dictionary
MEITES · Handbook of Analytical Chemistry
PERRY · Engineering Manual
PERRY, CHILTON, AND KIRKPATRICK · Chemical Engineers' Handbook
RICHEY · Agricultural Engineers' Handbook
ROTHBART · Mechanical Design and Systems Handbook
STREETER · Handbook of Fluid Dynamics
TERMAN · Radio Engineers' Handbook
TOULOUKIAN · Retrieval Guide to Thermophysical Properties Research Literature
TRUXAL · Control Engineers' Handbook
URQUHART · Civil Engineering Handbook
WOLMAN · Handbook of Clinical Psychology

HANDBOOK OF PROBABILITY AND STATISTICS WITH TABLES

Second Edition

Richard Stevens Burington, Ph.D.

Donald Curtis May, Jr., Ph.D.

Revised by Richard Stevens Burington

McGraw-Hill Book Company

New York   St. Louis   San Francisco   London
Sydney   Toronto   Mexico   Panama

PREFACE

The rapidly growing use of the methods of probability theory and statistics, the excellent reception of the First Edition of this book, and the valuable suggestions from its users for additions, improvements, extensions and modifications have resulted in the publisher's requesting a revision and enlargement of the First Edition.

As in the First Edition, this book has been designed to provide in a small convenient size a Handbook of probability and statistics, sufficiently comprehensive to fill a broad variety of needs, and yet simple enough in its structure to facilitate its use by people having wide differences in training, background, interest and experience. The book is intended to provide a convenient summary of theory, working rules, and tabular material useful in the study and solution of practical problems involving probability and statistics. The book has been constructed and arranged to meet the needs of students and workers in statistics, probability, mathematics, operations analysis, systems analysis, communications, engineering, physics, chemistry, biology, agriculture, science, education, economics, business, and other fields in which computations and analyses of a probabilistic or statistical nature are required. Considerable introductory material has been included so readers without detailed statistical training will find the volume a sufficient guide for the more commonly met statistical aspects of their studies. Those with a statistical background should find that the book offers a convenient summary of the material most often needed.

The content of and the manner in which mathematics, statistics, and related fields are taught and used have been undergoing many changes during the last few years. The languages of logic, set theory, and various mathematical structures have found their way into the elementary and secondary schools, into colleges, institutes, universities, and into many laboratories and industries where mathematical and statistical sciences are now widely used. The rapid and continuing growth of the use of various data processing machines, high speed digital and analog computers has had a marked impact on the teaching and practice of the statistical, physical, engineering, business, management, and social sciences. Statistical and probability analyses and computations, which only a few years ago would not have been undertaken because of the magnitude of the computations involved, now, with the availability of high-capacity computing devices, are well within the practical reach of many users. All this has resulted in the requirement that many more workers have a greater knowledge of probability and statistics, and in particular, have readily available the information they need. The existence of computers has not reduced the need for analyses of a statistical or probabilistic character; in fact it has increased the possibility and need for various types of analyses and computations, a large number of which are covered in this book.

The general arrangement of the First Edition has been retained. The first part of the book includes a comprehensive summary of the more important formulas, definitions, theorems, tests, and methods of elementary statistics and probability theory. The second part of the book consists of a carefully selected set of tables of distributions and other quantities of frequent use in the applications of statistics. New sections on order statistics, nonparametric methods, design of experiments, the usefulness and limitations of statistics, the analysis of variance, regression theory, orthogonal polynomials, acceptance sampling, and reliability theory have been added. Included are many numerical examples illustrating the principles outlined. Additional tables on tolerance intervals and random numbers have been included. Topics and tables which would be included in an extensive treatise on the subject have had to be kept short or omitted altogether. In order that the reader can readily find such additional topics and tables, an extensive bibliography of easily found reference material has been included. Throughout the text pertinent references are made to sources which include material too detailed or bulky for inclusion in this book.

This edition is designed to complement the *Handbook of Mathematical Tables and Formulas*, Fourth Edition, by Richard Stevens Burington, McGraw-Hill Book Company.

The entire book has been gone over, and many changes and additions made. Considerable effort has been taken to insure accuracy. Extensive efforts have been made to arrange the text and tables in such a manner that the user will interpret them readily, properly and accurately.

The authors are indebted to Professor Sir Ronald A. Fisher, Cambridge, and to Oliver & Boyd, Ltd., Edinburgh, for permission to reprint Tables XI, XII, XIV from their book *Statistical Methods for*

Research Workers; to Professor N. Arley of the Universitetets Institut for Teoretisk Fysik, Copenhagen, for Table 13.76.1; to the British Standards Institution for Table 13.58.1; to Professor G. W. Snedecor of the Iowa State College, and The Iowa State College Press, for Tables X, 14.57.1 and 14.57.2; to Professor E. S. Pearson and *Biometrika* for Table 13.86.1; and the American Society for Testing and Materials for Table 17.43.2.

The undersigned wishes to acknowledge the valuable suggestions made by the users of the book, and the many critical comments of Dr. D. C. May, who because of other commitments was unable to participate in this revision.

I am especially indebted to my wife, Jennet Mae Burington, for her continued encouragement, and her very able and considerable assistance in the preparation of the manuscript and in the reading of the proof.

Richard Stevens Burington

GREEK ALPHABET

A	α	alpha	N	ν	nu
B	β	beta	Ξ	ξ	xi
Γ	γ	gamma	O	o	omicron
Δ	δ	delta	Π	π	pi
E	ϵ	epsilon	P	ρ	rho
Z	ζ	zeta	Σ	σ	sigma
H	η	eta	T	τ	tau
Θ	θ	theta	Υ	υ	upsilon
I	ι	iota	Φ	ϕ	phi
K	κ	kappa	X	χ	chi
Λ	λ	lambda	Ψ	ψ	psi
M	μ	mu	Ω	ω	omega

CONTENTS

xi

The field of statistics

1.1. Introduction. Modern mathematical and statistical methods are playing a most important and growing role in many fields of endeavor, such as: engineering; agriculture; business, insurance, economics; the physical sciences such as chemistry, physics, astronomy; biological sciences, medical research; sociology and psychology; transportation; industrial studies, etc. In nearly all fields there arise problems where precise measurements or observations are impossible to make, or where events are not exactly reproducible or predictable. The analysis of such situations leads to considerable uncertainty. Statistical methods lend themselves to the analysis of such situations and furnish a means of describing and indicating trends, or expectations, often with an associated degree of reliability. Statistics makes use of scientific methods of collecting, analyzing, and interpreting data.

1.2. Modern *statistical method* is a science which deals with such problems as: (1) How to plan a program for obtaining data so that reliable conclusions can be made from the data so obtained. (2) How to analyze the data obtained. (3) What valid conclusions can be drawn properly from the data? (4) To what extent are the conclusions reliable?

To outline all known statistical methods in use today would lead deeply into theory, and into the subject matter of many fields, and would require several large volumes. However, the principal elementary fundamental concepts and methods can be outlined in a reasonably small space. It is the purpose of this Handbook to furnish such an outline, in elementary and compact form, and in a form that can be used by workers in many fields.

1.3. Quantitative statistical observations. A sequence or set of measurements, or observations, made on a set of objects in a specified set, or population, of objects is known as *quantitative statistical observations*. When the observations are made on only some of the objects of the population the set of observations is called a *sample*. Populations, or samples, may be *finite*, or *indefinitely large*.

To condense and describe samples of quantitative data, *frequency*

1

distributions are used, and certain parameters, called *statistics*, are calculated from the samples used to describe the frequency distributions. Likewise, populations and their distributions are described by *population parameters*.

As a rule, one only has a sample from a population and consequently does not have data for the entire population. A common procedure is to use statistics calculated from the sample frequency distribution to estimate likely values of the population parameters. For very large samples the statistics of properly chosen samples will have values quite close to those of the corresponding population parameters. For small samples the discrepancies are apt to be larger, and the business of predicting values of the population parameters from the sample statistics becomes involved and has to be resolved by the use of probability theory.

1.4. Qualitative statistical observations. A sequence or set of observations in which each observation in the set (and population) belongs to one of several *mutually exclusive* classes (perhaps non-numerical) is called *qualitative statistical observations*. (For example, an inspector at the ramp of an airplane equipped with a miniature television set at each seat, asking each disembarking passenger whether or not he liked the miniature set, might get a sequence of 50 answers, such as: yes, no, yes, yes, no comment, yes, no, · · · . The answers are qualitative.)

The problems of how to gather, analyze, and draw conclusions from qualitative observations are similar to those for quantitative observations.

In some cases the order of the observations is of principal interest; and time is often a leading variable. This is true both in qualitative and quantitative observations.

1.5. Probability theory and statistics. The theory of statistics is concerned with the mathematical description and analysis of observations that subsequently form the basis for prediction of the occurrence of events under given conditions. In certain respects the theory of statistics rests on the theory of probability, the latter often furnishing much of the basic or underlying structure for the former.

1.6. On the applications of statistics. The most useful and successful applications of statistics require knowledge of both statistical methods and the subject matter to which the methods are applied. Unless the methods used are well anchored in the field of application the results of the statistical analyses may prove to be misleading. In the applications of statistics and the theory of probability it is often wise to draw upon what might be termed the "dynamical approach" (e.g., the appropriate engineering, economical, biological, or physical approach) appropriate to the fields under study. This is particularly true if small samples are involved, and if to gain a sizable sample may require undue expense and effort.

The usefulness and limitations of statistics

1.7. **Usefulness.** Statistics is a powerful tool for analyzing data of many types. As with any analytical discipline, there are also definite limitations to the proper application of statistics that should be well understood.

Statistical theory has wide applications in many fields of endeavor, and its correct use can prove to be profitable. Statistics can be of assistance in management, industry, production, research, test, development and evaluations, and in fields such as agriculture, transportation, communications, public health, · · · . The proper applications of the methods of statistics have proved useful to the producer as well as the consumer in many diverse activities.

Statistics have helped to sharpen traditional tools in many fields, such as inspection, surveillance, quality control, market research, operational research, systems analysis, · · · . In the fields of research, test and development statistical methods have helped improve criteria for the assessment of data, and have helped make possible objective planning and implementation of experiments involving a large number and a considerable diversity of factors under test, and have helped to bring about successful results in these fields.

Wherever one's work calls for interpretation of data, one may possibly succeed without using statistical methods, but one usually can derive more information from the data by the proper use of statistics.

With the increase in the use of the scientific method there has been a considerable increase in the use of measurements of all kinds. Since much investigation is made without a good knowledge of the underlying structure of the field, and the measurements made and used are samples, the proper use of statistical techniques and the importance of well-designed experiments become more important than ever.

1.8. **Limitations.** There are limitations in the use of statistics that should be appreciated by the users. The computations used in statistics tend to summarize differences between items or individuals by the use of such parameters as average, standard deviation, etc. This loss of individual treatment of each piece of data occasionally precipitates an undesirable reaction in the eyes of some, although summarizing a large amount of information is one of the most useful results of statistical analysis.

While the arithmetic of statistics may be relatively easy, the interpretation of the results as they relate to some specific problem may not be simple. For example, it is not unusual for the enthusiast to examine data and come up with estimates of statistical parameters that appear to be far more accurate than the knowledge of the problem warrants.

Statistical analysis, properly and thoroughly done, requires a careful examination of the uncertainties and suppositions involved. The product obtained should be the result of objective, honest and careful inquiry, an inquiry that is fully conscious of the premises and limitations of the statistical and other methods used in reaching the results.

1.9. The interpretations of statistics. Statistics can be very useful when properly collected and interpreted. Skill, care, and objectivity in handling statistical data are necessary; without these, the data may have the appearance of supporting things which are not true.

It is often a simple matter to be well armed with numerical data and to become enamored with figures, though the data as presented may prove to be misleading (deliberately or otherwise). The ability to make proper legitimate use of data once in hand is sometimes scarce, though the zeal for collecting the data may be more than adequate.

There is an art to the proper handling of figures, an art that must be developed through the proper study of the mathematical and statistical principles involved. While much data may be collected relating to a specific field, it is a common result to find that little mature analysis of the data has been made, and some of the data have been interpreted incorrectly. One must learn when it is safe or legitimate to say definitely one thing or another, when to suspend judgment, etc. This is not to say that crude approximations are not often useful in arriving at well-founded answers.

Some of the common misuses of statistics are those traceable to:

(a) uncertain, ambiguous, or different *definitions* of terms used in statistical studies, or the absence of essential definitions, or the misinterpretation of definitions,

(b) inaccurate, uncertain, or biased *classifications* of cases involved,

(c) uncertain, inaccurate, or biased *measurements* of the pertinent quantities,

(d) ambiguous, uncertain, misleading, or biased *methods of selecting the cases for study*,

(e) misleading or inappropriate *comparisons* or associations of groups, classifications, or pertinent measured parameters,

(f) inadequate *planning and collection* of the data,

(g) the *changing make-up* of the *classifications* involved,

(h) false or *misleading association* or correlation between groups, classifications, cases, etc.

(i) neglect of the implications of the *dispersion* exhibited by the data,

(j) mistakes due to the improper use of *statistical techniques*,

(k) errors due to inappropriately *omitted data*, and associated improper use of the techniques of statistics,

(l) errors due to *misleading information* or *statements*,

(m) errors due to improper, poorly designed, or misleading *graphical pictures*,

(n) *insufficient understanding of the fields* in which the statistical analyses are being conducted.

Patience, care, good sense, skill and objectivity are essential if the pitfalls of statistics are to be avoided and the full usefulness of statistical analysis realized.

1.10. **Presentation of data.** There are many ways of presenting data. Graphical representations are particularly helpful since visual means for presenting data are easy to understand and remember. Graphs should be constructed in the particular manner needed to meet the needs of individuals who are to use them. Many forms of graphs appear in this book (e.g., as in §3.1 to §3.3, §10.5).

Graphs summarizing statistical data are generally effective in giving an overall view, but they may not include sufficient information to provide more detailed and accurate comparisons, or the means for deriving precise inferences from the available data. More powerful tools, some statistical in character, are needed for reaching reasonable inferences and related quantitative decisions. It is possible to design graphical representations of data which (even when technically correct) may mislead the not so critical or experienced viewer. The reader and user, as well as the designer of the graphical displays, should be alert to avoid such cases.

1.11. For those who make use of the fields of probability and statistics it is hoped that this short discussion of the usefulness and limitations of statistics will help to make clear the importance of knowing and understanding the conditions under which one may use the various methods and techniques outlined in this book.

Use of digital computers in statistical computations

1.12. Statistical computations make use of many of the numerical techniques and methods employed generally in the physical and engineering sciences. Many of the calculations of statistics involve the manipulations of sums of squares, the numerical evaluations of integrals, and the solution of systems of linear equations and other algebraic operations (e.g., regression and the analysis of variance often require the inversion of matrices). Such numerical manipulations may be extensive. In many cases digital computers can be of great help. For this reason many of the formulas presented herein have been written in a form suitable for use as part of a digital computer program.

These and related matters involving the theory of computational methods, the use of computers in statistical calculations and the related subject of computer programming have been kept in mind in preparing this work. However, it is beyond the scope of this report to treat these subjects here.

Organization of the book

1.13. **Descriptive statistics.**   In the study of samples it is desirable to define and calculate certain representative quantities which summarize pertinent information. Such values are called *descriptive statistics*. The principal descriptive statistics are those which serve as measures of (a) location, (b) spread (scatter, variation, or dispersion) of the individual values, (c) skewness, and (d) kurtosis. These statistics are defined in Chapter II.

1.14. **Frequency distributions.**   A common way for representing the results of a series of measurements of a quantity X is by means of frequency distributions. Such distributions may be given by tables or by graphs (e.g., histograms, cumulative frequency polygons). Chapter III is concerned with such distributions.

1.15. **Certain mathematical concepts.**   The theory of statistics is closely associated with the theory of probability, that is, the systematic description of chance events. Measurements of the uncertainties of statistics make use of this theory. Basic to the theory are certain concepts of the *theory of sets*.

In the study of outcomes (events, states) of experiments the *algebra of events* is of considerable importance.

Some of the earliest approaches to probability made use of the number of different arrangements and subgroups that can be formed from a given set. The mathematical approach to such matters comes under the subject matter of partitions, selections, permutations, and combinations. Brief sketches of these concepts are given in Chapter IV.

1.16. Probability theory. Statistics enters into real problems through experiment or observation. The actual result of each trial in a particular situation is a uniquely determined outcome. However, each result can be thought of as a member of the set of all possible results. A chance set-up may be described in part by stating the way in which the chances are distributed among all the possible outcomes (e.g., by histograms). The key to the study of chance is the notion of a distribution of outcomes.

Many of the statements (hypotheses) stemming from scientific investigations are based on inferences. Such statements commonly involve uncertainties of various kinds. Hence such statements must be tested for their validity. The theorems of statistics furnish logical methods for the measurement of these uncertainties.

1.17. Axioms of probability. The laws of probability may be introduced in various ways. Most formulations of these laws include certain features which are almost universally accepted and which have been motivated by experience with large samples of outcomes in numerous chance set-ups.

The chance (probability) of an event (outcome, result) is a measure of the likelihood of occurrence of the event. The laws and axioms for handling such probabilities are derived from idealizations of the properties of frequency ratios. From these axioms and definitions the rules of the calculus of probabilities are developed. These in turn permit the calculation of probabilities of combinations of events from specified probabilities for the individual events.

1.18. Random variables. In certain types of experiments it is not possible to predict what may happen in each individual trial. However, experience has shown that the data reveal certain statistical regularities such as the long run stability of frequency ratios, and averages. In such situations idealized mathematical probabilities are constructed to serve as a conceptual counterpart of the frequency ratios. This leads to the notion of a *chance* (or *random*) *variable* and its probability distribution. The theory is constructed to preserve the additive properties of the axioms treated in Chapter V.

In the theory of probability distributions *probability functions* are defined in much the same way as are the frequency functions of Chapter III.

1.19. Mathematical expectation. The concept of *mathematical ex-*

pectation plays an important role in the description of probability distributions. The expected value $E[\theta]$ of any function θ of a chance variable x is the average, weighted by the probability density of x, of the function θ over all possible values of x. (For example, the expected value of x is the weighted average of x.)

The underlying bases of these theories are discussed in Chapter V.

1.20. Probability distributions.   It is common to be concerned with the probabilities of the occurrence of a set of events, the events being described by numbers. In case an event is described by a single number, the theory is that of a *probability distribution in one dimension*. Chapter VI is concerned with the description of such distributions, both the discrete and continuous types.

In case an event requires two (or more) numbers for its description the probability distribution of concern is two-dimensional (or more). Such distributions are described in Chapter XI.

1.21. Generating and characteristic functions.   There are problems in which it is easier to describe a probability distribution in terms of generating or characteristic functions than to compute the probability distribution functions directly. From such functions, if known, the mean value, variance and moments can be computed by differentiation. Examples of such functions are treated in Chapter VII.

1.22. Binomial, Poisson and normal distributions.   There are certain probability distributions which have wide application. The most common of these are the *binomial, Poisson* and *normal* distributions. Related to these distributions are the *negative binomial*, the *exponential* and the *log normal* distributions. Chapters VIII, IX and X consider and illustrate by examples the characteristics of these distributions in considerable detail.

1.23. Regression theory.   Many problems are encountered in which information concerning two or more related variables is known. In such cases a way of mathematically expressing the form of the relationship between the variables may be sought. In addition one may wish to know how well the value of one variable can be predicted if one knows the values of the associated variables. *Regression methods* and *correlation methods* furnish means for accomplishing these objectives. One method for fitting data with polynomials is that involving the use of *orthogonal polynomials*. Chapter XII is concerned with such methods.

1.24. Sampling distributions.   In the study of sample statistics there are three distributions of great importance which together with the normal distribution are extensively used. They are the χ^2 distribution, the Student's t-distribution, and the F-distribution.

These and related distributions are considered in Chapter XIII. Applications of these distributions to the drawing of inferences about populations from samples are treated in Chapter XIV.

1.25. Random numbers. There are many situations where the techniques of random sampling are used. In some cases it is possible to use advantageously tables of *random sampling numbers*. Such a table consists of a sequence of digits designed to represent the result of a simple random sampling from a population consisting of the digits $0, 1, 2, \cdots, 9$. Table XXIII is such a table. (See §13.101.)

1.26. Estimation of population parameters. In statistical problems one is often concerned with a large lot or *population* of objects whose characteristics are only partially known. If a certain characteristic of each object were measured a distribution of these measurements might be made. But the measurements may not have been made. Perhaps the best that can be done is to estimate the distribution from data obtained from measurements made on a random sample of objects taken from the population. From the sample data it may then be possible to estimate the characteristics (i.e., the *population parameters*) of the distribution. For example, one may use the sample mean as an estimate of the population mean.

There are two important types of estimators for population parameters, *point estimators* and *interval estimators* (e.g., *confidence limits*). Such estimators are considered in Chapters XIII and XIV, along with the related topics of sampling distributions and the use of such distributions as a basis for drawing inferences about populations.

1.27. Statistical inference. A statistical hypothesis is a statement about a statistical population (e.g., a statement about the values of one or more parameters of the population). Often it is desirable to test the validity of such hypotheses. A significance test of a hypothesis consists in determining whether or not an observed result could reasonably have been expected if the hypothesis about the population were assumed to be true. These and related matters are treated in Chapter XIV.

1.28. Non-parametric tests. In recent times much effort has been placed on the development of statistical tests which are valid for situations in which little is known about the shape of the population distribution from which the samples are drawn. One class of such tests, valid for samples from continuous population distributions of any shape, is called *non-parametric tests* (or *distribution-free tests*). They are based on order statistics (considered in Chapter XIII). A few of the more common non-parametric tests are given in Chapter XIV.

1.29. Design of experiments. The analysis of any set of data is to a considerable extent dictated by the manner in which the data were obtained.

The *design of an experiment* is the sequence of steps taken ahead of time to insure that appropriate data will be obtained in a manner which permits an objective analysis from which valid inferences with respect to the problem under investigation may be made. The elements of this subject are considered in Chapter XV.

1.30. Analysis of variance. The term *analysis of variance* is an expression used to cover a wide range of statistical techniques. It is a flexible method of constructing statistical models for experimental material. The subject is a vast one. For example, in the analysis and design of experiments it is often useful to resolve the variance of a variable relative to its sample mean into separate components of variance, each corresponding to a specific source (real or otherwise) of variation. This process of resolution of a variance into components is an example of the analysis of variance. Chapter XV treats this subject.

1.31. Interpolation. In statistics much use is made of tables of mathematical functions. Tables of a function $y(x)$ can include values of $y(x)$ for only a finite number of values of the argument x. To find a value y_1 of y for a non-tabulated value of x_1, either y_1 must be calculated for x_1 or else a value must be interpolated from the tabulated values of y. The simplest method is that of *linear interpolation*. (See §16.9.) When more precision is desired more elaborate methods must be employed. Chapter XVI gives a brief outline of some of these methods. For example: to interpolate values slightly larger than a tabulated value y_0, formula (16.9.1) might be used; to interpolate near the middle of a tabulated set, *central difference formulas* such as listed in §16.11 to §16.15 could be used; to interpolate values slightly smaller than y_n, formula (16.10.1) might be used.

1.32. Sampling inspection. The use of statistical methods in industry has had a tremendous upswing during the last twenty years. This use of statistics has been very marked in *research and development* and in *reliability and quality control*. Two techniques especially useful in controlling and improving the reliability and quality of products are *acceptance sampling plans* and *control charts*. A brief summary of some of these methods is given in Chapter XVII.

1.33. Reliability theory and practice. The increasing need for higher reliability and safety in modern complex systems (e.g., aircraft, communication systems, power distribution systems) has resulted in a considerable expansion of the body of knowledge in reliability control. To meet this need a much better understanding of the characteristics of such systems is required and of the mathematical aspects of the systems. Work bearing on the reliability of such systems is very important because of the extremely serious results that can and do stem from breakdowns of such systems. The mathematical aspects of reliability theory and practice involve the use of the methods of logic, probability theory and statistics, coupled with the definition of certain parameters designed to measure the reliability, availability, and maintainability, etc., of the systems studied. Chapter XVII includes a section concerned with some of the quantities encountered in reliability theory and practice.

1.34. Tables. In applying the methods of statistics many tables of distributions and other related quantities are needed. The second part of the book consists of such tables. In order to keep the book at pocket size much care has been taken to include the most important statistical tables commonly needed and yet keep their size and extent consistent with the aim and scope of the book. For more extensive tables the reader is referred to the various references given herein. A short table of integrals is given in Chapter XVIII.

1.35. Bibliography and index of symbols. At the close of the book a bibliography of references is given which bear on the general topics considered. Included is a list of some of the more extensive numerical tables to be found in the literature. A short index of symbols encountered in the book is also given.

In various sections of the book references are made to specific books and articles which the reader may wish to pursue.

2.1. The term *variate* is often used in statistics and probability theory to denote a *variable*. Variates may be discrete or continuous.

Measures of location

2.2. Let $x_1, x_2, \cdots, x_k$ be a set of numbers or variates. An *arithmetic mean*, or *average*, $\bar{x}$ of this set is

$$(2.2.1) \qquad \bar{x} = (1/k) \sum_{i=1}^{k} x_i = (x_1 + x_2 + \cdots + x_k) \div k.$$

If the values of $x_1, \cdots, x_k$ occur with corresponding frequencies $f_1, f_2, \cdots, f_k$, respectively, the *weighted arithmetic mean*, or *weighted average*, $\bar{x}$ is

$$(2.2.2) \qquad \bar{x} = (1/n) \sum_{i=1}^{k} f_i x_i,$$

where $n = \sum_{i=1}^{k} f_i$. Each $f_i \geqq 0$. The values $f_1, \cdots, f_k$ are called *weighting factors*; n is the *total weight*.

The quantity $x_i - \bar{x}$ is called the *deviation*, or *error*, of x_i with respect to $\bar{x}$; $|x_i - \bar{x}|$ is the *absolute deviation*, or *absolute error* of x_i.

The *least*, L, and *greatest*, G, of the variates $x_1, \cdots, x_k$ define the *range* $(G - L)$ of the variate.

2.3. The *root-mean-square* (R.M.S.) value of $x_1, \cdots, x_k$ is

$$(2.3.1) \qquad \left[\left(\sum_{i=1}^{k} f_i x_i^2 \right) \div n \right]^{1/2}.$$

The *geometric mean* (G.M.) of n positive numbers $x_1, \cdots, x_n$ is the nth root of their product, that is,

$$(2.3.2) \qquad (x_1 \, x_2 \cdots x_n)^{1/n}.$$

The *harmonic mean* (H.M.) of n positive numbers $x_1, \cdots, x_n$ is the reciprocal of the arithmetic mean of the reciprocals of the values, that is,

$$(2.3.3) \qquad n \div \sum_{i=1}^{n} (1/x_i).$$

If the positive number x_i appears f_i times, $i = 1, \cdots, k$, and $n = \sum_{i=1}^{k} f_i$, then the *geometric mean* is

$$(2.3.4) \qquad [x_1^{f_1} x_2^{f_2} \cdots x_k^{f_k}]^{1/n}.$$

2.4. Mode (Mo). A value of the variable x which occurs most frequently is called a *mode*. A mode (Mo) for $x_1, x_2, \cdots, x_n$ is a value of x_i having a maximum frequency f_i. If $x_1, \cdots, x_n$ all have the same frequencies, a mode is a value about which $x_1, \cdots, x_n$ "cluster most densely". A frequency distribution may have one or more than one mode. In certain cases it is difficult to give a completely satisfactory definition of mode.

2.5. Median. The *median* is defined as a value which is greater than half the variates and less than the other half, the median value being selected as follows. Let $x_1, x_2, \cdots, x_n$ be a set of real numbers arranged in order of magnitude algebraically,

$$x_1 \leqq x_2 \leqq \cdots \leqq x_n.$$

When n is odd, the median is x_h, where $n = 2h - 1$. When n is even, the median is not uniquely defined unless $x_h = x_{h+1}$, when $n = 2h$, in which case the median is this common value. In case the median is not uniquely defined, it is conventional to take for the median $(x_h + x_{h+1}) \div 2$, when $n = 2h$.

Moments

2.6. Moments about the origin. The rth *moment about the origin* $x = 0$ of the numbers $x_1, \cdots, x_k$, having a weighted mean $\bar{x}$, where f_i is the frequency of x_i and $n = \sum_{i=1}^{k} f_i$, is

$$(2.6.1) \qquad \nu_r = (1/n) \sum_{i=1}^{k} f_i x_i^{\,r}. \qquad (r = 1, 2, \cdots.)$$

To indicate that ν_r is calculated for the variable x, the notation $\nu_{r:x}$ is often used.

2.7. Moments about the mean $\bar{x}$. The rth *(central) moment* μ_r about *the weighted mean* $\bar{x}$ of the numbers $x_1, x_2, \cdots, x_k$ is

$$(2.7.1) \qquad \mu_r = (1/n) \sum_{i=1}^{k} f_i (x_i - \bar{x})^r.$$

2.8. Relations between μ_r and ν_r.

$$(2.8.1) \qquad \mu_0 = 1, \qquad \mu_1 = 0,$$

$$\mu_2 = \nu_2 - \nu_1{}^2 = \sigma^2. \qquad (\mu_2 \text{ is called the } variance)$$

$$\mu_3 = \nu_3 - 3\nu_2\nu_1 + 2\nu_1{}^3.$$

$$\mu_4 = \nu_4 - 4\nu_3\nu_1 + 6\nu_2\nu_1{}^2 - 3\nu_1{}^4.$$

$$\vdots \qquad\qquad \vdots$$

$$\mu_k = \sum_{r=0}^{k} C_r{}^k \nu_{k-r}(-\nu_1)^r, \quad \text{where } C_r{}^k = k!/(k-r)!\, r!.$$

$$(2.8.2) \qquad \nu_0 = 1, \qquad \nu_1 = \bar{x}.$$

$$\nu_2 = \mu_2 + \nu_1{}^2.$$

$$\nu_3 = \mu_3 + 3\mu_2\nu_1{}^1 + \nu_1{}^3.$$

$$\nu_4 = \mu_4 + 4\mu_3\nu_1{}^1 + 6\mu_2\nu_1{}^2 + \nu_1{}^4.$$

$$\vdots \qquad\qquad \vdots$$

$$\nu_k = \sum_{r=0}^{k} C_r{}^k \mu_{k-r}\nu_1{}^r.$$

2.9. Moments about arbitrary point.

The rth *moment about an arbitrary value* x_0 of x is

$$(2.9.1) \qquad \mu_r(x_0) = (1/n) \sum_{i=1}^{k} f_i(x_i - x_0)^r.$$

$\mu_2(x_0)$ is called the *mean square deviation* from $x = x_0$, and $\sqrt{\mu_2(x_0)}$ is called the *root mean square deviation* from x_0. The *order* of the moment μ_r is r.

The function $\mu_2(x_0)$, treated as a function of x_0, has a minimum value σ^2 when $x_0 = \bar{x}$.

2.10. Absolute moments about x_0.

The rth *absolute moment about the arbitrary value* x_0 is

$$(2.10.1) \qquad (1/n) \sum_{i=1}^{k} f_i \mid x_i - x_0 \mid^r.$$

This function is commonly used with $x_0 = \bar{x}$ or $x_0 = 0$.

2.11. Variance. The *variance* of $x_1, \cdots, x_k$ is defined as

$$(2.11.1) \qquad \sigma^2 = \mu_2(\bar{x}) = \mu_2 = (1/n) \sum_{i=1}^{k} f_i(x_i - \bar{x})^2.$$

σ is the *standard deviation*. Also,

$$(2.11.2) \qquad \sigma^2 = (1/n) \sum_{i=1}^{k} f_i x_i^2 - \bar{x}^2 = \nu_2 - \bar{x}^2.$$

The relation

$$(2.11.3) \qquad \sigma^2 = \mu_2(x_0) - (\bar{x} - x_0)^2$$

is often useful since it enables one to calculate σ^2 from the second moment about any convenient value x_0 of x.

2.12. Mean deviation, M.D.$|_0$, from an arbitrary point x_0 is:

$$(1/n) \sum_{i=1}^{k} f_i \mid x_i - x_0 \mid.$$

The value of x_0 for which M.D.$|_0$ is a minimum is the median.

2.13. Mean (absolute) deviation, M.D., from the mean is:

$$(1/n) \sum_{i=1}^{k} f_i \mid x_i - \bar{x} \mid.$$

This is also called the *mean absolute error (m.a.e.)*.

2.14. Factorial moments. The rth *factorial moment about the origin* $x = 0$ of the numbers $x_1, \cdots, x_k$, is

$$(2.14.1) \qquad \nu_{[r]} = (1/n) \sum_{i=1}^{k} f_i x_i^{[r]}, \qquad (r = 1, 2, \cdots),$$

where

$$(2.14.2) \qquad x_i^{[r]} = x_i(x_i - 1) \cdots (x_i - r + 1). \qquad (r = 1, 2, 3, \cdots .)$$

The rth *central factorial moment about the weighted mean* $\bar{x}$ is

$$(2.14.3) \qquad \mu_{[r]} = (1/n) \sum_{i=1}^{k} f_i(x_i - \bar{x})^{[r]},$$

where

$$(2.14.4) \qquad (x_i - \bar{x})^{[r]} = (x_i - \bar{x})(x_i - \bar{x} - 1) \cdots (x_i - \bar{x} - r + 1).$$
$$(r = 1, 2, \cdots .)$$

The relation

$$(2.14.5) \qquad \sigma^2 = (1/n) \sum_{i=1}^{k} f_i x_i (x_i - 1) - \bar{x}(\bar{x} - 1) = \nu_{[2]} - \bar{x}(\bar{x} - 1)$$

is often more convenient to use than (2.11.2).

Factorial moments are particularly useful in those cases where the variable x takes on discrete values spaced at equal intervals (e.g., when the x_i are positive integers, and when $r \leq x_i$ in (2.14.1), and $r \leq x_i - \bar{x}$ in case (2.14.4)). Such moments play a role in the theory of interpolation and curve fitting. Some writers define $x^{[0]} = 1$.

Relations between $\nu_{[r]}$ and moments ν_r about $x = 0$.

$$(2.14.6) \qquad \begin{aligned} &\nu_1 = \nu_{[1]}, & &\nu_2 = \nu_{[2]} + \nu_{[1]}, \\ &\nu_3 = \nu_{[3]} + 3\nu_{[2]} + \nu_{[1]}, & &\nu_4 = \nu_{[4]} + 6\nu_{[3]} + 7\nu_{[2]} + \nu_{[1]}. \end{aligned}$$

2.15. Change of units. To indicate that μ_r is calculated for the variable x, the notation $\mu_{r:x}$ is sometimes used for μ_r. Thus,

$$(2.15.1) \qquad \mu_{r:x} = (1/n) \sum_{i=1}^{k} f_i (x_i - \bar{x})^r, \qquad \mu_{r:u} = (1/n) \sum_{i=1}^{k} f_i (u_i - \bar{u})^r.$$

If the change of variables, or units, is given by

$$(2.15.2) \qquad u = (x - x_0)/c,$$

then

$$(2.15.3) \qquad \mu_{r:x} = c^r \mu_{r:u}.$$

If $\sigma_x = \sqrt{\mu_{2:x}}$ is the standard deviation expressed in terms of x, and $\sigma_u = \sqrt{\mu_{2:u}}$ is the standard deviation in terms of u,

$$(2.15.4) \qquad \sigma_x = c\sigma_u.$$

Note: μ_r remains unchanged under any transformation of the form $u = x - x_0$.

2.16. Standard units. Deviations from the mean $\bar{x}$ are sometimes measured in units of the standard deviation σ_x, called *t units*, where

$$(2.16.1) \qquad t = (x - \bar{x})/\sigma_x.$$

The rth *moment in standard t units about $t = 0$* is

$$(2.16.2) \qquad \alpha_r = (1/n) \sum_{i=1}^{k} f_i t_i{}^r.$$

2.17. Relations between α_r and $\mu_{r:x}$.

(2.17.1) $\alpha_1 = 0,\ \alpha_2 = 1,\ \alpha_3 = \mu_{3:x}/\sigma_x^3,\ \cdots,\ \alpha_r = \mu_{r:x}/\sigma_x^r,\ \cdots$

The t-variate, (2.16.1), and α_r are independent of the units in which the original measurements were taken, e.g., if the change of variables given by (2.15.2) is made,

(2.17.2) $\mu_{r:x}/\sigma_x^r = \mu_{r:u}/\sigma_u^r, \qquad t = (x - \bar{x})/\sigma_x = (u - \bar{u})/\sigma_u.$

Measures of dispersion

2.18. Consider the ordered variates, $x_1 \leqq x_2 \leqq \cdots \leqq x_k$, where x_i has the frequency f_i, mean $\bar{x}$, and standard deviation σ. Let $n = \sum_{i=1}^{k} f_i$. Define cum $f|_i = \sum_{i=1}^{i} f_i$.

2.19. **p^{th} percentile.** Suppose cum $f|_i$ is plotted as a function of x, as in Fig. 3.3.1. A value x' of x for which cum $f|_i = p/100$ is called the pth *percentile*. Approximately p percent of the values of x are less than or equal to x', and $x_i \leqq x'$.

Lower (first) quartile, Q_1: the 25th percentile.
Median (second) quartile, Q_2: the 50th percentile.
Upper (third) quartile, Q_3: the 75th percentile.
Semi-interquartile range (quartile deviation), Q: $Q = (Q_3 - Q_1)/2.$
Note: Q is not necessarily equal to Q_2.
Coefficient of variation, V: $V = \sigma/\bar{x}$, or $100\sigma/\bar{x}\%$.
Standard deviation, σ, and variance, σ^2: See Eq. (2.11.1).

Measures of skewness

2.20. **Coefficient of skewness, γ_1:** $\gamma_1 = \mu_3/\sigma^3 = \alpha_3.$ (See §7.12.)
Momental skewness $= \mu_3/2\sigma^3$.
Pearson's measure of skewness, S_k:

$$S_k = (\bar{x} - \text{Mo})/\sigma = (\text{mean} - \text{mode})/\sigma.$$

Measures of kurtosis ("flatness")

2.21. **Coefficient of excess, γ_2:** $\gamma_2 = (\mu_4/\sigma^4) - 3 = \alpha_4 - 3.$
Kurtosis, β_2: $\beta_2 = \mu_4/\sigma^4 = \gamma_2 + 3 = \alpha_4.$
Some writers call $\beta_2/2$ the kurtosis. (See §7.12.)

3.1. Ungrouped measurements. Suppose a series of n measurements $X_1, X_2, \cdots, X_n$ are made of a quantity X. The results obtained may be represented by a *dot frequency diagram* as in Figure 3.1.1, where each dot represents an observation.

These results may also be represented by a *cumulative graph* as in Figure 3.1.2. The ordinate y in Figure 3.1.2 for any given abscissa x gives the *cumulative frequency* (or *cumulative per cent*); that is, the number of objects having measurements X less than or equal to x. In one sense, the only values of ordinates at which abscissas are defined are those corresponding to the left-hand end of the "horizontal step". An abscissa x' corresponding to a value p on the percent scale of Figure 3.1.2 is called a pth *percentile*. Approximately p per cent of the sample measurements X are less than or equal to x'.

The *cumulative frequency* is also called *accumulative*, or *cumulated*, or *total frequency*; or merely the *distribution function*.

EXAMPLE 1. The lengths of a sample of 60 rods (to the nearest 0.01 inch) were found to be as given in Table 3.1.1.

TABLE 3.1.1

Lengths (in inches) of 60 rods

2.53	2.73	2.78	2.63	2.54	2.55
2.68	2.71	2.63	2.56	2.56	2.60
2.56	2.70	2.70	2.63	2.60	2.56
2.42	2.57	2.65	2.70	2.62	2.44
2.56	2.46	2.72	2.67	2.53	2.48
2.57	2.56	2.63	2.55	2.72	2.59
2.73	2.44	2.65	2.57	2.70	2.61
2.53	2.48	2.72	2.56	2.65	2.63
2.63	2.56	2.70	2.48	2.56	2.70
2.75	2.49	2.67	2.44	2.58	2.55

Figures 3.1.1 and 3.1.2 give the dot frequency and cumulative graph, respectively, for Table 3.1.1.

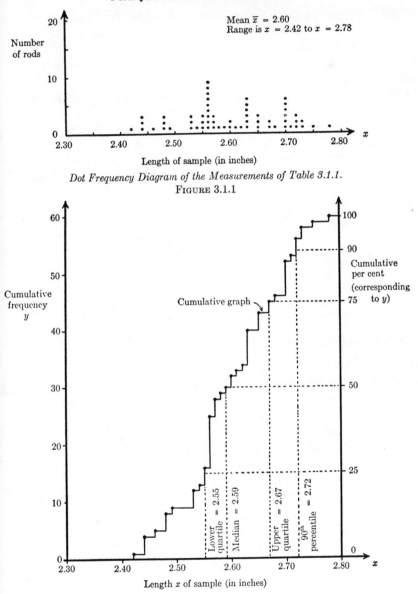

Dot Frequency Diagram of the Measurements of Table 3.1.1.

FIGURE 3.1.1

Cumulative Graph of Data in Table 3.1.1.

The ordinate y for any given abscissa x gives the cumulative frequency (or per cent), that is, the number of rods having lengths less than or equal to x.

FIGURE 3.1.2

Note: In one sense, the only values of ordinates at which abscissas are defined are those at which "horizontal steps" occur, and the abscissa for such an ordinate is the abscissa corresponding to the left hand end of the "horizontal step."

3.2. Grouped measurements. When the number of measurements $X_1, X_2, \cdots, X_n$ becomes large (say 25 or more), ungrouped measurements become too laborious to deal with easily. Grouping the data simplifies matters. In such a case, find the *least, L*, and *greatest, G*, values of the measurements; divide the *range* $(G - L)$ into some number k of equal intervals, selecting the interval length c to be a simple one in terms of the original units of measurements. Divide the x-axis up into intervals of length c, called *class intervals,* or *cells,* locating these intervals so as to have a convenient *mid-point.* Assign every measurement falling in a given class interval i a value equal to the value of x at the mid-point x_i of the interval. The boundaries (*class boundaries*) of the class interval are placed at $(x_i - c/2)$ and $(x_i + c/2)$. x_i is called a *class mark,* or *mid-value,* or *central value.* The class intervals are ordered as $1, 2, \cdots, k$ and the corresponding *class frequencies* $f_1, \cdots, f_k$ are listed. The *cumulative frequency* associated with the upper boundary of the ith class interval is $F_i = \sum_{j=1}^{i} f_j = \operatorname{cum} f|_i$.

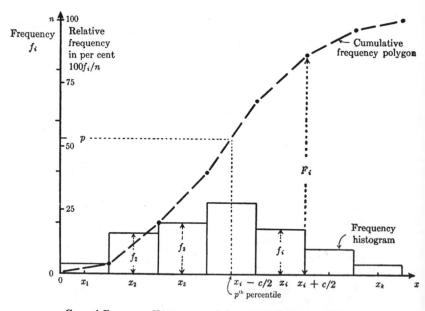

General Frequency Histogram and Cumulative Frequency Polygon

FIGURE 3.3.1

NOTE: f_i is the frequency of measurements X in the ith class interval. $F_i = \sum_{j=1}^{i} f_j = \operatorname{cum} f|_i$ is the cumulative frequency associated with $x_i + c/2$. Strictly speaking, F_i is only defined for the right end boundary of the ith cell; F_i is the number of measurements X less than or equal to $x_i + c/2$.

Ogive. A graph of F_i as a function of the x-values at the upper ends of the class intervals is called an *ogive.*

3.3. A convenient arrangement for exhibiting grouped frequencies is shown in Table 3.3.1. The frequencies f_i and cumulative frequencies F_i, when plotted give a *frequency histogram*, and a *cumulative polygon*, respectively, as illustrated in Figure 3.3.1, Figure 3.3.2, and Figure 3.3.3,

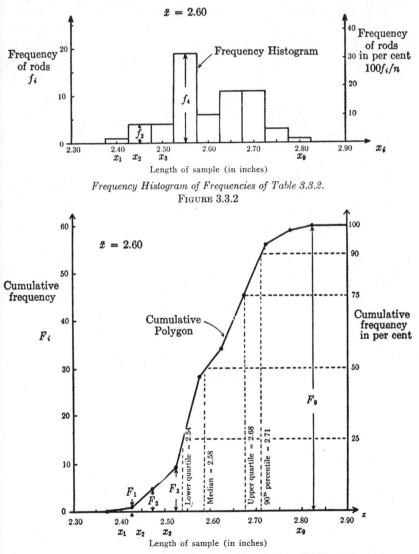

Frequency Histogram of Frequencies of Table 3.3.2.
FIGURE 3.3.2

Cumulative Polygon for the Cumulative Frequencies of Table 3.3.2.
FIGURE 3.3.3

TABLE 3.3.1

Frequency distribution of grouped measurements

	Class interval (cell)					
Number	Boundaries	Mid-point	Frequency	Relative frequency	Cumulative frequency	Relative cumulative frequency
1	$(x_1 - c/2)$, $(x_1 + c/2)$	x_1	f_1	f_1/n	$f_1 = F_1$	F_1/n
2	$(x_2 - c/2)$, $(x_2 + c/2)$	x_2	f_2	f_2/n	$f_1 + f_2 = F_2$	F_2/n
$\vdots$	$\vdots$	$\vdots$	$\vdots$	$\vdots$	$\vdots$	$\vdots$
i	$(x_i - c/2)$, $(x_i + c/2)$	x_i	f_i	f_i/n	$\sum_{j=1}^{i} f_j = F_i$	F_i/n
$\vdots$	$\vdots$	$\vdots$	$\vdots$	$\vdots$	$\vdots$	$\vdots$
k	$(x_k - c/2)$, $(x_k + c/2)$	x_k	f_k	f_k/n	$n = F_k$	1
Total			$n = \sum_{j=1}^{k} f_j$	1		

k = number of class intervals.

c = length of class interval.

x_i = midpoint of ith class interval.

The boundaries $(x_i - c/2)$ and $(x_i + c/2)$ to the ith class interval are usually given to one more decimal than original measurements.

f_i is the number, or frequency, of the measurements $X_1, \cdots, X_n$ in the ith class interval.

F_i is the cumulative frequency associated with the upper class interval boundary $(x_i + c/2)$ for the ith class interval.

f_i and F_i are plotted as shown in Figure 3.3.1.

EXAMPLE 1. In Example 1, §3.1, the least value is $L = 2.42$, the greatest value is $G = 2.78$, the range is $G - L = 0.36$. Divide the x-axis into intervals of length 0.05 with midpoints as indicated in Table 3.3.2. The frequency histogram for Table 3.3.2 is plotted on Figure 3.3.2; the cumulative polygon is plotted on Figure 3.3.3 above the upper class interval boundaries (*not* above the interval midpoints). Figure 3.3.3 for the grouped case approximates the cumulative graph for the ungrouped case (Figure 3.1.2).

TABLE 3.3.2

Frequency distribution for a certain grouping of measurements of lengths of 60 rods
(See Table 3.1.1)

Class interval							
Number	Boundaries	Midpoint	Tallied frequency	Frequency	Relative frequency	Cumulative frequency	Relative cumulative frequency
i		x_i		f_i	f_i/n	F_i	F_i/n
1	2.375 to 2.425	2.40	|	1	.017	1	.017
2	2.425 to 2.475	2.45	||||	4	.067	5	.083
3	2.475 to 2.525	2.50	||||	4	.067	9	.150
4	2.525 to 2.575	2.55	⊤⊤⊤ ⊤⊤⊤ ⊤⊤⊤ ||||	19	.317	28	.467
5	2.575 to 2.625	2.60	⊤⊤⊤ |	6	.100	34	.567
6	2.625 to 2.675	2.65	⊤⊤⊤ ⊤⊤⊤ |	11	.183	45	.750
7	2.675 to 2.725	2.70	⊤⊤⊤ ⊤⊤⊤ |	11	.183	56	.933
8	2.725 to 2.775	2.75	|||	3	.050	59	.983
9	2.775 to 2.825	2.80	|	1	.017	60	1.000
Total			$n = 60$		1.00		

3.4. Measures of "spread" and "middle" of a distribution.

The "middle" of a frequency distribution may be described in various ways; as for example, by the arithmetic mean, median, mode, etc., as defined in §2.2 to §2.5; the "spread", or "dispersion", by variance, standard deviation, p^{th} percentile, range, $\cdots$, as in §2.11 to §2.19.

3.5. Calculation of mean and standard deviation. For grouped data the mean and standard deviation of a sample may be calculated conveniently by the scheme shown in Table 3.5.1.

Table 3.5.1

Scheme for calculating mean $\bar{X}$ and standard deviation σ of sample of n items from grouped data

Class interval midpoint	Frequency		
x_i	f_i	$f_i x_i$	$f_i x_i^2$
x_1	f_1	$f_1 x_1$	$f_1 x_1^2$
x_2	f_2	$f_2 x_2$	$f_2 x_2^2$
$\vdots$	$\vdots$	$\vdots$	$\vdots$
x_i	f_i	$f_i x_i$	$f_i x_i^2$
$\vdots$	$\vdots$	$\vdots$	$\vdots$
x_k	f_k	$f_k x_k$	$f_k x_k^2$
Total $\rightarrow$	n	$S(X)$	$S(X^2)$

$$S(X) = \sum_{i=1}^{k} f_i x_i, \qquad\qquad S(X^2) = \sum_{i=1}^{k} f_i x_i^2.$$

$$S(X) = n\bar{X} = sample\ sum. \qquad n\sigma^2 = S(X^2) - (1/n)[S(X)]^2.$$

The mean $\bar{X}$ as calculated from the ungrouped measurements as a rule differs slightly from the $\bar{X}$ as calculated for grouped measurements, and may differ from the $\bar{X}$ as calculated from a different grouping of the measurements. A similar statement can be made for the calculated values of the standard deviation σ, and variance σ^2.

EXAMPLE 1. An actual calculation of $\bar{X}$ and σ is illustrated in Table 3.5.2 for the measurements of Table 3.1.1.

T<small>ABLE</small> 3.5.2

Calculations to obtain mean $\bar{X}$ and variance σ^2 of sample from grouped measurements of lengths of 60 rods

(See Table 3.3.2)

Class interval midpoint	Frequency		
x_i	f_i	$f_i x_i$	$f_i x_i{}^2$
2.40	1	2.40	5.7600
2.45	4	9.80	24.0100
2.50	4	10.00	25.0000
2.55	19	48.45	123.5475
2.60	6	15.60	40.5600
2.65	11	29.15	77.2475
2.70	11	29.70	80.1900
2.75	3	8.25	22.6875
2.80	1	2.80	7.8400
Total	$n = 60$	$S(X) = 156.15$	$S(X^2) = 406.8425$

$\bar{X} = S(X) \div n.$ $\bar{X} = 156.15 \div 60 = 2.60.$

$n\sigma^2 = S(X^2) - (1/n)[S(X)]^2.$ $60\sigma^2 = 406.8425 - (1/60)(156.15)^2 = 0.4621.$

$\sigma^2 = 0.0077.$ $\sigma = 0.088 =$ standard deviation.

The more commonly used unbiased estimator of variance is found from $59\,\sigma^2 = 0.4621$ to be 0.0078. (See §13.12.)

3.6. Sheppard's corrections.   In calculating moments (e.g., standard deviations) from grouped measurements there is introduced an error due to the grouping. Under certain conditions, this error can be reduced by using Sheppard's corrections. This is discussed in §6.47.

3.7. Simplified schemes for calculating mean and standard deviations. For ease of computation it is sometimes well to calculate the mean and standard deviation after changing the measurements to a new origin and scale. Suppose Y measurements are related to X measurements by the formula

$$(3.7.1) \qquad\qquad X_i = a + bY_i, \qquad\qquad (b \neq 0)$$

where a and b are constants. (The Y measurements are sometimes called *coded* values of the X measurements.) Let $\bar{Y}$ and σ_Y be the mean and standard deviation, respectively, of the measurements Y. Then

$(3.7.2)\ \ \bar{X} = a + b\bar{Y},\ \sigma_X{}^2 = b^2\sigma_Y{}^2,\ \sigma_X = b\sigma_Y,$ (use positive square root).

In using coded values with grouped data, it is often well to select b to be equal to the length of class interval, and a to be at a class interval midpoint near the middle of the grouped distribution.

EXAMPLE 1. An actual calculation of $\bar{X}$ and σ_X using coded values with grouped data is illustrated in Table 3.7.1 for the measurements used in Table 3.5.2. In this example, $a = 2.60$, $b = 0.05$.

TABLE 3.7.1

Calculations to obtain mean $\bar{X}$ and variance σ_X^2 of sample from grouped measurements of lengths of 60 rods using coded computation scheme

(See Table 3.3.2 and Table 3.5.2)

Class interval midpoint x_i	Frequency f_i	y_i	$f_i y_i$	$f_i y_i^2$	$(y_i + 1)^2$	$f_i(y_i + 1)^2$
2.40	1	−4	−4	16	9	9
2.45	4	−3	−12	36	4	16
2.50	4	−2	−8	16	1	4
2.55	19	−1	−19	19	0	0
2.60	6	0	0	0	1	6
2.65	11	1	11	11	4	44
2.70	11	2	22	44	9	99
2.75	3	3	9	27	16	48
2.80	1	4	4	16	25	25
Total	60		$S(Y) = 3$	$S(Y^2) = 185$,	$S[(Y + 1)^2] = 251$

Let $\quad x_i = 2.60 + 0.05 y_i$

$\bar{Y} = S(Y) \div n = 3 \div 60 = 0.05.$ $n\sigma_Y^2 = S(Y^2) - (1/n)[S(Y)]^2.$

$60\sigma_Y^2 = 185 - (1/60)(3)^2 = 184.85.$ $\sigma_Y^2 = 3.0808,$ $\sigma_Y = 1.76.$

$\bar{X} = 2.60 + 0.05\bar{Y} = 2.60 + (0.05)(0.05) = 2.603.$

$\sigma_X = 0.05\sigma_Y = (0.05)(1.76) = 0.088.$

Partial check: $S[(Y + 1)^2] = S(Y^2) + 2S(Y) + n.$

$$251 = 185 + 2(3) + 60 = 251.$$

3.8. Mean and variance of a linear combination of variables.
Let $u = c_1 x_1 + c_2 x_2 + \cdots + c_m x_m,$ where the c's are constants and the x's

are arbitrary random variables. Let $\bar{x}_i$ and σ_i^2 be the mean and variance, respectively, of the distribution of the variable x_i. Let ρ_{ij} be the correlation coefficient between the frequency distributions of x_i and x_j. (§5.37). (The frequency distribution functions of x_i and x_j need not be the same.) Then the mean $\bar{u}$ of the variable u is

$$(3.8.1) \qquad \bar{u} = c_1 \bar{x}_1 + c_2 \bar{x}_2 + \cdots + c_m \bar{x}_m,$$

and the variance σ_u^2 of the variable u is

$$(3.8.2) \qquad \sigma_u^2 = \sum_{i=1}^{m} c_i^2 \sigma_i^2 + \sum_{\substack{i \neq j \\ i,j=1}}^{m} c_i c_j \rho_{ij} \sigma_i \sigma_j.$$

In the special case when the variables x_i are statistically independent, $\rho_{ij} = 0$, and

$$(3.8.3) \qquad \sigma_u^2 = \sum_{i=1}^{m} c_i^2 \sigma_i^2.$$

3.9. Mean and variance of a set in terms of mean and variance of subsets. Suppose sets of n_i values have mean $\bar{x}_i$ and variance σ_i^2, $i = 1, 2, \cdots, m$. Then the whole set of $N = n_1 + n_2 + \cdots + n_m$ values has

$$(3.9.1) \qquad \text{mean} = \bar{x} = (1/N) \sum_{i=1}^{m} n_i \bar{x}_i,$$

and

$$(3.9.2) \qquad \text{variance} = \sigma^2 = (1/N) \sum_{i=1}^{m} n_i \sigma_i^2 + (1/N) \sum_{i=1}^{m} n_i d_i^2,$$

where

$$d_i = \bar{x}_i - \bar{x}.$$

It may be noted that σ^2 is the sum of the weighted mean of the variances of the individual subsets plus the variance of the means of the subsets.

3.10. Measures of skewness and excess ("flatness") of a distribution. Parameters commonly used to measure *skewness* or *asymmetry* in a frequency distribution are the third moment μ_3 about the mean, the *coefficient of skewness (skewness)* γ_1, and other parameters as defined in §2.20. When $\gamma_1 > 0$, the distribution has a "long tail" on the right hand side and is said to have *positive skewness*. When $\gamma_1 < 0$, the "long tail" is on the left, and the distribution has *negative skewness*. If a distribution is symmetric, every odd order moment about the mean (if it exists) is zero.

To measure the degree of "flattening" (*kurtosis*) of a distribution near its center, the fourth moment μ_4 about the mean, the *coefficient of excess* (*excess*) γ_2, $\cdots$, and other parameters as defined in §2.21 are sometimes used.

3.11. Frequency distributions in two, or more, dimensions.   Suppose a series of measurements, (X_1, Y_1), (X_2, Y_2), $\cdots$, (X_n, Y_n), are made of a pair (X, Y) of quantities X and Y. The terminology and schemes for describing frequency distributions in two or more variables follow those described above for one dimension. Two-dimensional distributions may be represented by rectangular tables.

4.1. **Sets.**   A *set* (*class, aggregate, ensemble*) S is a well-defined collection of objects, or symbols, called *elements* or *members* of the set. No restriction is placed on the nature of the objects.

Set A is a *subset* of set B, written $A \subset B$ or $B \supset A$, if and only if each element of A is also an element of B. A is a *proper subset* of B if at least one member of B is not a member of A. $A \not\subset B$ means that at least one element of A is not included in B. Some writers use $A \subseteq B$ to indicate that A is a subset of B and reserve $A \subset B$ to indicate that A is a proper subset of B.

Set A is said to *equal* set B, written $A = B$, if every element of A is an element of B and every element of B is an element of A. Two equal sets are said to be *identical*. If A and B are not equal, we write $A \neq B$.

A *universal set I* (or *universe* or *space*) is the term given to an overall set which includes all the sets of concern in a given study.

An *empty* (*null*) set, written $\emptyset$, is a set that has no elements. It is usually assumed that for every set A, $\emptyset \subset A$.

The *complement A'* of a set A with respect to a given universal set I is the set of elements in I that are not in A. [Some writers use $\bar{A}$ or $\mathcal{C}(A)$ or $I - A$ to denote the complement A'.]

Notation. The elements of a set are usually denoted by small letters. Capital letters are used as the names of sets. The elements of a set are often listed by enclosing them in braces. The notation $x \in S$ means that x is a member of set S; $x \notin S$ means that x is not a member of S.

To indicate a set of objects x having the property P, the symbol $\{x \mid x$ has the property $P\}$ is used. $\{x \mid \ \}$ is called a *set builder*. The bar "$\mid$" is read "such that."

EXAMPLE. $S = \{1,2,4,6\}$ is a set S consisting of the elements 1, 2, 4, 6.

EXAMPLE. $\{(x,y) \mid (x - y) > 2\}$ denotes the set of number pairs (x,y) such that $(x - y)$ exceeds 2.

4.2. **Union and intersection.**   The *union* (*join*) of sets A and B, written $A \cup B$ or "A cup B," is the set of all elements that are in A or in B or in both A and B.

The *intersection* (*meet*) of the sets A and B, written $A \cap B$, is the set of all elements which are in both A and B.

The sets A and B are said to be *disjoint* or *mutually exclusive* if they have no elements in common, that is, if their intersection is the empty set. A and B *overlap* if they have some elements in common.

4.3. Venn diagram. One scheme for representing a subset V of ordered pairs (x,y) consists in using a rectangle to represent the universal set, and sets of points inside the rectangle to represent V. For example, Figs. 4.3.1 and 4.3.2 illustrate the commutative law; Fig. 4.3.3, the complement.

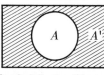

Union of A and B

Crosshatched Section Illustrates $A \cup B = B \cup A$.
FIGURE 4.3.1

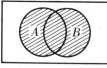

Intersection of A and B

Crosshatched Section Illustrates $A \cap B = B \cap A$.
FIGURE 4.3.2

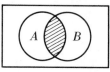

Complement A' of A

Crosshatched Section Illustrates A'.
FIGURE 4.3.3

4.4. Operations on sets. Consider the ordered n-tuple $X = \{x_1, x_2, \cdots, x_n\}$ of n elements, where the elements x_i are members of set S_i. An *operation* on X is a *rule* which assigns to each X a uniquely defined element in some set S. If $n = 1$, the operation is *unary*; if $n = 2$, the operation is *binary*; if $n = m$, the operation is *m-ary*.

4.5. Algebra of sets. If A, B, C are any three subsets of a universe

I, then, for the operations $\cup$ and $\cap$, the following relationships (laws) hold:

1. Closure laws
   - (a) There exists a unique set $A \cup B$
   - (b) There exists a unique set $A \cap B$

2. Commutative laws
   - (a) $A \cup B = B \cup A$ (b) $A \cap B = B \cap A$

3. Associative laws
   - (a) $(A \cup B) \cup C = A \cup (B \cup C)$
   - (b) $(A \cap B) \cap C = A \cap (B \cap C)$

4. Distributive laws
   - (a) $A \cup (B \cap C) = (A \cup B) \cap (A \cup C)$
   - (b) $A \cap (B \cup C) = (A \cap B) \cup (A \cap C)$

5. Idempotent laws
   - (a) $A \cup A = A$ (b) $A \cap A = A$

6. Properties of universe I. Identity laws
   - (a) $A \cup I = I$ (b) $A \cap I = A$

7. Properties of null set $\emptyset$. Identity laws
   - (a) $A \cup \emptyset = A$ (b) $A \cap \emptyset = \emptyset$

8. Properties of inclusion $\subset$
   - (a) $A \subset I$ (b) $\emptyset \subset A$
   - (c) If $A \subset B$, then $A \cup B = B$
   - (d) If $B \subset A$, then $A \cap B = B$
   - (e) $A \subset (A \cup B)$ (f) $(A \cap B) \subset A$

9. Properties of complement
   - (a) For every set A there exists a unique set A' with respect to I.
   - (b) $A \cup A' = I$
   - (c) $A \cap A' = \emptyset$
   - (d) $(A \cup B)' = A' \cap B'$ De Morgan's law (dualization)
   - (e) $(A \cap B)' = A' \cup B'$ De Morgan's law (dualization)
   - (f) $\emptyset' = I$
   - (g) $I' = \emptyset$
   - (h) $(A')' = A$ (involution)

10. Duality principle
    If in any "correct" formula involving the symbols $\cup$, $\cap$, $\emptyset$, I, $\subset$, and $\supset$ we interchange $\cup$ and $\cap$, $\emptyset$ and I, $\subset$ and $\supset$ throughout the formula, the resulting formula is correct.

11. Absorption laws
    - (a) $A \cup (A \cap B) = A$ (b) $A \cap (A \cup B) = A$

12. Consistency property

The conditions $A \subset B$, $A \cap B = A$, and $A \cup B = B$ are mutually equivalent.

4.6. One-to-one correspondence between sets. Two sets A and B are said to be in *one-to-one (reciprocal) correspondence* when a pairing of the elements of A with the elements of B is such that each element of A corresponds to exactly one element of B and each element of B corresponds to exactly one element of A. ("Exactly" means "one and only one.")

4.7. Equivalent sets. Two sets A and B are said to be *equivalent*, written $A \backsim B$, if and only if they can be placed in one-to-one correspondence.

4.8. Cardinal numbers. Two sets are said to have the same *cardinal number* if and only if they are equivalent.

A set A is said to have the cardinal number n if and only if there exists a one-to-one correspondence between the elements of A and the natural numbers $1, 2, 3, \cdots, n$.

4.9. Finite and infinite sets. A nonempty set is *finite* if and only if its cardinal number is one of the cardinal numbers $1, 2, 3, \cdots$. An *infinite set* is a nonempty set which is *not finite*.

An *n-set* is a set of n elements.

4.10. Product set. Let $A_1, A_2, \cdots, A_n$ be n sets. Let $a_i \epsilon A_i$, for $i = 1, 2, \cdots, n$. The collection of all ordered n-ples $(a_1, a_2, \cdots, a_n)$ is called the *product set* $A_1 \times A_2 \times \cdots \times A_n$.

The set of all ordered pairs (x,y), where $x \epsilon X$ and $y \epsilon Y$, is called the *cartesian product* of X and Y, is denoted by $X \times Y$, and is read "X cross Y."

Variable, relation, function, inverse

4.11. Variable. A symbol in a sentence, phrase, or expression is a *variable* (or *placeholder*) if and only if it is replaceable by the name of an element (number) of a set called the *replacement set* (or *domain* or *universal set*) specifically defined for the variable. If the set contains only a single element, the symbol used to represent this element is called a *constant*. Variables are used as placeholders for numbers and other mathematical quantities in algebraic sentences, such as equations and inequalities.

4.12. Relation. A *relation R* in a set U is a subset of ordered pairs

(x,y) from $U \times U$. The *domain* of the relation R is the subset of U for which x is a placeholder; the *range* of the relation is the subset of U for which y is a placeholder.

4.13. Solution set. Suppose x belongs to the universal set I. Consider the set $\{x|S\}$, where S is a statement involving x, or a condition placed on x. Any member of I for which the statement S is true is said to *satisfy* S, or to be a *solution set* of S. S is called a *defining* condition, or *set selector*.

EXAMPLE. If x belongs to the set of real numbers, the solution set of $\{x|2x = 3\}$ is $\{3/2\}$.

EXAMPLE. If $I = \{1,3,5\}$ and x belongs to I, the solution set of $\{x|x > 2\}$ is $\{3,5\}$.

4.14. Function. A *function* f in a set U is a subset of ordered pairs (x,y) from $U \times U$ having exactly one y for each x. In other words, a *function* f in U is a relation R in U such that for every x in the domain of R there corresponds exactly one y in the range of R. (This definition of function corresponds to the notion of a single-valued function given by many writers.)

The notation $f:(x,y)$ is sometimes used to mean "the function f whose ordered pairs are (x,y)."

The notation $f(x)$, read "f at x," is often used to denote the second element y of the ordered pair (x,y) whose first element is x. $f(x)$ is called "the value of the function f at x." It is common to write $y = f(x)$, meaning y is the value of f at x.

EXAMPLE. The relation $\{(x,y)|y = x^2 + 1\}$ in the set U of real numbers defines a function F. $F(1) = 2$ is the value of $F(x) = x^2 + 1$ at $x = 1$. The domain of the function is U; the range of F is the set of all real numbers equal to or greater than 1.

4.15. Inverse relation. The set R^{-1} obtained by interchanging the coordinates of all the ordered pairs of a given relation R is called the *inverse relation of R*.

EXAMPLE. The inverse R^{-1} of the relation $R = \{(1,2), (1,3), (2,4)\}$ is $R^{-1} = \{(2,1), (3,1), (4,2)\}$. R is not a function; R^{-1} is a function.

4.16. Inverse function. The inverse of a function is obtained by interchanging the coordinates of each of its ordered pairs; the inverse is a relation, but may or may not be a function.

A function F is a set of ordered pairs (x,y) no two of which have the same first component x with different second components. If a function

F is such that no two of its ordered pairs have the same second component with different first components, then the set of ordered pairs obtained from F by interchanging in each ordered pair the first and second elements is called the *inverse function* and is denoted by F^{-1}.

The range of F is the domain of F^{-1}, and the domain of F is the range of F^{-1}. We write $F:(a,b)$ and $F^{-1}:(b,a)$.

EXAMPLE.   If the function F has elements (ordered pairs) of the form (a_1,b_1), (a_2,b_2), $\cdots$, the inverse function F^{-1}, if it exists, has elements (ordered pairs) of the form (b_1,a_1), (b_2,a_2), $\cdots$.

EXAMPLE.   Let x and y be real numbers. The inverse to the relation $R = \{(x,y)\,|\,y = 2x\}$ is $R^{-1} = \{(y,x)\,|\,x = y/2\}$. R^{-1} and R are each functions. The inverse function is then sometimes written $R^{-1} = \{(x,y)\,|\,y = x/2\}$.

EXAMPLE.   Let x and y be real numbers. The inverse relation to the relation $R = \{(x,y)\,|\,y = 3x^2\}$ is $R^{-1} = \{(y,x)\,|\,x = \pm\sqrt{y/3}\}$. R is a function; R^{-1} is not a function. The inverse relation may be written $R^{-1} = \{(x,y)\,|\,y = \pm\sqrt{x/3}\}$. If the domain of R is restricted so that $x \geqq 0$, then $R^{-1} = \{(x,y)\,|\,y = \sqrt{x/3}\}$ is a function.

EXAMPLE.   The inverse relation to $R = \{(x,y)\,|\,y - 2 = |x| - x\}$ is $R^{-1} = \{(x,y)\,|\,x - 2 = |y| - y\}$. R^{-1} is not a function.

For a fuller treatment of sets, relations and functions see R. S. Burington, *Handbook of Mathematical Tables and Formulas*, 4th ed., McGraw-Hill Book Company, New York, 1964.

Partitions, selections, samples, permutations, combinations

4.17. Partition of a set.   The subsets A_1, A_2, ..., A_r of A form a *partition* of A provided

$$A = A_1 \cup A_2 \cup \cdots \cup A_r \qquad \text{and} \qquad A_i \cap A_j = \emptyset$$

for $i \neq j$, $(i, j = 1, 2, \cdots, r)$.

If for two partitions of a set $A = A_1 \cup A_2 \cup \cdots \cup A_r$ and $A = A_1' \cup A_2' \cup \cdots \cup A_r'$ means that $A_i = A_i'$ $(i = 1, 2, \cdots, r)$, then the partitions are said to be *ordered* and *equal*; the partitions are *unordered* if equality means that each A_i is equal to some A_i'.

If

(4.17.1) $$S = A_1 \cup A_2 \cup \cdots \cup A_r$$

is a partition of an n-set S into subsets A_1, A_2, $\cdots$, A_r containing k_1, k_2,

$\cdots$, k_r elements, then $n = k_1 + k_2 + \cdots + k_r$, and (4.17.1) is called a $(k_1, k_2, \cdots, k_r)$-*partition* of S.

4.18. Ordered selections. *Samples.* Let

$$(4.18.1) \qquad X = (x_1, x_2, \cdots, x_r)$$

be an ordered r-tuple of not necessarily distinct elements of set S. X is a *sample* of S. The sample is of *size* r and is called an *r-sample* of S.

THEOREM. The number of r-samples of an n-set S is n^r.

4.19. Permutations. If S is an n-set and the components x_i of the r-sample $(x_1, x_2, \cdots, x_r)$ are distinct, and if $r \leqq n$, then the r-sample is an *r-permutation* of n elements.

An n-permutation is called a *permutation* (or *arrangement* or *ordering*) of n elements.

Where there is no ambiguity introduced the term permutation (or arrangement) is often used to mean r-permutation.

THEOREM. The number of different arrangements of r different elements selected from an n-set (i.e., the number of r-permutations of n different elements) is

$$(4.19.1) \quad P(n,r) \equiv P_r^n = n(n-1) \cdots (n-r+1) = \frac{n!}{(n-r)!}$$

EXAMPLE 1. The permutations of a, b, c taken 2 at a time are: ab, ac, bc, ba, ca, cb. There are 6 of these permutations, $P_2^3 = 3!/(3-2)! = 6$. However, there are only 3 combinations of these 3 letters taken 2 at a time, $C_2^3 = 3!/2!(3-2)! = 3$.

EXAMPLE 2. The number of permutations of n different things is $P_n^n = n!$.

EXAMPLE 3. If the letters a, b, c, d, e are all different (and the repetition of a letter within a permutation or combination is not allowed), the number of permutations of the five letters taken 3 at a time is $P_3^5 = 5!/(5-3)! = 60$ and the number of combinations of 5 letters taken 3 at a time is $C_3^5 = 5!/3!(5-3)! = 10$.

4.20. Circular permutations. The number of different orders in which n different things taken r at a time can be arranged about the circumference of a circle is P_r^n/r.

4.21. Permutations of different things when repetitions are allowed.

With n different items there can be formed n^r arrangements or permutations of r items each, provided the repetition of items within a permutation is allowed.

4.22. The number of permutations of n things not all of which are different.

The number N of distinguishable permutations of n things taken n at a time, of which p are alike, q others alike, $\cdots$, z others alike, is given by $N = n!/p!q! \cdots z!$. Here $p + q + \cdots + z = n$. This is sometimes written

$$(4.22.1) \qquad N = P^{(n)}(p, q, \cdots, z) = n!/p!q! \cdots z!.$$

EXAMPLE. From four a's, four b's, three c's, and two d's the number of 11-letter words that can be formed is

$$11!/4!\,4!\,3!\,2!.$$

4.23. Unordered selections. Let

$$(4.23.1) \qquad X = (x_1, x_2, \cdots, x_r)$$

be an *unordered r-tuple* (*selection* or *collection*) of r not necessarily distinct elements of S. The number of occurrences of an element in (4.23.1) is called the *multiplicity* of the element.

The unordered selection (4.23.1) of S is of size r and is called an *r-selection* of S.

Two collections X and X^* such as (4.23.1) are said to be *equal* if the elements with their respective multiplicities are the same for both collections.

If each element in (4.23.1) is of multiplicity 1, the r-selection is an *r-subset* of S.

4.24. Combinations.

An *r-combination* of n elements is an r-subset of an n-set.

Where there is no ambiguity introduced, the term *combination* is often used to mean r-combinations.

THEOREM 1. The number of combinations of n different elements (things) taken r at a time (i.e., the number r subsets of an n-set) is $C(n,r) \equiv C_r^n$, where

$$(4.24.1) \qquad C_r^n = \binom{n}{r} = \frac{P_r^n}{r!} = \frac{n(n-1) \cdots (n-r+1)}{r(r-1) \cdots (1)} = \frac{n!}{r!(n-r)!},$$

where

(4.24.2) $$n! = n(n-1)(n-2) \cdots (1).$$

(4.24.3) *Note*: $0! = 1$. $C(n,0) = 1$, $C(0,0) = 1$.

$C(n,r)$ is defined when r is positive and n is real. If r is a negative integer, by definition $C(n,r) = 0$. $C(n,r)$ is not defined when r is negative and not an integer. If $n > 0$,

(4.24.4) $$C(-n,r) = (-1)^r C(n+r-1, r).$$
(See § 8.12.)

Many different symbols are used for C_r^n, e.g., C_r^n, ^nC_r, $\binom{n}{r}$, $_nC_r$, $C(n,r)$, $C_{n,r}$; likewise for P_r^n.

EXAMPLE 1. There are $C_{13}^{52} = 52!/(13!\, 39!)$ different bridge hands, each consisting of a selection of 13 cards from a full deck of 52 cards.

THEOREM 2. (a) The total number of ways in which one or more items may be selected from n items is $2^n - 1 = \sum_{j=1}^{n} C_j^n$.

(b) The number of ways in which r items may be selected from n distinguishable items, without regard to order, and with repetitions permitted, is equal to the number of combinations of $(n+r-1)$ different things taken r at a time without repetitions, that is, C_r^{n+r-1}.

(c) The number of r-selections of an n-set is $C_r^{n+r-1} = C_{n-1}^{n+r-1}$.

EXAMPLE 2. The number of different throws possible with four dice (with 6 faces each) is $C_4^{6+3} = 126$.

THEOREM 3. If a certain thing A can be done in m ways, and if when A has been done, a certain other thing B can be done in n ways, the entire number of ways in which both things can be done in the order stated is mn.

EXAMPLE 3. A set of n elements contains a red elements and b black elements, and $a + b = n$. From the set a group G of r elements is picked (without replacement or regard to order) containing exactly k red elements and $r - k$ black elements. The total number of ways G can be selected is $C_k^a C_{r-k}^{n-a}$. This is in contrast to the number of ways C_r^n a set of r elements can be picked from a set of n elements.

EXAMPLE 4. The number of different possible samples of size r from a population of n elements is n^r if sampling is done with replacement, and $n(n-1) \cdots (n-r+1)$ if sampling takes place without replacement.

THEOREM 4.   The number of ordered $(k_1, k_2, \cdots, k_r)$-partitions of an n-set is $n!/k_1! \, k_2! \cdots k_r!$. That is,

$$(4.24.5) \qquad C_{k_1}{}^{n} C_{k_2}{}^{n-k_1} \cdots C_{k_r}{}^{n-k_1-\cdots-k_{r-1}} = \frac{n!}{k_1! \, k_2! \cdots k_r!}$$

where $n = k_1 + k_2 + \cdots + k_r$.

EXAMPLE 5.   The 52 cards of a deck are distributed among 4 players; so the number of possible different situations at a bridge table is $52!/(13!)^4$.

4.25. Various formulas relating to combinations and permutations.

$$r! \times C_r{}^{n} = P_r{}^{n}. \qquad C_{n-r}{}^{n} = C_r{}^{n}.$$

$$C_0{}^{n} = C_n{}^{n} = 1. \qquad C_r{}^{n-1} + C_{r-1}{}^{n-1} = C_r{}^{n}. \qquad P^{(n)}(r, n-r) = C_r{}^{n}.$$

$$C_0{}^{n} + C_1{}^{n} + C_2{}^{n} + \cdots + C_n{}^{n} = 2^n.$$

$$C_0{}^{n} - C_1{}^{n} + C_2{}^{n} - \cdots + (-1)^n \, C_n{}^{n} = 0.$$

$$C_r{}^{n} - C_{r-1}{}^{n} + C_{r-2}{}^{n} - \cdots + (-1)^{r-1} \, C_1{}^{n} + (-1)^r \, C_0{}^{n} = C_r{}^{n-1},$$
$$\text{where } \ 0 \leqq r < n.$$

$$C_0{}^{n}C_k{}^{n} - C_1{}^{n}C_{k-1}{}^{n-1} + C_2{}^{n}C_{k-2}{}^{n-2} - \cdots + (-1)^k \, C_k{}^{n}C_0{}^{n-k} = 0,$$
$$\text{where } \ 0 \leqq k \leqq n.$$

$$C_0{}^{m}C_r{}^{n} + C_1{}^{m}C_{r-1}{}^{n} + C_2{}^{m}C_{r-2}{}^{n} + \cdots + C_{r-1}{}^{m}C_1{}^{n} + C_r{}^{m}C_0{}^{n} = C_r{}^{m+n},$$
$$\text{where } \ 0 \leqq r \leqq n, \qquad 0 \leqq r \leqq m.$$

$$C_0{}^{n}C_k{}^{n} + C_1{}^{n}C_{k-1}{}^{n-1} + C_2{}^{n}C_{k-2}{}^{n-2} + \cdots + C_k{}^{n}C_0{}^{n-k} = 2^k C_k{}^{n}, \quad k \leqq n.$$

$$C_1{}^{n} + 2C_2{}^{n} + 3C_3{}^{n} + \cdots + nC_n{}^{n} = n2^{n-1}.$$

$$1 \cdot 2C_2{}^{n} + 2 \cdot 3C_3{}^{n} + 3 \cdot 4C_4{}^{n} + \cdots + n(n-1)C_n{}^{n} = n(n-1)2^{n-2}.$$

$$C_0{}^{m-1} + C_1{}^{m} + C_2{}^{m+1} + C_3{}^{m+2} + \cdots + C_n{}^{m+n-1} = C_n{}^{m+n}, \, m \geqq 1$$

$$C_0{}^{m}C_0{}^{n} + C_1{}^{m}C_1{}^{n} + C_2{}^{m}C_2{}^{n} + \cdots + C_n{}^{m}C_n{}^{n} = C_n{}^{m+n}, \qquad n \leqq m.$$

4.26. Binomial expansion.

$$(4.26.1) \qquad (q + p)^n = q^n + C_1{}^{n}pq^{n-1} + \cdots + C_r{}^{n}p^rq^{n-r} + \cdots + p^n,$$

where

$$(4.26.2) \qquad C_r^{\,n} = \frac{n!}{r!(n-r)!} = \binom{n}{n-r}, \qquad \binom{n}{0} = 1.$$

The quantities $C_r^{\,n}$ are called *binomial coefficients*.

Table IV gives values of the binomial coefficients for $n = 1, \cdots, 20$.

4.27. For a fuller treatment of combinatorial mathematics see

Feller, W., *Probability Theory and Its Applications*, Vol. 1, 3d ed., John Wiley & Sons, Inc., New York, 1968.

Riordan, J., *An Introduction to Combinatorial Analysis*, John Wiley & Sons, Inc., New York, 1958.

Ryser, H. J., *Combinatorial Mathematics*, Mathematical Association of America, distributed by John Wiley & Sons, Inc., New York, 1963.

Algebra of events

4.28. Algebra of events associated with a given experiment. Let S denote a set (space) of sample descriptions of all possible *outcomes* (states, events) of a given experiment. An *event E* is any subset of S. To say that the event E *is realized*, or *has occurred*, is to say that the outcome of a specific experiment under consideration has a description that is a member of subset E.

Suppose the events $E_1, E_2, \cdots$ are such as to permit the following definitions:

1. The *union* (*logical sum*, or *join*) of a set of events $E_1, E_2, \cdots$, written $E_1 \cup E_2 \cup E_3 \cup \cdots$, is the outcome of realizing *at least one* of the events $E_1, E_2, \cdots$.

2. The *intersection* (*logical product*, or *meet*) of two events E_1 and E_2, written $E_1 \cap E_2$, is the joint outcome of realizing *both* E_1 and E_2.

3. The (*logical*) *complement* of an event E is the outcome E' of not realizing E. $E' \cup E = S$. $E \cap E' = \emptyset$.

4. The *certain event I* is the set S (i.e., at least one of the outcomes of S is realized in a specific experiment).

5. The *impossible event $\emptyset$* is an event that contains no descriptions in S, and therefore cannot occur in the experiment. $S' = \emptyset$.

The *algebra of events* associated with the given experiment under the hypothesis that event E_1 occurs is the set B_1 of joint events $E \cap E_1$. In this case the certain event in set B_1 is $E_1 \cap E_1 = E_1$.

NOTE: $E_1 \cap E_2$ is often written $E_1 E_2$; and $E_1 \cup E_2$, written $E_1 + E_2$.

4.29. Events E_1 and E_2 are said to be *mutually exclusive*, or *disjoint*, if and only if $E_1 \cap E_2$ is $\emptyset$, that is, if and only if the joint event of simultaneously realizing *both* E_1 and E_2 is impossible.

If E_1 and E_2 are events of the set S, we say $E_1 \subset E_2$ if and only if $E_1 \cup E_2 = E_2$. ($E_1 \subset E_2$ is read "E_1 is included in E_2.") The symbol $=$ here means "is". In other words, the event E_1 is included in the event E_2 if and only if the event of realizing at least one of the two events E_1 and E_2 is the realization of the event E_2.

THEOREM. $E_1 \subset E_2$ if and only if $E_1 \cap E_2 = E_1$. In other words, the event E_1 is included in the event E_2 if and only if the event of realizing both events E_1 and E_2 is the realization of the event E_1.

THEOREM. For any event E, $\emptyset \subset E \subset I$.

4.30. Event E is *equal* to event F if and only if $E \subset F$ and $F \subset E$. (Written $E = F$.)

Let E and F be any two events. Then:

(a) $E'F'$ is the event that *exactly none* of the events E and F will occur.

(b) $EF' \cup E'F$ is the event that *exactly one* of the events E and F will occur.

(c) $E \cup F$ is the event that *at least one* of the events E or F will occur.

(d) $(EF)' = E' \cup F'$ is the event that *at most one* of the events E and F will occur.

The various laws and formulas given for sets in §4.5 hold for events.

4.31. **Event algebras and two-valued logic.** In two-valued (Aristotelian) logic an algebra $\mathfrak{B}$ of events (logical propositions, assertions) is related to a simpler Boolean algebra $\mathfrak{I}$ of *truth-values*. A truth-value $T[E]$ is taken to be equal to 1 if an event E is realized (i.e., if E is true) and equal to 0 if the event E is not realized (i.e., if E is false). Let I denote a certain event and $\emptyset$ denote the impossible event. The relation between the algebras $\mathfrak{B}$ and $\mathfrak{I}$ is given by the homomorphism

$$(4.31.1) \qquad T[I] = 1 \qquad\qquad T[\emptyset] = 0$$
$$T[E_1 \cup E_2] = T[E_1] \oplus T[E_2] \qquad T[E_1 \cap E_2] = T[E_1] \otimes T[E_2]$$

The truth tables for the events $E_1 \cup E_2$ and $E_1 \cap E_2$ and the truth-values T, with associated arithmetic, are shown in Tables 4.31.1 and 4.31.2.

TABLE 4.31.1
Truth table

E_1	E_2	$E_1 \cup E_2$	$E_1 \cap E_2$
T	T	T	T
T	F	T	F
F	T	T	F
F	F	F	F

TABLE 4.31.2
Arithmetic of truth-values

$T[E_1] \oplus T[E_2] = T[E_1 \cup E_2]$					$T[E_1] \otimes T[E_2] = T[E_1 \cap E_2]$			
1	$\oplus$	1	=	1	1	$\otimes$	1	=	1
1	$\oplus$	0	=	1	1	$\otimes$	0	=	0
0	$\oplus$	1	=	1	0	$\otimes$	1	=	0
0	$\oplus$	0	=	0	0	$\otimes$	0	=	0

4.32, Let $E = F(E_1, E_2, \cdots)$ be any proposition logically related to a set of events $E_1, E_2, \cdots$. The truth-value $T[E]$ of E is $T = T[G\{T(E_1), T(E_2), \cdots\}]$, where G is the image of F obtained by replacing $\cup$ and $\cap$ by $\oplus$ and $\otimes$, respectively. In the arithmetic of truth-values the complement of 0 is 1, $\cdots$, that is, $0' = 1$, $1' = 0$. In view of the assumptions made here for the algebras involved, any proposition E (i.e., an event) is either true or false.

5.1. The problem of describing a given sample of quantitative measurements is outlined in earlier sections. Interest is usually centered in making inferences about the populations from which the samples come, such as intervals within which *population parameters* (i.e., population mean, variance, $\cdots$) are likely to lie. To make such inferences one must learn something about how *sample statistics* (i.e., sample mean, variance, $\cdots$) vary from sample to sample when repeated samples are drawn from the same population. Sample to sample variations of sample statistics are studied by *experimental* and *mathematical* methods.

Experimental method.   In this method one determines by repeated experiments how a given sample statistic (e.g., mean of a sample) is distributed. Such experiments are often carried out by repeated drawing of samples. The method is often very costly and time consuming.

Mathematical method.   Mathematical methods can be used to find theoretical sampling laws for some sampling statistics (e.g., sample means). These methods involve *the theory of probability*.

5.2. **Chance and probability.**   The terms *chance* and *probability* are used quite loosely in ordinary conversation and writing, the interest usually lying in a future *event*. In mathematical probability the attempt is to lay out: (a) conditions under which sensible numerical statements concerning uncertainties can be made; (b) methods for finding numerical values of expectations and probabilities.

5.3. **Random phenomenon.**   Whenever a certain group of events is described in a statistical way, it may be said to be treated as a *random phenomenon*. In other words, for practical reasons or from principles it is considered impossible to predict the precise final state of the phenomenon from the initial state and the known laws of nature.

5.4. **Definitions of probability.**   Various definitions of probability have been proposed, though no single particular definition has as yet met with general acceptance as completely adequate and rigorous. Since rigorous definitions are beyond the scope of this book, merely a listing of certain definitions will be given.

The definition to use in a specific case depends on the breadth of application and order of simplicity required, and on the closeness of the definition to certain intuitions in which the concept of probability originates for the case of interest.

5.5. Definition A. (Involving an enumeration of cases.) If an event E can occur in m cases and E can not occur in $n - m$ cases out of a total of n possible cases (all of which are agreed to be considered in advance as "equally likely", *a priori*), then the probability p of the event E occurring is m/n, that is, $p = P(E) = m/n$.

The failure of the event E is denoted by $\bar{E}$, and is called the *complementary event*. The probability of $\bar{E}$ is $(n - m)/n$, namely $1 - p$. $q = 1 - p$ is called the *complementary probability*, $p + q = 1$.

Definition A agrees with certain intuitive notions, and is useful in simple games of chance problems. It can not be used when it is impossible to make a simple enumeration of cases considered to be equally likely.

EXAMPLE 1. An unbiased "ideal" die is rolled. Being unbiased, the occurrence of any face up is considered equally likely with any other face. The probability of an "ace" is $1/6$; the probability of a "deuce" is $1/6$, etc.

EXAMPLE 2. An unbiased coin is tossed n times. The probability p that exactly r heads are obtained is found as follows: The total number of different sequences is 2^n. Each is considered equally likely. The number of ways of permuting n symbols of which r are heads and $n - r$ are tails without regard to order is $C_r^n = n!/r!(n - r)!$. The probability p is

$$p = P(\text{of obtaining exactly } r \text{ heads in } n \text{ tosses}) = [n!/r!(n - r)!] \div 2^n.$$

5.6. Definition B. (Involving the stability of relative frequency.) Suppose that: (a) whenever a series of many trials is made, the number m of trials in which the event E occurred divided by the total number n of trials is approximately p, and (b) m/n as a general rule becomes nearer to p as more extensive trials are made. Then the probability of the event E is defined to be p, that is, $P(E) = p$.

This definition gives a method of estimating probabilities from experimental results.

Probabilities determined from measurements of the corresponding relative frequencies are known as *a posteriori* probabilities (i.e., determined afterwards).

EXAMPLE 1. A symmetric coin is tossed 500 times, and 243 heads are observed. One would expect $m = 250$. The experiment is repeated a number of times, the heads obtained being 243, 252, 249, 257, $\cdots$. The ratios $243/500 = 0.486$, $495/1000 = 0.495$, $744/1500 = 0.496$, $1001/2000 = 0.501$, $\cdots$ appear usually to come nearer to $p = 0.5$ as the experiment is continued. $P(\text{heads}) = 0.5$.

5.7. If n is finite and if E must happen in all of the n possible cases, then $p = 1$ and E is *certain*, while $q = 1 - p = 0$ and $\overline{E}$ is *impossible*.

5.8. Definition C. (Probability as a measure of a sub-aggregate.) Let S be a set of circumstances (such as the many physical circumstances surrounding the spinning of a real symmetric die on a rough surface):

(a) events E (such as fall of "ace") are considered as associated with, or caused by, *phases S_i* of circumstances;

(b) each S_i results unambiguously either in E or $\overline{E}$ (not E);

(c) the totality of phases S_i form a set or an aggregate S, of which the phases favorable to E, and those favorable to $\overline{E}$ form complementary subsets;

(d) if M is a measure given to the whole set S and pM is the measure of the subset favorable to E, then the probability $p(E;S)$ of the occurrence of the event E under circumstances S is p;

(e) the question of "equally likely" phases is the same abstractly as the question of specifying the aggregate S and its measure.

The term *fundamental probability set* is used by some writers for the set S of phases S_i. The probability $p(E;S)$ is sometimes written as $Pr(E;S)$, or merely as $Pr(E)$, or $P(E)$.

Definition C is useful where other definitions may fail to furnish the needed concepts. Such is often the case where the systems involved depend on non-finite sets. A truly rigorous definition of probability involves the theory of abstract measure and is beyond the scope of this book.

If one tries to explain the effect (e.g., the relative frequency of E) by analysis of the cause (e.g., the circumstances S), one is led into the *a priori* points of view (i.e., determined in advance) in formulating a definition of probability. The *a priorist* point of view directs attention to the invariant part of the circumstances S involved.

5.9. Geometric probability. In some types of work the notion of geometric probability is useful.

Let the term *content* mean length, or area, or volume, $\cdots$, of a *region* depending on whether the regions under consideration are one-, or two-, or three-, $\cdots$ dimensional regions.

Definition D. (*Geometric probability*). Suppose an event E (or the non-occurrence of the event) is associated with the selecting of a point in a region R. Suppose all points in the region R are taken to be equally likely selections. If an event E can happen by the selection of a point in a region R_E within the region R, (and E can not happen by the selection of any other points in R except those in R_E) then the probability of the event E occurring is defined to be

$$P(E) = k_E/k,$$

where k_E is the content of R_E, and k is the content of R.

5.10. Probability in a continuous case. Suppose for a set of circumstances S an event E is described by the value of a continuous variable x. Suppose the probability that x falls between $x - \Delta x/2$ and $x + \Delta x/2$ is denoted by

$$\Delta p(x;S) = p(x - \Delta x/2, x + \Delta x/2; S).$$

If as the number n of phases in S increases without bound, Δx and Δp tend to zero, then Δp is sometimes called the *probability differential* and is often written

$$dp(x;S) = p(x - dx/2, x + dx/2; S).$$

5.11. Elementary foundations of probability. Let $E, F, \cdots$ be events. The probability of the event E occurring is denoted by $P(E)$, the probability of the event E not occurring is denoted by $P(\overline{E})$. $\overline{E}$ means "not E"; i.e., that the event E does not occur. $\overline{E}$ is the *opposite* or *complementary* event of E.

AXIOM I. $\qquad\qquad 0 \leq P(E) \leq 1.$

A *certain event* is an event which occurs in every observation.
An *impossible event* is one which can not occur in any observation.

AXIOM II. The probability of a certain event is 1.

AXIOM III. The probability of an impossible event is zero.

If an event has the probability 1, the event is *practically certain* to occur in every observation, and the event is said to be *statistically* (or *stochastically*) *certain*.

If an event has the probability 0, the event is *practically impossible* in every observation, and the event is said to be *statistically* (or *stochastically*) *impossible*.

5.12. Frequencies associated with two events. Consider two events E and F and a series of n observations in which in each observation one and only one of the following possibilities occurs: E and F, E and $\overline{F}$, $\overline{E}$

and F, $\bar{E}$ and $\bar{F}$. The number of times each of these possibilities has occurred in n observations is as indicated below:

	F	$\bar{F}$	Total
E	n_{11}	n_{12}	$n_{11} + n_{12}$
$\bar{E}$	n_{21}	n_{22}	$n_{21} + n_{22}$
Total	$n_{11} + n_{21}$	$n_{12} + n_{22}$	n

Let $f(X)$ denote the relative frequency of event X,

$E \cup F$ denote the occurrence of either E, or F, or both,

EF, or $E \cap F$, denote the occurrence of both E and F, and

$F|E$ denote the occurrence of F under the condition that E has occurred.

Then

$$f(E) = (n_{11} + n_{12})/n, \qquad\qquad f(F) = (n_{11} + n_{21})/n,$$

$$f(E \cup F) = (n_{11} + n_{12} + n_{21})/n, \qquad f(EF) = n_{11}/n,$$

$$f(F|E) = n_{11}/(n_{11} + n_{12}), \qquad\qquad f(E|F) = n_{11}/(n_{11} + n_{21}).$$

$$(5.12.1) \qquad f(E \cup F) = f(E) + f(F) - f(EF).$$

$$(5.12.2) \qquad f(EF) = f(E)\, f(F|E) = f(F)\, f(E|F).$$

Notation. In this chapter, the symbol $P(E \cup F)$, sometimes written $P(E$ or $F)$, or $P(E + F)$, denotes the probability (chance) of occurrence of either E, or F, or both. The symbol $P(EF)$, sometimes written $P(E \cap F)$, or as $P(E$ and $F)$, denotes the probability of the simultaneous occurrence of both E and F.

5.13. **Axiom IV. (Addition law of probability).** The probability that at least one of two events occurs is the sum of the probabilities of each event minus the probability of the simultaneous occurrence of both events:

$$(5.13.1) \qquad P(E \cup F) = P(E) + P(F) - P(EF).$$

EXAMPLE 1. Two separate well-shuffled decks of cards are given. A card is drawn from each deck. The probability that either one or the other, or both, of the cards drawn is an ace of spades is

$$(1/52) + (1/52) - 1/(52)(52) = 103/2704.$$

Axiom IV corresponds to the frequency relation (5.12.1).

5.14. Conditional probabilities. The probability that F occurs on the condition that E has already occurred is written $P(F|E)$, and is known as the *conditional* (or *conditioned*, or *relative*) *probability* of F under condition E. In contrast $P(F)$ is called the *absolute probability* of F.

5.15. Axiom V. (Multiplication law, or law of compound probabilities). The probability that both of two events occur is the product of the absolute probability of one event by the conditional probability of the other under the condition of the first event:

(5.15.1) $\qquad P(EF) = P(E) \, P(F|E) = P(F) \, P(E|F).$

Axiom V corresponds to the frequency relation (5.12.2).

Remark. An alternate approach used by some writers is to define the *conditional probability* $P(F|E)$ as

$$P(F|E) = P(EF)/P(E), \qquad \text{if } P(E) \neq 0.$$

Then, relation (5.15.1) is a theorem.

EXAMPLE 1. All the spades from a deck of cards are removed. The remaining cards are well-shuffled. Two cards are then drawn successively from the remaining deck without replacing either card. The probability of both cards being diamonds is $(13/39)(12/38) = 2/19$.

5.16. A scheme for representing various combinations of events in case the events are not mutually exclusive. One can represent by regions of a rectangle the various possible events EF, $E\overline{F}$, $\overline{E}F$, $\overline{E}\,\overline{F}$, as illustrated in Fig. 5.16.1. Such a diagram is called an *"Euler diagram"*.

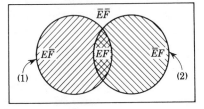

FIGURE 5.16.1

The scheme can be used to represent situations related to three or more events.

In such a diagram a point in the rectangle represents a possible situation resulting from a possible trial. Associated with any point inside circle (1) is a situation in which event E occurs, inside circle (2) a situation in which event F occurs, inside the region covered by circles (1) and (2) a situation in which either E or F or both E and F occur. The points falling inside the region common to the two circles represent situations in which both event E and event F occur; that is, EF occurs. The region outside both circles represents situations in which both $\overline{E}$ and $\overline{F}$ occur, etc.

5.17. Mutually exclusive events. If the events E and F can not occur simultaneously (symbolized by $EF = 0$), they are said to be *mutually exclusive*. In this case $P(EF) = 0$.

THEOREM 1.   If the events are mutually exclusive, then the probability of the occurrence of either one or the other is the sum of the probabilities of each occurring separately:

(5.17.1) $$P(E \cup F) = P(E) + P(F).$$

EXAMPLE 1.   The probability in a throw with an 8-sided die of obtaining either a 2 or a 3 is $(1/8) + (1/8) = 1/4$.

THEOREM 2.   If E is an event and $\overline{E}$ the complementary event,

(5.17.2) $$P(E) = 1 - P(\overline{E}).$$

Consider an experiment involving repeated independent trials. If for each trial there are only two possible outcomes, E and $\overline{E}$, called *success* and *failure*, respectively, and if throughout the trials the probability $P(E)$ remains the same, then the repeated trials are called *Bernoulli trials*.

5.18. Statistical independence. If the probability of the occurrence of F is independent of whether or not E has occurred, the event F is said to be *statistically (stochastically) independent* of the event E. In this case $P(F|E) = P(F)$. Then from Axiom V, $P(E) = P(E|F)$ and E is statistically (stochastically) independent of F. E and F are said to be *statistically (stochastically) independent*.

THEOREM 1.   If two events are statistically independent the probability of both occurring is the product of the probabilities of the occurrence of each event separately:

(5.18.1) $$P(EF) = P(E)\ P(F).$$

EXAMPLE 1.   One card is drawn from each of two decks from which all the spades have been removed. The probability that both cards are hearts is $(13/39)(13/39) = 1/9.$[cf. §5.15, Ex. 1.]

In the practice of statistics the term *independent* is used commonly to mean *statistical independence*. It should be realized that *statistical independence* and *causal independence* do not necessarily mean the same thing. Likewise *functional independence* means still something different. (See §18.10.)

5.19. Many events.   Let $E_1 \cup E_2 \cup \cdots \cup E_k$ indicate that at least one of k events $E_1, E_2, \cdots, E_k$ occurs. Let $E_1 E_2 \cdots E_k$ indicate that the k events $E_1, E_2, \cdots, E_k$ occur simultaneously.

THEOREM 1. (THEOREM OF ADDITION).   If $E_1, E_2, \cdots$ is any infinite sequence of events which exclude each other two-by-two, that is, $E_i E_j = 0$, $i \neq j$, then the probability that at least one of the events $E_1, E_2, \cdots$ occurs is the sum of the probabilities of each occurring separately; that is,

(5.19.1) $$P\left(\sum_{i=1}^{\infty} E_i\right) = \sum_{i=1}^{\infty} P(E_i).$$

THEOREM 2. (THEOREM OF MULTIPLICATION).   If each of the events $E_1, \cdots, E_k$ is statistically independent of each combination of all the other events, then the probability of the simultaneous occurrence of all the events is the product of the probabilities of each event occurring separately; that is,

(5.19.2) $$P(E_1 E_2 \cdots E_k) = P(E_1)\ P(E_2) \cdots P(E_k).$$

It is to a large extent upon the addition and multiplication theorems of probability theory that the mathematical theory of statistics rests.

Some writers use the following definition for statistical independence.

5.20. Definition.   Two events E and F are said to be *statistically independent* if Eq. (5.18.1) holds,

(5.20.1) $$P(EF) = P(E)\ P(F).$$

This definition is also accepted when $P(F) = 0$, in which case $P(E|F)$ is not defined.

5.21. Definition.   The events $E_1, E_2, \cdots, E_n$ are called *mutually independent* if for all combinations $1 \leq i < j < k \cdots \leq n$ the following multiplication rules apply

$$P(E_iE_j) = P(E_i)\,P(E_j), \qquad (C_2{}^n \text{ equations})$$

$$P(E_iE_jE_k) = P(E_i)\,P(E_j)\,P(E_k), \qquad (C_3{}^n \text{ equations})$$

(5.21.1)

$$\vdots$$

$$P(E_1 \cdots E_n) = P(E_1)\,P(E_2)\cdots P(E_n). \qquad (C_n{}^n \text{ equations})$$

Axioms and theorems for several events

5.22. Axiom IVa. The probability of the occurrence of at least one of the three events E, F, G is

$$(5.22.1) \quad P(E \cup F \cup G) = P(E) + P(F) + P(G)$$
$$- P(EF) - P(EG) - P(FG) + P(EFG).$$

Axiom IVb. The probability of the occurrence of at least one among n events $E_1, \cdots, E_n$ is

$$(5.22.2) \quad P(E_1 \cup \cdots \cup E_n) = S_1 - S_2 + S_3 - S_4 + \cdots + (-1)^{n-1}S_n,$$

where

$$S_1 = \sum P(E_i), \quad S_2 = \sum P(E_iE_j),$$
$$S_3 = \sum P(E_iE_jE_k), \cdots, \quad S_n = P(E_1\,E_2\cdots E_n).$$

The symbol $P(E_iE_j \cdots E_q)$ is sometimes written $p_{i,j,\ldots,q}$. In these sums each combination appears once and only once, and $i, j, k, \cdots$ run from 1 to n, with $i < j < k \cdots \leqq n$, S_r has $C_r{}^n$ terms.

5.23. Axiom Va. The probability of the occurrence of all of the three events E, F, G is

$$(5.23.1) \qquad P(EFG) = P(G)\,P(F|G)\,P(E|FG).$$

Axiom Vb. The probability of the occurrence of all of the n events $E_1, \cdots, E_n$ is

$$(5.23.2) \quad P(E_1\,E_2 \cdots E_n)$$
$$= P(E_1)\cdot P(E_2|E_1)\cdot P(E_3|E_1E_2) \cdots P(E_n|E_1E_2 \cdots E_{n-1}).$$

Here $P(E_3|E_1E_2)$ denotes the probability of E_3 after E_1 and E_2 have happened, and so on.

5.24. All general theorems on probabilities are valid also for conditional probabilities with respect to any particular hypothesis H. For example,

$$(5.24.1) \qquad P((E \cup F)|H) = P(E|H) + P(F|H) - P(EF|H).$$

(5.24.2) $$P(E) = P(EF) + P(E\overline{F})$$

(5.24.3) $$P(E \cup F) = 1 - P(\overline{E}\overline{F}) = 1 - P(\overline{E})\,P(\overline{F}\,|\,\overline{E})$$

(5.24.4) $$P(E \cup F) = 1 - \{1 - P(E)\}\{1 - P(F\,|\,\overline{E})\}.$$

If E and F are statistically independent,

(5.24.5) $$P(E \cup F) = 1 - \{1 - P(E)\}\{1 - P(F)\}.$$

EXAMPLE 1. A device is assembled from two parts. The chance that part A is defective is p_1, that part B is defective is p_2. For the assembled device the chance that both parts are defective is $p_1 p_2$, that A is defective and B is not defective is $p_1(1 - p_2)$, that A is not defective and B is defective is $(1 - p_1)p_2$ and that both A and B are without defects is $(1 - p_1)(1 - p_2)$. Of these mutually exclusive events one must occur. Hence, the sum of the chances must be unity, that is,

$$1 = p_1 p_2 + p_1(1 - p_2) + (1 - p_1)p_2 + (1 - p_1)(1 - p_2).$$

5.25. Bayes' theorem. Consider the k events $E_1, \cdots, E_k$ which exclude each other two-by-two, and of which at least one occurs with certainty. An event X is observed and is known to have occurred in conjunction with or as a consequence of one of the events E_i. The probability of the occurrence of the event E_i, on the hypothesis that the event X has occurred, is

(5.25.1) $$P(E_i\,|\,X) = \frac{P(XE_i)}{P(X)} = \frac{P(E_i)\,P(X\,|\,E_i)}{P(X)}$$

where

(5.25.2) $$P(X) = \sum_{j=1}^{k} P(E_j)\,P(X\,|\,E_j)$$

The events $E_1, \cdots, E_k$ are called "hypotheses" or "causes" of X. Eq. (5.25.1) is known as "Bayes' Rule for the probability of causes".

5.26. Mathematical expectation of gain. Let $E_1, E_2, \cdots, E_n$ denote mutually exclusive and exhaustive events. Suppose $p_1, p_2, \cdots, p_n$ are the respective probabilities of these events occurring. Suppose that by agreement or otherwise, an individual I receives money to the amount of m_1 if E_1 happens, m_2 if E_2 happens, $\cdots$, m_n if E_n happens. By I's *mathematical expectation E of gain (winnings)* is meant

(5.26.1) $$E = E(m) = m_1 p_1 + m_2 p_2 + \cdots + m_n p_n.$$

Non-finite case; case of events expressible in terms of a continuous variable

5.27. The above axioms and theorems have been stated as if an event E can result from the phases of a system S having n different phases and no more. They actually hold for more general cases S involving non-finite phases (sets), and for events characterized by continuous variables. A careful treatment of these cases is beyond the scope of this book.

When n is finite, if E must inevitably happen in all of the n cases, then the probability p of the event E occurring is $p = 1$, and E is "certain", and $\overline{E}$ is "impossible". But, if the circumstances S depend on a non-finite set, or result in events expressible by a continuous variable, $p = 1$ does not imply "certainty", or $p = 0$ "impossibility".

EXAMPLE 1. A point B is selected at random on a line segment containing a particular point A. The chance that B is the point A is 0, though since some point is selected the chance $p = 0$ does not mean that selecting A is impossible.

In considering the continuous case some writers speak of "almost mutually exclusive" and "almost statistically independent" events, etc., just as some refer to "almost impossible" events as events for which the probability $p = 0$.

5.28. **The term "random".** A universally acceptable definition of *random* is probably not possible. In certain fields or types of experiments (as in casting a coin) it is not possible to predict satisfactorily what may happen (head or tail) in each individual experiment, even though the experiment be carefully controlled, the results varying in such an irregular fashion as to elude all attempts as to prediction. In such cases the results are described as *random* and the experiments called *random experiments*. However, when attention is drawn to the results of extensive sequences of random experiments, experience has shown that the data reveal a certain statistical regularity, as in the averages of the results obtained, and as in the long run stability of frequency ratios. In the mathematical theory constructed to serve such situations the fact of statistical regularity is idealized by assuming the existence of "mathematical probabilities" as conceptual counterparts of the frequency ratios. This idealization is carried further in the mathematical theory by constructing the theory so the additive probability theorems hold for an enumerable sequence of events. This leads to the concept of a *random variable* and its probability distribution.

5.29. **Probability distributions (one dimension).** A random experiment is repeated a number of times. Each trial leads to an event E represented by a real number called a chance quantity X. Suppose all possible values of X are $x_1, x_2, \cdots$. Suppose there is a non-negative function f such that the chance that X takes the value x is $f(x)$. Then $f(x)$ is called the *probability function* (p.f.) of the chance (random) variable X and is written $P(X = x) = f(x)$.

Discrete case. Consider the case when X takes on the values $x_1, x_2, \cdots, x_k$, where $x_1 < x_2 < \cdots < x_k$, and where x_i corresponds to the event E_i. Then

$$P(E_i) = P(X = x_i) = f(x_i).$$

The probability that $X \leqq x'$ is

$$(5.29.1) \qquad F(x') = P(X \leqq x') = \sum_{i=1}^{j} f(x_i),$$

where the summation is carried out over values of i such that $x_i \leqq x'$. $F(x')$ is called the *cumulative discrete probability distribution function* of X. This is written d.f. of X or c.d.f. of X.

$F(x')$ and $f(x_i)$ have the following properties:

$$F(x_k) = \sum_{i=1}^{k} f(x_i) = 1, \qquad\qquad f(x_i) \geqq 0.$$

$(5.29.2)$

$$P(a < X \leqq b) = F(b) - F(a) = \sum_{a < x_i \leqq b} f(x_i).$$

Continuous case. In case the chance variable X can take on any real number value, the probability $P(X \leqq x)$ is denoted by $F(x)$, and $F(x)$ is called the *cumulative distribution function* of the variable X.

The function $f(x) = dF(x)/dx$ is called the *probability density function* (p.d.f.) of the chance variable X.

$F(x)$ and $f(x)$ have the following properties:

$$P(X \leqq x) = F(x) = \int_{-\infty}^{x} f(x)\, dx, \qquad\qquad f(x) \geqq 0.$$

$(5.29.3)$

$$F(-\infty) = 0, \qquad 0 \leqq F(x) \leqq 1, \qquad F(\infty) = 1.$$

$$P(a < X \leqq b) = F(b) - F(a) = \int_{a}^{b} f(x)\, dx.$$

(See Chapter VI for further details.)

5.30. **Probability distributions** (**two dimensions**). If X and Y are continuous chance variables having the joint *probability density function* $f(x,y)$, and *cumulative distribution function* $F(x,y)$ the probability that $X \leqq x$ and $Y \leqq y$ is

(5.30.1) $P(X \leqq x', Y \leqq y') = F(x',y') = \int_{-\infty}^{x'} \int_{-\infty}^{y'} f(x,y)\, dx\, dy.$

$F(x,y)$ and $f(x,y)$ have the following properties:

$$F(-\infty,y) = F(x,-\infty) = F(-\infty,-\infty) = 0. \qquad F(\infty,\infty) = 1.$$

$$f(x,y) \geqq 0. \qquad 0 \leqq F(x,y) \leqq 1. \qquad f(x,y) = \frac{\partial^2 F(x,y)}{\partial x\, \partial y}.$$

$F(\infty,y) = F_2(y)$ is the *marginal distribution function* of y.
$F(x,\infty) = F_1(x)$ is the *marginal distribution function* of x.

(5.30.2) $P(a < X \leqq b, c < Y \leqq d) = \int_a^b \int_c^d f(x,y)\, dx\, dy.$
   $= F(b,d) - F(b,c) - F(a,d) + F(a,c).$

The two-dimensional discrete case is treated similarly, with the integral being replaced by the corresponding sum. (See Chapter XI.)

5.31. **Probability distributions** (*n***-dimensions**). In the following sections generalizations of the notions of probability and distribution functions to cases involving many variables are given. For specific examples involving two variables see Chapter XI.

Probability function. A random experiment C is repeated a number of times under uniformly controlled conditions. Suppose the results of each particular experiment are given by n real numbers $X_1, X_2, \cdots, X_n$. Let $X = (X_1, \cdots, X_n)$. Each experiment C yields observations $X = (X_1, \cdots, X_n)$. Let S be any sub-set of points in R, where R is the n-dimensional space within which X lies. Suppose that when the point X falls in S the event E takes place. Let the probability P that X falls in S be assumed to exist, to be uniquely defined for all suitably defined sets S, and be written in any of the following ways

(5.31.1) $P = P(E) = P(X \subset S) = Pr(X \subset S) = P(S).$

Here $X \subset S$ means X is included in S. Thus the probability P is a *set function*. $P(S)$ is assumed to be non-negative with $P(R) = 1$. $P(S)$ is also assumed to obey the addition theorem as in §5.19, Th. 1. $P(S)$ is called the *probability function* of X, often written *pr.f.* of X.

Any set function $P(S)$ of the sort described above defines a *distribution* in R, such that any set S has associated with it the number $P(S)$.

Cumulative distribution function. Corresponding to $P(S)$ is a point function $F(x) = F(x_1, \cdots, x_n)$ defined as follows, and known as the *cumulative (probability) distribution function* of X, (written *d.f.* of X, or *c.d.f.* of X). [The term *cumulative distribution function* is also called the *cumulated,* or *accumulative,* or *accumulated,* or *total, distribution function.*] This distribution is uniquely defined. $F(x)$ is a non-decreasing function in each variable x_i.

$$(5.31.2) \qquad F(x) = F(x_1, \cdots, x_n) = P(X_1 < x_1, \cdots, X_n < x_n),$$
$$0 \le F(x) \le 1,$$
$$F(-\infty, x_2, \cdots, x_n) = \cdots = F(x_1, \cdots, x_{n-1}, -\infty) = 0,$$
$$F(+\infty, \cdots, +\infty) = 1.$$

The variable X ranges over a set R and has the distribution (5.31.2). X is commonly called a *random variable,* and R the *spectrum* of X.

5.32. **Combined random variables.**

Suppose random experiments $C_1, C_2, \cdots, C_n$ lead to the random variables $X_1, \cdots, X_n$, of dimensions $k_1, \cdots, k_n$, respectively. A combined experiment $C = (C_1, \cdots, C_n)$ in which each experiment C_i is carried out and all the results of all of the experiments are observed jointly results in a combined variable $(X_1, \cdots, X_n)$ of dimensions $k_1 + \cdots + k_n$. In such a case it is assumed that *when $X_1, \cdots, X_n$ are random, so is any combined variable $(X_1, \cdots, X_n)$ a random variable*. This combined variable has a joint probability distribution involving variables $X_1, \cdots, X_n$.

Any function $\eta(X)$ of the random variable X, with suitable restrictions, is itself a random variable with a probability distribution uniquely determined from the distribution of X.

5.33. **Probability density, or frequency function.**

If the partial derivative

$$(5.33.1) \qquad f(x_1, x_2, \cdots, x_n) = \frac{\partial^n F}{\partial x_1 \, \partial x_2 \, \cdots \, \partial x_n}$$

exists, (5.33.1) is called the *probability density* or *frequency function* of the distribution F defined in (5.31.2).

The *conditional frequency* function of $x_1, \cdots, x_k$ relative to the hypothesis $X_{k+1} = x_{k+1}, \cdots, X_n = x_n$, is defined to be

$$(5.33.2) \qquad f(x_1, \cdots, x_k | X_{k+1}, \cdots, X_n)$$

$$= \frac{f(x_1, \cdots, x_n)}{\int_{-\infty}^{\infty} \cdots \int_{-\infty}^{\infty} f(x_1, \cdots, x_k, X_{k+1}, \cdots, X_n) \, dx_1 \cdots dx_k}.$$

5.34. Marginal distribution. When all of the variables in Eq. (5.31.2) except x_i tend to $+\infty$, F will tend to a limit $F_i(x_i)$, the *marginal distribution* of x_i,

$$(5.34.1) \qquad F_i(x_i) = F(+\infty, \cdots, +\infty, x_i, +\infty, \cdots, +\infty).$$

When $n - k$ of the variables in (5.31.2) tend to $+\infty$, the limit of F defines a *k-dimensional marginal distribution* of the remaining k variables.

5.35. Statistically independent random variables. k random variables are *statistically (mutually) independent* (or, *stochastically independent*) if their c.d.f. is of the form

$$(5.35.1) \qquad F(x_1, x_2, \cdots, x_k) = F_1(x_1) \, F_2(x_2) \cdots F_k(x_k),$$

where $F_i(x_i)$ is the marginal distribution of X_i.

EXAMPLE 1. Let $F(x_1, x_2)$ be the distribution function (c.d.f.) of X_1, X_2. Then $F_1(x_1) = F(x_1, +\infty)$, $F_2(x_2) = F(+\infty, x_2)$ are the marginal distributions of X_1 and X_2, respectively. The random variables X_1, X_2 are *independent in the probability sense*, or *statistically independent*, if

$$F(x_1, x_2) = F_1(x_1) \, F_2(x_2).$$

THEOREM 1. A necessary and sufficient condition for the statistical independence of X_1 and X_2 is that their joint c.d.f. be factorable into the product of a function of x_1 alone and a function of x_2 alone, that is,

$$F(x_1, x_2) = H_1(x_1) \, H_2(x_2).$$

Mathematical expectation

5.36. Expected value (one dimension). Many of the properties of probability distributions are associated with the concept of mathematical expectation.

If $g(x)$ is a function of the chance variable x having the probability (density) function $f(x)$ and cumulative distribution $F(x)$, the *expected value* of g is defined to be

$$E[g(x)] = \int_{-\infty}^{\infty} g(x) f(x) \, dx, \qquad \text{if } x \text{ is continuous}$$

(5.36.1)

$$= \sum_{\substack{-\infty \\ \text{all } x_i}}^{\infty} g(x_i) f(x_i), \qquad \text{if } x \text{ is discrete.}$$

The kth moment about the origin of x is $\mu_k' = E(x^k)$. The expected value of x is $E(x) = \mu_1' = \mu$. μ is called the *mean* of x.

The kth moment about the mean μ is $\mu_k = E[(x - \mu)^k]$. μ_k is called the kth *central moment*. μ_2 is the *variance*, and $+\sqrt{\mu_2} = \sigma$ is the *standard deviation* of x. $\sigma^2 = E(x^2) - [E(x)]^2$. (See Chapter VI for further details.)

5.37. **Expected value (two or more dimensions).** If $g(x,y)$ is a function of the chance variables x and y having probability density function $f(x,y)$, the *expected value*, or *mean*, of the function $g(x,y)$ is defined to be

(5.37.1) $$E[g(x,y)] = \int_{-\infty}^{\infty} \int_{-\infty}^{\infty} g(x,y) f(x,y) \, dx \, dy.$$

The *product moments of x and y* about the origin are

(5.37.2) $$E[x^i y^j] = \int_{-\infty}^{\infty} \int_{-\infty}^{\infty} x^i y^j f(x,y) \, dx \, dy.$$

The mean of the x's is $\bar{x} = \mu_x = E[x]$, and the mean of the y's is $\bar{y} = \mu_y = E[y]$.

The *central product moments* are

(5.37.3) $$\mu_{ij} = E[(x - \mu_x)^i (y - \mu_y)^j].$$

The *variance* of x is $\mu_{20} = \sigma_x^2$; the *variance* of y is $\mu_{02} = \sigma_y^2$. μ_{11} is the *covariance* of x and y.

The above definitions hold for the discrete cases when the integral is replaced by the corresponding sum. Further details may be found in Chapters VI and XI.

5.38. If $g(x_1, \cdots, x_n)$ is a function of the chance variables $x_1, \cdots, x_n$, having the joint probability density function $f(x_1, \cdots, x_n)$ and cumulative distribution $F(x_1, \cdots, x_n)$, the *expected value* (or *mean value*) of $g(x_1, \cdots, x_n)$ is defined to be

(5.38.1) $$E[g(x_1, \cdots, x_n)]$$

$$= \int_{-\infty}^{\infty} \cdots \int_{-\infty}^{\infty} g(x_1, \cdots, x_n) f(x_1, \cdots, x_n) \, dx_1 \cdots dx_n.$$

This definition holds for the discrete case when the integral is replaced by the corresponding appropriate sum.

The *moments* (about the origin) of the distribution are the expected values

$$\nu_{v_1 \cdots v_n} = E(x_1^{v_1} \cdots x_n^{v_n}),$$

where $v_1 + \cdots + v_n$ is the *order* of the moment.

The *means* are

$$m_i = E(x_i). \qquad\qquad (i = 1, \cdots, n.)$$

The *central moments* (about the point $(m_1, m_2, \cdots, m_n)$) are

$$\mu_{v_1 \cdots v_n} = E[(x_1 - m_1)^{v_1} \cdots (x_n - m_n)^{v_n}].$$

The *variance* λ_{ii} of x_i and *covariance* λ_{ij} of x_i and x_j are

$$\lambda_{ii} = \sigma_i^2 = E[(x_i - m_i)^2],$$

$$\lambda_{ij} = \rho_{ij}\sigma_i\sigma_j = E[(x_i - m_i)(x_j - m_j)].$$

Here ρ_{ij} is the *correlation coefficient* and σ_i is the *standard deviation* of x_i.

Moment matrix Λ:

$$\Lambda = \begin{bmatrix} \lambda_{11} & \cdots & \lambda_{1n} \\ \cdots\cdots\cdots\cdots\cdots \\ \lambda_{n1} & \cdots & \lambda_{nn} \end{bmatrix}.$$

Correlation matrix P:

$$P = \begin{bmatrix} \rho_{11} & \cdots & \rho_{1n} \\ \cdots\cdots\cdots\cdots\cdots \\ \rho_{n1} & \cdots & \rho_{nn} \end{bmatrix}.$$

Special cases of these definitions, as for the case of a single variable, are considered in other chapters.

5.39. Many theorems relating to mathematical expectation are known. The following elementary cases are typical:

THEOREM 1.   The expected value (or mean value) of a sum of two functions $g_1(x_1)$ and $g_2(x_2)$ of the variables x_1 and x_2 is equal to the sum of the expected values (or mean values), if they exist, of the functions, that is,

$$E[g_1(x_1) + g_2(x_2)] = E[g_1(x_1)] + E[g_2(x_2)].$$

(The variables x_1 and x_2 need not be statistically independent.)

THEOREM 2.   The expected value (or mean value) of a product of two functions $g_1(x_1)$ and $g_2(x_2)$ is equal to the product (if they exist) of the expected values (or mean values) of the functions, provided the variables x_1 and x_2 are statistically independent [and consequently $g_1(x_1)$ and $g_2(x_2)$ are statistically independent]. (This theorem is not necessarily true if the variables are not statistically independent.)

THEOREM 3. $E(ax + b) = a\,E(x) + b.$
$E(ax + by) = a\,E(x) + b\,E(y).$
$E(x + y) = E(x) + E(y).$
$E(xy) = E(x)\,E(y),$ if x and y are
statistically independent.

Description of distributions by Stieltjes integrals

5.40. Discrete and continuous distributions may be treated in a unified manner by the use of *Stieltjes integrals*.

The *Stieltjes integral* of a function $g(x)$ with respect to $F(x)$ over the interval $a \leq x \leq b$ is defined to be

$$(5.40.1) \qquad \int_a^b g(x)\,dF(x) = \lim \sum_{j=1}^k g(\xi_j)[F(x_{j+1}) - F(x_j)],$$

where $x_j < \xi_j < x_{j+1}$, and where the limit is taken as the maximum length of the intervals (x_j, x_{j+1}) tends to zero.

If $g(x)$ is continuous on $a \leq x \leq b$, and $F(x)$ is monotone, integral (5.40.1) exists. When $F(x) = x$, (5.40.1) is a *Riemann integral*.

If $F(x)$ is a cumulative distribution function of X and $g(x)$ is continuous on $a \leq x \leq b$, then

(a) if $F(x)$ is continuous, (5.40.1) is an ordinary (Riemann) integral.

$$(5.40.2) \qquad \int_a^b g(x)\,f(x)\,dx, \qquad\qquad f(x) = F'(x);$$

(b) if $F(x)$ is discrete, (5.40.1) is a simple sum

$$(5.40.3) \qquad \sum_{j=1}^k g(x_j)\,p_j, \qquad\qquad a < x_j < b,$$

where

$$(5.40.4) \qquad p_j = P(X = x_j) = F(x_{j+1}) - F(x_j).$$

The integral (5.40.1) is a representation of both (5.40.2) and (5.40.3). The familiar properties of the ordinary (Riemann) integrals extend to Stieltjes integrals. For example, $\int_a^b dF(x) = F(b) - F(a)$.

5.41. Any distribution function $F(x)$ can be written as a Stieltjes integral

(5.41.1)
$$F(x) = \int_{-\infty}^{x} dF(t).$$

(a) if $F(x)$ is continuous,

(5.41.2)
$$F(x) = \int_{-\infty}^{x} f(t)\, dt, \qquad\qquad f(t) = F'(t);$$

(b) if $F(x)$ is discrete,

(5.41.3)
$$F(x) = \sum_{j=1}^{k} F(x_{j+1}) - F(x_j) = \sum_{j=1}^{k} p_j.$$

EXAMPLE. The causal distribution (6.23.2) may be expressed as a Stieltjes integral in the form

(5.41.4)
$$F(x') = \sum_{t=-\infty}^{x'} d\epsilon(t - u) = \epsilon(x' - u).$$

Also, from (5.40.2)

(5.41.5)
$$\int_{t=-\infty}^{\infty} g(t)\, d\epsilon(t - u) = g(u),$$

(5.41.6)
$$\int_{t=-\infty}^{\infty} d\epsilon(t - u) = 1.$$

5.42. **δ-function.** As a shorthand for rigorous statements involving Stieltjes integrals and the causal distribution function $\epsilon(x - u)$, an entity known as the δ-function was defined by Dirac. (See §6.23.) It is widely used in theoretical physics.

Let a be a constant. The δ-function is defined as a "function" $\delta(x - a)$ such that

(5.42.1)
$$\int_{-\infty}^{\infty} \delta(t - a)\, dt = 1$$

and

(5.42.2)
$$\delta(x - a) = 0 \qquad\qquad \text{if } x \neq a.$$

By analogy with (5.41.5),

(5.42.3)
$$\int_{-\infty}^{\infty} g(t)\, \delta(t - a)\, dt = g(a).$$

The "derivative" of the δ-function with respect to x is defined as an entity $\delta'(x - a)$, such that

(5.42.4)
$$\int_{-\infty}^{\infty} g(t)\, \delta'(t - a)\, dt = -g'(a).$$

The δ-function, though useful, is not a true mathematical function.

Many rules for manipulating the δ-function have been formalized. To avoid pitfalls in the use of the δ-function requires much care and attention to details. Many writers consider it wise to use the more rigorous approaches available in the literature, which include a mathematically rigorous treatment of the delta function.

5.43. **On applications of the theory of probability.** In applying probability theory one should bear in mind the manner in which the quantities in the theory arise and the mode of measuring them. Measurements of pertinent quantities may take the form of relative frequencies which suggest assumptions as to the corresponding probabilities in the theory under consideration. The "verification" of the theory hinges around the comparison of the probabilities obtained from the theory and corresponding experimentally measured relative frequencies.

A theory of probability can never tell us anything about the actual course that a single event may follow. The theory begins from given or assumed probabilities and the results are stated in the form of probabilities.

six

PROBABILITY DISTRIBUTIONS IN ONE DIMENSION

6.1. Frequently one is interested in the probabilities for a set of events, the events being described by numbers. For example, $x_1, x_2, \cdots$ may be possible values of a chance quantity X. If X can take on only certain isolated values, X is called a *discrete chance quantity*. With each value x that X can take on there is a probability $f(x)$ that $X = x$. This is written $P(X = x) = f(x)$. Here $f(x)$ is called the *probability distribution*, or *probability density*, or *frequency function*, of X.

Discrete probability distributions

6.2. Suppose X can take on only the possible discrete values $x_1, x_2, \cdots, x_k$, with $x_1 < x_2 < \cdots < x_k$. Let the probability that X takes the value x_i be

(6.2.1) $$P(X = x_i) = f(x_i). \qquad (i = 1, \cdots, k)$$

Then

(6.2.2) $$\sum_{i=1}^{k} f(x_i) = 1.$$

Such a distribution $f(x)$ is said to be *discrete*. The distribution $f(x)$ can be represented by a *probability bar chart* as in Fig. 6.2.1, where the

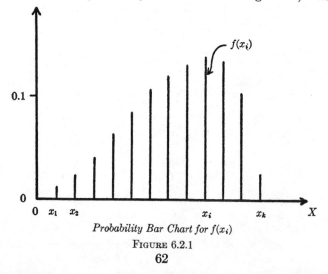

Probability Bar Chart for $f(x_i)$

FIGURE 6.2.1

62

ordinates at x_i represent $f(x_i)$; or by a *probability histogram* as in Fig. 6.2.2, where the rectangles of height $f(x_i)$ are drawn centered at x_i. In

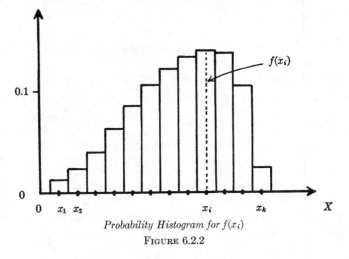

Probability Histogram for $f(x_i)$

FIGURE 6.2.2

the case of a probability histogram it is customary to make the rectangles of equal width so that the areas of the rectangles are in proportion to their heights.

6.3. By the *cumulative discrete probability* $F(x')$ is meant the probability that $X \leqq x'$. If x_i is the largest discrete value of X less than or equal to x',

$$(6.3.1) \qquad F(x') = P(X \leqq x') = \sum_{i=1}^{i=j} f(x_i).$$

Note: $F(x_k) = 1$.

The probability that $x' < X \leqq x''$ is

$$(6.3.2) \qquad P(x' < X \leqq x'') = F(x'') - F(x').$$

EXAMPLE 1. The probability $f(x)$ of obtaining x dots when two dice are rolled is given in Table 6.3.1.

TABLE 6.3.1

x	2	3	4	5	6	7	8	9	10	11	12
$f(x)$	1/36	2/36	3/36	4/36	5/36	6/36	5/36	4/36	3/36	2/36	1/36
$F(x)$	1/36	3/36	6/36	10/36	15/36	21/36	26/36	30/36	33/36	35/36	36/36

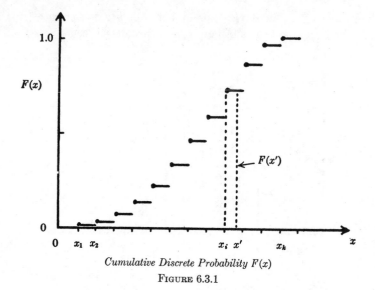

Cumulative Discrete Probability F(x)

FIGURE 6.3.1

The graphs of $f(x)$ and the cumulative probability $F(x)$ are given in Figs. 6.3.2, 6.3.3, and 6.3.4, respectively. From Table 6.3.1

$$P(X = 5) = 4/36, \qquad P(X \leqq 5) = 10/36,$$

$$P(X \leqq 8.7) = 26/36, \qquad P(X \leqq 3.5) = 3/36,$$

$$P(3.5 < X \leqq 8.7) = F(8.7) - F(3.5) = 26/36 - 3/36 = 23/36,$$

$$P(4 < X \leqq 6) = F(6) - F(4) = 15/36 - 6/36 = 9/36.$$

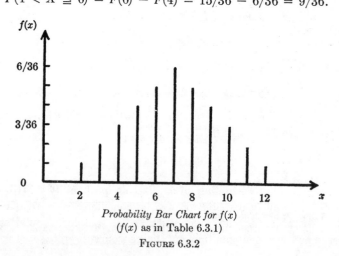

Probability Bar Chart for f(x)
(f(x) as in Table 6.3.1)

FIGURE 6.3.2

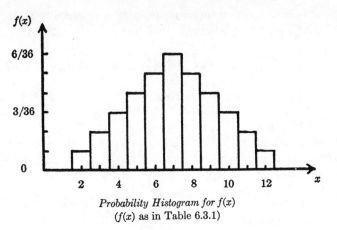

Probability Histogram for f(x)
(f(x) as in Table 6.3.1)

FIGURE 6.3.3

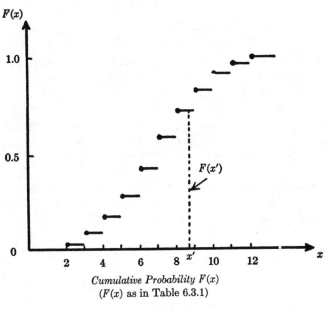

Cumulative Probability F(x)
(F(x) as in Table 6.3.1)

FIGURE 6.3.4

6.4. Statistical interpretation of a discrete probability distribution.

A discrete probability distribution can be considered as a relative frequency distribution for an indefinitely large sample (population) in which each measurement can take on a value from a discrete set of values.

Means, variances, $\cdots$ are calculated from a probability distribution in a manner similar to the case of a relative frequency distribution.

Parameters which describe a discrete probability distribution

6.5. In the following sections the discrete chance variable X is assumed to have a probability distribution $f(x)$ with a cumulative probability $F(x') = P(X \leqq x')$.

The principal features of the distribution $f(x)$ may be described by means of certain parameters. These parameters are defined in very much the same manner as in Chapter II for frequency distributions.

6.6. **Mean and variance.** The mean μ and the variance σ^2 of the discrete chance quantity X are, respectively:

$$(6.6.1) \qquad \mu = \sum_{i=1}^{k} x_i\, f(x_i),$$

$$(6.6.2) \qquad \sigma^2 = \sum_{i=1}^{k} (x_i - \mu)^2 f(x_i), \qquad \text{or} \qquad \sigma^2 = \sum_{i=1}^{k} x_i^2 f(x_i) - \mu^2.$$

$\sigma = +\sqrt{\sigma^2}$ is called the *standard deviation*, (or *root mean square error*, or *standard error*).

The mean μ is also called the *mean of distribution* $f(x)$, or the *mathematical expectation* $E(X)$ of X, or *the mean of* X, or the *expected value of* X. σ^2 is also called the *variance of the probability distribution* $f(x)$.

The quantity $(x - \mu)$ is called the *error* or *deviation* of the variable x; $|x - \mu|$, the *absolute error*, or *absolute deviation* of x.

6.7. **Mathematical expectation.** If $y = g(X)$ is a function of the discrete chance variable X, and X has the probability distribution $f(x)$, the *expected value* (or *mean value*) of $g(X)$ is defined to be

$$(6.7.1) \qquad E[g(X)] = \sum_{i=1}^{k} g(x_i)\, f(x_i).$$

6.8. r^{th} **moment** μ_r **about the mean** μ **(or central moments).** The mean value of $(X - \mu)^r$, or the expected value of $(X - \mu)^r$, is

$$(6.8.1) \qquad E[(X - \mu)^r] = \mu_r = \sum_{i=1}^{k} (x_i - \mu)^r f(x_i).$$

μ_r is called the rth *moment* of the distribution $f(x)$ about the mean μ. For a given $f(x)$, $\mu_0 = 1$, $\mu_1 = 0$, $\mu_2 = \sigma^2 = $ variance.

6.9. r^{th} **moment ν_r about the origin** $X = 0$. The mean value of X^r, or the expected value of X^r, is

$$(6.9.1) \qquad E[X^r] = \nu_r = \sum_{i=1}^{k} x_i^{\,r} f(x_i).$$

ν_r is called the rth *moment* of the distribution $f(x)$ about the origin $X = 0$. For any given $f(x)$, $\nu_0 = 1$, $\nu_1 = E[X] = \mu$, the mean.

6.10. **Relations between μ_r and ν_r.** These are as given in Eqs. (2.8.1).

6.11. r^{th} **absolute moment β_r about** $X = 0$. By the rth *absolute moment β_r* of the distribution $f(x)$ about $X = 0$ is meant

$$(6.11.1) \qquad E[|\,X\,|^r] = \beta_r = \sum_{i=1}^{k} |\,x_i\,|^r f(x_i).$$

6.12. **Moments about** $X = c$. By the rth *moment of the distribution* $f(x)$ *about* $X = c$ is meant

$$(6.12.1) \qquad E[(X - c)^r] = \sum_{i=1}^{k} (x_i - c)^r f(x_i).$$

The second moment about $X = c$ becomes a minimum when c is the mean μ.

6.13. **Factorial moments.** By the rth *factorial moment of the distribution* $f(x)$ *about* $X = 0$ is meant

$$(6.13.1) \qquad \nu_{[r]} = E[X^{[r]}] = \sum_{i=1}^{k} x_i^{\,[r]} f(x_i),$$

where the symbol $x^{[r]}$ denotes the factorial

$$(6.13.2) \qquad x^{[r]} = x(x - 1) \cdots (x - r + 1). \qquad (r = 1, 2, 3, \cdots)$$

The relation

$$(6.13.3) \qquad \sigma^2 = \nu_{[2]} - \mu(\mu - 1) = \sum_{i=1}^{k} x_i(x_i - 1)f(x_i) - \mu(\mu - 1)$$

is sometimes more convenient to use than Eq. (6.6.2).

6.14. **Measures locating a discrete distribution.** Three common parameters used to locate a distribution $f(x)$ are:

Mean: The *mean* $\mu = E[X]$ is the "center of gravity" of the distribution. The mean does not always exist in the sense that the mean is one of the x_i of the distribution.

Median: Any root of the equation $F(x) = 1/2$, where $F(x)$ is the cumulative probability defined in §6.3, is called a *median* of the distribution. In some cases the median is not uniquely defined. Every distribution has at least one median. The first absolute moment $E[|X - c|]$ about a point c becomes a minimum when c is the median.

Mode: Any maximum point x_0 of the distribution $f(x)$ is called a *mode* of the distribution. For a discrete distribution the values of the chance variable X are arranged in increasing order of magnitude, $x_1, x_2, \cdots, x_{i-1}, x_i, x_{i+1}, \cdots$. The number x_i is called a *mode* of the distribution $f(x)$ if $f(x_i) > f(x_{i-1})$, $f(x_i) > f(x_{i+1})$.

Unimodal distributions are those having a single maximum.

Multimodal distributions are those having several maximums.

6.15. Measures of dispersion. The most common measures of the "spread" of a distribution are the *variance, standard deviation*, and *mean deviation*. Other measures are defined as in Chapter II.

6.16. Variance. The *variance* is $\mu_2 = \sigma^2$, which is the "moment of inertia" about a line normal to the X axis through the center of gravity μ. Let $\sigma = \sigma(X)$ denote the standard deviation of X. Then $\sigma^2(X) = E[(X - \mu)^2] \equiv \text{var } X$ are equivalent representations for the variance of X.

THEOREM 1.   If a and b are constants

$$\sigma(aX + b) = |a| \sigma(X).$$

THEOREM 2.   If X_1 and X_2 are random variables, and c_1 and c_2 are constants,

$$\sigma^2(c_1X_1 + c_2X_2) = c_1{}^2\sigma_1{}^2 + c_2{}^2\sigma_2{}^2 + 2c_1c_2 \text{ cov } (X_1,X_2),$$

where cov (X_1,X_2) is the *covariance* of X_1 and X_2. (See §12.6.) If X_1 and X_2 are (statistically) independent,

$$\sigma^2(c_1X_1 + c_2X_2) = c_1{}^2\sigma_1{}^2 + c_2{}^2\sigma_2{}^2$$

When the standard variable $t = (X - \mu)/\sigma$ is used, the mean and variance of t are $E[t] = 0$ and $E[t^2] = 1$, respectively.

6.17. Mean deviation. The *mean (absolute) deviation* of X with respect to the median m is defined to be

$$E[|X - m|],$$

where m is the *median* of the distribution $f(x)$.

Some writers define "mean deviation" (or "average deviation") of X to be $E[|X - \mu|]$, where μ is the mean. This is not preferred.

The *mean (absolute) deviation, M.D.*, with respect to the mean μ, or *mean absolute error (m.a.e.)*, is defined to be $E[|X - \mu|]$.

6.18. Quartile of order p. A number ζ_p such that $F(\zeta_p) = p$, where p is a number such that $0 < p < 1$, is called the *percentile of order p*, or *p percentile*, for the distribution $f(x)$. $\zeta_{1/4}$ is the *lower quartile*, $\zeta_{3/4}$ the *upper quartile*, and $\zeta_{0.1}$, $\zeta_{0.2}$, $\cdots$ the *deciles*. The *semi-interquartile range* is $(\zeta_{3/4} - \zeta_{1/4})/2$.

6.19. Range of distribution $f(x)$. Let the smallest interval on X including the entire distribution $f(x)$ be (L,U). The length of this interval $U - L$ is the *range* of the distribution $f(x)$. Some writers call the interval (L,U) the *range*.

6.20. Measures of "skewness" and "flatness". The most common measures of "skewness" in a probability distribution are the third moment μ_3 and the coefficient of skewness γ_1; of "flatness" or "kurtosis", the fourth moment μ_4 and coefficient of excess γ_2. Other measures are defined in §§2.20, 2.21.

6.21. Similarity between probability distributions and sample frequency distributions.

Sample frequency distribution.	$\rightarrow$	*Probability distribution.*
f_i/n	$\rightarrow$	$f(x_i)$
$\bar{x} = S(x)/n = \sum_{i=1}^{k} x_i(f_i/n)$	$\rightarrow$	$\mu = E[X] = \sum_{i=1}^{k} x_i f(x_i)$
$\sigma_x^2 = \sum_{i=1}^{k} (x_i - \bar{x})^2(f_i/n)$	$\rightarrow$	$\sigma^2 = \sum_{i=1}^{k} (x_i - \mu)^2 f(x_i)$

Some examples of discrete distributions

6.22. The most common discrete distributions are the binomial and Poisson distributions discussed in Chapters VIII and IX. The following special distributions are also frequently used.

6.23. Causal distribution. If a discrete variable X is such that the probability $f(x)$ that X takes the value x is

$$(6.23.1) \qquad P(X = x) = f(x) = \begin{cases} 0 & \text{for} \quad x \neq u, \\ 1 & \text{for} \quad x = u \end{cases}$$

with the understanding that X assumes with certainty only one definite value u, the variable X is said to possess a *causal (unitary) distribution*. The *cumulative causal distribution*, for any value x', is

$$(6.23.2) \qquad F(x') = P(X \leq x') = \epsilon(x' - u) \equiv \begin{cases} 0 & \text{for} \quad x' < u \\ 1 & \text{for} \quad x' \geq u \end{cases}.$$

Properties. The mean and variance of the variable X having distribution (6.23.1) are u and 0, respectively. The characteristic function for distribution (6.23.1) is

$$(6.23.3) \qquad \chi(t) = \exp\,[iut], \qquad\qquad i^2 = -1.$$

This is a very useful distribution in the development of the theory of probability. Also, it is a sort of limiting distribution corresponding to a causal description of some phenomenon as contrasted to the more usually encountered statistical description with its "random spread". (See §7.18.)

6.24. Special discrete case.

A variable X ranges over the two points 1 and 0. Suppose that the probability that X takes the value x is

$$(6.24.1) \qquad P(X = x) = f(x) = \begin{cases} p & \text{when} \quad x = 1 \\ q = 1 - p & \text{when} \quad x = 0 \end{cases}.$$

The corresponding cumulative distribution for any value x' is

$$(6.24.2) \qquad F(x') = P(X \leq x') = p \cdot \epsilon(x' - 1) + q \cdot \epsilon(x'),$$

where the $\epsilon(x' - 1)$ and $\epsilon(x')$ are defined as in (6.23.2).

Properties. The mean and variance of the variable X having distribution (6.24.1) are p and pq, respectively. The characteristic function for distribution (6.24.1) is

$$(6.24.3) \qquad \chi(t) = pe^{it} + q, \qquad\qquad i^2 = -1.$$

6.25. Discrete distribution of equal probability.

If a discrete variable X, ranging over $x_1, x_2, \cdots, x_n$, is such that the probability that X takes the value x_i is

$$(6.25.1) \qquad P(X = x_i) = f(x_i) = 1/n, \qquad (i = 1, 2, \cdots, n)$$

the distribution is said to be a *discrete equal probability distribution*. Let $x_1 < x_2 < \cdots < x_n$. The corresponding *cumulative equal probability distribution*, for any value x' is

$$(6.25.2) \qquad F(x') = P(X \leq x') = k(1/n),$$

where k is the largest integer such that $x_k \leq x'$.

Properties. The mean and variance of the variable X having distribution (6.25.1) are, respectively,

$$(6.25.3) \qquad \mu = (1/n) \sum_{i=1}^{n} x_i, \qquad \sigma^2 = (1/n) \sum_{i=1}^{n} x_i^2 - \mu^2.$$

The characteristic function for distribution (6.25.1) is

$$(6.25.4) \qquad \chi(t) = (1/n) \sum_{j=1}^{n} e^{itx_j}. \qquad i^2 = -1.$$

Discrete distribution. General case. For the discrete probability distribution treated in §6.2 to §6.21, with

$$(6.25.5) \qquad P(X = x_i) = f(x_i), \qquad (i = 1, \cdots, k)$$

the cumulative distribution may be written [see (6.23.3) and §7.10]

$$(6.25.6) \qquad F(x') = P(X \leq x') = \sum_{i=1}^{k} f(x_i) \, \epsilon(x' - x_i).$$

The characteristic function for distribution (6.25.5) is

$$(6.25.7) \qquad \chi(t) = \sum_{j=1}^{k} f(x_j) \, e^{itx_j}.$$

6.26. Pascal's (or Furry's) distribution. If a discrete variable X is such that

$$(6.26.1) \qquad P(X = x) = f(x) = m^x \div (1 + m)^{x+1},$$

$$m > 0, \qquad x = 0, 1, 2, \cdots$$

X is said to possess a *Pascal* (or *Furry*) *distribution.*

Properties. The mean, factorial moment of order 2, and variance of X having distribution (6.26.1) are m, $2m^2$, $m^2 + m$, respectively. The characteristic function for distribution (6.26.1) is

$$(6.26.2) \qquad \chi(t) = [1 + m(1 - e^{it})]^{-1}, \qquad i^2 = -1.$$

6.27. Pólya's distribution. If a discrete variable X is such that

$$(6.27.1) \qquad P(X = x) = f(x)$$

$$= \left[\frac{m}{1 + Bm} \right]^x \left[\frac{1(1 + B) \cdots (1 + \{x - 1\}B)}{x!} \right] f(0),$$

where

$$f(0) = (1 + Bm)^{-1/B}, \qquad B \geq 0, \qquad m > 0,$$

and where x can take the values $0, 1, 2, \cdots$, with $P(X = 0) = f(0)$, X is said to have a *Pólya distribution,* or *negative binomial distribution.*

Properties. The mean, factorial moment of order 2, and variance of X having distribution (6.27.1) are m, $(1 + B)m^2$, and $m(1 + Bm)$, respectively. The characteristic function for distribution (6.27.1) is

$$(6.27.2) \qquad \chi(t) = [1 + Bm(1 - e^{it})]^{-1/B}, \qquad i^2 = -1.$$

The Poisson distribution is obtained by taking the limit of (6.27.1) as $B \to 0$. When $B = 1$, (6.27.1) is Pascal's distribution.

Probability distributions continuous on an interval

6.28. Let X be a chance variable which is certain to fall in the interval (α, β). Here α and β are any finite numbers; or α could be $-\infty$, β could be $+\infty$. By a *continuous cumulative distribution function $F(x)$* is meant a function which gives the probability that the chance variable X is less than or equal to any particular value x which may be selected, that is,

$$(6.28.1) \qquad P(X \leqq x) = F(x).$$

The function $P(X \leqq x)$ is called the *probability function* of the variable X. The probability that X is greater than x' and less than or equal to x'' is

$$(6.28.2) \qquad P(x' < X \leqq x'') = P(X \leqq x'') - P(X \leqq x')$$
$$= F(x'') - F(x').$$

In such a continuous distribution the probability that X has a particular value x_0 is zero, that is, $P(X = x_0) = 0$.

In many practical problems, to plot $F(x)$ one may need a large number of sample measurements from which one can construct a cumulative frequency polygon, and in turn from which a smooth curve $F(x)$ may be drawn which approximates the polygon. (See Fig. 6.28.1). This curve $F(x)$ then yields a method for obtaining "estimated" probabilities.

6.29. **Probability density function.** If the derivative $F'(x)$ of $F(x)$ with respect to x exists, the function $f(x) = F'(x)$ is called the *continuous probability density function*, or *distribution*, or *frequency function*. Then

$$P(x' < X \leqq x'') = \int_{x'}^{x''} f(x) \, dx = F(x'') - F(x'),$$

$$(6.29.1) \qquad F(x') = \int_{\alpha}^{x'} f(x) \, dx = P(X \leqq x'),$$

$$F(\beta) = \int_{\alpha}^{\beta} f(x) \, dx = 1, \qquad F(\alpha) = 0, \qquad 0 \leqq F(x) \leqq 1.$$

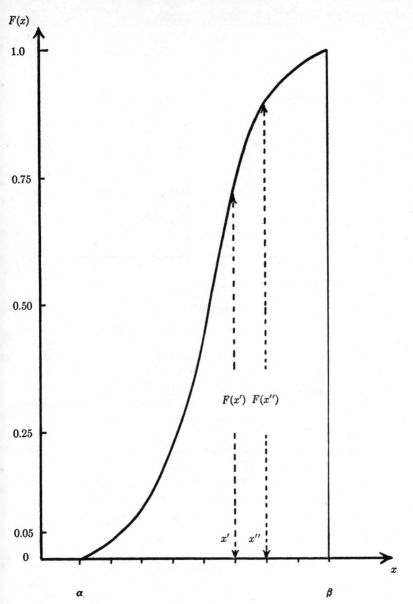

FIGURE 6.28.1

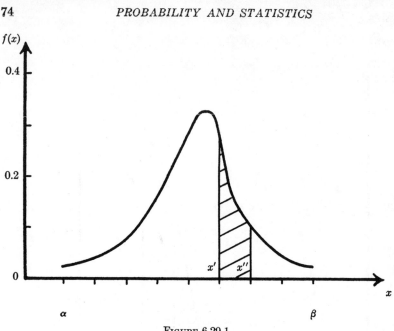

FIGURE 6.29.1

The function $f(x) = F'(x)$ can be constructed from $F(x)$ by plotting the derivative $F'(x)$, (that is, the slope of the curve $F(x)$), as a function of x. The total area under the histogram for $f(x)$ is 1. (See Fig. 6.29.1).

The total area under $f(x)$ from $x = \alpha$ to $x = x'$ is $F(x')$, the probability that $X \leq x'$.

6.30. If in (6.29.1) $x'' = x' + \Delta x$

$$(6.30.1) \qquad P(x' < X \leq x' + \Delta x) = \int_{x'}^{x' + \Delta x} f(x)\, dx.$$

By the theorem of the mean,

$$(6.30.2) \qquad P(x' < X \leq x' + \Delta x) = f(\xi)\, \Delta x,$$

where ξ lies somewhere between x' and $x' + \Delta x$. Thus, for Δx sufficiently small, the product $f(x')\Delta x$ gives approximately the probability that X assumes a value in the interval x' to $x' + \Delta x$. For this reason the probability density is often expressed in the form

$$(6.30.3) \qquad\qquad dF = f(x)\, dx.$$

This is called the *probability differential*.

Parameters which describe a continuous probability distribution

6.31. In the following sections the continuous variable X is assumed to range over the interval (α, β) with a probability density $f(x) = F'(x)$, and a cumulative distribution function $F(x)$.

The principal features of a distribution $f(x)$ may be described by means of certain parameters. These parameters are defined in very much the same manner as for discrete probability distributions.

6.32. **Mean and variance.** The mean μ and the variance σ^2 of the variable X are defined to be

$$\mu = \int_\alpha^\beta x\, f(x)\, dx,$$

(6.32.1)

$$\sigma^2 = \int_\alpha^\beta (x - \mu)^2 f(x)\, dx, \qquad \text{or,} \qquad \sigma^2 = \int_\alpha^\beta x^2 f(x)\, dx - \mu^2.$$

$\sigma = +\sqrt{\sigma^2}$ is called the *standard deviation* (or, *root mean square deviation*, or, *standard error*). μ and σ^2 are also called the *mean* and *variance*, respectively, of the distribution $f(x)$.

6.33. **Probable error.** The *probable error* is a number $(p.e.)$ such that

(6.33.1)
$$\int_{\mu-(p.e.)}^{\mu+(p.e.)} f(x)\, dx = 0.5.$$

The term "probable error" is a somewhat misleading term. "Equally likely deviation" might be more appropriate. Some call this term "probable deviation".

6.34. **Mathematical expectation and moments.** If $y = g(X)$ is a function of the continuous variable X, and X has the probability distribution $f(x)$, X being certain to fall in (α, β), the *expected value* (or, *mean value*) *of* $g(X)$ is

(6.34.1)
$$E[g(X)] = \int_\alpha^\beta g(x) f(x)\, dx.$$

6.35. r^{th} **moment μ_r about the mean μ (or central moments).** The mean value of $(X - \mu)^r$, or the expected value of $(X - \mu)^r$, is

(6.35.1)
$$E[(X - \mu)^r] = \mu_r = \int_\alpha^\beta (x - \mu)^r f(x)\, dx.$$

μ_r is called the rth *moment* of the distribution *about the mean* μ. For a given $f(x)$, $\mu_0 = 1$, $\mu_1 = 0$, $\mu_2 = \sigma^2 = $ variance.

6.36. r^{th} **moment** ν_r **about the origin** $X = 0$. The mean value of X^r, or the expected value of X^r, is

(6.36.1) $$E[X^r] = \nu_r = \int_\alpha^\beta x^r f(x)\ dx.$$

ν_r is called the rth *moment* of the distribution $f(x)$ *about the origin* $X = 0$. For any given X, $\nu_0 = 1$, $\nu_1 = E[X] = \mu$, the mean.

6.37. r^{th} **absolute moment** β_r **about** $X = 0$. By the rth *absolute moment* β_r of the distribution $f(x)$ about $X = 0$ is meant

(6.37.1) $$E[|\ X\ |^r] = \beta_r = \int_\alpha^\beta |\ x\ |^r f(x)\ dx.$$

6.38. Moments about $X = c$**.** By the rth moment of the distribution $f(x)$ about $X = c$ is meant

(6.38.1) $$E[(X - c)^r] = \int_\alpha^\beta (x - c)^r f(x)\ dx.$$

The second moment about $X = c$ becomes a minimum when c is the mean μ.

6.39. Relations between μ_r **and** ν_r**.** These are given in Eqs. (2.8.1).

6.40. r^{th} **factorial moment.** The rth *factorial moment* $\nu_{[r]}$ of the distribution $f(x)$ about $X = 0$ is the mean value of $x^{[r]}$,

(6.40.1) $$E[x^{[r]}] = \nu_{[r]} = \int_\alpha^\beta x^{[r]} f(x)\ dx,$$

where $x^{[r]}$ is a symbol for the factorial

(6.40.2) $\qquad x^{[r]} = x(x - 1) \cdots (x - r + 1). \qquad (r = 1, 2, 3, \cdots .)$

6.41. Measures of location, dispersion, skewness, and flatness. The same parameters used to describe a discrete distribution (§§6.15 to 6.20) are also used for continuous distributions.

Some examples of continuous distributions

6.42. The most common continuous distribution is the normal distribution discussed in Chapter X. The following special distributions are frequently encountered.

6.43. Rectangular distribution. If a continuous variable X is such that the (cumulative) probability that X is less than or equal to x' is

$$(6.43.1) \qquad P(X \leq x') = F(x') = \int_{-\infty}^{x'} f(x) \, dx,$$

where the probability density is

$$(6.43.2) \qquad f(x) = \begin{cases} 0 & \text{for} \quad x < a \\ 1/(b - a) & \text{for} \quad a \leq x \leq b \\ 0 & \text{for} \quad b < x, \end{cases}$$

and a and b are two constants, the distribution is said to be *uniformly distributed* in (a,b); (6.43.2) is called a *rectangular distribution*.

Properties. The mean and variance of X having distribution (6.43.2) are $(b + a)/2$ and $(b - a)^2/12$, respectively.

If a be fixed and $b \to a$, $F(x') \to \epsilon(x' - a)$, the causal distribution defined in §6.23, with $a = u$.

6.44. Cauchy's distribution. If X is such that

$$(6.44.1) \qquad P(X \leq x') = F(x') = \int_{-\infty}^{x'} f(x) \, dx,$$

with

$$(6.44.2) \qquad f(x) = \frac{1}{\pi a[1 + (x - u)^2/a^2]}, \qquad (-\infty < x < \infty)$$

in which a and u are constants, X is said to have a *Cauchy distribution*. It is encountered in physics.

Properties. The mean value of X is indeterminant from the definition, but it is defined to be equal to the parameter u. The mode and median are both equal to u. The variance does not exist. As a measure of dispersion a is used, which is equal to the semi-interquartile range. The characteristic function for the distribution (6.44.2) is

$$(6.44.3) \qquad \chi(t) = \exp\left[iut - a\,|\,t\,|\right], \qquad i^2 = -1.$$

If u be fixed and $a \to 0$, $F(x') \to \epsilon(x' - u)$, the causal distribution (6.23.1).

6.45. Laplace's distribution. If X is such that

$$(6.45.1) \qquad P(X \leq x') = F(x') = \int_{-\infty}^{x'} f(x) \, dx,$$

where

$$(6.45.2) \qquad f(x) = (1/2a)e^{-|x-u|/a}, \qquad (-\infty < x < \infty)$$

in which a and u are constants, X is said to have a *Laplace distribution*.

Properties. The mean and variance of X are u and $2a^2$, respectively. The characteristic function for the distribution (6.45.2) is

$$(6.45.3) \qquad \chi(t) = [\exp{(iut)}] \div (1 + a^2t^2), \qquad i^2 = -1.$$

If u be fixed and $a \to 0$, $F(x') \to \epsilon(x' - u)$, the causal distribution (6.23.1).

6.46. On the statistical interpretation of continuous probability distributions.
Just as in the discrete case, a continuous probability distribution ·is used as a population distribution "model". A continuous probability density may be regarded as a relative frequency distribution for an indefinitely large sample (population) when the measurements can take on any value within a given interval. In practice, if one introduces a smooth curve to represent approximately the distribution of relative frequencies (for large enough samples of the population under consideration) one frequently obtains a fairly simple and accurate model to use in calculating means, variances, frequencies and other quantities related to the population distribution.

6.47. Sheppard's corrections.
In calculating moments from grouped measurements, there is an error due to the grouping. Under some conditions, this error can be reduced by using *Sheppard's corrections.*

Let $f(x)$ be the continuous distribution function of x in a population. Let $x_1', x_2', \cdots, x_n'$ be a sample from this population. Suppose the range of x is divided into intervals of length c with midpoints $x_i, i = 1, 2, \cdots, k$. Let the class frequencies of the sample, for these intervals, be f_i; $\sum_{i=1}^{k} f_i = n$. Then

$$\nu_r = \int_{-\infty}^{\infty} x^r f(x) \, dx \qquad \text{is the } r^{\text{th}} \text{ moment (about the origin) of the population.}$$

$$M_r = (1/n) \sum_{j=1}^{n} x_j'^r \qquad \text{is the } r^{\text{th}} \text{ moment (about the origin) of the sample.}$$

$$(6.47.1)$$

$$M_{r,c} = (1/n) \sum_{i=1}^{k} f_i x_i^r \qquad \text{is the } r^{\text{th}} \text{ moment (about the origin) of the grouped sample.}$$

$$\lim_{c \to 0} M_{r,c} = M_r. \qquad\qquad E(M_r) = \nu_r.$$

If c is sufficiently small, and if $f(x)$ has a high order of contact with the x-axis for very large positive and negative values of x (i.e., if the distribution curve $f(x)$ has "tails" which are very nearly tangent to the x-axis) then the following approximations are valid.

$$M_1 \approx M_{1,c}.$$

(6.47.2)
$$M_2 \approx M_{2,c} - c^2/12.$$

$$M_3 \approx M_{3,c} - (c^2/4)M_{1,c}.$$

$$M_4 \approx M_{4,c} - [(c^2/2)M_{2,c} - (7c^4/240)].$$

The quantities subtracted from $M_{r,c}$ on the right in these expressions are *Sheppard's corrections.*

If M_r' denotes the rth moment (about the mean) of the sample and $M_{r,c}'$ the corresponding moment of the grouped sample, the expressions corresponding to (6.47.2) are

$$M_1' \approx M_{1,c}', \qquad M_1' = 0.$$

(6.47.3)
$$M_2' \approx M_{2,c}' - c^2/12.$$

$$M_3' \approx M_{3,c}'.$$

$$M_4' \approx M_{4,c}' - [(c^2/2)M_{2,c}' - (7c^4/240)].$$

The expected value of M_r is ν_r; of M_r' is μ_r. In case the sample is identical with the population, then $M_r = \nu_r$ and $M_r' = \mu_r$.

The normal distribution function has a high order of contact with the *x*-axis for large positive and negative values of x.

6.48. Many statisticians use the corrections (6.47.2) and (6.47.3) for grouping in the case of frequency distributions, the assumption being that the moments thus corrected are a better representation of the moments of the underlying probability distribution. Such a step requires considerable justification. Hence, some discretion is required to determine whether Sheppard's corrections are worthwhile in any given calculation of moments, taking account of the magnitude of the corrections as compared to the magnitude of the moments, and taking account of the accuracy of the observed data.

6.49. **Folded distributions.** Sometimes observations are recorded without indicating whether the values are positive or negative. In such cases the measurements may be regarded as all plus. This modifies the distribution by *"folding"* over the negative part and adding it to the positive part.

EXAMPLE. The probability density function for the folded normal distribution is

(6.49.1) $\quad f(x) = [e^{-(x-\mu)^2/2\sigma^2} + e^{-(x+\mu)^2/2\sigma^2}] \div \sqrt{2\pi}\, \sigma, \qquad\qquad x \geqq 0$

where μ and σ^2 are the mean and variance of Eq. (10.1.1). The mean μ_f and variance σ_f^2 of a variable x distributed as in (6.49.1) are

(6.49.2)

$$\mu_f = \sigma \sqrt{\frac{2}{\pi}} \, e^{-\frac{1}{2}(\mu/\sigma)^2} + \frac{1}{\sqrt{2\pi}} \int_{-\mu/\sigma}^{\mu/\sigma} e^{-t^2/2} \, dt$$

$$\sigma_f^2 = \sigma^2 + \mu^2 - \mu_f^2$$

6.50. Truncated and censored distributions.   Limitations in measuring instruments, or practical considerations may result in the observation of only values of x less than a fixed value x_0. The distribution of x is then said to be *truncated* at x_0.

In some cases, the number of observations greater than x_0 may be known, but practical considerations indicate the use of only values of x less than x_0. The distribution of x is then called *censored*, in contrast to truncated, in which it is presumed that nothing is known about values of x greater than x_0.

A truly truncated distribution is one *say* for which there are no values of $x > x_0$ or for $x < x_1$.

If the p.d.f. for x is $f(x)$, the p.d.f. $f^*(x)$ for the truncated distribution in which only values of x less than x_0 are considered is

(6.50.1) $$f^*(x) = \begin{cases} f(x)/F(x_0), & x < x_0 \\ 0, & x \geqq x_0 \end{cases}$$

where

$$F(x_0) = \int_{-\infty}^{x_0} f(x) \, dx$$

EXAMPLE.   If $f(x)$ is normal $N(\mu, \sigma)$, the mean and variance of the truncated distribution $f^*(x)$ in which $x \leqq x_0$ are

(6.50.2) $\mu^* = \mu - \sigma W, \qquad \sigma^{*2} = \sigma^2[1 - W^2 - t_0 W]$

where

$$W = Z/F, \qquad t_0 = (x_0 - \mu)/\sigma$$

(6.50.3) $Z = (1/\sqrt{2\pi}) \exp[-t_0^2/2].$

$$F = (1/\sqrt{2\pi}) \int_{-\infty}^{t_0} e^{-t^2/2} \, dt.$$

If the distribution is truncated to consider only values of x greater than x_0, then the p.d.f. $g^*(x)$ for the truncated distribution is

(6.50.4) $$g^*(x) = \begin{cases} f(x)/[1 - F(x_0)], & x > x_0 \\ 0 & x \leqq x_0 \end{cases}$$

seven

GENERATING AND CHARACTERISTIC FUNCTIONS

Generating functions

7.1. Moment generating function. Let $F(x)$ be a cumulative distribution function (c.d.f.) of the variable x with probability density $f(x) = F'(x)$. The *moment generating function* (m.g.f.) of the variable x is defined as

$$(7.1.1) \qquad M(\theta) = E[e^{\theta x}] = \int_{-\infty}^{\infty} e^{\theta x} \, dF(x) = \int_{-\infty}^{\infty} e^{\theta x} f(x) \, dx,$$

where $E[e^{\theta x}]$ is the expected value of $e^{\theta x}$. If $M(\theta)$ and its derivatives exist, $|\theta| \leq h,\ h > 0$, the rth moment of $F(x)$ about the origin is

$$(7.1.2) \qquad\qquad \nu_r = M^{(r)}(0), \qquad\qquad r = 0, 1, 2, \cdots,$$

where $M^{(r)}(0)$ is the rth derivative $M^{(r)}(\theta)$ of $M(\theta)$ with respect to θ, at $\theta = 0$. If $M(\theta)$ can be expanded into the form

$$(7.1.3) \qquad M(\theta) = 1 + \nu_1 \theta + \nu_2(\theta^2/2!) + \nu_3(\theta^3/3!) + \cdots,$$

the moments ν_r appear as coefficients of $\theta^r/r!$. Thus, $M(\theta)$ may be regarded as generating the moments ν_r. Moment generating functions are often used to facilitate manipulations. For example, if $M(\theta)$ is known, the moments ν_r can be found from (7.1.2).

7.2. Generating function for $F(x)$. The expected value of t^x is called the *generating function* (g.f.) for the variable x having the cumulative distribution $F(x)$, that is,

$$(7.2.1) \qquad G(t) = E[t^x] = \int_{-\infty}^{\infty} t^x \, dF(x) = \int_{-\infty}^{\infty} t^x f(x) \, dx.$$

Note that if one sets $t = e^{\theta}$, $G(t) = M(\theta)$.

In case the distribution is discrete, with probability density $f(x)$, similar definitions for $M(\theta)$ and $G(t)$ are made: namely,

$$(7.2.2) \qquad\qquad M(\theta) = E[e^{\theta x}] = \sum_i f(x_i) e^{\theta x_i},$$

$$(7.2.3) \qquad\qquad G(t) = E[t^x] = \sum_i f(x_i) t^{x_i}.$$

81

THEOREM 1.   If the cumulative distribution functions $F_1(x)$ and $F_2(x)$ have the same m.g.f. $M(\theta)$, for $|\theta| \leq h$, $h > 0$, then $F_1(x) = F_2(x)$.

THEOREM 2.   Suppose $F_{(n)}(X_n)$ and $M_{(n)}(\theta)$ are respectively the c.d.f and m.g.f. of a random variable X_n, $n = 1, 2, \cdots$. If for all n and for $|\theta| < h$, $M_{(n)}(\theta)$ exists, and if a function $M(\theta)$ exists such that $\text{limit}_{n \to \infty} M_{(n)}(\theta) = M(\theta)$ for θ sufficiently small numerically, then $\text{limit}_{n \to \infty} F_{(n)}(X_n) = F(x)$, where $F(x)$ is the c.d.f. of a random variable X with m.g.f. $M(\theta)$.

7.3. Factorial moment generating function.
Let $t = 1 + \theta$. Then the generating function $G(t)$ becomes the *factorial moment generating function* (f.m.g.f.) (in case the values of x are discrete and spaced at unit intervals).

$$(7.3.1)\qquad G(1 + \theta) = \sum_j f(x_i)(1 + \theta)^{x_i}$$

$$= 1 + \nu_{[1]}\theta + \nu_{[2]}(\theta^2/2!) + \nu_{[3]}(\theta^3/3!)$$

$$+ \cdots + \nu_{[r]}(\theta^r/r!) + \cdots .$$

Here $\nu_{[r]} = \sum_i x_i^{[r]} \cdot f(x_i)$ are *factorial moments*, where

$$(7.3.2)\qquad x^{[r]} = x(x - 1)(x - 2) \cdots (x - r + 1).$$

7.4. Semi-invariant generating function.
Let $M(\theta)$ be a m.g.f. If $\log_e M(\theta)$ can be expanded in the form

$$(7.4.1)\qquad L(\theta) \equiv \log_e M(\theta) = \kappa_1\theta + \kappa_2(\theta^2/2!) + \kappa_3(\theta^3/3!)$$

$$+ \cdots + \kappa_r(\theta^r/r!) + \cdots ,$$

then $L(\theta)$ is called the *semi-invariant generating function* (s.g.f.), and κ_r are called the *semi-invariants*, or *cumulants** of the distribution $F(x)$.

7.5. Factorial s.g.f. and factorial semi-invariants
may be defined in a similar way by taking the logarithm of the f.m.g.f.

7.6. Change of origin.
If the origin be changed from $x = 0$ to $x = a$ by the change of variable $x - a = z$, the m.g.f. for the variable z is $e^{-a\theta}$ times the m.g.f. (7.1.1) for the variable x; the f.m.g.f. for z is $(1 + \theta)^{-a}$ times the f.m.g.f. (7.3.1) for x; the s.g.f. for z is $-a\theta$ plus the s.g.f. (7.4.1) for x.

7.7. Change of scale.
If a change of scale be made by the change of variable $z = kx$, the g.f. for the variable z is obtained from (7.2.1) by substituting t^k for t in (7.2.1); the m.g.f. for z is obtained from (7.1.1) by substituting $k\theta$ for θ in (7.1.1); the rth moment for the variable z is $k^r \nu_r$,

* Some writers use only the term *cumulant* in this setting.

where ν_r is the rth moment (7.1.2) for the variable x; the rth semi-invariant for the variable z is $k^r \kappa_r$ where κ_r is the rth semi-invariant for x.

EXAMPLE 1. If all the points of a straight line from $x = 0$ to $x = 1$ are equally likely, the g.f. is $\int_0^1 t^x \, dx = (t - 1)/\log_e t$.

EXAMPLE 2. The binomial distribution has probability density function $f(x) = C_x^n p^x q^{n-x}$, $q = 1 - p$, The moment generating function is

$$(7.7.1) \qquad M(\theta) = E[e^{\theta x}] = \sum_{x=0}^{n} C_x^n (pe^\theta)^x q^{n-x} = (q + pe^\theta)^n$$

The rth moment about $x = 0$ is found from the rth derivative of $M(\theta)$, $M^{(r)}(0) = \nu_r$. Thus,

$$\nu_1 = M^{(1)}(0) = np, \qquad \nu_2 = M^{(2)}(0) = np + n(n-1)p^2$$

In (7.7.1) set $e^\theta = t$. Then $M(\theta)$ becomes the generating function (g.f.)

$$(7.7.2) \qquad G(t) = E[t^x] = M(\theta) = (q + pt)^n$$

$G(t)$ is called the *probability generating function* (p.g.f.).

In (7.7.1) set $\theta = it$. Then $M(\theta)$ becomes the characteristic function

$$(7.7.3) \qquad \chi(t) = E[e^{itx}] = (q + pe^{it})^n$$

The moment ν_r may be found from the rth derivative $\chi(t)$, $\chi^{(r)}(0) = i^r \nu_r$. Thus, $\chi^{(1)}(0) = i\nu_1 = inp$. (See §7.8.)

EXAMPLE 3. Suppose the chance that the discrete variable X takes on the integer value j is $P(X = j) = p_j$, $j = 0, 1, 2, \cdots$. The probability generating function (p.g.f.) is $G(t) = \sum_{j=0}^{\infty} p_j t^j$, for $|t| \leq 1$. Denote first and second derivatives of $G(t)$ by $G^{(1)}(t)$ and $G^{(2)}(t)$, respectively. Then

$$G(1) = \sum_{j=0}^{\infty} p_j = 1, \qquad G^{(1)}(1) = \sum_{j=0}^{\infty} j p_j = E(X)$$

$$G^{(2)}(1) = \sum_{j=0}^{\infty} j(j-1) p_j = E[X(X-1)]$$

$$(7.7.4) \qquad G^{(v)}(1) = E[X(X-1) \cdots (X - v + 1)]$$

Mean of $X = E(X) = \bar{x}$

Variance of $X = G^{(2)}(1) + \bar{x} - \bar{x}^2$
$$= G^{(2)}(1) + G^{(1)}(1) - [G^{(1)}(1)]^2$$
$$= \sigma^2.$$

Characteristic functions

7.8. **Continuous case.** The *characteristic function* $\chi(t)$ corresponding to a distribution function $F(x)$ is the expected value of e^{itx}, namely,

$$(7.8.1) \qquad \chi(t) = E[e^{itx}] = \int_{-\infty}^{\infty} e^{itx} \, dF(x) = \int_{-\infty}^{\infty} e^{itx} f(x) \, dx.$$

Here t is a real number and $e^{itx} = \cos tx + i \sin tx$, and $F'(x) = f(x)$. $\chi(t)$ is a complex-valued function of t.

$$\chi(0) = 1, \qquad\qquad\qquad i^2 = -1.$$

$$(7.8.2)$$

$$| \chi(t) | \leq \int_{-\infty}^{\infty} dF(x) = 1, \quad \chi(-t) = \overline{\chi(t)}, \quad \text{for all values of } t.$$

By differentiation, the rth derivative of $\chi(t)$ is found to be

$$(7.8.3) \qquad \chi^{(r)}(t) = i^r \int_{-\infty}^{\infty} x^r e^{itx} \, dF(x).$$

At $t = 0$,

$$(7.8.4) \qquad \chi^{(r)}(0) = i^r \int_{-\infty}^{\infty} x^r \, dF(x) = i^r \nu_r,$$

where ν_r is the rth moment about $x = 0$ (if it exists) of the distribution $F(x)$.

By MacLaurin Series, if the moments ν_r exist,

$$(7.8.5) \qquad \chi(t) = 1 + \sum_{r=1}^{k} (\nu_r/r!)(it)^r + o(t^k),$$

where the error term $o(t^k)$, divided by t^k, approaches zero as $t \to 0$.

7.9. The characteristic functions for various important distributions are listed in this book along with the properties of these distributions.

EXAMPLE 1. If the distribution function $F(x)$ is the normal distribution function

$$(7.9.1) \qquad \Psi(x) = (1/\sqrt{2\pi}) \int_{-\infty}^{x} e^{-x^2/2} \, dx,$$

the characteristic function corresponding to $\Psi(x)$ is

$$(7.9.2) \qquad \chi(t) = \int_{-\infty}^{\infty} e^{itx} \, d\Psi(x) = (1/\sqrt{2\pi}) \int_{-\infty}^{\infty} e^{(itx - x^2/2)} \, dx = e^{-t^2/2}.$$

The characteristic function corresponding to $\Psi^{(n)}(x)$, the nth derivative of $\Psi(x)$, is

$$(7.9.3) \qquad \int_{-\infty}^{\infty} e^{itx} \, d\Psi^{(n)}(x) = (-it)^n e^{-t^2/2}.$$

EXAMPLE 2.   The characteristic function of any function $g(x)$ of the variable x is the expected value of $e^{itg(x)}$.

EXAMPLE 3.

7.9.4)
$$E[e^{it(ax+b)}] = e^{bit}\chi(at).$$
$$E[e^{it(x-m)/\sigma}] = e^{-(mit/\sigma)}\chi(t/\sigma).$$

7.10. Discrete case.   In case the distribution is discrete, with probability density $f(x)$, a similar definition for the characteristic function can be made: namely (see §6.25),

7.10.1)
$$\chi(t) = E[e^{it}] = \sum_j f(x_j)e^{itx_j}, \qquad i^2 = -1.$$

7.11. Semi-invariants.   It can be shown that

7.11.1)
$$\log \chi(t) = \sum_{r=1}^{\infty} \kappa_r (it)^r/r! \equiv K,$$

or

7.11.2)
$$\chi(t) = e^K,$$

where κ_r are the *semi-invariants* of the distribution $F(x)$. (See §7.4.)

7.12. Relations between semi-invariants and moments.   From (7.11.1) and (7.11.2) the following relations between κ_r and the moments ν_r about the origin and the central moments μ_r are found:

(7.12.1)
$$\kappa_1 = \nu_1 = \mu,$$
$$\kappa_2 = \nu_2 - \nu_1^2 = \sigma^2,$$
$$\kappa_3 = \nu_3 - 3\nu_1\nu_2 + 2\nu_1^3,$$
$$\kappa_4 = \nu_4 - 3\nu_2^2 - 4\nu_1\nu_3 + 12\nu_1^2\nu_2 - 6\nu_1^4,$$
$$\vdots$$

(7.12.2)
$$\nu_1 = \kappa_1,$$
$$\nu_2 = \kappa_2 + \kappa_1^2,$$
$$\nu_3 = \kappa_3 + 3\kappa_1\kappa_2 + \kappa_1^3,$$
$$\nu_4 = \kappa_4 + 3\kappa_2^2 + 4\kappa_1\kappa_3 + 6\kappa_1^2\kappa_2 + \kappa_1^4,$$
$$\vdots$$

$$\kappa_1 = \mu,$$

$$\kappa_2 = \mu_2 = \sigma^2,$$

(7.12.3) $$\kappa_3 = \mu_3,$$

$$\kappa_4 = \mu_4 - 3\mu_2^2,$$

$$\kappa_5 = \mu_5 - 10\mu_2\mu_3,$$

$$\kappa_6 = \mu_6 - 15\mu_2\mu_4 - 10\mu_3^2 + 30\mu_2^3.$$

$$\vdots$$

The coefficients of skewness and excess are, respectively,

(7.12.4) $$\gamma_1 = \kappa_3/\kappa_2^{3/2}, \qquad \gamma_2 = \kappa_4/\kappa_2^2.$$

Multivariable case

7.13. The moment generating function of a cumulative distribution $F = F(x_1, x_2, \cdots, x_k)$ is defined as:

(7.13.1) $$M(\theta_1, \theta_2, \cdots, \theta_k) = E[e^A] = \int_S e^A \, dF,$$

where

$$A = \sum_{i=1}^{k} \theta_i x_i,$$

and S indicates the domain of integration for the distribution F.

THEOREM 1. If $F_j(x_j)$ are the c.d.f.'s of the statistically independent variables X_j, $(j = 1, \cdots, k)$, and $M_j(\theta_j)$ is the m.g.f. for $F_j(x_j)$, then

(7.13.2) $$M(\theta_1, \theta_2, \cdots, \theta_k) = \prod_{j=1}^{k} M_j(\theta_j),$$

and

(7.13.3) $$F(x_1, x_2, \cdots, x_k) = \prod_{j=1}^{k} F_j(x_j).$$

THEOREM 2.   If the c.d.f.'s $F_j(x_j)$ for the variables X_j, $(j = 1, 2, \cdots, k)$ have m.g.f. $M_j(\theta_j)$, with $|\theta_j| \leq h$, $h > 0$, then the variables X_j are statistically independent if and only if the m.g.f. M of the joint distribution F of the variables X_j, $(j = 1, \cdots, k)$, can be factored into the form (7.13.2).

7.14. The generalization to k variables of the concept of a *characteristic unction* $\chi(t_1, \cdots, t_k)$ *of a cumulative distribution* F is made by replacing t_i by it_i in definition (7.13.1).

Likewise, the *generating function* $G(t_1, \cdots, t_k)$ for the k variables *of the cumulative distribution* F is defined by replacing $e^{\theta i}$ by t_i in definition (7.13.1).

EXAMPLE 1. The g.f. for the throw of an unsymmetrical coin in which the probability of heads is p, of tails q, is $G(t) = pt + qt^0$.

EXAMPLE 2. The g.f. for the simultaneous throw of n different unsymmetrical coins is $G(t) = (p_1 t + q_1 t^0)(p_2 t + q_2 t^0) \cdots (p_n t + q_n t^0)$, where $(p_i t + q_i t^0)$ is the g.f. for the ith coin. The f.m.g.f. is $(1 + p_1\theta)(1 + p_2\theta) \cdots (1 + p_n\theta)$.

EXAMPLE 3. If $G_1(t)$ is a g.f. of the variable x, and $G_2(t)$ a g.f. of a statistically independent variable y, then the g.f. of $x + y$ is $G_1(t) \cdot G_2(t)$.

EXAMPLE 4. The g.f. for the variable X with probability distribution

$$P[X = 0] = a_0, \qquad P[X = 1] = a_1,$$

where $a_0 = q$, $a_1 = p = 1 - q$, is

$$G_1(t) = \sum_{j=0}^{1} a_j t^j = q + pt$$

EXAMPLE 5. If X and Y each have the same probability distribution as in Example 4, the generating function for $X + Y$ is

$$G(t) = c_0 + c_1 t + c_2 t^2,$$

with
$$c_0 = a_0 b_0, \qquad c_1 = a_0 b_1 + a_1 b_0, \qquad c_2 = a_0 b_2 + a_1 b_1 + a_2 b_0.$$

Since $a_0 = b_0 = q$, $a_1 = b_1 = p$, $a_2 = b_2 = 0$,

$$G(t) = q^2 + 2qpt + p^2 t^2 = (q + pt)^2 = \sum_{k=0}^{2} \binom{2}{k}(pt)^k q^{2-k}.$$

EXAMPLE 6. If $Z = \sum_{j=1}^{n} X_j$, each X_j having the probability distribution as in Example 4, Z has the generating function

$$(q + pt)^n = \sum_{k=0}^{n} \binom{n}{k}(pt)^k q^{n-k}.$$

THEOREM 1. The characteristic function $\chi(t)$ of a sum of n statistically independent variables is equal to the product of the characteristic functions $\chi_i(t)$ of the individual variables, that is,

(7.14.1)
$$\chi(t) = \chi_1(t) \cdot \chi_2(t) \cdot \ \cdots \ \cdot \chi_n(t).$$

THEOREM 2. The cumulative distribution function $F(x)$ of the sum x of two statistically independent variables y and z is given by

(7.14.2)
$$F(x) = \int_{-\infty}^{\infty} F_1(x - w) \, dF_2(w) = \int_{-\infty}^{\infty} F_2(x - w) \, dF_1(w),$$

where $F_1(y)$ and $F_2(z)$ are cumulative distribution functions of the variables. In case differentiation is permissible,

(7.14.3)
$$f(x) = \int_{-\infty}^{\infty} f_1(x - w) f_2(w) \, dw = \int_{-\infty}^{\infty} f_2(x - w) f_1(w) \, dw,$$

where $f(x) = F'(x)$, $f_1(y) = F_1'(y)$, $f_2(z) = F_2'(z)$, $x = y + z$.
Here $F(x) = P(y + z \leq x)$.

The integrals in (7.14.2) and (7.14.3) are called the *convolution* of F_1 and F_2. (7.14.2) and (7.14.3), respectively, are sometimes written

$$F(x) = F_1 * F_2, \qquad f(x) = f_1 * f_2$$

THEOREM 3. If $x_1, x_2, \cdots , x_n$ are n statistically independent variables, then the mean, variance, and third moment about the mean of the sum $\sum_{j=1}^{n} x_j$ are, respectively,

(7.14.4)
$$m = \sum_{j=1}^{n} m_j, \qquad \sigma^2 = \sum_{j=1}^{n} \sigma_j^2, \qquad \mu_3 = \sum_{j=1}^{n} \mu_3^{(i)},$$

where m_j, σ_j^2, $\mu_3^{(i)}$ are the mean, second, and third moments about the mean, respectively, of the variable x_j. Furthermore, the rth semi-invariant κ_r of the sum is related to the rth semi-invariant $\kappa_r^{(i)}$ of x_j by $\kappa_r = \sum_{j=1}^{n} \kappa_r^{(i)}$.

Remark. If c_i are constants, the rth cumulant of the linear function $L = \sum_{j=1}^{n} c_j x_j$ is $\kappa_r = \sum_{j=1}^{n} c_j^r \kappa_r^{(j)}$. It is this additive property of semi-invariants (cumulants) that makes them important.

THEOREM 4. The characteristic function $\chi(t)$ is uniquely determined by the cumulative distribution function $F(x)$, and conversely. In fact, in the case of a continuous distribution,

(7.14.5)
$$\chi(t) = \int_{-\infty}^{\infty} e^{itx} \, dF(x) = \int_{-\infty}^{\infty} e^{itx} f(x) \, dx$$

and

(7.14.6) $$f(x) = F'(x) = (1/2\pi) \int_{-\infty}^{\infty} e^{-itx} \chi(t) \, dt.$$

Sometimes it is difficult to obtain the density function for the probability distribution by inverting the characteristic function [i.e., by using (7.14.6)]. However, the characteristic function can be used to obtain the moments and cumulants of the distribution. In this connection Theorem 3 is useful.

THEOREM 5. If $F_1(t)$, $F_2(t)$, $\cdots$ is a sequence of cumulative distribution functions, with corresponding characteristic functions $\chi_1(t)$, $\chi_2(t)$, $\cdots$, then a necessary and sufficient condition for the convergence of $F_n(t)$ to a cumulative distribution $F(t)$ is that for every t, $\chi_n(t)$ converge to a limit $\chi(t)$, which is continuous at $t = 0$. $\chi(t)$ is the characteristic function of $F(t)$.

THEOREM 6. (*Inversion Theorem*). The characteristic function $\chi(t)$ uniquely determines the distribution function $F(x)$.

(7.14.7) $$F(x) - F(0) = (1/2\pi) \int_{-\infty}^{\infty} \chi(t) \frac{1 - e^{-ixt}}{it} \, dt$$

THEOREM 7. Let x_1, x_2, $\cdots$ be independent random variables having the same distribution function. Consider the sum $z = \sum_{j=1}^{N} x_j$. Suppose N is a discrete random variable which takes on values $0, 1, 2, \cdots$. If the generating functions of x_j and N are $G_x(t)$ and $G_N(t)$, respectively, the distribution of the sum z is given by the (*compound*) *generating function* $G_z(t) \equiv G_N[G_x(t)]$. Distributions of this sort are called *compound distributions*.

EXAMPLE 7. Let x_k be a random variable which assumes the value 1 with probability p and the value 0 with probability $q = 1 - p$. Suppose N has the Poisson distribution with parameter m. The generating functions for x_k and N are $G_x(t) = q + pt$ and $G_N(t) = e^{-m+mt}$. By Theorem 7, the generating function for the sum

$$z = \sum_{j=1}^{N} x_j \text{ is}$$

(7.14.8) $$G_z(t) = \exp\left[-m + m(q + pt)\right] = \exp\left[-mp + mpt\right].$$

(7.14.8) is the generating function of a Poisson distribution with parameter mp.

7.15. Convolution. The sequence $\{c_x\} = c_0, c_1, c_2, \cdots$ defined by

$$(7.15.1) \qquad c_x = \sum_{j=0}^{j=x} a_j\, b_{x-j}, \qquad\qquad x = 0, 1, 2, \cdots$$

is the *convolution* (or *faltung*, *composition*) of the sequences $\{a_j\} = a_0,$ $a_1, a_2, \cdots$ and $\{b_k\} = b_0, b_1, b_2, \cdots$
The sequence $\{c_x\}$ is often written

$$\{c_x\} = \{a_x\} * \{b_x\}$$

The generating functions g.f. of $\{a_j\}$ and $\{b_k\}$ are

$$A(t) = \sum_{j=0} a_j t^i, \qquad B(s) = \sum_{k=0} b_k t^k$$

The generating function of $\{c_x\}$ is $C(t) = \sum_{x=0} c_x t^x$, and

$$(7.15.2) \qquad\qquad C(t) = A(t)\, B(t)$$

(7.15.2) is the generating function of the sum $Z = X + Y$. [Relation (7.15.2) corresponds to $G_1(t) \cdot G_2(t)$ in Example 3 of §7.14.]

The convolution is an associative and commutative operation (as is the sum of chance variables). For example

$$\{a_k\} * \{\{b_k\} * \{c_k\}\} = \{\{a_k\} * \{b_k\}\} * \{c_k\}$$

$$(7.15.3)$$

$$\{a_k\} * \{b_k\} * \{c_k\} = \{\{b_k\} * \{a_k\}\} * \{c_k\}, \text{ etc.}$$

EXAMPLE. Let X and Y be non-negative integral-valued independent chance variables having probability (density) distributions $P[X = j] = a_j$, $P[Y = k] = b_k$. The event $[X = j, Y = k]$ has the probability $a_j b_k$. Let $X + Y = Z$. The event $Z = x$ is a union of the mutually exclusive events

$$(7.15.4) \qquad [X = 0, Y = x], [X = 1, Y = x - 1], \cdots, [X = x, Y = 0].$$

The probability that $Z = x$ is the sum of the probabilities of the events in (7.15.4), that is,

$$(7.15.5) \qquad c_x = P[Z = x] = \sum_{j=0}^{j=x} a_j\, b_{x-j}, \qquad x = 0, 1, 2, \cdots$$

Here c_x is the probability (density) distribution of the chance variable $Z = X + Y$. Relation (7.15.5) corresponds to relation (7.14.3), with c_x corresponding to $f(x)$.

7.16. The distribution of the sum of independent variables may be obtained directly without use of characteristic or generating functions.

EXAMPLE. Let $x = y + z$ be the sum of two jointly related random variables. The probability that x will be no greater than some fixed number x_0 is

$$(7.16.1) \qquad P(y + z \leqq x_0) = \int_{-\infty}^{x_0} f(x)\, dx$$

where $f(x)$ is the joint density function of y and z over the region of y, z plane for which $y + z \leqq x_0$. (7.16.1) can be written

$$(7.16.2) \qquad P(x \leqq x_0) = \int_{-\infty}^{\infty} dy \int_{-\infty}^{x_0-y} g(y,z)\, dz$$

where $g(y,z)$ is the joint density function expressed in terms of y and z. By differentiation

$$(7.16.3) \qquad f(x_0) = \int_{-\infty}^{\infty} g(y, x_0 - y)\, dy$$

If y and z are statistically independent variables the integrand can be written

$$(7.16.4) \qquad g(y, x_0 - y) = f_1(y)\, f_2(x_0 - y)$$

where $f_1(y)$ and $f_2(z)$ are the density functions for y and z, respectively; and (7.16.3) becomes

$$(7.16.5) \qquad f(x_0) = \int_{-\infty}^{\infty} f_1(y)\, f_2(x_0 - y)\, dy$$

The right-hand side of (7.16.5) is called the *convolution operation*. (See §7.14, Theorem 2.)

7.17. Regenerative property of distributions. Suppose a family of distributions has the property that the convolutions of two such distributions with different parameters produce a distribution that is also in the family. The distribution of the sum of independent random variables described by such a distribution would then belong to the same family. This property is called the regenerative (reproductive) property of distributions. The normal, Poisson, chi-square distributions are examples of distributions which are regenerative. The exponential distribution is not regenerative, nor is the binomial distribution regenerative unless the parameter p is fixed.

eight

BINOMIAL, NEGATIVE BINOMIAL, HYPERGEOMETRIC AND RELATED DISTRIBUTIONS

Binomial distribution

8.1. If a discrete variable X is such that the probability $f(x)$ that X takes the value x is

$$(8.1.1) \qquad P(X = x) = f(x) = C_x^{\,n} p^x q^{n-x}, \qquad (x = 0, 1, \cdots, n)$$

where p is a constant, $0 \leqq p \leqq 1$,

$$q = 1 - p, \qquad \text{and} \qquad C_x^{\,n} = n!/x!(n - x)!,$$

the variable X is said to possess a *binomial probability distribution* (or, *Bernoulli distribution*). This discrete distribution is sometimes called the *point binomial* since a variable X so distributed assumes only integer values from 0 to n, and hence the probabilities are concentrated at "points".

The *cumulative binomial distribution* for any value x' less than or equal to n is given by

$$(8.1.2) \qquad P(X \leqq x') = F_B(x')$$

$$= C_0^{\,n} p^0 q^n + C_1^{\,n} p^1 q^{n-1} + \cdots + C_{x''}^{\,n} p^{x''} q^{n-x''},$$

where x'' is the largest integer which does not exceed x'. In other words,

$$F_B(x') = \sum_{x \leqq x'} C_x^{\,n} p^x q^{n-x}.$$

$$(8.1.3) \qquad F_B(n) = \sum_{x=0}^{x=n} C_x^{\,n} p^x q^{n-x} = 1.$$

The expression on the right hand side of Eq. (8.1.1) is the $(x + 1)^{\text{th}}$ term in the binomial expansion of $(q + p)^n$:

$$(8.1.4) \qquad (q + p)^n = q^n + C_1^{\,n} p q^{n-1} + \cdots + C_x^{\,n} p^x q^{n-x} + \cdots + p^n.$$

8.2. **Properties of binomial distribution.** The mean μ, variance σ^2, $\cdots$ of the variable X having distribution (8.1.1) are:

$$\text{Mean} = \mu = np = \text{expected value of } X.$$

$$(8.2.1) \qquad \text{Variance} = \sigma^2 = np(1 - p) = npq.$$

$$\text{Standard deviation} = \sigma = \sqrt{npq}.$$

r^{th} moment μ_r about the mean μ:

$$(8.2.2) \quad \mu_1 = 0, \qquad \mu_2 = \sigma^2 = npq, \qquad (q = 1 - p)$$
$$\mu_3 = npq(q - p), \qquad \mu_4 = npq[1 + 3(n - 2)pq].$$

r^{th} moment ν_r about $x = 0$:

$$\nu_1 = np, \qquad \nu_2 = np + n(n - 1)p^2,$$
$$(8.2.3) \quad \nu_3 = n(n - 1)(n - 2)p^3 + 3n(n - 1)p^2 + np,$$
$$\nu_4 = n(n - 1)(n - 2)(n - 3)p^4 + 6\nu_3 - 11\nu_2 + 6\nu_1.$$

r^{th} factorial moment $\nu_{[r]}$ about $x = 0$:

$$(8.2.4) \quad \nu_{[r]} = n^{[r]}p^r,$$

where

$$n^{[r]} = n(n - 1) \cdots (n - r + 1), \qquad n^{[0]} = 1.$$

Coefficient of *skewness* $= \gamma_1 = \mu_3/\sigma^3 = (q - p)/\sqrt{npq}$.

$(8.2.5)$ Coefficient of *excess* $= \gamma_2 = (\mu_4/\sigma^4) - 3 = (1 - 6pq)/npq$.

Mode $=$ positive integral value (or values) of x for which

$$np - q \leqq x \leqq np + p.$$

The characteristic function for distribution (8.1.1) is

$$(8.2.6) \quad \chi(t) = (pe^{it} + q)^n = [1 + p(e^{it} - 1)]^n, \qquad i^2 = -1.$$

8.3. Repeated trials. Let p be the probability of occurrence of an event E ("success") on each of n trials, so that the probability of non-occurrence of E ("failure") is $q = 1 - p, 0 \leqq p \leqq 1$.

Such a case can be written in a functional form. Let $\omega = 1$ denote a success, $\omega = 0$ a failure, and

$$g(\omega) = \begin{cases} p & \text{when} \quad \omega = 1 \\ q & \text{when} \quad \omega = 0. \end{cases}$$

Then $g(\omega)$ is the probability of obtaining ω successes in a single trial. Suppose n mutually independent trials are carried out. The probability of x successes followed by $n - x$ failures, in the particular order $\omega_1 = 1$, $\omega_2 = 1, \cdots, \omega_x = 1, \omega_{x+1} = 0, \cdots, \omega_n = 0$, is

$$(8.3.1) \quad g(\omega_1) g(\omega_2) \cdots g(\omega_n) = [g(1)]^x[g(0)]^{n-x} = p^x q^{n-x}.$$

If the order in which the successes occur is of no consequence the number of orders in which x successes and $n - x$ failures can occur is $C_x^n = n!/x!(n - x)!$. These orders are mutually exclusive events, so the probability $f(x)$ of exactly x successes (in n independent trials) irrespective of order is the sum of the probabilities for all the C_x^n orders, namely,

$$(8.3.2) \qquad P(X = x) = f(x) = C_x^n p^x q^{n-x}.$$

The probability of *at least* x occurrences (i.e., of $X = x$ or more occurrences) in n trials is

$$(8.3.3) \qquad P(X \geq x) = \sum_{i=x}^{n} C_i^n p^i q^{n-i}, \qquad \text{for} \qquad x = 0, 1, 2, \cdots, n.$$

Equation (8.3.3) is the sum of the last $n - x + 1$ terms of the binomial expansion (8.1.4) of $(q + p)^n$.

The probability of *at most* x occurrences (i.e., of $X = x$ or less occurrences) in n trials is

$$(8.3.4) \qquad P(X \leq x) = \sum_{i=0}^{i=x} C_i^n p^i q^{n-i}, \qquad \text{for} \qquad x = 0, 1, 2, \cdots, n.$$

The mean number of occurrences of event E is $\mu = np$, the mean number of non-occurrences is nq.

Some writers denote the function (8.3.4) by $B_n(x;p)$.

8.4. Case when p varies with the trial. Suppose that p_i is the probability of occurrence of an event E ("success") in the ith trial ($i = 1, 2, \cdots, n$), and $q_i = 1 - p_i$, the corresponding failure probability. Suppose the n trials are mutually independent. For the ith trial, let $\omega_i = 1$ denote a success, and $\omega_i = 0$ a failure. Define

$$g(\omega_i) = \begin{cases} p_i & \text{when} \quad \omega_i = 1 \\ q_i & \text{when} \quad \omega_i = 0 \end{cases} \qquad (i = 1, \cdots, n).$$

$g(\omega_i)$ is the probability of obtaining ω_i successes in the ith trial. Suppose n independent trials are carried out. The probability of x successes followed by $n - x$ failures, in the particular order $\omega_1 = 1$, $\omega_2 = 1$, $\cdots$, $\omega_x = 1$, $\omega_{x+1} = 0$, $\cdots$, $\omega_n = 0$, is

$$(8.4.1) \qquad g(\omega_1) \cdots g(\omega_n) = p_1 p_2 \cdots p_x q_{x+1} \cdots q_n.$$

If the order in which the successes occur is of no consequence the number of orders in which x successes and $n - x$ failures can occur is $C_x^n = n!/x!(n - x)!$. With each of these orders θ_j there is an expression correponding to (8.4.1), here called $g(\theta_i)$. These orders are mutually exclusive events, so that the probability $f(x)$ of exactly x successes (in n independent trials) irrespective of order is the sum of the probabilities for all the C_x^n orders, namely

$$8.4.2) \qquad P(X = x) = f(x) = \sum_i g(\theta_i).$$

Let $X = \sum_{i=1}^{n} \omega_i$, that is, X is the total number of successes in n mutually independent trials. The expected (or mean) value of X is

$$8.4.3) \qquad E(\omega_1 + \cdots + \omega_n) = \sum_{i=1}^{n} E(\omega_i) = \sum_{i=1}^{n} p_i.$$

The variance of ω_i is $p_i q_i$, so for n independent trials the variance of X is (see Eq. 7.14.4)

$$8.4.4) \qquad \sum_{i=1}^{n} p_i q_i = npq - \sum_{i=1}^{n} (p_i - p)^2,$$

where

$$8.4.5) \qquad p = (1/n) \sum_{i=1}^{n} p_i, \qquad q = 1 - p.$$

When the probability p_i remains constant from trial to trial, the distribution (8.4.2) reduces to (8.3.2) and is called the *Bernoulli case*; when p_i varies (8.4.2) is called the *Poisson case*.

In the Bernoulli case (8.4.2) is a certain general term of the binomial expansion $(p + q)^n$, while in the Poisson case, (8.4.2) is a certain sum of typical terms in the expansion of the product $(p_1 + q_1)(p_2 + q_2) \cdots (p_n + q_n)$. Distribution (8.4.2) is sometimes called the *generalized binomial distribution of Poisson*. The characteristic function for (8.4.2) is

$$8.4.6) \qquad \chi(t) = \prod_{j=1}^{n} (p_i e^{it} + q_i), \qquad i^2 = -1.$$

EXAMPLE 1. Gun A has, for each shot, a probability p_1, gun B a probability p_2. A fires three shots and B fires two shots. Find the probabilities of exactly $0, 1, \cdots, 5$ hits.

This problem may be solved by expanding the left member of the identity

$$(p_1 + q_1)^3 (p_2 + q_2)^2 = 1.$$

The probability of obtaining exactly five hits is the term containing only p_1 and p_2; the probability of obtaining exactly four hits is the sum of all the terms containing four p's and one q, etc.

8.5. Calculations with binomial distribution. In using the binomial distribution to calculate probabilities one can not always know in advance the value of p. In such cases p must usually be estimated experimentally. To do this set the mean $\mu = np$ of the theoretical distribution equal to the mean $\bar{x}$ of the (observed) frequency distribution, and calculate p. This value of p can then be used to calculate the binomial distribution function $f(x)$ given in (8.1.1).

EXAMPLE 1.   If n "unbiased" coins are tossed once, the probability of obtaining x heads and $n - x$ tails is

$$[n!/x!(n - x)!](1/2)^n, \qquad \text{for} \qquad x = 0, 1, \cdots, n.$$

EXAMPLE 2.   The probability of obtaining a head in one toss of a "biased" coin is $p = 0.1$. The probability of getting 3 "heads" if the coin is tossed 4 times is

$$C_3{}^4(0.1)^3(0.9) = 0.0036.$$

EXAMPLE 3.   An unbiased die is rolled 5 times. The probability of obtaining "four" exactly x times is

$$C_x{}^5(1/6)^x(5/6)^{5-x}, \qquad \text{for} \qquad x = 0, 1, \cdots, 5.$$

EXAMPLE 4.   A set of six cards was dropped on the floor 100 times. The number x of cards falling face up was distributed with frequency as shown below.

x	0	1	2	3	4	5	6
f	2	13	21	32	22	6	4

If the probability of a card falling face up is p, the theoretical probability of exactly x cards of the 6 falling face up is

$$(8.5.1) \qquad f(x) = C_x{}^6 p^x q^{6-x}, \qquad \text{for} \qquad x = 0, 1, \cdots, 6.$$

The mean of this theoretical distribution is $\mu = 6p$, in contrast to the mean $\bar{x} = 2.93$ of the observed distribution. Set $\mu = \bar{x}$, hence $p = 0.488$. With this value of p, the binomial distribution (8.1.1) could be "fitted" to the observed data, thus obtaining an approximation to the observed relative frequency. This is illustrated in Table 8.5.1.

TABLE 8.5.1

Binomial distribution (8.5.1) fitted to data listed in the first two columns

Observed frequency	Observed relative frequency	Fitted binomial probability distribution	Fitted expected frequency	Cumulative observed frequency	Cumulative expected frequency	Fitted normal probability distribution
x	f	$f(x)$				$\psi(t) \div (1.22)$
0	2	0.02	0.018	1.8	2	1.8	0.018
1	13	0.13	0.103	10.3	15	12.1	0.094
2	21	0.21	0.245	24.5	36	36.6	0.246
3	32	0.32	0.312	31.2	68	67.8	0.327
4	22	0.22	0.223	22.3	90	90.1	0.223
5	6	0.06	0.085	8.5	96	98.6	0.078
6	4	0.04	0.014	1.4	100	100	0.014
Total	100	1.00	1.000	100			1.000

$$\bar{x} = \sum_{i=1}^{6} f_i x_i / 100 = 2.93 = \text{mean}$$

of observed distribution.
$6p$ = mean of theoretical distribution (8.5.1).

Set $6p = 2.93$. Then $p = 0.488$.

$$f(x) = C_x^6 (0.488)^x (0.512)^{6-x}.$$

Expected frequency of exactly x cards of the six falling face up in a set of N drops $= N \cdot f(x)$, $N = 100$.

$$f(x) \cong \psi(t) \div 1.22$$
$$= (1/\sqrt{2\pi}) e^{-t^2/2} \div 1.22.$$

EXAMPLE 5. To fit the normal distribution $\psi(t) = (1/\sqrt{2\pi}) e^{-t^2/2}$, where $t = (x - \mu)/\sigma$, to the binomial distribution $f(x) = C_x^n p^x q^{6-x}$, with $p = 0.488$, $q = 1 - p$, set $n = 6$, $\mu = 6(0.488) = 2.928$, $\sigma = [6(0.488)(0.512)]^{1/2} = 1.22$. Then

$$C_x^6 (0.488)^x (0.512)^{6-x} \cong (1/\sqrt{2\pi}) e^{-t^2/2} \div (1.22) = \psi(t) \div (1.22).$$

The results of this fit are shown in the last column of Table 8.5.1.

The following recursion formula is sometimes useful in calculations involving Eq. (8.1.1), especially when n is very much larger than x.

$$(8.5.2) \qquad f(x + 1) = [(n - x)/(x + 1)][p/q]f(x),$$

where $f(x) = C_x^n p^x q^{n-x}$. To use this formula, begin by finding $f(0) = q^n$, $f(1)$, $f(2)$, and so on.

The expression (8.3.3) can be calculated with the aid of the identity

$$8.5.3) \qquad \sum_{i=x}^{n} C_i{}^n p^i q^{n-i} = \frac{\int_0^p t^{x-1}(1-t)^{n-x}\,dt}{\int_0^1 t^{x-1}(1-t)^{n-x}\,dt} = \frac{B_p(x, n-x+1)}{B(x, n-x+1)}$$

$$= I_p(x, n-x+1),$$

where B_p is the Incomplete Beta-function, B the Beta-function. (See §13.25, §18.5, §18.6, §18.7; Table III.)

The expression (8.1.1) can be calculated with the aid of

$$(8.5.4) \qquad C_x{}^n p^x q^{n-x} = I_p(x, n-x+1) - I_p(x+1, n-x).$$

The sum $\sum C_x{}^n p^x q^{n-x}$ for $x = a$ to $x = b$, $p + q = 1$, is approximately (if n is large enough, and certain other conditions are met),

$$(8.5.5) \qquad \sum_{x=a}^{x=b} C_x{}^n p^x q^{n-x}$$

$$= \int_{t_1}^{t_2} \psi(t)\,dt + [(-\gamma_1/3!)\psi^{(2)}(t) + (\gamma_2/4!)\psi^{(3)}(t)]_{t_1}^{t_2},$$

where

$$t_1 = [a - (1/2) - np] \div \sigma, \qquad t_2 = [b + (1/2) - np] \div \sigma,$$

$$(8.5.6) \qquad \gamma_1 = (q - p) \div \sigma, \qquad \gamma_2 = (1 - 6pq) \div \sigma^2, \qquad \sigma^2 = npq,$$

$$\psi(t) = (1/\sqrt{2\pi})e^{-t^2/2},$$

$\psi^{(2)}(t)$ and $\psi^{(3)}(t)$ being the second and third derivatives of $\psi(t)$. (See §10.10.) (For values of these derivatives see Table IX.) Here the bracket has the same meaning as in calculus, i.e.,

$$(8.5.7) \qquad f(t)]_{t_1}^{t_2} \equiv f(t_2) - f(t_1).$$

8.6. DeMoivre's theorem for binomial distribution. If $n \to \infty$ in such a way that $(x - np)/\sqrt{npq} \to t$, t finite, $p + q = 1$, then $\sqrt{npq}\,C_x{}^n p^x q^{n-x} \to (1/\sqrt{2\pi})e^{-t^2/2} = \psi(t)$. Also, for every fixed T and p,

$$(8.6.1) \qquad \lim_{n\to\infty} \sum_{x \le M} C_x{}^n p^x q^{n-x} = \int_{-\infty}^{T} \psi(t)\,dt,$$

where

$$M = np + T\sqrt{npq}.$$

For any fixed values of T_1, T_2, and p,

$$(8.6.2) \qquad \lim_{n\to\infty} \sum_{M_1 < x \le M_2} C_x{}^n p^x q^{n-x} = \int_{T_1}^{T_2} \psi(t)\,dt,$$

where

$$M_i = np + T_i \sqrt{npq}, \qquad (i = 1, 2).$$

Approximating the binomial with the normal curve. The normal probability distribution (see §10.3) gives a good approximation to the binomial distribution when n is large enough. To fit the normal curve $y = \psi(t)/\sigma$ to the binomial probability distribution $f(x) = C_x^n p^x q^{n-x}$, set $\mu = np$, $\sigma = \sqrt{npq}$, $t = (x - \mu)/\sigma$. (This follows from DeMoivre's limit theorem given above.)

8.7. Correction for continuity.

When x is a discrete variable having a binomial distribution (8.1.1) it is often necessary to compute the probability that x lies between a and b, inclusive, that is,

$$(8.7.1) \qquad P_B(a \leqq x \leqq b) = \sum_{x=a}^{x=b} f_B(x).$$

Note that in computing (8.7.1), one adds areas of rectangles centered at $x = a, \cdots, x = b$. If a normal distribution is used to estimate (8.7.1), one should approximate (8.7.1) by

$$(8.7.2) \qquad P_N(\alpha \leqq x \leqq \beta), \qquad \alpha = a - \tfrac{1}{2}, \qquad \beta = b + \tfrac{1}{2},$$

in order not to omit the half of the first and the half of the last rectangles in the approximation, and thus underestimate the required probability. Corrections of this sort are called *half-integer corrections for continuity*. The following example is illustrative.

EXAMPLE. A machine mass-produced a certain object. Records indicate that 20 percent of a production run are defective. A random sample of 100 is drawn from a run. The probability that exactly x items are defective is

$$f_B(x) = C_x^{100} p^x q^{100-x}, \qquad p = 0.2, \qquad q = 0.8.$$

The probability that not more than 10 defective items occur in the sample is

$$(8.7.3) \qquad P_B(x \leqq 10) = \sum_{x=0}^{x=10} f_B(x).$$

The probability that between 10 and 25, inclusive, are defective is

$$(8.7.4) \qquad P_B(10 \leqq x \leqq 25) = \sum_{x=10}^{x=25} f_B(x).$$

Use the normal distribution to approximate (8.7.3). The mean is

$\mu = np = 100(0.2) = 20$, the variance $\sigma^2 = npq = 16$. Let $t = (x - \mu)/\sigma$. Then, from §10.1, §10.3, §10.11, §8.5, and Table IX,

$$P_B(x \leq 10) \cong P_N(x \leq 10.5) = P_N(t \leq (10.5 - 20)/4)$$
$$= P_N(t \leq -2.38) = 0.009.$$

Similarly,

$$P_B(10 \leq x \leq 25) \cong P_N(9.5 \leq x \leq 25.5) = P_N(-10.5/4 \leq t \leq 5.5/4)$$
$$= P_N(t \leq 1.38) - P_N(t \leq -2.62)$$
$$= 0.9155 - 0.0044 = 0.9111.$$

The probability that exactly 10 defective objects are in the sample is

$$P_B(x = 10) = C_{10}{}^{100}(0.2)^{10}(0.8)^{90} \cong P_N(9.5 \leq x \leq 10.5)$$
$$= P_N[(9.5 - 20)/4 \leq t \leq (10.5 - 20)/4]$$
$$= P_N(t \leq -2.375) - P_N(t \leq -2.625)$$
$$= 0.0088 - 0.0043 = 0.0045$$

8.8. Reproductive property of binomial distribution.

Suppose X_1 and X_2 are statistically independent variables, having binomial distributions with the same value of the parameter p, and

$$(8.8.1) \quad P(X_1 = x_1) = C_{x_1}{}^{n_1}p^{x_1}q^{n_1-x_1}, \qquad P(X_2 = x_2) = C_{x_2}{}^{n_2}p^{x_2}q^{n_2-x_2}.$$

Then the sum $Z = X_1 + X_2$ has a binomial distribution with the parameter p, and

$$(8.8.2) \quad P(Z = z) = C_z{}^{n}p^{z}q^{n-z}, \qquad \text{where} \quad z = x_1 + x_2,\, n = n_1 + n_2.$$

This result is known as the *reproductive property of the binomial distribution.*

8.9. Multinomial distribution.

Suppose a set of discrete variables $X_1, \cdots, X_n$ is such that the probability that the set takes the value $x_1, \cdots, x_n$ is

$$(8.9.1) \quad P(X_1 = x_1, \cdots, X_n = x_n) = f(x_1, \cdots, x_n)$$

$$= \frac{N!}{x_1! \cdots x_n!}\, p_1{}^{x_1} \cdots p_n{}^{x_n},$$

where the x_i are positive integers and each $p_i > 0$ for $i = 1, \cdots, n$, and

$$\sum_{j=1}^{n} p_i = 1, \qquad \sum_{j=1}^{n} x_i = N.$$

The joint distribution (8.9.1) of $x_1, \cdots, x_n$ is called the *multinomial distribution*. (8.9.1) is the general term of the expansion of $(p_1 + \cdots + p_n)^N$.

8.10. **Properties.** The expected or mean value of $\bar{x}_i$ is $m_i = Np_i$; the variance of x_i is $Np_i(1 - p_i)$; the covariance of x_i and x_k is $-Np_ip_k$. When $n = 2$, (8.9.1) reduces to the *binomial distribution*.

EXAMPLE 1.   Transit tokens of a certain type are produced the diameters of which should be 0.500 inch. The tokens are classified as *undersize, acceptable* and *oversize* if their diameters measure less than 0.495, between 0.495 and 0.505 inch, or more than 0.505 inch, respectively.

The production process is such that for a large lot 3, 89, and 8 percent are undersize, acceptable, and oversize, respectively. From a production lot 100 of the tokens are picked at random. The probability of getting x_1 undersize, x_2 acceptable, and x_3 oversize tokens from the sample of 100 is

$$f\,(x_1, x_2, x_3) = \frac{100!}{x_1!\,x_2!\,x_3!}\,(.03)^{x_1}(.89)^{x_2}(.08)^{x_3}$$

where $\sum_{i=1}^{3} x_i = 100$, and $0 \leqq x_i \leqq 100$ for $i = 1, 2, 3$.

EXAMPLE 2.   Suppose a certain random experiment can produce any of n mutually exclusive events $E_1, \cdots, E_n$. Let the probability of event E_i be $p_i > 0$, and $\sum_{i=1}^{n} p_i = 1$.

In a series of N mutually independent trials (in which the p_i do not vary with the trial) the probability of the events taking place in the order: x_1 occurrences of E_1, followed by x_2 occurrences of E_2, $\cdots$, x_n occurrences of E_n; is

(8.10.1) $p_1^{x_1} p_2^{x_2} \cdots p_n^{x_n}.$

The number of different orders in which N objects can be permuted when x_1 are of type E_1, x_2 of type E_2, $\cdots$ is $N!/x_1!\,x_2! \cdots x_n!$. These orders are mutually exclusive events. Hence, the probability $f(x_1, \cdots, x_n)$ of x_1 occurrences of E_1, x_2 occurrences of E_2, $\cdots$, x_n occurrences of E_n, irrespective of the order in which they occur in the trials, is found by adding the probabilities for the various possible orders. The result is given in Eq. (8.9.1).

In case the p_i vary from trial to trial a method similar to that given in §8.4 can be used.

8.11. **Tables.** Table I gives values of $C_x^n p^x (1 - p)^{n-x}$. Table II gives values of $\sum_{x=x'}^{x=n} C_x{}^n p^x (1 - p)^{n-x}$, for values of $x' = 1, 2, \cdots, n$. More extensive tables of these quantities are given in *Tables of the Binomial Probability Distribution*, National Bureau of Standards, Applied Mathematics Series 6, Washington, D. C., 1950.

8.12. Negative binomial distribution. If a discrete variable X is such that the probability $f(x)$ that X takes the value x is

$$(8.12.1) \qquad P(X = x) = f(x) = C(x + r - 1, r - 1)p^r q^x, \quad x = 0, 1, \cdots$$

where

$$C(x + r - 1, r - 1) = (x + r - 1)!/(r - 1)! \, x!, \qquad q = 1 - p$$

the variable X is said to have a *negative binomial (Pascal) distribution.*

Properties. The mean and variance of the variable X with distribution (8.12.1) are:

$$\text{Mean} = rq/p, \qquad \text{variance} = rq/p^2$$

(8.12.1) may be written

$$(8.12.2) \qquad f(x) = f(x; r,p) = C(r + x - 1, x)p^r q^x$$

$$= \binom{-r}{x} p^r (-q)^x, \qquad x = 0, 1, \cdots$$

The terms $f(x; r,p)$ of distribution (8.12.2) can be calculated from tables of binomial distributions, using the identities

$$\begin{aligned}
f(x; r,p) &= p \cdot b(x; r + x - 1, q) \\
&= p \cdot b(r - 1; r + x - 1, p) \\
&= [r/(r + x)] \cdot b(r; r + x, p) \\
(8.12.3) \qquad &= [r/(r + x)] \cdot b(x; r + x, q)
\end{aligned}$$

where

$$b(x; n,p) = C(n,x)p^x q^{n-x} = \binom{n}{x} p^x q^{n-x}$$

is the binomial probability density (8.1.1).

EXAMPLE 1.   Let p be the probability of a defective item appearing in a production lot, and $q = 1 - p$. A random sample is taken from the lot. (The trials are Bernoulli.) The probability that the rth defective unit will occur on the $(x + r)^{\text{th}}$ unit sampled is

$$(8.12.4) \qquad f(x) = C(x + r - 1, r - 1)p^r q^x, \qquad x = 0, 1, \cdots$$

Or, if $m = x + r$, the probability of the rth defective unit occurring on the mth unit sampled is

$$(8.12.5) \qquad g(m) = C(m - 1, r - 1)p^r q^{m-r}, \qquad m = r, r + 1, \cdots$$

The probability that the rth defective unit occurs on the rth or $(r + 1)^{st}$ or $\cdots$ or n^{th} unit sampled is

$$(8.12.6) \qquad \sum_{x=0}^{x=n-r} C(x + r - 1, r - 1)p^r q^x = \sum_{x=0}^{x=n-r} f(x)$$

$$(8.12.7) \qquad \sum_{x=0}^{x=n-r} C(x + r - 1, r - 1)p^r q^x = \sum_{m=r}^{m=n} C(m - 1, r - 1)p^r q^{m-r}.$$

Distribution (8.12.1) should not be confused with Pólya's distribution of §6.27 (which is also known as a negative binomial distribution).

In case $r = 1$, the distribution (8.12.1) is known in some applications as a *geometric distribution*. In this case

$$(8.12.8) \qquad f(x) = pq^x, \qquad\qquad x = 0, 1, \cdots$$

The mean is q/p; the variance is q/p^2.

EXAMPLE 2.   The probability that the first defective unit will occur on the $(x + 1)^{th}$ unit sampled is pq^x.

The probability that the first defective unit will occur after at most x trials is

$$(8.12.9) \qquad F(x) = 1 - (1 - p)^{x+1}.$$

Equation (8.12.9) is the *cumulative geometric distribution* for (8.12.8).

8.13. **Hypergeometric distribution.**   A sample of r items out of a lot of n items is selected. The sampling is done without replacement. The lot contains n_1 items having property A (i.e., defective). Let x be the number of items in the sample with property A. The chance of selecting r items that have exactly x items with property A is

$$(8.13.1) \qquad f(x) = \binom{n_1}{x}\binom{n - n_1}{r - x} \div \binom{n}{r}.$$

Here $x = 0, 1, \cdots, b$, $b =$ smaller of n_1 and r,

$$n_1 < n, r < n.$$

The distribution whose density function is (8.13.1) is called the *hypergeometric distribution*.

The probability of selecting r items that have x' defective items, where $x' \leqq x$, is

$$(8.13.2) \qquad P(x \leqq x') = F(x) = \sum_{i=0}^{x} \binom{n_1}{i}\binom{n - n_1}{r - i} \div \binom{n}{r}$$

$F(x)$ is called the *cumulative hypergeometric distribution*.

Properties. The mean and variance of the chance variable X having probability density distribution (8.13.1) are, respectively,

$$(8.13.3) \qquad \begin{aligned} \mu &= rn_1/n, \\ \sigma^2 &= rn_1(n - n_1)(n - r)/n^2(n - 1). \end{aligned}$$

The hypergeometric distribution occurs in problems of acceptance sampling by attributes where an item is classified as either defective or non-defective.

When the population size n is large, the hypergeometric distribution function (8.13.1) may be approximated by the corresponding binomial distribution function (8.1.1).

For tables see

Lieberman, G. J., and Owen, D. B., *Tables of Hypergeometric Distributions*, Stanford University Press, Stanford, Calif., 1960.

nine

POISSON, EXPONENTIAL AND WEIBULL DISTRIBUTIONS

Poisson distribution

9.1. If a discrete chance variable X is such that the probability $f(x)$ that X takes the value x is

(9.1.1) $P(X = x) = f(x) = e^{-m}m^x/x!, \quad m > 0$, for $x = 0, 1, 2, \cdots$,

the variable X is said to possess a *Poisson distribution*. When $x = 0$, $f(x) = e^{-m}$, since $0! = 1$. In (9.1.1) m is a constant.

9.2. The probability that X is greater than or equal to x' is

(9.2.1) $$P(X \geqq x') = \sum_{x=x'}^{x=\infty} e^{-m}m^x/x!.$$

Note: $P(X \geqq 0) = 1$.
The probability that X is less than or equal to x'' is

(9.2.2) $$P(X \leqq x'') = \sum_{x=0}^{x=x''} e^{-m}m^x/x!.$$

9.3. **Properties of Poisson distribution (9.1.1).** The mean μ, variance $\sigma^2, \cdots$, of the variable X having distribution (9.1.1) are:

Mean $= \mu = m$.

(9.3.1) Variance $= \sigma^2 = m$. $\quad \sum_{x=0}^{\infty} e^{-m}m^x/x! = 1$.

Standard deviation $= \sigma = \sqrt{m}$.

r^{th} moment μ_r about mean μ: $\qquad r^{\text{th}}$ moment ν_r about $x = 0$:

$$\mu_2 = \sigma^2 = m. \qquad\qquad \nu_1 = m.$$

$$\mu_3 = m. \qquad\qquad\qquad \nu_2 = m(m + 1).$$

(9.3.2) $\mu_4 = 3m^2 + m. \qquad\quad \nu_3 = m(m^2 + 3m + 1).$

$$\qquad\quad \nu_4 = m(m^3 + 6m^2 + 7m + 1).$$

105

r^{th} factorial moment $\nu_{[r]}$ about $x = 0$: $\nu_{[2]} = m^2$.

(9.3.3)   Coefficient of skewness $= \gamma_1 = 1/\sqrt{m}$.

Coefficient of excess $= \gamma_2 = 1/m$.

Characteristic function of distribution (9.1.1) is

(9.3.4) $\chi(t) = \exp\,[m(e^{it} - 1)]$.

Figures 9.3.1 and 9.3.2 give a representation of (9.1.1) for certain values of m.

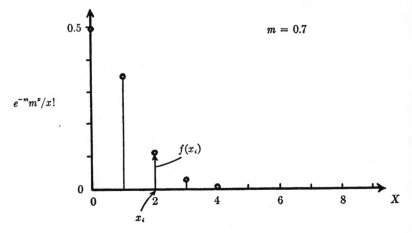

FIGURE 9.3.1

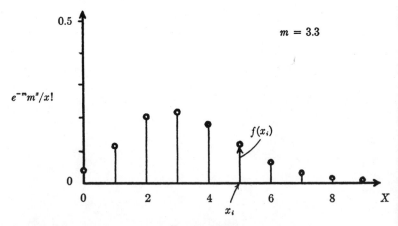

FIGURE 9.3.2

9.4. Reproductive property. If X_1 and X_2 are independent Poisson distributed variables with parameters m_1 and m_2, respectively, the sum $X_1 + X_2$ has a Poisson distribution with parameter $m_1 + m_2$.

9.5. Approximation to binomial distribution. The binomial distribution $C_x{}^n p^x q^{n-x}$ (see §8.1) approaches the Poisson distribution (9.1.1) as a limit when $n \to \infty$ and $p \to 0$ in such a way that $np = m$ is a constant. The Poisson distribution (9.1.1) is often used to approximate the binomial distribution when n is large enough $(n > 50)$, p is small enough $(p < 0.1)$, and np is between, say, 0 and 10.

EXAMPLE 1.   The probability of obtaining five heads x times when five coins are tossed 128 times is (by the binomial distribution)

$$(9.5.1) \qquad C_x{}^{128}(1/32)^x(31/32)^{128-x}.$$

An approximate value to (9.5.1) is given by the Poisson distribution (9.1.1) with $m = np = 128(1/32) = 4$; that is by

$$(9.5.2) \qquad 4^x e^{-4}/x!.$$

When $x = 2$, the values of (9.5.1) and (9.5.2) are 0.1453, and 0.1465, respectively.

9.6. Fitting a Poisson distribution to a sample frequency.   In fitting (9.1.1) to a sample frequency distribution m may not be known. In such cases, to make the fit, set the mean np of the sample distribution equal to m in (9.1.1).

EXAMPLE 1.   A certain screw making machine produces an average of 2 defective screws out of 100, and packs the screws in boxes of 500. The expected number of defectives in a box of 500 is 10. Set $m = np = 10$. The probability that a box of 500 contains x defective screws is approximately $10^x e^{-10}/x!$; and $C_x{}^{500}(10/500)^x(490/500)^{500-x}$ by (8.3.2).

9.7. Applications.   The Poisson distribution has found many applications where a large number n of observations is involved and the probability p of an event occurring in any specific observation is very small, as in problems concerned with the rare occurrence of events in a fixed time interval (e.g., the number born deaf per year in N. Y. C., the frequency of certain "peaks" per interval of time in telephone and other traffic problems, bacteria count, disintegration of atoms in a given radioactive source, etc.)

Poisson's distribution arises in practice, whenever a process is such that the probability of exactly x events occurring within an interval t is approximately, or exactly, given by (9.1.1), where $m = at$ is proportional to t.

EXAMPLE 1.   Let m be the expected (average) number of pulses striking an object in t seconds, the pulses being independent and random, and the chance of a hit rare. The probability that x of the pulses hit the object in t seconds is often taken to be approximately $e^{-m}m^x/x!$.

EXAMPLE 2.   A wire is coated by a certain process. The probability that a break in the coating in a piece of wire Δl units long is assumed to be $\lambda \Delta l$, where λ is a constant independent of the position on the wire and the number of breaks elsewhere on the wire. Suppose the probability that two or more breaks in a piece of wire Δl units in length is denoted by $o(\Delta l)$, where as $\Delta l \to 0$, $o(\Delta l)/\Delta l \to 0$.

Let $p_x(l)$ denote the probability that x breaks will occur in the coating along a wire of length l. Then

$$(9.7.1) \quad p_x(l + \Delta l) = p_x(l) \cdot (1 - \lambda \Delta l) + p_{x-1}(l) \cdot \lambda \Delta l + o(\Delta l).$$

From (9.7.1), with $p_0(0) = 1$, $p_x(0) = 0$ for $x > 0$.

$$(9.7.2) \quad \begin{cases} \dfrac{dp_x(l)}{dl} = \lambda[p_{x-1}(l) - p_x(l)] \\[2ex] \dfrac{dp_0(l)}{dl} = -\lambda p_0(l), \end{cases}$$

and

$$(9.7.3) \quad p_x(l) = (\lambda l)^x e^{-\lambda l}/x!$$

where λ is the mean number of coating breaks per unit length of the wire.

9.8. **Tables.**   The following tables are useful in dealing with the Poisson distribution: Table VII gives values of (9.1.1). Table VIII gives values of (9.2.1) for $x' = 0, 1, 2, \cdots$. More extensive tables of these quantities are given in Molina, E. C., *Poisson's Exponential Binomial Limit*, D. Van Nostrand, New York, 1942.

Other exponential type distributions

9.9. **Exponential distribution.**   If the chance variable X is such that the (cumulative) probability that X is less than or equal to x' is

$$(9.9.1) \quad P(X \le x') = F(x') = \int_{-\infty}^{x'} f(x)\, dx,$$

where, with $\lambda > 0$, the probability density function is

(9.9.2) $\qquad f(x) = 0$ for $x < 0, \qquad f(x) = \lambda e^{-\lambda x}$ for $x \geqq 0$,

the variable X is said to possess an *exponential distribution*.

Properties. The mean and variance of X are, respectively, $\mu = 1/\lambda$, $\sigma^2 = 1/\lambda^2$. The distribution does not have the reproductive property.

9.10. Weibull distribution. If a continuous chance variable X is such that the (cumulative) probability that X is less than or equal to x' is

(9.10.1) $\qquad P(X \leqq x') = F(x') = \int_{-\infty}^{x'} f(x) \, dx,$

where, with $a > 0$, $b > 0$, the probability density function is

(9.10.2) $\qquad f(x) = 0$ for $x \leqq 0, \qquad f(x) = abx^{b-1} \exp[-ax^b]$ for $x > 0$,

the variable X is said to possess a *Weibull distribution*.

Properties. The mean and variance of X are, respectively,

(9.10.3) $\qquad \mu = \alpha^\beta \Gamma(\beta + 1), \qquad \sigma^2 = \alpha^{2\beta}\{\Gamma(2\beta + 1) - [\Gamma(\beta + 1)]^2\},$

where $\alpha = 1/a$, $\beta = 1/b$. b is known as the *shape* parameter, α the *scale* parameter. The distribution does not have the reproductive property.

In certain fields, a generalization of this distribution is used in which x is replaced by $x - g$, where g represents a location or threshold parameter (e.g., a minimum life, or "guarantee time" in reliability studies).

The Weibull distribution is a generalization of the exponential distribution. It has found considerable use in the study of the aging of equipment, that is, of in time variations of the mean time between failures.

10.1. The normal distribution is the most important continuous probability distribution. It is widely used in probability and statistics.

If a continuous chance variable X is such that the (cumulative) probability that X is less than or equal to x is

$$(10.1.1) \qquad P(X \leqq x) = F_N(x) = (1/\sqrt{2\pi}\ \sigma) \int_{-\infty}^{x} e^{-(X-\mu)^2/2\sigma^2}\ dX,$$

where μ and σ are constants, the variable X is said to be distributed *normally*. The mean and standard deviation of X are μ and σ, respectively.

The (random) variable x is said to have the *normal* distribution $N(\mu,\sigma^2)$.

$F_N(x)$ is called the *cumulative normal* (or *Gaussian*) *distribution*. The normal probability density function $f_N(x)$ is the derivative with respect to x of $F_N(x)$,

$$(10.1.2) \qquad f_N(x) = (1/\sqrt{2\pi}\ \sigma)e^{-(x-\mu)^2/2\sigma^2}.$$

The probability that X falls between x' and x'' is

$$(10.1.3) \qquad P(x' < X \leqq x'') = F_N(x'') - F_N(x').$$

10.2. **Values of** $F_N(x)$. Values of $F_N(x)$ for values of x ranging from $\mu - 3\sigma$ to $\mu + 3\sigma$ are given in Table 10.2.1. For a more comprehensive table see Table 10.8.3 and Table IX.

EXAMPLE 1. The chance variable X has a normal distribution with mean 40 and standard deviation 3. The probability that X lies between 34 and 43 is $F_N(43) - F_N(34)$. Here $t = (x - 40)/3$. When $x = 43$, $t = 1$, when $x = 34$, $t = -2$. From Table 10.2.1

$$P(34 < X \leqq 43) = P(-2 < t \leqq 1) = F_N(t = 1) - F_N(t = -2)$$

$$= 0.8413 - 0.0228 = 0.8185.$$

Table 10.2.1

Values of cumulative normal distribution function $F_N(x)$

$$F_N(x) = (1/\sqrt{2\pi}\,\sigma) \int_{-\infty}^{x} e^{-(X-\mu)^2/2\sigma^2}\,dX = (1/\sqrt{2\pi}) \int_{-\infty}^{(x-\mu)/\sigma} e^{-\tau^2/2}\,d\tau,$$

$$t = (x - \mu)/\sigma$$

t	x	$F_N(x)$	t	x	$F_N(x)$
−3.0	$\mu - 3.0\sigma$	0.0013	0	μ	0.5000
−2.9	$\mu - 2.9\sigma$	0.0019	0.1	$\mu + 0.1\sigma$	0.5398
−2.8	$\mu - 2.8\sigma$	0.0026	0.2	$\mu + 0.2\sigma$	0.5793
−2.7	$\mu - 2.7\sigma$	0.0035	0.3	$\mu + 0.3\sigma$	0.6179
−2.6	$\mu - 2.6\sigma$	0.0047	0.4	$\mu + 0.4\sigma$	0.6554
−2.5	$\mu - 2.5\sigma$	0.0062	0.5	$\mu + 0.5\sigma$	0.6915
−2.4	$\mu - 2.4\sigma$	0.0082	0.6	$\mu + 0.6\sigma$	0.7257
−2.3	$\mu - 2.3\sigma$	0.0107	0.7	$\mu + 0.7\sigma$	0.7580
−2.2	$\mu - 2.2\sigma$	0.0139	0.8	$\mu + 0.8\sigma$	0.7881
−2.1	$\mu - 2.1\sigma$	0.0179	0.9	$\mu + 0.9\sigma$	0.8159
−2.0	$\mu - 2.0\sigma$	0.0228	1.0	$\mu + 1.0\sigma$	0.8413
−1.9	$\mu - 1.9\sigma$	0.0287	1.1	$\mu + 1.1\sigma$	0.8643
−1.8	$\mu - 1.8\sigma$	0.0359	1.2	$\mu + 1.2\sigma$	0.8849
−1.7	$\mu - 1.7\sigma$	0.0446	1.3	$\mu + 1.3\sigma$	0.9032
−1.6	$\mu - 1.6\sigma$	0.0548	1.4	$\mu + 1.4\sigma$	0.9192
−1.5	$\mu - 1.5\sigma$	0.0668	1.5	$\mu + 1.5\sigma$	0.9332
−1.4	$\mu - 1.4\sigma$	0.0808	1.6	$\mu + 1.6\sigma$	0.9452
−1.3	$\mu - 1.3\sigma$	0.0968	1.7	$\mu + 1.7\sigma$	0.9554
−1.2	$\mu - 1.2\sigma$	0.1151	1.8	$\mu + 1.8\sigma$	0.9641
−1.1	$\mu - 1.1\sigma$	0.1357	1.9	$\mu + 1.9\sigma$	0.9713
−1.0	$\mu - 1.0\sigma$	0.1587	2.0	$\mu + 2.0\sigma$	0.9772
−0.9	$\mu - 0.9\sigma$	0.1841	2.1	$\mu + 2.1\sigma$	0.9821
−0.8	$\mu - 0.8\sigma$	0.2119	2.2	$\mu + 2.2\sigma$	0.9861
−0.7	$\mu - 0.7\sigma$	0.2420	2.3	$\mu + 2.3\sigma$	0.9893
−0.6	$\mu - 0.6\sigma$	0.2743	2.4	$\mu + 2.4\sigma$	0.9918
−0.5	$\mu - 0.5\sigma$	0.3085	2.5	$\mu + 2.5\sigma$	0.9938
−0.4	$\mu - 0.4\sigma$	0.3446	2.6	$\mu + 2.6\sigma$	0.9953
−0.3	$\mu - 0.3\sigma$	0.3821	2.7	$\mu + 2.7\sigma$	0.9965
−0.2	$\mu - 0.2\sigma$	0.4207	2.8	$\mu + 2.8\sigma$	0.9974
−0.1	$\mu - 0.1\sigma$	0.4602	2.9	$\mu + 2.9\sigma$	0.9981
0	μ	0.5000	3.0	$\mu + 3.0\sigma$	0.9987

EXAMPLE 2. The probability that X will differ from μ by not more than σ is

$$P(\mu - \sigma \leqq X \leqq \mu + \sigma) = P(|X - \mu| \leqq \sigma) = P(-1 < t \leqq 1)$$
$$= F_N(\mu + \sigma) - F_N(\mu - \sigma)$$
$$= 0.8413 - 0.1587 = 0.6826.$$

10.3. **Notation.** Various symbols are used for the relations (10.1.1) and (10.1.2). One set of symbols commonly used is as follows:

$$f_N(x) = \varphi(x) = (1/\sqrt{2\pi}\,\sigma)\exp[-(x-\mu)^2/2\sigma^2].$$

$$F_N(x) = \Phi(x) = \int_{-\infty}^{x}\varphi(X)\,dX.$$

$$\psi(t) = (1/\sqrt{2\pi})e^{-t^2/2}.$$

(10.3.1)
$$\Psi(t) = (1/\sqrt{2\pi})\int_{-\infty}^{t}e^{-\tau^2/2}\,d\tau.$$

$$\Theta(z) = (2/\sqrt{\pi})\int_{0}^{z}e^{-v^2}\,dv = \operatorname{erf} z.$$

$$\Theta(z) = 2\cdot\Psi(\sqrt{2}\,z) - 1.$$

$$\Theta(a/\sqrt{2}) = 2\cdot\Psi(a) - 1.$$

$\Theta(z)$ is sometimes called the *error integral,* or *error function,* or *probability integral,* or *integrale de Gauss.*

10.4. **Values of** $\psi(t)$, $\Psi(t)$. Table (10.8.3) gives values of $\psi(t)$, $\Psi(t)$, $\Theta(t/\sqrt{2}) = 2\,\Psi(t) - 1$ for various values of t. Table IX gives values of $\Psi(t) - 1/2$, $\psi(t)$, and derivatives of $\psi(t)$.

10.5. **Properties of the normal distribution.** The mean, variance, $\cdots$, of the variable X having distribution (10.1.1) are:

Mean $= \mu$. $\qquad\qquad F_N(-\infty) = 0.$

(10.5.1) Variance $= \sigma^2$. $\qquad\quad F_N(+\infty) = 1.$

Standard deviation $= \sigma$.

r^{th} moment μ_r about the mean μ: $\qquad r^{\mathrm{th}}$ moment ν_r about $x = 0$:

$$\mu_1 = 0. \qquad\qquad\qquad \nu_1 = \mu.$$

$$\mu_2 = \sigma^2. \qquad\qquad\qquad \nu_2 = \mu^2 + \sigma^2.$$

$$\mu_3 = 0. \qquad\qquad\qquad \nu_3 = \mu(\mu^2 + 3\sigma^2).$$

(10.5.2)
$$\mu_4 = 3\sigma^4. \qquad\qquad\quad \nu_4 = \mu^4 + 6\mu^2\sigma^2 + 3\sigma^4.$$

$$\vdots \qquad\qquad\qquad\qquad \vdots$$

$$\mu_{2r-1} = 0.$$

$$\mu_{2r} = 1\cdot3\cdots(2r-1)\sigma^{2r}.$$

Coefficient of skewness $= \gamma_1 = \mu_3/\sigma^3 = 0.$

Coefficient of excess $= \gamma_2 = (\mu_4/\sigma^4) - 3 = 0.$

Mode $= \mu.$ Median $= \mu.$

(10.5.3) Mean (absolute) deviation from the mean $=$ MD $=$ m.a.e.
$$= \sigma\sqrt{2/\pi} = 0.79788\sigma.$$

Modulus of precision $= h = 1/\sigma\sqrt{2} = 0.70711/\sigma.$

Probable error $=$ p.e. $= 0.67449\sigma.$

The upper and lower quartiles are: $Q_3 = \mu +$ p.e.,
$$Q_1 = \mu -$$ p.e.

The characteristic function for distribution (10.1.1) is

(10.5.4) $$\chi(t) = \exp[i\mu t - \sigma^2 t^2/2], \qquad i^2 = -1.$$

Semi-invariants:

(10.5.5) $$\kappa_1 = \mu, \ \kappa_2 = \sigma^2, \ \kappa_3 = \kappa_4 = \cdots = 0.$$

Figure 10.5.1 shows the relative shapes of the normal probability density function $f_N(x)$ for various values of σ. Further properties of the normal distribution are shown in Figures 10.5.2 and 10.5.3.

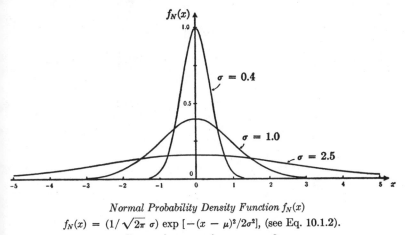

Normal Probability Density Function $f_N(x)$
$$f_N(x) = (1/\sqrt{2\pi}\,\sigma)\exp[-(x-\mu)^2/2\sigma^2], \text{ (see Eq. 10.1.2)}.$$
Curves shown are for case $\mu = 0.$

FIGURE 10.5.1

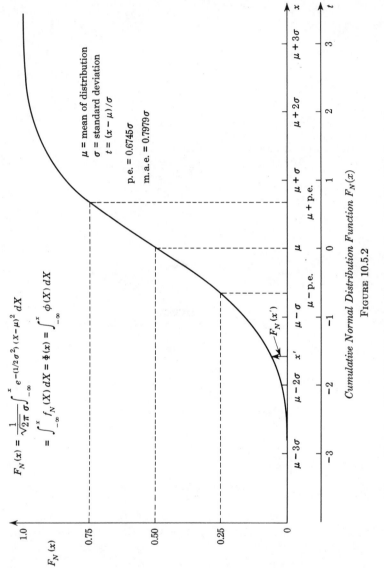

$$F_N(x) = \frac{1}{\sqrt{2\pi}\,\sigma} \int_{-\infty}^{x} e^{-(1/2\sigma^2)\,(X-\mu)^2}\, dX$$

$$= \int_{-\infty}^{x} f_N(X)\, dX = \bar\Phi(x) = \int_{-\infty}^{x} \phi(X)\, dX$$

μ = mean of distribution
σ = standard deviation
$t = (x - \mu)/\sigma$

p.e. = 0.6745σ
m.a.e. = 0.7979σ

Cumulative Normal Distribution Function $F_N(x)$

FIGURE 10.5.2

NOTE: The value of the ordinate $F_N(x')$ at x' is equal to the numerical value of the shaded area in Figure 10.1.1. (See Eq. 10.5.3.) $F_N(x)$ is the probability that X is less than or equal to x.

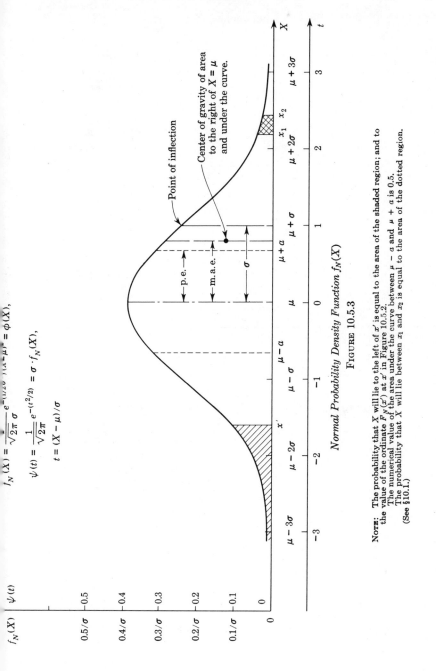

Normal Probability Density Function $f_N(X)$

FIGURE 10.5.3

NOTE: The probability that X will lie to the left of x' is equal to the area of the shaded region; and to the value of the ordinate $F_N(x')$ at x' in Figure 10.5.2.

The numerical value of the area under the curve between $\mu - a$ and $\mu + a$ is 0.5.

The probability that X will lie between x_1 and x_2 is equal to the area of the dotted region.

(See §10.1.)

10.6. **Relationships between σ, (p.e.), (m.a.e.).**

$$\sigma = 1.2533 \text{ (m.a.e.)} = 1.4826 \text{ (p.e.)}$$

$$\text{(p.e.)} = 0.6745\sigma \qquad = 0.8453 \text{ (m.a.e.)}$$

$$\text{(m.a.e.)} = 0.7979\sigma \qquad = 1.1829 \text{ (p.e.)}$$

10.7. **Reproductive property of normal distribution.** Let X_1, X $\cdots$, X_n be independent normally distributed variables with mean $\mu_1, \mu_2, \cdots , \mu_n$ and variances $\sigma_1{}^2, \sigma_2{}^2, \cdots , \sigma_n{}^2$. The sum $X = X_1 + X_2$ $\cdots + X_n$ of these variables is itself normally distributed with mean $\mu = \mu_1 + \mu_2 + \cdots + \mu_n$, and variance $\sigma^2 = \sigma_1{}^2 + \sigma_2{}^2 + \cdots + \sigma_n{}^2$.

Furthermore, if $a_1, \cdots , a_n$ are arbitrary constants, then $Z = a_1X_1$ $a_2X_2 + \cdots + a_nX_n$ is also normally distributed with the mean $\mu_z = \sum_{i=1}^{n} a_i\mu_i$ and variance $\sigma_z{}^2 = \sum_{i=1}^{n} a_i{}^2\sigma_i{}^2$.

10.8. **Probability of occurrence of certain deviations.** The probability P' that measurements x in a normally distributed population deviate less than $\lambda\sigma$ from the mean μ is

$$(10.8.1) \qquad P' = P(|x - \mu| < \lambda\sigma)$$

$$= (1/\sqrt{2\pi}) \int_{-\lambda}^{\lambda} \exp\left[-t^2/2\right] dt = 2\,\Psi(\lambda) - 1.$$

Values of the probability (10.8.1) are given in Table 10.8.3. P' can also be found from Table 10.8.1.

Table 10.8.2 gives values of the probability P' as a function of λ and or p.e., or m.a.e.

The probability P that x will deviate more than $\lambda\sigma$ from the mean μ

$$(10.8.2) \qquad P = P(|x - \mu| > \lambda\sigma)$$

$$= (2/\sqrt{2\pi}) \int_{\lambda}^{\infty} \exp\left[-t^2/2\right] dt = 2[1 - \Psi(\lambda)].$$

Values of this probability (10.8.2) are given in Table 10.8.1 as a function of λ. The sum of the probabilities given in (10.8.1) and (10.8.2), for given μ, λ, and σ, is 1. Table 10.8.3 can also be used to find P since $P = 1 - P'$.

The quantity $(x - \mu)/\sigma$ is sometimes called the *normal deviate*.

EXAMPLE 1.   For measurements x in a normally distributed population with mean μ and standard deviation σ, then, of the measurements,

68.26% deviate less than σ from μ,

$$\text{(that is, } P(|\,x - \mu\,| < \sigma) = 0.6826)$$

95.45% deviate less than 2σ from μ,

99.74% deviate less than 3σ from μ.

Tolerance limits.   The limits within which x lies with a given probability P', as given by (10.8.1), (when the interval defined by these limits is assumed symmetric about the mean μ) are called *tolerance limits*.

In practice, it is usually more convenient to use the value of λ corresponding to the probability P of the complementary event, namely, $P = P(|\,x - \mu\,| \geqq \lambda\sigma)$.

EXAMPLE 1.   The 0.1%, 0.5%, 1%, 2.5%, 5%, and 10% limits, respectively, are given as follows:

P	0.001	0.005	0.01	0.025	0.05	0.10
λ	3.29	2.81	2.58	2.24	1.96	1.64

TABLE 10.8.1

The probability of occurrence of deviations (for a normally distributed variable)

Let x be a normally distributed variable with mean μ and standard deviation σ. The probability P that an observed value of x differs from μ in either direction by more than λ times σ is

$$P = P(|x - \mu| > \lambda\sigma) = (2/\sqrt{2\pi}) \int_\lambda^\infty e^{-t^2/2}\, dt = 2[1 - \Psi(\lambda)].$$

The *p percent value of a normal deviate* $t = (x - \mu)/\sigma$ is the value λ_p of λ corresponding to $P = p/100$.

The probability P' that x differs from μ in either direction by less than λ times σ is

$$P' = P(|x - \mu| < \lambda\sigma) = 1 - P = 2\,\Psi(\lambda) - 1.$$

Note: $\qquad\qquad\qquad 100P' = 100 - p.$

λ_p	$p = 100P$	$p = 100P$	λ_p
0.0	100.000	100	0.0000
0.2	84.148	95	0.0627
0.4	68.916	90	0.1257
0.6	54.851	85	0.1891
0.8	42.371	80	0.2533
1.0	31.731	75	0.3186
1.2	23.014	70	0.3853
1.4	16.151	65	0.4538
1.6	10.960	60	0.5244
1.8	7.186	55	0.5978
2.0	4.550	50	0.6745
2.2	2.781	45	0.7554
2.4	1.640	40	0.8416
2.6	0.932	35	0.9346
2.8	0.511	30	1.0364
3.0	0.270	25	1.1503
3.2	0.137	20	1.2816
3.4	0.067	15	1.4395
3.6	0.032	10	1.6449
3.8	0.014	5	1.9600
4.0	0.006	1	2.5758
5.0	5.73×10^{-5}	0.1	3.2905
6.0	2.0×10^{-7}	0.01	3.8906
7.0	2.6×10^{-10}	0.001	4.4172
		0.0001	4.8916
		0.00001	5.3267

TABLE 10.8.2

Table whose entries give (when x is normally distributed) the probability P' that the absolute deviation $|x - \mu|$ is not greater than λ times the parameter ω indicated, that is,

$$P' \equiv P(|x - \mu| \leqq \lambda\omega).$$

λ	when:	$\omega = \sigma$	$\omega = $ p.e.	$\omega = $ m.a.e.
1		0.683	0.500	0.575
2		0.955	0.823	0.889
3		0.997	0.957	0.983
4		0.9999	0.993	0.9985

TABLE 10.8.3

The normal distribution

t	$\psi(t)$	$\Psi(t)$	$2\Psi(t) - 1$	t	$\psi(t)$	$\Psi(t)$	$2\Psi(t) - 1$
0.0	0.39894	0.50000	0.00000	2.0	0.05399	0.97725	0.95450
0.1	0.39695	0.53983	0.07966	2.1	0.04398	0.98214	0.96427
0.2	0.39104	0.57926	0.15852	2.2	0.03547	0.98610	0.97219
0.3	0.38139	0.61791	0.23582	2.3	0.02833	0.98928	0.97855
0.4	0.36827	0.65542	0.31084	2.4	0.02239	0.99180	0.98360
0.5	0.35207	0.69146	0.38292	2.5	0.01753	0.99379	0.98758
0.6	0.33322	0.72575	0.45149	2.6	0.01358	0.99534	0.99068
0.7	0.31225	0.75804	0.51607	2.7	0.01042	0.99653	0.99307
0.8	0.28969	0.78814	0.57629	2.8	0.00792	0.99744	0.99489
0.9	0.26609	0.81594	0.63188	2.9	0.00595	0.99813	0.99627
1.0	0.24197	0.84134	0.68269	3.0	0.00443	0.99865	0.99730
1.1	0.21785	0.86433	0.72867	3.1	0.00327	0.99903	0.99806
1.2	0.19419	0.88493	0.76986	3.2	0.00238	0.99931	0.99863
1.3	0.17137	0.90320	0.80640	3.3	0.00172	0.99952	0.99903
1.4	0.14973	0.91924	0.83849	3.4	0.00123	0.99966	0.99933
1.5	0.12952	0.93319	0.86639	3.5	0.00087	0.99977	0.99953
1.6	0.11092	0.94520	0.89040	3.6	0.00061	0.99984	0.99968
1.7	0.09405	0.95543	0.91087	3.7	0.00042	0.99989	0.99978
1.8	0.07895	0.96407	0.92814	3.8	0.00029	0.99993	0.99986
1.9	0.06562	0.97128	0.94257	3.9	0.00020	0.99995	0.99990
				4.0	0.00013	0.99997	0.99994

$$P' = P(|x - \mu| < \lambda\sigma) = 2\Psi(\lambda) - 1 = \Theta(\lambda/\sqrt{2}). \qquad \text{(See §10.8)}$$

$$\psi(t) = (1/\sqrt{2\pi})e^{-t^2/2}, \qquad \Psi(t) = \int_{-\infty}^{t} \psi(\tau)\, d\tau.$$

10.9. **Fitting a cumulative normal distribution to a cumulative frequency polygon.** It is very often possible to approximate a cumulative distribution (frequency polygon) of measurements in a sample by a cumulative normal distribution $F_N(x)$. This is done by replacing μ and σ in (10.1.1) by the sample mean $\bar{x}$ and the sample standard deviation σ_x, respectively. The use of the "dimensionless" number $t = (x - \bar{x})/\sigma$ simplifies the calculation.

EXAMPLE 1.   Fig. 10.9.1 gives the cumulative polygon for the cumulative frequencies of Table 3.3.2. The mean and standard deviation of the sample are $\bar{x} = 2.60$, $\sigma_x = 0.088$, respectively. Let $t = (x - 2.60)/0.088$. Fig. 10.9.1 shows the "fitted" cumulative normal distribution curve and the sample cumulative polygon. Table 10.9.1 below indicates the "goodness of fit". The "fitted" cumulative frequencies are read from the "fitted" curve at the upper boundaries of the class intervals. The "fitted" frequency is calculated from the "fitted" cumulative frequency. The "observed" frequencies are obtained from Table 3.3.2.

TABLE 10.9.1

Comparison of fitted and observed frequencies for Fig. 10.9.1

Midpoint of class interval	Cumulative frequency		Frequency	
	Observed	Fitted	Observed	Fitted
2.35	0	0.3	0	0
2.40	1	1.3	1	1.0
2.45	5	4.5	4	3.2
2.50	9	11.5	4	7.0
2.55	28	22.8	19	11.3
2.60	34	36.2	6	13.4
2.65	45	47.8	11	11.6
2.70	56	55.1	11	7.3
2.75	59	58.5	3	3.4
2.80	60	59.9	1	1.4
2.85	60	60.0	0	0.1
Total		60.0	60	59.7

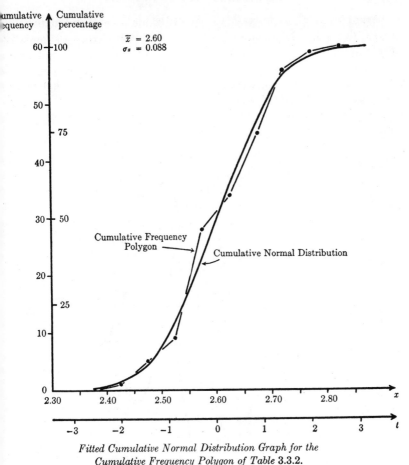

Fitted Cumulative Normal Distribution Graph for the
Cumulative Frequency Polygon of Table 3.3.2.

FIGURE 10.9.1

10.10. Approximation to arbitrary distributions. Consider a random variable x with a continuous distribution (not necessarily normal) having a mean μ, a standard deviation σ, and r^{th} order central moment μ_r. Introduce the standardized variable $t = (x - \mu)/\sigma$. Denote the cumulative distribution function and probability density by $F(t)$ and $f(t) = F'(t)$, respectively. Under certain conditions $f(t)$ may be approximated by

$$(10.10.1) \qquad f(t) = \psi(t) - (\gamma_1/3!)\psi^{(3)}(t)$$
$$+ (\gamma_2/4!)\psi^{(4)}(t) + (10\gamma_1^{2}/6!)\psi^{(6)}(t),$$

where $\gamma_1 = \mu_3/\sigma^3$, $\gamma_2 = (\mu_4/\sigma^4) - 3$ are the coefficients of skewness and excess defined in §10.5 and $\psi^{(r)}(t)$ is the rth derivative with respect to t of $\psi(t)$.

The expansion for $F(t)$ is obtained from (10.10.1) by replacing $\psi(t)$ by $\Psi(t)$, and $f(t)$ by $F(t)$. These expansions are related to the *Gram-Charlier A series* and *Edgeworth's series*.

Graphs of the derivatives $\psi^{(3)}(t)$, $\psi^{(4)}(t)$, $\psi^{(6)}(t)$ are shown in Fig 10.10.1.

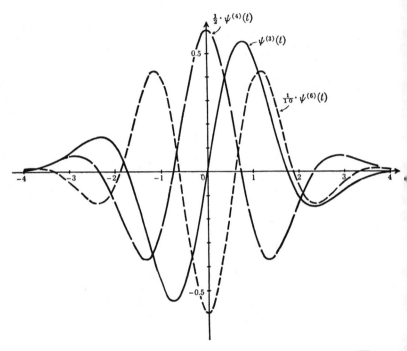

Third, Fourth and Sixth Derivatives of the Normal Function $\psi(t) = (1/\sqrt{2\pi})e^{-t^2/2}$.

FIGURE 10.10.1

10.11. **Standard normal distribution.** If x has normal distribution $N(\mu,\sigma^2)$ then $t = (x - \mu)/\sigma$ has the *standard (unit) normal distribution* $N(0,1)$. t is called the *standard (unit) normal variable*. The p.d.f. and c.d.f. of t are, respectively, $\psi(t) = (1/\sqrt{2\pi})e^{-t^2/2}$, $\Psi(t) = \int_{-\infty}^{t} \psi(\tau)\, d\tau$ The probability that $x' \leqq x$ is

(10.11.1) $P(x' \leqq x) = P(t' \leqq t) = \Psi(t) = \Psi((x - \mu)/\sigma).$

See Figures 10.5.2, 10.5.3.)

Note:

(10.11.2) $$\Psi(-t) = 1 - \Psi(t).$$

EXAMPLE. Experience with the production of a $\frac{1}{2}$ inch rod indicates that the diameters of the rods produced have a normal distribution with mean $\mu = 0.500$ inch and standard deviation $\sigma = 0.0005$ inch. The chance P' that a particular rod taken at random from the production line has a diameter x which lies between 0.4990 and 0.5010 inch is $P' \equiv P(0.4990 \leq x \leq 0.5010) = P(-2 \leq t \leq 2)$, where $t = (x - 0.500)/0.0005$. By (10.11.1), (10.11.2), and Table 10.8.3

$$P' = P(t \leq 2) - P(t \leq -2) = \Psi(2) - \Psi(-2)$$
$$= 2\Psi(2) - 1 = 2(0.97725) - 1 = 0.9545.$$

Thus, 95% of the production lies between 0.4990 and 0.5010 inch.

10.12. Tables. Table IX gives values of $\Psi(t) - 1/2$, $\psi(t)$, and the first 6 derivatives of $\psi(t)$. More extensive tables are available, such as those for $\psi(t)$ and $2\Psi(t) - 1$ given to 15 decimal places in *Tables of Probability Functions*, Vol. II, Federal Works Agency, sponsored by National Bureau of Standards, New York, 1942.

10.13. Random sampling from a normal distribution. If $x_1, x_2, \ldots, x_n$ are independent random variables with means μ_i and variances σ_i^2, $i = 1, \cdots, n$, then any linear combination of the x_i, say $A = \sum_{i=1}^{n} c_i x_i$, has the mean $\mu_A = E(A) = \sum_{i=1}^{n} c_i \mu_i$ and variance

$$\sigma_A^2 = E[A - \mu_A]^2 = \sum_{i=1}^{n} c_i^2 \sigma_i^2.$$

EXAMPLE. The difference $D = x_1 - x_2$ has the mean $\mu_D = \mu_1 - \mu_2$ and variance $\sigma_D^2 = \sigma_1^2 + \sigma_2^2$. If x_1 and x_2 are distributed normally, so is D. D is called a *contrast*.

If $x_1, \cdots, x_n$ is a random sample from a population having the normal distribution $N(\mu, \sigma^2)$: the mean $\bar{x} = (1/n) \sum_{i=1}^{n} x_i$ is a random variable with distribution $N(\mu, \sigma^2/n)$; and the sum $B = \sum_{i=1}^{n} x_i$ is a random variable with distribution $N(n\mu, n\sigma^2)$.

10.14. Log normal distribution. If a continuous chance variable X is such that the (cumulative) probability that X is less than or equal to x' is

(10.14.1) $P(X \leq x') = F(x') = \int_{-\infty}^{x'} f(x)\, dx,$

where the probability density function is

(10.14.2)
$$\begin{cases} f(x) = 0 & \text{for } x \leq 0, \\ f(x) = \dfrac{1}{\sqrt{2\pi}\,\sigma x} \exp\left[-\dfrac{(\ln x - \mu)^2}{2\sigma^2} \right] & \text{for } x > 0 \end{cases}$$

and

$$\ln x \equiv \log_e x, \qquad -\infty < \mu < \infty, \qquad \sigma > 0,$$

the variable X is said to possess a *log normal distribution*. (In other words a variable X has a logarithmic normal distribution if $\log_e X$ is normall distributed with mean $\mu \equiv \ln \xi$ and variance σ^2.)

Properties.   The chance variable X having distribution (10.14.1) ha the following properties:

(10.14.3)
$$\begin{aligned} &\text{Mode of } X = \exp[\mu - \sigma^2] = e^{\mu - \sigma^2} \\ &\text{Median of } X = e^\mu = \xi \\ &\text{Mean of } X = \exp[\mu + \sigma^2/2] = e^{\mu + \sigma^2/2} \\ &\text{Variance of } X = e^{2\mu + \sigma^2}(e^{\sigma^2} - 1) \end{aligned}$$

The kth moment of X about $x = 0$ is $\exp[k\mu + \frac{1}{2}k^2\sigma^2]$. The distributio is not reproductive. The larger the value of σ, the sharper the skewness

When a cumulative distribution function of the form (10.14.1) i plotted on *logarithmic probability paper* (i.e., a graph paper with a loga rithmic scale as the abscissa and a normal probability scale as the ordi nate), the cumulative distribution curve appears as a straight line whos slope depends on the parameter σ. The smaller σ, the greater the slope

The log normal distribution has many important applications in eco nomics and engineering.

**PROBABILITY DISTRIBUTIONS IN
TWO OR MORE DIMENSIONS**

The two-dimensional case

11.1. The two-dimensional case is of concern where an event requires
two numbers for its description. This case is of frequent application in
military problems.

Let X and Y be real chance variables ranging over the real X,Y-plane.
Suppose the probability that $X \leq x'$ and $Y \leq y'$ is given by $P(X \leq x',$
$Y \leq y') = F(x',y')$. $F(x',y')$ is called the *joint (cumulative) distribution
function of X and Y.*

The *cumulative distribution function of the marginal distribution of X* is

11.1.1) $$F_1(X) = P(X \leq x') = F(x', \infty).$$

Likewise, the *marginal distribution of Y* is

11.1.2) $$F_2(Y) = P(Y \leq y') = F(\infty, y').$$

11.2. **Discrete case.** Suppose (X,Y) can take on the values (x_i,y_j),
$j = 1, \cdots, k$. Let $P(X = x_i, Y = y_i) = p_{ij}$. Then $\sum_{i,j=1}^{k} p_{ij} = 1$.
The marginal distributions of X and Y are, respectively,

11.2.1) $$p_{i\bullet} = P(X = x_i) = \sum_{j=1}^{k} p_{ij},$$

11.2.2) $$p_{\bullet j} = P(Y = y_i) = \sum_{i=1}^{k} p_{ij},$$

11.2.3) $$\sum_{i=1}^{k} p_{i\bullet} = 1, \qquad \sum_{j=1}^{k} p_{\bullet j} = 1.$$

A necessary and sufficient condition for the statistical independence
of X and Y is that for all i and j, $p_{ij} = p_{i\bullet}.p_{\bullet j}$.

The *conditional probability* that the event $Y = y_i$ occurs, relative to
the hypothesis $X = x_i$, is

$$(11.2.4) \qquad P(Y = y_i \mid X = x_i) = \frac{P(X = x_i,\ Y = y_i)}{P(X = x_i)} = \frac{p_{ij}}{p_{i\cdot}}.$$

$$(11.2.5) \qquad \sum_{j=1}^{k} p_{ij}/p_{i\cdot} = 1.$$

The conditional probabilities (for a fixed x_i) for the various possible values of y_j define the *conditional distribution* of Y, relative to the hypothesis $X = x_i$.

The *conditional mean* value of a function $g(X,Y)$, relative to the hypothesis $X = x_i$, is

$$(11.2.6) \qquad E[g(X,Y)|X = x_i] = [\sum_j p_{ij} g(x_i,y_j)] \div \sum_j p_{ij}.$$

The *conditional mean* of Y relative to the hypothesis $X = x_i$ is

$$(11.2.7) \qquad E[Y|X = x_i] = m_2^{(i)} = [\sum_j p_{ij} y_j] \div \sum_j p_{ij};$$

the *conditional variance* of Y relative to the hypothesis $X = x_i$ is Eq. (11.2.6) with g replaced by $(Y - m_2^{(i)})^2$.

Similar definitions hold with respect to the hypothesis $Y = y_j$.

11.3. Continuous case.   Suppose $F(x,y)$ is everywhere continuous and

$$f(x,y) = \frac{\partial^2 F(x,y)}{\partial x\ \partial y}$$

exists, is non-negative, single-valued and continuous almost everywhere. $f(x,y)$ is called the *probability density* or *frequency function of the dis-*

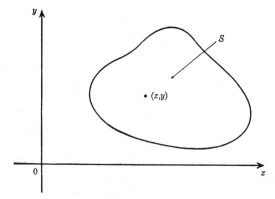

FIGURE 11.3.1

ribution $F(x,y)$. Let S be a region of the X,Y-plane. The probability $P(S)$ hat the point (X,Y) falls in the region S is

$$(11.3.1) \qquad P(S) = \iint\limits_{S} f(x,y) \, dx \, dy,$$

where the integration is carried out over the entire region S. Then

$$(11.3.2) \qquad \int_{-\infty}^{\infty} \int_{-\infty}^{\infty} f(x,y) \, dx \, dy = 1.$$

The *marginal distribution of X* has the distribution function

$$(11.3.3) \qquad P(X \leqq x') = \int_{x=-\infty}^{x=x'} \int_{y=-\infty}^{y=\infty} f(x,y) \, dx \, dy = \int_{-\infty}^{x'} f_1(x) \, dx,$$

where

$$(11.3.4) \qquad f_1(x) = \int_{y=-\infty}^{y=\infty} f(x,y) \, dy.$$

The *marginal distribution of Y* has the distribution function

$$(11.3.5) \qquad P(Y \leqq y') = \int_{y=-\infty}^{y=y'} \int_{x=-\infty}^{x=\infty} f(x,y) \, dy \, dx = \int_{-\infty}^{y'} f_2(y) \, dy,$$

where

$$(11.3.6) \qquad f_2(y) = \int_{x=-\infty}^{x=\infty} f(x,y) \, dx.$$

A necessary and sufficient condition for the statistical independence of X and Y is that for all x and y,

$$(11.3.7) \qquad f(x,y) = f_1(x) \, f_2(y).$$

11.4. Parameters used to describe a probability distribution. The parameters used to describe a probability distribution $F(X,Y)$ with frequency function $f(X,Y)$ are defined in a manner similar to that used for the single variable case. Only the *continuous case* is written out in detail. The discrete case is similar with the integral signs replaced by appropriate sums.

The *mean*, or *expected value*, of the function $g(X,Y)$ is defined to be

$$(11.4.1) \qquad E[g(X,Y)] = \int_{-\infty}^{\infty} \int_{-\infty}^{\infty} g(x,y) \, f(x,y) \, dx \, dy.$$

The *general product moments* (about the origin) of order $i + j$ of the distribution are

$$\nu_{ij} = E[X^i Y^j] = \int_{-\infty}^{\infty} \int_{-\infty}^{\infty} x^i y^j \, f(x,y) \, dx \, dy.$$

(11.4.2)

$$\nu_{10} = E[X] = \mu_x = \text{mean of } x\text{'s} = \bar{x},$$

$$\nu_{01} = E[Y] = \mu_y = \text{mean of } y\text{'s} = \bar{y}.$$

The *general product (central) moments* about $X = \mu_x$, $Y = \mu_y$ are

$$\mu_{ij} = E[(X - \mu_x)^i (Y - \mu_y)^j]$$

$$= \int_{-\infty}^{\infty} \int_{-\infty}^{\infty} (x - \mu_x)^i (y - \mu_y)^j f(x,y) \, dx \, dy.$$

$$\mu_{10} = \mu_{01} = 0. \qquad \mu_{20} = \sigma_x^2, \qquad \mu_{02} = \sigma_y^2.$$

(11.4.3)

$$\mu_{20} = \nu_{20} - \mu_x^2, \qquad \mu_{11} = \nu_{11} - \mu_x \mu_y, \qquad \mu_{02} = \nu_{02} - \mu_y^2.$$

$\sigma_x = $ *standard deviation* of X, $\quad \sigma_y = $ *standard deviation* of Y

$\mu_{20} = $ *variance* of X, $\qquad\qquad \mu_{02} = $ *variance* of Y.

$\mu_{11} = $ *covariance* of the joint distribution of X and Y

$\qquad = $ *product moment* or *mixed moment*.

(11.4.4) $r = \mu_{11}/\sigma_x \sigma_y = $ *correlation coefficient* of X and Y.

Other parameters may be defined along the lines given in §10.5 for the single variable case.

11.5. Conditional distributions.   Let

(11.5.1) $$P(x < X < x + h) = \int_{x}^{x+h} \int_{-\infty}^{\infty} f(x,y) \, dx \, dy.$$

The *conditional probability* of the occurrence of the event $Y \leqq y$ relative to the hypothesis $x < X < x + h$ is

(11.5.2) $$P(Y \leqq y | x < X < x + h) = \frac{P(x < X < x + h, \, Y \leqq y)}{P(x < X < x + h)}$$

If $f_1(x) > 0$,

(11.5.3) $$\lim_{h \to 0} P(Y \leqq y | x < X < x + h) = \int_{-\infty}^{y} f(x,y) \, dy \div f_1(x)$$

Eq. (11.5.3) is the *conditional distribution function* of Y, relative to the hypothesis $X = x$. The *conditional frequency function* of Y is

$$11.5.4) \qquad f(y|x) = f(x,y) \div f_1(x).$$

The *conditional mean value* of a function, $g(X,Y)$, relative to the hypothesis $X = x$, is

$$11.5.5) \qquad E[g(X,Y)|X = x] = \int_{-\infty}^{\infty} g(x,y) \, f(x,y) \, dy \div f_1(x).$$

Also,

$$11.5.6) \qquad E[g(X,Y)] = \int_{-\infty}^{\infty} \int_{-\infty}^{\infty} g(x,y) \, f(x,y) \, dx \, dy$$

$$= \int_{-\infty}^{\infty} E[g(X,Y)|X = x] \, f_1(x) \, dx.$$

The *conditional mean* and *conditional variance* of Y are

$$11.5.7) \qquad E[Y|X = x] = m_2(x) = \int_{-\infty}^{\infty} y \, f(x,y) \, dy \div \int_{-\infty}^{\infty} f(x,y) \, dy,$$

$$11.5.8) \qquad E[(Y - m_2(x))^2|X = x]$$

$$= \int_{-\infty}^{\infty} (y - m_2(x))^2 f(x,y) \, dy \div \int_{-\infty}^{\infty} f(x,y) \, dy.$$

Similar definitions hold with respect to the hypothesis $Y = y$.

11.6. **Normal distribution.** In case the probability density is expressible in the form

$$11.6.1) \qquad f(x,y) = \frac{1}{2\pi\sigma_x\sigma_y \sqrt{1 - r^2}} e^{-G/2},$$

where

$$11.6.2) \qquad G = \frac{1}{(1 - r^2)} \left[\frac{(x - \bar{x})^2}{\sigma_x^2} - \frac{2r(x - \bar{x})(y - \bar{y})}{\sigma_x\sigma_y} + \frac{(y - \bar{y})^2}{\sigma_y^2} \right],$$

the distribution is said to be *normal*, or *Gaussian*. The mean of this distribution is $(\bar{x},\bar{y})$. r is the correlation coefficient.

The probability that the point (X,Y) falls in the region S of the x,y-plane is

$$11.6.3) \qquad P(S) = \iint_S f(x,y) \, dx \, dy.$$

11.7. Probability ellipse.   The locus of $G = c^2$, where c is a constant, is an equi-probability ellipse. On this ellipse $f(x,y)$ is a constant. The probability p that a point (x,y) taken at random will fall within the

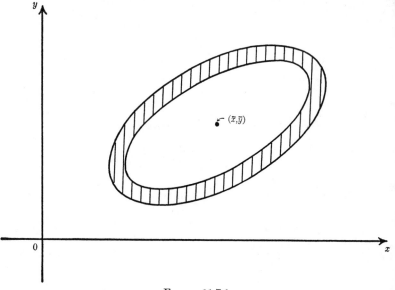

FIGURE 11.7.1

ellipse corresponding to a particular value of c is $p = 1 - e^{-c^2/2}$. When $p = 0.5$, $c = 1.1774$ and the corresponding ellipse is called the *50% probability ellipse*. When $c = 1$, $p = 0.39347$. Values of c for other values of $P = 100p$ are given in Table 11.7.1.

TABLE 11.7.1

Values of c for $P\%$ probability ellipses

P	c
25	0.7585
50	1.177
75	1.665
90	2.146
95	2.448
99	3.035

The probability p' that a point taken at random will fall in the elliptic ring formed by the ellipses with parameters c and $c + \Delta c$, respectively, is approximately, $p' = c\ e^{-c^2/2}\ \Delta c$. For a given value of Δc, the probability p' has a maximum when $c = 1$. The corresponding ellipse is called the *ellipse of maximum probability*.

In many applications $r = 0$. In case $r \neq 0$, an appropriate rotation of the x and y axes will reduce $f(x,y)$ to a similar form $f'(x',y')$ in which the terms containing the product $x'y'$ vanish. This is often desirable for purposes of calculation.

11.8. **Circular case.** In case $\sigma_x = \sigma_y = \sigma$, and $r = 0$, the normal distribution (11.6.1) is called *circular*. Then the probability density corresponding to $f(x,y)$ is

$$(11.8.1) \qquad f(\rho) = (\rho/\sigma^2)e^{-\rho^2/2\sigma^2},$$

where

$$(11.8.2) \qquad \rho^2 = (x - \bar{x})^2 + (y - \bar{y})^2.$$

ρ is called the *radial error*. The probability ellipses are circles.

The probability p' that a point (X,Y) taken at random will fall within a circular ring, whose center is at $(\bar{x},\bar{y})$ and whose inner and outer radii are ρ and $\rho + \Delta\rho$, is

$$(11.8.3) \qquad p' = f(\rho)\ \Delta\rho.$$

The probability p that a point (x,y) taken at random will fall within the circle

$$(11.8.4) \qquad (x - \bar{x})^2 + (y - \bar{y})^2 = (c\sigma)^2,$$

for a particular value of c, is

$$(11.8.5) \qquad p = 1 - e^{-c^2/2}.$$

When $c = 1.1774$, $p = 0.5$, and the circle of radius 1.1774σ is called the *50 per cent probability circle*. The radius of this circle is called the *circular probable error* (c.p.e.), or CEP, or CPE, and

$$(11.8.6) \qquad \text{c.p.e.} = 1.1774\sigma.$$

When $\sigma_x = \sigma_y = \sigma$, some writers use the term *radial probable error* (r.p.e.) to mean

$$(11.8.7) \qquad \text{r.p.e.} = \text{c.p.e.} = 1.1774\sigma = 1.7456(\text{p.e.}),$$

where

$$(11.8.8) \qquad \text{p.e.} = 0.6745\sigma.$$

Values of the radius $c\sigma$ for other values of p are given in Table 11.8.1, where $P = 100p$.

TABLE 11.8.1

Radius of $P\%$ probability circles.

$(\sigma_x = \sigma_y = \sigma)$

P	$c\sigma$ =	Radius	
25	0.7585σ	0.6052 (m.r.e.)	= 0.6442 (c.p.e.)
39.3	1.0000σ	0.7979 (m.r.e.)	= 0.8493 (c.p.e.)
50	1.177σ	0.9394 (m.r.e.)	= 1.0000 (c.p.e.)
54.4	1.253σ	1.0000 (m.r.e.)	= 1.064 (c.p.e.)
75	1.665σ	1.329 (m.r.e.)	= 1.414 (c.p.e.)
90	2.146σ	1.712 (m.r.e.)	= 1.823 (c.p.e.)
95	2.448σ	1.953 (m.r.e.)	= 2.079 (c.p.e.)
99	3.035σ	2.421 (m.r.e.)	= 2.578 (c.p.e.)

The *mean radial error* (m.r.e.) of the circular normal distribution (11.8.1) is the expected value of ρ, namely,

$$(11.8.9) \qquad \text{m.r.e.} = E[\rho] = \int_0^\infty \rho\, f(\rho)\, d\rho = \sigma\sqrt{\pi/2} = 1.2533\sigma.$$

The expected value of ρ^2 is $E[\rho^2] = 2\sigma^2$.

Various relationships between σ, c.p.e., and m.r.e. are:

$$\sigma = 0.8493 \text{ (c.p.e.)} = 0.7979 \text{ (m.r.e.)}$$

$$(11.8.10) \qquad \text{(c.p.e.)} = 1.1774\sigma \qquad = 0.9394 \text{ (m.r.e.)}$$

$$\text{(m.r.e.)} = 1.2533\sigma \qquad = 1.0645 \text{ (c.p.e.)}$$

Some military organizations use the term *dispersion* to mean the diameter δ of a circle about $(\bar{x}, \bar{y})$ within which 75% of the observations fall;

$$(11.8.11) \qquad\qquad \delta = 3.330\sigma.$$

11.9. Circular case. Polar form. Mean at origin.   In case $\sigma_x = \sigma_y = \sigma$, $\bar{x} = \bar{y} = 0$, $r = 0$, the integral (11.6.3) can be written (in polar coordinates) as

$$(11.9.1) \qquad P = (1/2\pi\sigma^2) \iint\limits_S e^{-G/2}\, dx\, dy = (1/2\pi) \iint\limits_S g(R)\, dR\, d\theta,$$

where

(11.9.2) $$R^2 = (x^2 + y^2)/\sigma^2, \qquad \tan\theta = y/x,$$

(11.9.3) $$g(R) = Re^{-R^2/2}, \qquad G = R^2.$$

11.10. Values of normal probability integral over a circular disk, mean at (0,0).

When S is a circle of radius R_1, center at $(0,0)$, (11.9.1) becomes

(11.10.1) $$P = \int_0^{R_1} g(R)\,dR = 1 - e^{-R_1^2/2}.$$

Table 11.10.1 gives values of P and $g(R)$ as a function of R.

TABLE 11.10.1

R	$g(R) = Re^{-R^2/2}$	$P = 1 - e^{-R^2/2}$
0	0	0.00
0.3203	0.3043	0.05
0.4590	0.4131	0.10
0.5701	0.4846	0.15
0.6680	0.5344	0.20
0.7585	0.5689	0.25
0.8446	0.5912	0.30
0.9282	0.6033	0.35
1.0108	0.6065	0.40
1.0935	0.6014	0.45
1.1774	0.5887	0.50
1.2637	0.5687	0.55
1.3537	0.5415	0.60
1.4490	0.5072	0.65
1.5518	0.4655	0.70
1.6651	0.4163	0.75
1.7941	0.3588	0.80
1.9479	0.2922	0.85
2.1460	0.2146	0.90
2.4477	0.1224	0.95
∞	0	1.0

$$P = \int_0^R g(r)\,dr. \qquad \text{See §§11.9 and 11.10.}$$

11.11. Values of normal probability integral over a circular disk, mean at $(\bar{x}, \bar{y})$.

The problem of calculating the normal probability integral

(11.6.3) over a circular disk C frequently arises. In case $r = 0$, $\sigma_x = \sigma_y = \sigma$, the integral (11.6.3) reduces to

$$(11.11.1) \qquad P(C) = \iint_C f(x,y) \; dx \; dy,$$

where

$$(11.11.2) \qquad f(x,y) = e^{-G/2} \div 2\pi\sigma^2, \quad G = [(x - \bar{x})^2 + (y - \bar{y})^2] \div \sigma^2.$$

Let R be the radius of disk C whose center is at $(0,0)$; d, the distance from center of disk to mean $(\bar{x}, \bar{y})$ of the distribution; $\sigma_x = \sigma_y = \sigma$. The values of the integral $P(C)$ for various values of the ratios d/σ and R/σ are given in Table 11.11.1. The distance d is known as the "bias" or "systematic error" in some applications.

Table 11.11.1

Values of the normal probability integral $P(C)$ over a circular disk C as a function of d/σ and R/σ

$$P = P(C) = \iint_C f(x,y) \; dx \; dy,$$

where

$$f(x,y) = e^{-G/2} \div 2\pi\sigma^2, \quad G = [(x - \bar{x})^2 + (y - \bar{y})^2] \div \sigma^2,$$

$$\sigma_x = \sigma_y = \sigma, \quad R = \text{radius of disk with center at } (0,0),$$

$$(\bar{x}, \bar{y}) = \text{mean of distribution } f(x,y),$$

$$d = \text{distance from } (0,0) \text{ to } (\bar{x}, \bar{y}).$$

Entries in table give values of P.

d/σ	R/σ 0.1	0.2	0.3	0.4	0.5	0.6	0.7	0.8	0.9	1.0
					P					
0.0	.0050	.020	.044	.077	.118	.165	.217	.274	.333	.393
0.1	.0050	.020	.044	.077	.117	.164	.216	.273	.332	.392
0.2	.0049	.019	.043	.075	.115	.162	.213	.269	.328	.387
0.3	.0048	.019	.042	.074	.113	.158	.209	.264	.321	.380
0.4	.0046	.018	.041	.071	.109	.153	.202	.256	.312	.370
0.5	.0044	.018	.039	.068	.104	.147	.195	.246	.301	.357
0.6	.0042	.017	.037	.065	.099	.140	.185	.235	.288	.342
0.7	.0039	.016	.035	.061	.093	.132	.175	.223	.273	.326
0.8	.0036	.014	.032	.057	.087	.123	.164	.209	.257	.307
0.9	.0033	.013	.030	.052	.080	.114	.152	.194	.240	.288

TABLE 11.11.1 (*Continued*)

d/σ	\multicolumn{10}{c}{R/σ}									
	0.1	0.2	0.3	0.4	0.5	0.6	0.7	0.8	0.9	1.0
	\multicolumn{10}{c}{P}									
1.0	.0030	.012	.027	.048	.073	.104	.140	.179	.222	.267
1.1	.0027	.011	.024	.043	.067	.095	.127	.164	.203	.246
1.2	.0024	.010	.022	.038	.060	.085	.115	.148	.185	.225
1.3	.0022	.0086	.019	.034	.053	.076	.103	.133	.167	.204
1.4	.0019	.0075	.017	.030	.047	.067	.091	.119	.150	.184
1.5	.0016	.0065	.015	.026	.041	.059	.080	.105	.133	.164
1.6	.0014	.0056	.013	.022	.035	.051	.070	.092	.117	.145
1.7	.0012	.0047	.011	.019	.030	.044	.060	.080	.102	.127
1.8	.0010	.0040	.0090	.016	.026	.037	.052	.069	.088	.111
1.9	.0008	.0033	.0075	.014	.022	.032	.044	.059	.076	.096
2.0	.0007	.0027	.0062	.011	.018	.026	.037	.050	.064	.082
2.1	.0006	.0022	.0051	.0092	.015	.022	.031	.042	.054	.070
2.2	.0004	.0018	.0041	.0075	.012	.018	.025	.034	.045	.059
2.3	.0004	.0014	.0033	.0060	.0098	.015	.021	.028	.038	.049
2.4	.0003	.0011	.0026	.0048	.0078	.012	.017	.023	.031	.040
2.5	.0002	.0009	.0021	.0038	.0062	.0094	.014	.019	.025	.033
2.6	.0002	.0007	.0016	.0030	.0049	.0074	.011	.015	.020	.027
2.7	.0001	.0005	.0012	.0023	.0038	.0058	.0085	.012	.016	.022
2.8	.0001	.0004	.0010	.0018	.0029	.0045	.0066	.0094	.013	.017
2.9	.0001	.0003	.0007	.0014	.0022	.0035	.0051	.0073	.010	.014
3.0	.0001	.0002	.0005	.0010	.0017	.0026	.0039	.0057	.0079	.011
3.2		.0001	.0003	.0006	.0010	.0014	.0024	.0034	.0048	.0066
3.4		.0001	.0001	.0003	.0005	.0008	.0013	.0020	.0028	.0039
3.6				.0001	.0002	.0004	.0008	.0011	.0016	.0022
3.8					.0001	.0002	.0004	.0006	.0009	.0012
4.0					.0001	.0001	.0002	.0003	.0005	.0007
4.2						.0001	.0001	.0002	.0002	.0004
4.4							.0001	.0001	.0001	.0002
4.6									.0001	.0001

For an extensive 5-place table see *Table of Circular Normal Probabilities*, Bell Aircraft Corporation, Report No. 02-949-106, Buffalo, New York, 1956.

TABLE 11.11.1 (*Continued*)

d/σ	R/σ									
	1.2	1.4	1.6	1.8	2.0	2.2	2.4	2.6	2.8	3.0
					P					
0.0	.513	.625	.722	.802	.865	.911	.944	.966	.980	.989
0.1	.512	.623	.720	.800	.863	.910	.943	.965	.980	.989
0.2	.506	.617	.715	.796	.859	.907	.941	.964	.979	.988
0.3	.498	.608	.706	.788	.852	.901	.937	.961	.977	.987
0.4	.486	.596	.694	.777	.843	.894	.931	.956	.974	.985
0.5	.471	.580	.678	.763	.831	.884	.923	.951	.970	.982
0.6	.454	.561	.660	.745	.816	.872	.914	.944	.965	.979
0.7	.434	.540	.639	.726	.799	.857	.902	.936	.959	.975
0.8	.412	.516	.615	.703	.778	.840	.889	.925	.952	.970
0.9	.388	.490	.588	.678	.756	.821	.873	.913	.943	.964
1.0	.364	.463	.560	.650	.731	.800	.856	.900	.933	.956
1.1	.338	.434	.530	.621	.704	.776	.836	.884	.921	.948
1.2	.312	.404	.498	.590	.674	.750	.814	.866	.907	.937
1.3	.285	.374	.466	.557	.643	.721	.789	.846	.891	.926
1.4	.259	.344	.433	.523	.610	.691	.763	.824	.873	.912
1.5	.234	.314	.400	.489	.576	.659	.734	.799	.853	.896
1.6	.210	.284	.367	.454	.541	.625	.703	.772	.831	.878
1.7	.186	.256	.334	.419	.505	.590	.671	.744	.806	.859
1.8	.164	.229	.303	.384	.469	.554	.637	.713	.780	.837
1.9	.144	.203	.272	.349	.433	.518	.601	.680	.751	.812
2.0	.125	.178	.243	.316	.396	.481	.565	.646	.720	.786
2.1	.107	.156	.215	.284	.361	.443	.528	.610	.687	.757
2.2	.092	.135	.189	.253	.327	.407	.490	.573	.653	.726
2.3	.078	.116	.165	.224	.294	.370	.452	.536	.617	.693
2.4	.065	.099	.143	.197	.262	.335	.415	.498	.580	.659
2.5	.054	.084	.123	.172	.232	.301	.378	.460	.542	.623
2.6	.045	.070	.105	.149	.204	.269	.342	.422	.504	.586
2.7	.037	.059	.089	.128	.178	.238	.308	.385	.466	.548
2.8	.030	.048	.074	.109	.154	.210	.275	.348	.428	.510
2.9	.024	.040	.062	.093	.133	.183	.244	.313	.390	.471
3.0	.019	.032	.051	.078	.113	.159	.215	.280	.353	.433
3.2	.012	.021	.034	.053	.080	.117	.163	.219	.284	.358
3.4	.007	.013	.022	.036	.055	.083	.119	.166	.222	.288
3.6	.004	.008	.014	.023	.037	.057	.085	.122	.169	.225
3.8	.002	.004	.008	.014	.024	.038	.059	.087	.124	.171

TABLE 11.11.1 (*Continued*)

| d/σ | \multicolumn{10}{c}{R/σ} |
	1.2	1.4	1.6	1.8	2.0	2.2	2.4	2.6	2.8	3.0
					P					
4.0	.0013	.0025	.0046	.0085	.015	.024	.039	.060	.088	.126
4.2	.0007	.0014	.0026	.0049	.0087	.015	.025	.040	.061	.090
4.4	.0004	.0007	.0014	.0027	.0051	.0090	.016	.026	.041	.062
4.6	.0002	.0004	.0008	.0015	.0028	.0052	.0093	.016	.026	.041
4.8	.0001	.0002	.0004	.0008	.0016	.0029	.0054	.0095	.016	.027
5.0			.0002	.0004	.0008	.0016	.0030	.0055	.0098	.017
5.2			.0001	.0002	.0004	.0008	.0016	.0030	.0056	.0100
5.4				.0001	.0002	.0004	.0009	.0016	.0031	.0058
5.6						.0002	.0005	.0009	.0017	.0033
5.8						.0001	.0002	.0005	.0009	.0017
6.0								.0002	.0005	.0009

11.12. A useful formula. The following integral is sometimes useful:

$$(11.12.1) \qquad P_1 = \int_{-\infty}^{\infty} \int_{-\infty}^{\infty} g(x,y)\, f(x,y)\, dx\, dy = (c^2/s^2)e^{-d^2/2s^2},$$

where

$$(11.12.2) \qquad g(x,y) = e^{-(x^2+y^2)/2c^2}, \qquad s^2 = \sigma^2 + c^2, \qquad d^2 = \bar{x}^2 + \bar{y}^2,$$

and $f(x,y)$ is as defined in (11.11.2). In certain applications Eq. (11.12.1) appears with $2c^2$ being replaced by b^2, and $2s^2$ by w^2.

The three dimensional case

11.13. The development of the probability theory for three chance variables X, Y, Z follows the same pattern as outlined above for one or two variables. The three-dimensional case is of concern where an event requires three numbers to describe it. The following example is of interest.

Let S be a region in X, Y, Z space. Suppose that the probability $P(S)$ that the point (X,Y,Z) falls in the region S is given by

$$(11.13.1) \qquad P(S) = \iiint_{S} f(x,y,z)\, dx\, dy\, dz,$$

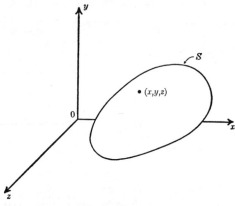

FIGURE 11.13.1

where the integration is carried out over the entire region S. Then

$$(11.13.2) \qquad \int_{-\infty}^{\infty} \int_{-\infty}^{\infty} \int_{-\infty}^{\infty} f(x,y,z) \, dx \, dy \, dz = 1.$$

The function $f(x,y,z)$ is the *probability density* or *frequency function* for the joint distribution of X, Y, and Z.

The *marginal distribution of X* has the distribution function

$$(11.13.3) \qquad P(X \leqq x') = \int_{x=-\infty}^{x=x'} \int_{y=-\infty}^{y=\infty} \int_{z=-\infty}^{z=\infty} f(x,y,z) \, dx \, dy \, dz$$

$$= \int_{-\infty}^{x'} f_1(x) \, dx,$$

where

$$(11.13.4) \qquad f_1(x) = \int_{y=-\infty}^{y=\infty} \int_{z=-\infty}^{z=\infty} f(x,y,z) \, dy \, dz.$$

Likewise, the marginal distributions of Y and Z may be defined.

A necessary and sufficient condition for the statistical independence of X, Y, and Z is that for all x, y, and z,

$$(11.13.5) \qquad f(x,y,z) = f_1(x) \, f_2(y) \, f_3(z).$$

The *mean*, or *expected value* of the function $g(X,Y,Z)$ is

$$(11.13.6) \qquad E[g(X,Y,Z)] = \int_{-\infty}^{\infty} \int_{-\infty}^{\infty} \int_{-\infty}^{\infty} g(x,y,z) \, f(x,y,z) \, dx \, dy \, dz.$$

The *moments* (about the origin) of the distribution are

$$\nu_{ijk} = E[X^i Y^j Z^k] = \int_{-\infty}^{\infty} \int_{-\infty}^{\infty} \int_{-\infty}^{\infty} x^i y^j z^k \, f(x,y,z) \, dx \, dy \, dz,$$

(11.13.7)
$$\nu_{100} = E[X] = \mu_x = \bar{x}, \qquad \nu_{010} = E[Y] = \mu_y = \bar{y},$$

$$\nu_{001} = E[Z] = \mu_z = \bar{z}.$$

The *(central) moments* about $X = \bar{x}$, $Y = \bar{y}$, $Z = \bar{z}$ are

$$\mu_{ijk} = E[(X - \bar{x})^i (Y - \bar{y})^j (Z - \bar{z})^k]$$

(11.13.8)
$$= \int_{-\infty}^{\infty} \int_{-\infty}^{\infty} \int_{-\infty}^{\infty} (x - \bar{x})^i (y - \bar{y})^j (z - \bar{z})^k \, f(x,y,z) \, dx \, dy \, dz.$$

$$\mu_{100} = \mu_{010} = \mu_{001} = 0, \quad \mu_{200} = \sigma_x^2, \quad \mu_{020} = \sigma_y^2, \quad \mu_{002} = \sigma_z^2.$$

11.14. **Normal distribution.** In case the probability density is expressible in the form

(11.14.1) $$f(x,y,z) = f(x_1,x_2,x_3) = e^{-\phi/2} \div (2\pi)^{3/2} \sigma_1 \sigma_2 \sigma_3 \sqrt{|R|},$$

where

(11.14.2) $$\phi = \sum_{i,j=1}^{3} [R_{ij}(x_i - \bar{x}_i)(x_j - \bar{x}_j) \div |R| \cdot \sigma_i \sigma_j] \geqq 0,$$

(11.14.3) $$R = \begin{vmatrix} 1 & r_{12} & r_{13} \\ r_{21} & 1 & r_{23} \\ r_{31} & r_{32} & 1 \end{vmatrix}, \qquad |R| = \text{determinant of } R,$$

R_{ij} = cofactor of the ith row and j^{th} column of R, the distribution is said to be *normal*, or *Gaussian*. Here the r_{ij} are correlation coefficients, and $x = x_1$, $y = x_2$, $z = x_3$. (See §12.6.)

The probability that the point (X,Y,Z) falls in the region S is

(11.14.4) $$P(S) = \iiint\limits_{S} f(x,y,z) \, dx \, dy \, dz.$$

11.15. **Special case.** Here, only a special case of (11.14.1) is considered, namely,

(11.15.1) $$f(x,y,z) = e^{-\phi/2} \div (2\pi)^{3/2} \sigma_x \sigma_y \sigma_z,$$

where

$$(11.15.2) \qquad \phi = \frac{(x - \bar{x})^2}{\sigma_x^{\,2}} + \frac{(y - \bar{y})^2}{\sigma_y^{\,2}} + \frac{(z - \bar{z})^2}{\sigma_z^{\,2}},$$

$$(11.15.3) \qquad E[X] = \bar{x}, \qquad E[Y] = \bar{y}, \qquad E[Z] = \bar{z}.$$

11.16. Probability ellipsoid. The normal density function (11.15.1) is constant for points of the ellipsoid $\phi = C^2$, where C is a constant. The probability that a random point (X,Y,Z) will fall inside the ellipsoid is

$$(11.16.1) \qquad p = \sqrt{2/\pi} \int_0^C t^2 \, e^{-t^2/2} \, dt.$$

When $C = 1.5382$, $p = 0.5$, and the corresponding ellipsoid $\phi = 1.5382$ is called the *50% probability ellipsoid*.

TABLE 11.16.1

Value of C for $P\%$ probability ellipsoids. ($P = 100p$).

$P\%$	C	$P\%$	C
25	1.101	90	2.500
50	1.538	95	2.795
75	2.027	99	3.368

11.17. Spherical case. In case $\sigma_x = \sigma_y = \sigma_z = \sigma$, the normal distribution is called *spherical*. Then the probability density corresponding to $f(x,y,z)$, Eq. (11.15.1), is

$$(11.17.1) \qquad f(\rho) = (2/\pi)^{1/2}(\rho^2/\sigma^3)e^{-\rho^2/2\sigma^2},$$

where

$$(11.17.2) \qquad \rho^2 = (x - \bar{x})^2 + (y - \bar{y})^2 + (z - \bar{z})^2.$$

The probability ellipsoids are spheres.

The probability that a point (X,Y,Z) taken at random will fall in a spherical shell, whose center is at $(\bar{x},\bar{y},\bar{z})$ and whose inner and outer radii are ρ and $\rho + \Delta\rho$, is $f(\rho)\Delta\rho$.

The probabilities p that a point (X,Y,Z) taken at random will fall within the sphere $\rho^2 = (C\sigma)^2$, for particular values of C are shown in Table 11.17.1.

When $C = 1.5382$, $p = 0.5$, and the sphere of radius 1.5382σ is called the *50% probability sphere*. The radius of this sphere is called the *spherical probable error* (s.p.e.), and (s.p.e.) $= 1.5382\sigma$.

TABLE 11.17.1

Radius of $P\%$ probability spheres, of radius $C\sigma$

$(P = 100p)$ $(\sigma_x = \sigma_y = \sigma_z \equiv \sigma)$

$P\%$	$C\sigma$	$=$	Radius		
25	1.101σ		0.7159 (s.p.e.)	$= 0.6900$ (m.s.r.e.)	
50	1.538σ		1.000 (s.p.e.)	$= 0.9639$ (m.s.r.e.)	
75	2.027σ		1.318 (s.p.e.)	$= 1.270$ (m.s.r.e.)	
90	2.500σ		1.625 (s.p.e.)	$= 1.567$ (m.s.r.e.)	
95	2.795σ		1.817 (s.p.e.)	$= 1.752$ (m.s.r.e.)	
99	3.368σ		2.190 (s.p.e.)	$= 2.111$ (m.s.r.e.)	

The *mean spherical radial error* (m.s.r.e.) of the spherical normal distribution (11.17.1) is the expected value of ρ, namely,

$$(11.17.3) \qquad \text{m.s.r.e.} = E[\rho] = \int_0^\infty \rho\, f(\rho)\, d\rho = 2\sigma\sqrt{2/\pi} = 1.5958\sigma.$$

Various relationships between s.p.e., σ, and m.s.r.e. are:

$$0.6501 \text{ (s.p.e.)} = \sigma = 0.6267 \text{ (m.s.r.e.)}$$

$$(11.17.4) \qquad \text{(s.p.e.)} = 1.5382\sigma = 0.9639 \text{ (m.s.r.e.)}$$

$$1.0374 \text{ (s.p.e.)} = 1.5958\sigma = \qquad \text{(m.s.r.e.)}$$

Some graphical methods

11.18. In calculating with normal distributions a number of graphical schemes are useful. Several such schemes are given below.

11.19. **Normal probability graph paper.** The cumulative normal distribution function $F_N(x)$ as given in Eq. (10.1.1), can be plotted from Table 10.2.1 as a function of x so as to appear as a straight line. The values of x are set off equidistant as shown on Fig. 11.19.1. A straight line is drawn, to represent $F_N(x)$, which passes through the center C of the rectangular sheet and the top edge of the sheet. The point C is given the abscissa μ, and ordinate 0.50. The ordinate y of a point (x,y) on the line is the cumulative normal probability $F_N(x)$ corresponding to x, and the scale on the ordinate is thus constructed with the aid of Table 10.2.1. When $x = \mu$ is the mean, y is 0.5; when $x = \mu + \sigma$, $y = 0.8413$ where σ is the standard deviation of the distribution, etc.

A *normal probability graph paper*, so constructed, is shown in Fig. 11.19.1. It is often quite useful for approximate work.

The cumulative polygon of Table 3.3.2, Fig. 10.9.1, is shown plotted on probability paper in Fig. 11.19.1 together with the corresponding cumulative normal distribution fitted to the polygon.

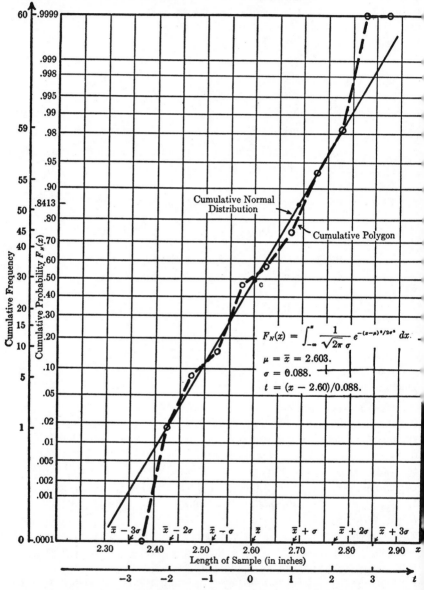

$$F_N(x) = \int_{-\infty}^{x} \frac{1}{\sqrt{2\pi}\ \sigma}\ e^{-(x-\mu)^2/2\sigma^2}\ dx.$$

$$\mu = \bar{x} = 2.603.$$

$$\sigma = 0.088.$$

$$t = (x - 2.60)/0.088.$$

Cumulative Polygon of Table 3.3.2, Figure 10.9.1, Plotted on Probability Graph Paper

FIGURE 11.19.1

If data plotted on such paper permit a straight line approximation, the data may be considered to be approximately normally distributed. The abscissa corresponding to the ordinate $y = 0.5$ gives the mean, μ, and the abscissa corresponding to the ordinate $y = 0.8413$ gives the value of $\mu + \sigma$.

11.20. Normal probability ruler.

The problem frequently arises to calculate crudely the integral

$$(11.20.1) \quad I = F_N(b) - F_N(a)$$

$$= (1/\sqrt{2\pi}\ \sigma) \int_a^b e^{-(x-\mu)^2/2\sigma^2}\ dx;$$

or, by Eq. (10.3.1),

$$(11.20.2) \quad I = \Psi(t_2) - \Psi(t_1)$$

$$= (1/\sqrt{2\pi}) \int_{t_1}^{t_2} e^{-t^2/2}\ dt,$$

where

$$t = (x - \mu)/\sigma,$$

$$t_2 = (b - \mu)/\sigma,$$

$$t_1 = (a - \mu)/\sigma.$$

This can be done easily with the aid of a *probability ruler*. (Fig. 11.20.1).

Suppose the ruler is constructed for the variable t. The t-scale of the ruler is drawn uniformly spaced in distance so that $t = 0$ is at the center C of the t-scale; $t = 1$ is one unit of length (e.g., 1 inch) above C, $t = 2$ is two units above C, etc.; $t = 0$ corresponds to $x = \mu$; $t = 1$ to $x = \mu + \sigma$.

Scale $\Psi(t)$ on the ruler is drawn so that to the value of t on scale t there corresponds the value of the cumulative probability $\Psi(t)$ as found in the normal probability Table 10.8.3.

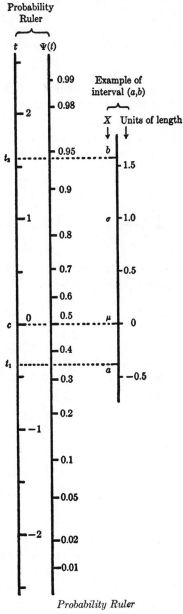

Probability Ruler
Figure 11.20.1

Suppose x, μ, a, and b are measured, say, in yards. Suppose these quantities are drawn on a piece of paper to the scale σ yards equals one unit of length on the ruler. To calculate the value of the integral I, place the ruler on the paper drawing so that the 0.5 point on the $\Psi(t)$ scale falls on the mean μ. The values of $\Psi(t_1)$ and $\Psi(t_2)$ corresponding to $x = a$ and $x = b$, respectively, can then be read off the ruler directly. The difference $\Psi(t_2) - \Psi(t_1)$ is the value of I sought.

11.21. Graphical scheme for calculating two-dimensional normal probability integral over an area A. The integral

$$(11.21.1) \qquad I = (1/2\pi) \iint\limits_{A} e^{-R^2/2} \, dX \, dY,$$

where

$$R^2 = X^2 + Y^2, \qquad X = x/\sigma_x, \qquad Y = y/\sigma_y,$$

over the area A in the X,Y-plane can be approximated quickly by means of a *chart* constructed as follows:

On a sheet of paper draw the rectangular coordinate axes, X and Y, intersecting at point C. Along the X-axis lay off the graduations of a normal probability ruler (see Fig. 11.20.1) to some convenient scale, say $\sigma_x = 2$ inches on sheet. In other words, the X scale is drawn uniformly spaced in distance so that $X = 0$ is at the center C of the X scale, $X = 1$ is 2 inches to the right of C; etc.; $X = 0$ corresponds to $x = 0$; $X = 1$ corresponds to $x = \sigma_x$; etc. Along the Y-axis do likewise, $\sigma_y = 2$ inches on sheet. Through the graduations selected on the X-axis draw lines parallel to the Y axis. Similarly for the Y-axis. If the adjacent parallel lines drawn parallel to the Y axis are so selected that the probability I for the area A_x between adjacent lines is p, and if, furthermore, adjacent parallel lines drawn parallel to the X-axis are so selected that the probability for the area A_y between these adjacent lines is p, then the probability I for the area A_{xy} common to the two strips A_x and A_y is p^2.

Select the graduation on the axes so that $p^2 = 0.001$, (or some other convenient number like 0.01, or 0.0001). Then $p = 0.031623$ (or 0.1, or 0.01). Draw lines through the corresponding graduations parallel to the two axes. A checkered design is formed made up of a number of rectangular "cells", forming the *chart*.

Draw on a transparent piece of paper the area A, choosing the same scale as before, in the direction of the X-axis, $\sigma_x = 2$ inches, and in the direction of the Y-axis $\sigma_y = 2$ inches. [*Note*: Unless $\sigma_x = \sigma_y$, the scales are different in the X and Y directions, and the area A will appear distorted when drawn on the X,Y-plane.]

Lay the transparent drawing of A on the *chart* so that the point C lies at the origin O of the transparent drawing and the X-axis of the drawing is parallel to the X-axis of the chart. The value of the integral I is equal to $(0.001)N$, where N is the number of "cells" included within the area A drawn on the transparent sheet.

Chart I shows such a chart designed from the principles described above, each "cell" having a probability of 0.001, but drawn to a scale σ_x and σ_y equal to one unit of length in the directions of the X- and Y-axes, respectively, i.e., $X = 0$ at C, $X = 1$ is drawn one unit of length to the right of C; and $X = 1$ corresponds to $x = \sigma_x$, etc.

The values of X in Chart I for consecutive rulings along the X-axis are: $X = 0, 0.08, 0.16, 0.24, 0.32, 0.41, 0.495, 0.59, 0.68, 0.79, 0.90, 1.03,$ $1.17, 1.35, 1.58, 1.95, \cdots$. Precisely the same numerical values are used for consecutive rulings along the Y-axis.

Various versions of this chart can be constructed, the best form and scale to use depending upon the nature of the problem and the accuracy with which the integral (11.21.1) is to be approximated.

11.22. Polar form.

The polar form of Chart I is shown in Chart II. The number of cells contained in the area A times 0.0025 is the approximate value of (11.21.1).

The values of the radius R from the origin C to the consecutive circular rulings of Chart II are $R = 0, 0.3203, 0.4590, 0.5701, 0.6680, 0.7585,$ $0.8446, 0.9282, 1.0108, 1.0935, 1.1774, 1.2637, 1.3537, 1.4490, 1.5518,$ $1.6651, 1.7941, 1.9479, 2.1460, 2.4477, \cdots$. Plot the values of R to a scale so that $R = 0, 1, 2, \cdots$, appear at uniform radial distances (e.g., $0, 1, 2, \cdots$ inches) from the origin C.

The value of the angles with vertex at C between consecutive radial rulings is $360 \sqrt{0.0025}$ degrees $= 360(.05)$ degrees $= 18°$.

Chart II is drawn so that the radius of C_1 is 1 unit; the scale used on the chart being such that $\sigma_x = 1$ unit along the direction of the X axis and $\sigma_y = 1$ unit along the direction of the Y axis. The circle C_1 is the so called "40% circle" so named since integral (11.21.1) over this circle has the value 0.3935.

The probability for the area between consecutive circles is 0.05; for the area between consecutive radial lines, is 0.05.

The integral (11.21.1) appears in many two-dimensional gunnery problems and the scheme outlined herein is of considerable practical use. This scheme can also be used to approximate (11.21.1) when

$$X = (x - \bar{x})/\sigma_x, \qquad Y = (y - \bar{y})/\sigma_y,$$

by the appropriate translation of coordinates.

Chart 1

Rectangular normal probability integral chart drawn for one quadrant

Scale: one unit of length on chart in X direction equals σ_x.

one unit of length on chart in Y direction equals σ_y.

The value of the integral I over the area enclosed in each rectangular "cell" is 0.001. The value of I over an arbitrary area A is $0.001N$, where N is the approximate number of cells enclosed within A. In the case illustrated, $I = 0.001(65) = 0.065$.

$$I = (1/2\pi) \iint\limits_{A} e^{-R^2/2}\ dX\ dY,$$

where

$$X = x/\sigma_x, \qquad Y = y/\sigma_y, \qquad R^2 = X^2 + Y^2.$$

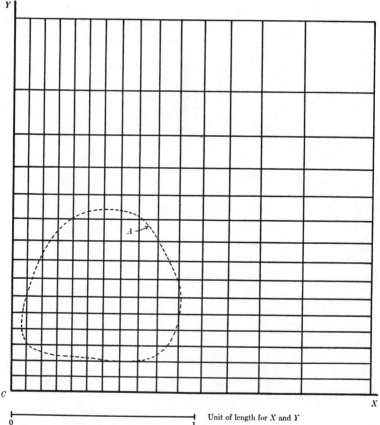

Unit of length for X and Y

0   1

Chart II

Polar form of normal probability integral chart drawn for four quadrants

Scale: one unit of length on chart in X direction equals σ_x.
   one unit of length on chart in Y direction equals σ_y.

The value of the integral I over the area enclosed in each "cell" is 0025. The value of I over an arbitrary area A is $0.0025N$, where N is the approximate number of cells enclosed within A. In the case illustrated, $I = .0025(16) = 0.04$.

$$I = (1/2\pi) \iint_A e^{-R^2/2} R\, dR\, d\theta = (1/2\pi) \iint_A e^{-R^2/2}\, dX\, dY,$$

$$X = x/\sigma_x, \qquad Y = y/\sigma_y, \qquad R^2 = X^2 + Y^2.$$

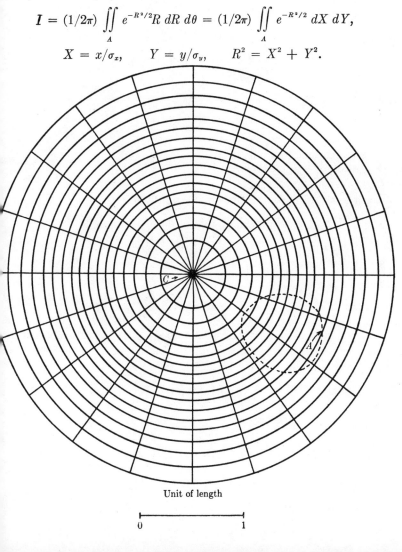

Unit of length

0   1

11.23. **Transformation of variables.** Suppose the joint density distribution of two random variables x_1, x_2 is $f(x_1, x_2)$. The joint distribution of the random variables y_1, y_2, where

$$(11.23.1) \qquad y_1 = y_1(x_1, x_2), \qquad y_2 = y_2(x_1, x_2)$$

are expressions which indicate a "one to one" transformation from (x_1, x_2) to (y_1, y_2), (i.e., there is exactly one set of values of the y's corresponding to a given set of values of the x's, and conversely), is found as follows.

If the event that (x_1, x_2) falls in the region R_x is ξ_x, then ξ_x is equivalent to the event ξ_y that (y_1, y_2) falls in the region R_y, where region R_y corresponds to region R_x. The probability that ξ_x occurs is equal to the probability that ξ_y occurs. That is,

$$(11.23.2) \qquad P[\xi_x] = P[\xi_y] = \iint\limits_{R_x} f(x_1, x_2) \, dx_1 \, dx_2$$

$$= \iint\limits_{R_y} g(y_1, y_2) \, dy_1 \, dy_2,$$

where

$$(11.23.3) \qquad g(y_1, y_2) = f(x_1, x_2) \cdot J\left(\frac{x_1, x_2}{y_1, y_2}\right).$$

Here $g(y_1, y_2)$ is the joint density distribution of the random variables y_1, y_2, and

$$(11.23.4) \qquad J\left(\frac{x_1, x_2}{y_1, y_2}\right) = \begin{vmatrix} \dfrac{\partial x_1}{\partial y_1} & \dfrac{\partial x_1}{\partial y_2} \\[2ex] \dfrac{\partial x_2}{\partial y_1} & \dfrac{\partial x_2}{\partial y_2} \end{vmatrix}$$

is the *Jacobian* of x_1, x_2 with respect to y_1, y_2. (See §18.9.)

The principles sketched here apply to the case of joint distributions of n random variables.

EXAMPLE 1. Suppose the joint probability density function for x_1 and x_2 is

$$(11.23.5) \qquad f(x_1, x_2) = 4x_1 x_2, \quad (0 \leqq x_1 \leqq 1, \, 0 \leqq x_2 \leqq 1)$$

By (11.23.3), the joint probability function for $y_1 = x_1{}^2$, $y_2 = x_2{}^2$ is

$$(11.23.6) \qquad g(y_1, y_2) = 4\sqrt{y_1 y_2} \begin{vmatrix} 1/(2\sqrt{y_1}), & 0 \\ 0 & , 1/(2\sqrt{y_2}) \end{vmatrix} = 1,$$

since $x_1 = +\sqrt{y_1}$, $x_2 = +\sqrt{y_2}$ for $0 \leq y_1 \leq 1$, $0 \leq y_2 \leq 1$.

The marginal probability density functions for y_1 and y_2 are

$$(11.23.7) \qquad \begin{cases} g(y_1) = \int_0^1 g(y_1, y_2)\, dy_2 = \int_0^1 dy_2 = 1 \\ g(y_2) = 1. \end{cases}$$

EXAMPLE 2. If the joint probability density function of x and y is

$$f(x,y) = \exp -(x+y), \qquad (x \geq 0, y \geq 0)$$

the joint probability density function for $u = x$, $v = (x+y)/2$ is

$$g(u,v) = e^{-2v} \begin{vmatrix} 1 & 0 \\ -1 & 2 \end{vmatrix} = 2e^{-2v},$$

since $x = u$, $y = 2v - u$, $0 \leq u \leq 2v$.

The marginal probability density functions for u and v are, respectively,

$$(11.23.8) \qquad \begin{cases} g(u) = \int_{v=u/2}^{v=\infty} g(u,v)\, dv = \int_{v=u/2}^{v=+\infty} 2e^{-2v}\, dv = e^{-u} \\ g(v) = \int_{u=0}^{u=2v} g(u,v)\, du = \int_{u=0}^{u=2v} 2e^{-2v}\, du = 4ve^{-2v} \end{cases}$$

The probability that $v = (x+y)/2$ is less than X is

$$P\left[\frac{x+y}{2} < X\right] = \int_{v=0}^{v=X} g(v)\, dv = 1 - e^{-2X} - 2Xe^{-2X}, \quad y \geq 0.$$

twelve

**ANALYSIS OF PAIRS OF
MEASUREMENTS, REGRESSION
THEORY, ORTHOGONAL
POLYNOMIALS, TIME SERIES**

Analysis of pairs of measurements

12.1. Fitting a polynomial to data represented by a set of points
Consider data represented by the set of points (x_1, y_1), (x_2, y_2), $\cdots$, (x_k, y_k)
Suppose it is required to find among all curves of the form

$$(12.1.1) \qquad y = y(x) = a_0 + a_1 x + \cdots + a_n x^n, \qquad (k > n)$$

a curve C which gives a "best possible fit" to the given data. Suppose
that the error lies only in the ordinates y_i. The method of least square
assumes that the "best" system of values for the coefficients in (12.1.1)
(namely, $A_0, A_1, \cdots, A_n$), is one that renders the sum $\sum_{i=1}^{k} d_i^2 = M$
of the squares of the deviations d_i a minimum. Here, $d_i = y_i - y(x_i)$
the *deviation* of y_i from the value $y(x_i)$. The curve (12.1.1) with coeff
cients $a_j = A_j, j = 0, 1, \cdots, n$, is the "best fitting curve" C. C is called
a *mean square regression curve*. When $n = 1$, C is called a *mean square
regression line*.

The coefficients of C may be found by solving the $n + 1$ "normal"
equations $\partial M / \partial a_j = 0$, $(j = 0, 1, \cdots, n)$, for a_j. Specifically, the coeff
cients $A_0, \cdots, A_n$ of C are the roots of the equations:

$$(12.1.2)$$
$$\sum y_i = A_0 k + A_1 \sum x_i + A_2 \sum x_i^2 + \cdots + A_n \sum x_i^n,$$
$$\sum x_i y_i = A_0 \sum x_i + A_1 \sum x_i^2 + A_2 \sum x_i^3 + \cdots + A_n \sum x_i^{n+1}$$
$$\vdots \qquad\qquad\qquad\qquad\qquad\qquad \vdots$$

$$\sum x_i^n y_i = A_0 \sum x_i^n + A_1 \sum x_i^{n+1} + A_2 \sum x_i^{n+2} + \cdots + A_n \sum x_i^{2n}$$

where the summations run from $i = 1$ to $i = k$.

$$
A_0 = \begin{vmatrix}
\sum y_i & \sum x_i & \sum x_i^2 & \cdots & \sum x_i^n \\
\sum x_i y_i & \sum x_i^2 & \sum x_i^3 & \cdots & \sum x_i^{n+1} \\
\vdots & \vdots & \vdots & & \vdots \\
\sum x_i^n y_i & \sum x_i^{n+1} & \sum x_i^{n+2} & \cdots & \sum x_i^{2n}
\end{vmatrix} \div L
$$

$(12.1.3)$

$$A_1 = \begin{vmatrix} k & \sum y_i & \sum x_i^2 & \cdots & \sum x_i^n \\ \sum x_i & \sum x_i y_i & \sum x_i^3 & \cdots & \sum x_i^{n+1} \\ \vdots & \vdots & \vdots & & \vdots \\ \sum x_i^n & \sum x_i^n y_i & \sum x_i^{n+2} & \cdots & \sum x_i^{2n} \end{vmatrix} \div D,$$

nd similarly for $A_2, \cdots, A_n$, where

$$(12.1.4) \qquad D = \begin{vmatrix} k & \sum x_i & \cdots & \sum x_i^n \\ \sum x_i & \sum x_i^2 & \cdots & \sum x_i^{n+1} \\ \vdots & \vdots & & \vdots \\ \sum x_i^n & \sum x_i^{n+1} & \cdots & \sum x_i^{2n} \end{vmatrix}.$$

12.2. Fitting curves of the form $y' = Ab^x$. Such curves may be fitted o data in a manner similar to that given in §12.1 after a suitable change f coordinates. Thus, to fit $y' = Ab^x$, set $y = \log y'$, $A_0 = \log A$, $l_1 = \log b$, and proceed as in §12.1.

12.3. Fitting a line to points. Errors in ordinates, y_i, only. Suppose ; is required to find the best linear fit to the points (x_i, y_i), $i = 1, \cdots, k$, y means of a linear expression in x, namely, $y = a_0 + a_1 x$, assuming the rrors to be in the ordinates y_i, only. A good fit may often be given by ye. The method of least squares can also be used to find a "best linear t", whose coefficients are $a_0 = A_0$, $a_1 = A_1$. The values of A_0 and A_1 o obtained may be found from the normal equations

$$(12.3.1) \qquad \begin{aligned} \sum y_i &= A_0 k + A_1 \sum x_i, \\ \sum x_i y_i &= A_0 \sum x_i + A_1 \sum x_i^2, \end{aligned}$$

vhere the summations run from $i = 1$ to $i = k$.

$$(12.3.2) \qquad \begin{aligned} A_0 &= [(\sum y_i)(\sum x_i^2) - (\sum x_i)(\sum x_i y_i)] \div D, \\ A_1 &= [k(\sum x_i y_i) - (\sum y_i)(\sum x_i)] \div D, \end{aligned}$$

vhere

$$D = k(\sum x_i^2) - (\sum x_i)^2.$$

'or these values A_0 and A_1 the sum $\sum_{i=1}^{k} d_i$ of deviations

$$d_i = [y_i - (A_0 + A_1 x_i)]$$

is zero. The deviation d_i is also called a *residual*. The "best linear fit" is found is

$$(12.3.3) \qquad y = A_0 + A_1 x.$$

12.4. Errors assumed to be in abscissae, x_i, only. This case similar to the case above with the role of x and y interchanged.

12.5. Errors assumed to be in both x_i and y_i. In this case if the principle of least squares is used the sum of the perpendicular distance from the points to the line is made a minimum. The resulting line is

$$(12.5.1) \qquad y - \bar{y} = m(x - \bar{x}),$$

where

$$\bar{x} = \left(\sum x_i\right) \div k, \qquad \bar{y} = \left(\sum y_i\right) \div k,$$

$$(12.5.2) \qquad \sigma_x{}^2 = \sum (x_i - \bar{x})^2 \div k = \left[\sum x_i{}^2 - k\bar{x}^2\right] \div k,$$

$$\sigma_y{}^2 = \sum (y_i - \bar{y})^2 \div k = \left[\sum y_i{}^2 - k\bar{y}^2\right] \div k,$$

$$r\sigma_x\sigma_y = \sum (x_i - \bar{x})(y_i - \bar{y}) \div k = \left(\sum x_i y_i - k\bar{x}\bar{y}\right) \div$$

and

$$(12.5.3) \qquad m = \{(\sigma_y{}^2 - \sigma_x{}^2) \pm [(\sigma_y{}^2 - \sigma_x{}^2)^2 + 4r^2\sigma_y{}^2\sigma_x{}^2]^{1/2}\} \div 2r\sigma_y$$

In all cases the summation runs from $i = 1$ to $i = k$. The proper sign must be selected from (12.5.3) to correspond to the given data.

Line (12.5.1) is an appropriate one to fit the given data when there are errors in both x and y.

12.6. Simple (linear) correlation. Consider data represented by (x_1, y_1), (x_2, y_2), $\cdots$, (x_k, y_k). One measure of the degree of association or *correlation*, between the x's and the corresponding y's is given by the ratio

$$(12.6.1) \qquad r = \operatorname{cov}(x,y)/\sigma_x\sigma_y,$$

where $\bar{x}$ and σ_x are the mean and standard deviation, respectively, of the x's; $\bar{y}$ and σ_y are the corresponding parameters for the y's; and

$$(12.6.2) \quad \operatorname{cov}(x,y) = \sum (x_i - \bar{x})(y_i - \bar{y}) \div k = \left[\sum x_i y_i - k\bar{x}\bar{y}\right] \div$$

is the *covariance* between the x's and the y's. The ratio r is called the *correlation coefficient*, (or, *product-moment about the mean*) of the variable x and y. Here $\bar{x}$, $\bar{y}$, σ_x, σ_y are calculated from (12.5.2).

The following notation is in common use:

12.6.3) $$\operatorname{var} x = \operatorname{cov}(x,x) = \sigma_x{}^2.$$

12.7. Regression line of y on x. Suppose a line is fitted to the points $x_i,y_i)$, $i = 1, \cdots, k$, so as to minimize the sum M (defined as in §12.1) of the squares of the vertical distances from the line (assuming the errors o be in the y-coordinates only, x treated as the independent variable). The line so obtained (by method of §12.3) has the equation

12.7.1) $$\frac{y - \bar{y}}{\sigma_y} = r \frac{(x - \bar{x})}{\sigma_x},$$

and is known as the (*mean*) *regression line of y on x*. The slope m_1 of this ine is called the *regression coefficient of y on x*, and

12.7.2) $$m_1 = r\sigma_y/\sigma_x = \sum_{i=1}^{k}(x_i - \bar{x})(y_i - \bar{y}) \div \sum_{i=1}^{k}(x_i - \bar{x})^2.$$

Let $d_i = y_i - y(x_i)$ be the difference between an observed value y_i and he value $y(x_i)$ of y on the approximating line $y(x_i) = a_0 + a_1x_i$. The quantity $(1/k)\sum d_i{}^2$ has a minimum when the line is the regression line of y on x. The value of this minimum is

12.7.3) $$S_y{}^2 = \sigma_y{}^2(1 - r^2).$$

S_y is the standard deviation of the errors of estimate in y and is called the *standard error of estimate in y*. $S_y{}^2$ is the *residual variance of y*.

12.8. Regression line of x on y. If a line is fitted to the points (x_i,y_i), $= 1, \cdots, k$, so as to minimize the sum of the squares of the horizontal distances from the line (assuming the errors to be in the x-coordinates only, y treated as the independent variable), the line so obtained is

12.8.1) $$\frac{y - \bar{y}}{\sigma_y} = \frac{1}{r}\left[\frac{x - \bar{x}}{\sigma_x}\right],$$

and is called the (*mean*) *regression line of x on y*. The slope of this line is

12.8.2) $$m_2 = \sigma_y/r\sigma_x.$$

The *regression coefficient of x on y* is defined to be $r\sigma_x/\sigma_y$ which is the reciprocal of m_2.

Let $\delta_i = x_i - x(y_i)$ be the difference between the observed value x_i and the value $x(y_i)$ of x on the approximating line $x(y_i) = b_0 + b_1x_i$. The quantity $(1/k)\sum \delta_i{}^2$ has a minimum when the line is the regression ine of x on y. The value of this minimum is

12.8.3) $$S_x{}^2 = \sigma_x{}^2(1 - r^2).$$

S_x is the standard deviation of the errors of estimate in x and is called the *standard error of estimate in* x; S_x^2 is the *residual variance of* x.

12.9. Properties of the two lines of regression.

Both lines (12.7.1) and (12.8.1) intersect at $(\bar{x},\bar{y})$. The lines coincide if and only if $r = \pm 1$. When $r = 0$ the lines are at right angles.

When all the points (x_i,y_i), $i = 1, \cdots , k$, lie on the regression line S_y^2 (or S_x^2) vanishes, and $r = \pm 1$. In this case there is said to be *perfect correlation*; when one variable is known, the other is uniquely determined and conversely.

When $r \neq 0$ the variables are said to be *correlated*; the correlation is *positive* when $r > 0$, *negative* when $r < 0$. When $r \neq 0$, a certain part of the variance of y can be removed by the subtraction of a linear function of x, the maximum amount being indicated by (12.7.3), or (12.8.3) according to the value of r. When y and x are statistically independent $r = 0$. When $r = 0$, the variables are said to be *uncorrelated*. Two uncorrelated variables (with $r = 0$) are not necessarily statistically independent.

EXAMPLE 1. The data given are indicated by (x_i,y_i) in Table 12.9.1 below. The regression lines are calculated as shown below and plotted in Fig. 12.9.1.

TABLE 12.9.1

Pair	x_i	y_i	x_i^2	y_i^2	x_iy_i
1	52	46	2704	2116	2392
2	48	27	2304	729	1296
3	43	24	1849	576	1032
4	39	22	1521	484	858
5	79	59	6241	3481	4661
6	61	53	3721	2809	3233
7	46	16	2116	256	736
8	56	47	3136	2209	2632
9	63	39	3969	1521	2457
10	45	41	2025	1681	1845
11	70	52	4900	2704	3640
12	74	55	5476	3025	4070
13	57	34	3249	1156	1938
Total	733	515	43211	22747	30790

$$k = 13, \qquad \sum x_i^2 = 43211,$$
$$\sigma_x^2 = (1/13)[43211 - 13(56.38)^2] = 145.2,$$
$$\sum x_i = 733, \qquad \sum y_i^2 = 22747,$$

$$\sigma_y{}^2 = (1/13)[22747 - 13(39.61)^2] = 180.8,$$

$$\sum y_i = 515, \qquad \sum x_i y_i = 30790, \qquad \sigma_x = 12.1, \qquad \sigma_y = 13.4.$$

$$\bar{x} = 733/13 = 56.38,$$

$$\text{cov}(x,y) = (1/13)[30790 - 13(2233.2)] = 135.2,$$

$$\bar{y} = 515/13 = 39.61,$$

$$\bar{x}\bar{y} = 2233.2, \qquad r = \text{cov}(x,y)/\sigma_x\sigma_y = 135.2/(12.1)(13.4) = 0.83.$$

Regression line of y on x. $y - 39.6 = 0.92(x - 56.4),$

$$n_1 = (0.83)(13.4)/(12.1) = 0.92, \quad S_y = \{(13.4)^2[1 - (0.83)^2]\}^{1/2} = 7.5.$$

Regression line of x on y. $y - 39.6 = 1.33(x - 56.4).$

$$n_2 = 13.4/(0.83)(12.1) = 1.33, \quad S_x = \{(12.1)^2[1 - (0.83)^2]\}^{1/2} = 6.7.$$

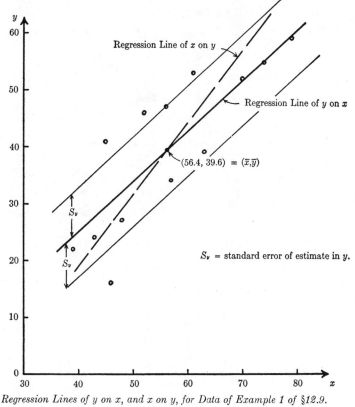

Regression Lines of y on x, and x on y, for Data of Example 1 of §12.9.

FIGURE 12.9.1

12.10. Computation of regression line using coded scheme. When the number n of pairs of measurements is under 50 the computation scheme of §12.9, Ex. 1 is satisfactory, but when the number of measurements is larger some scheme of grouping and choice of variable is often desirable. A method involving the selection of a convenient working origin (x_0, y_0), and units of measure h, k is often useful. Such a scheme is shown in the following example.

EXAMPLE 1. Consider the points (x, y) listed in Table 12.10.1. A change of variable is used, namely, $x = x_0 + hz$, $y = y_0 + kw$. In the example a two-way frequency table is used with $h = 5$, $k = 3$, with cell midpoints at 30, 35, $\cdots$, 60 for x; 18, 21, $\cdots$, 39 for y. The origin is selected so that $x_0 = 50$, $y_0 = 30$. A table giving the frequency with which combinations of values of x and y fall into the various cells is constructed as shown (bounded by heavy border) in Table 12.10.2. The frequency of occurrence of a point in cell (i, j) will be indicated by α_{ij}, the coordinates of cell midpoints being (x_i, y_i).

In Table 12.10.2, the following notation and formulas are used:

$$f_i = \sum_j \alpha_{ij}, \qquad g_j = \sum_i \alpha_{ij}, \qquad n = \sum_j g_j = \sum_i f_i,$$

(12.10.1) $$u_i = \sum_j \alpha_{ij} w_j, \qquad v_j = \sum_i \alpha_{ij} z_i.$$

$$\sum_i u_i z_i = \sum_j v_j w_j = \sum_i \sum_j \alpha_{ij} z_i w_j.$$

$$x = x_0 + hz, \qquad\qquad\qquad y = y_0 + kw.$$

$$\bar{x} = x_0 + h\bar{z}, \qquad\qquad\qquad \bar{y} = y_0 + k\bar{w}.$$

$$\bar{z} = (1/n) \sum f_i z_i, \qquad\qquad\qquad \bar{w} = (1/n) \sum g_j w_j.$$

$$\sigma_z^2 = (1/n)[\sum f_i z_i^2 - n\bar{z}^2], \qquad \sigma_w^2 = (1/n)[\sum g_j w_j^2 - n\bar{w}^2].$$

$$\sigma_x^2 = h^2 \sigma_z^2, \qquad\qquad\qquad \sigma_y^2 = k^2 \sigma_w^2.$$

$$\text{cov } (z, w) = (1/n)[\sum u_i z_i - n\bar{z}\bar{w}], \qquad \text{cov } (x, y) = hk \text{ cov } (z, w).$$

$$r = \text{cov } (x, y)/\sigma_x \sigma_y, \qquad\qquad m_1 = \text{cov } (x, y)/\sigma_x^2,$$

$$y - \bar{y} = m_1(x - \bar{x}).$$

TABLE 12.10.1

Measurements (x, y)

x	y	x	y	x	y	x	y	x	y
32	19	38	27	46	31	48	33	56	34
34	20	39	31	46	32	51	34	53	35
36	21	40	29	44	33	49	35	57	36
36	23	41	29	43	37	50	36	56	37
33	24	40	32	46	36	48	37	55	36
36	25	41	36	47	39	51	36	53	38
37	24	45	20	48	23	49	37	57	39
33	26	46	22	50	24	52	35	56	40
35	27	43	26	52	25	49	38	54	39
42	20	45	28	51	24	52	39	55	38
40	21	47	27	50	29	48	40	59	36
39	22	46	26	50	32	55	26	61	37
39	24	43	29	49	34	53	33	61	38

TABLE 12.10.2

Example of computation scheme to determine regression line using coded scheme

	z_i	-4	-3	-2	-1	0	1	2	g_i	g_iw_i	$g_iw_i^2$	v_i	v_iw_i
	x_i	30	35	40	45	50	55	60					
w_j	y_j												
3	39				1	3	5	1	10	30	90	6	18
2	36			1	2	6	4	2	15	30	60	4	8
1	33			1	2	4	2		9	9	9	-2	-2
0	30			3	2	1			6	0	0	-8	0
-1	27		2	1	4		1		8	-8	8	-11	11
-2	24		4	1		4			9	-18	36	-14	28
-3	21		2	3	2				7	-21	63	-14	42
-4	18	1							1	-4	16	-4	16
f_i		1	8	10	13	18	12	3	$n = 65$		$18 = \Sigma\, g_iw_i,$		
											$282 = \Sigma\, g_iw_i^2,$		
											$121 = \Sigma\, v_iw_i$		
f_iz_i		-4	-24	-20	-13	0	12	6	$-43 = \Sigma\, f_iz_i$	$x_0 = 50,\, y_0 = 30,$			
										$h = 5,\, k = 3,$			
$f_iz_i^2$		16	72	40	13	0	12	12	$165 = \Sigma\, f_iz_i^2$	$n = \Sigma\, f_i =$			
										$\Sigma\, g_j = 65$			
u_i		-4	-16	-9	-1	17	24	7		Entries in heavy			
										bordered part			
										are α_{ij}			
u_iz_i		16	48	18	1	0	24	14	$121 = \Sigma\, u_iz_i$				

The calculations from Table 12.10.2 are as follows:

$$\bar{z} = -43/65 = -0.6615, \qquad \bar{w} = 18/65 = 0.2769.$$

$$\bar{x} = 50 + 5(-0.6615) = 46.69, \qquad \bar{y} = 30 + 3(0.2769) = 30.83.$$

$$\sigma_z^2 = (1/65)[165 - 65(-0.6615)^2] = 2.538 - 0.4376 = 2.101.$$

$$\sigma_w^2 = (1/65)[282 - 65(0.2769)^2] = 4.3385 - 0.07667 = 4.262.$$

$$\sigma_x^2 = (5)^2(2.101) = 52.52, \qquad \sigma_y^2 = (3)^2(4.262) = 38.36.$$

$$\sigma_x = 7.25, \qquad \sigma_y = 6.19.$$

$$\text{cov } (z,w) = (1/65)[121 - 65(-0.6615)(0.2769)] = 2.045.$$
$$\text{cov } (x,y) = (5)(3)(2.045) = 30.67.$$

$$r = 30.67/(7.25)(6.19) = 0.683. \qquad m_1 = 30.67/52.52 = 0.584.$$

Line of regression of y on x is: $y - 30.8 = 0.58(x - 46.7)$.

Note: There is an approximation involved in grouping the (x,y) pairs at the centers of the (z,w) cells. When $n > 50$, this approximation is adequate in practice, provided the (z,w) cells are not too large.

For a treatment of regression lines when y is distributed normally see §12.22 to §12.25.

General regression and correlation

12.11. **General regression curves.** Let X and Y be variables with a joint continuous distribution of density $f(x,y)$. Consider those values of Y corresponding to X. Among all possible functions of the variable X, that function which gives a "best estimate" (in the least squares sense) of the variable Y is a function $g(X)$ for which

$$(12.11.1) \quad E[Y - g(X)]^2 = \int_{-\infty}^{\infty} \int_{-\infty}^{\infty} [y - g(x)]^2 f(x,y) \, dx \, dy$$

is a minimum. The function desired is $g(X) = m_2(X)$, where $m_2(X)$ is the *conditional mean*,

$$(12.11.2) \qquad m_2(X) = E[Y|X = x] = \int_{-\infty}^{\infty} y \, f(x,y) \, dy \div \int_{-\infty}^{\infty} f(x,y) \, dy.$$

The graph of $Y = m_2(X)$ is called a *regression curve of Y on X*.

If instead of all possible functions $g(X)$ one is restricted to some particular class of functions (e.g., polynomials of degree n, as in §12.1, or lines, as in §12.7), then the "mean square regression curve" obtained is generally different from that obtained above.

When the function $m_2(X)$ is substituted for $g(x)$ in (12.11.1), the latter expression takes on its minimum value. This minimum value is called the (*residual*) *variance* $S_Y{}^2$ *of* Y *about the regression curve of* Y *on* X. For example, in case the regression curve is a line, the variance is given by Eq. (12.7.3).

If Y is considered as the independent variable a *regression curve of* X *on* Y, $X = m_1(Y)$, is obtained in a similar manner, and

$$(12.11.3) \qquad m_1(Y) = E[X \,|\, Y = y] = \int_{-\infty}^{\infty} x\, f(x,y)\, dx \div \int_{-\infty}^{\infty} f(x,y)\, dx.$$

The curves $Y = m_2(X)$ and $X = m_1(Y)$ are not usually identical.

The theory of regression curves for discrete distributions follows in an analogous way. (See §12.27.)

12.12. Regression surfaces. Let $X_1, \cdots, X_n$ be variables with a joint continuous distribution of density $f(x_1, \cdots, x_n)$. The *conditional mean value* of X_1, relative to the hypothesis $X_2 = x_2, \cdots, X_n = x_n$ is

$$E[X_1 | X_2 = x_2, \cdots, X_n = x_n] = m_1(x_2, \cdots, x_n) = m_1$$

$$(12.12.1) \qquad = \int_{-\infty}^{\infty} x_1\, f(x_1 | x_2, \cdots, x_n)\, dx_1$$

$$= \int_{-\infty}^{\infty} x_1\, f(x_1, \cdots, x_n)\, dx_1 \div \int_{-\infty}^{\infty} f(x_1, \cdots, x_n)\, dx_1.$$

The locus $x_1 = m_1$ for all possible values of $x_2, \cdots, x_n$ is called a *regression surface for the mean* of X_1.

The variance of X_1 relative to $X_2 = x_2, \cdots, X_n = x_n$ is

$$(12.12.2) \qquad S_{1 \cdot x_2 x_3 \cdots x_n}^2 = \int_{-\infty}^{\infty} (x_1 - m_1)^2\, f(x_1 | x_2, \cdots, x_n)\, dx_1.$$

The (*residual*) *variance of* X_1 *about the regression surface* m_1 is the mean value of $S_{1 \cdot x_2 x_3 \cdots x_n}^2$ with respect to $x_2, \cdots, x_n$, that is,

$$(12.12.3) \quad S_{1 \cdot 23 \cdots n}^2 = \int_{-\infty}^{\infty} \cdots \int_{-\infty}^{\infty} S_{1 \cdot x_2 x_3 \cdots x_n}^2\, f_1(x_2, \cdots, x_n)\, dx_2 \cdots dx_n$$

$$= \int_{-\infty}^{\infty} \cdots \int_{-\infty}^{\infty} (x_1 - m_1)^2\, f(x_1, \cdots, x_n)\, dx_1 \cdots dx_n.$$

Here $f_1(x_2, \cdots, x_n)$ is the density of the marginal distribution of $X_2, \cdots, X_n$ corresponding to $f(x_1, \cdots, x_n)$.

12.13. Regression for normal distribution. When the distribution $f(x,y)$ is normal,

$$(12.13.1) \qquad f(x,y) = \frac{1}{2\pi\sigma_x\sigma_y\sqrt{1-r^2}} \exp\left(-G/2\right),$$

where

$$(12.13.2) \qquad G = \frac{1}{(1-r^2)}\left[\frac{(x-\bar{x})^2}{\sigma_x^2} - \frac{2r(x-\bar{x})(y-\bar{y})}{\sigma_x\sigma_y} + \frac{(y-\bar{y})^2}{\sigma_y^2}\right].$$

The *marginal frequency function for x* is

$$(12.13.3) \qquad f_1(x) = \frac{1}{\sigma_x\sqrt{2\pi}} \exp\left[-\frac{(x-\bar{x})^2}{2\sigma_x^2}\right].$$

The *conditional frequency function of y*, relative to a fixed value of x is

$$(12.13.4) \quad f(y|x) = f(x,y) \div f_1(x)$$

$$= \frac{1}{\sigma_y\sqrt{2\pi(1-r^2)}} \exp\left[-\frac{1}{2\sigma_y^2(1-r^2)}(y-m_2(x))^2\right],$$

where

$$(12.13.5) \qquad m_2(x) = \bar{y} + (r\sigma_y/\sigma_x)(x-\bar{x}).$$

$f(y|x)$ is a normal frequency function in y with mean $m_2(x)$ and standard deviation $\sigma_y\sqrt{1-r^2}$. Thus, the regression of y on x for the normal distribution is *linear*, and the conditional variance of y is independent of the value assumed by x.

This linear property of the regression of y on x holds for a n-dimensional normal distribution, also.

12.14. Partial correlation. Sometimes the degree of correlation between two variables X_1 and X_2 is wanted under the hypothesis that the other variables have assigned values $X_3 = x_3, \cdots, X_n = x_n$. X_1 and X_2 are considered as being estimated by means of regression curves m_1 and m_2 which are estimated in terms of $X_3, \cdots, X_n$. For this purpose the *partial correlation coefficient* between X_1 and X_2, with respect to $X_3, \cdots, X_n$ is used, namely,

$$(12.14.1) \qquad \rho_{12\cdot34\cdots n} = A \div \sqrt{BC},$$

where

$$A = E[(x_1 - m_1)(x_2 - m_2)]$$

$$= \int_{-\infty}^{\infty} \cdots \int_{-\infty}^{\infty} (x_1 - m_1)(x_2 - m_2) f(x_1, \cdots, x_n) \, dx_1 \cdots dx_n$$

(12.14.2)
$$B = E[(x_1 - m_1)^2]$$

$$= \int_{-\infty}^{\infty} \cdots \int_{-\infty}^{\infty} (x_1 - m_1)^2 f(x_1, \cdots, x_n) \, dx_1 \cdots dx_n,$$

$$C = E[(x_2 - m_2)^2]$$

$$= \int_{-\infty}^{\infty} \cdots \int_{-\infty}^{\infty} (x_2 - m_2)^2 f(x_1, \cdots, x_n) \, dx_1 \cdots dx_n,$$

$$m_1 = m_1(x_3, \cdots, x_n) = \int_{-\infty}^{\infty} x_1 \, f_2(x_1|x_3, \cdots, x_n) \, dx_1$$

$$= \int_{-\infty}^{\infty} x_1 \, f_2(x_1, x_3, \cdots, x_n) \, dx_1 \div \int_{-\infty}^{\infty} f_2(x_1, x_3, \cdots, x_n) \, dx_1,$$

$$m_2 = m_2(x_3, \cdots, x_n) = \int_{-\infty}^{\infty} x_2 \, f_1(x_2|x_3, \cdots, x_n) \, dx_2$$

$$= \int_{-\infty}^{\infty} x_2 \, f_1(x_2, x_3, \cdots, x_n) \, dx_2 \div \int_{-\infty}^{\infty} f_1(x_2, x_3, \cdots, x_n) \, dx_2,$$

and where the density of the joint distribution of $X_1, X_2, X_3, \cdots, X_n$ is $f(x_1, x_2, x_3, \cdots, x_n)$, and the density of the corresponding marginal distribution of $X_1, X_3, \cdots, X_n$ is $f_2(x_1, x_3, \cdots, x_n)$, of $X_2, X_3, \cdots, X_n$ is $f_1(x_2, x_3, \cdots, x_n)$.

12.15. Multiple correlation. Let $X_1, \cdots, X_n$ be a set of random variables with joint probability (or frequency) density $f(x_1, \cdots, x_n)$. A "best fitting" linear regression function

$$(12.15.1) \qquad\qquad L = b_1 + \sum_{i=2}^{n} b_i x_i$$

(in the sense of least squares) is a function L such that

$$(12.15.2) \qquad S = E[(X_1 - L)^2]$$

$$= \int_{-\infty}^{\infty} \cdots \int_{-\infty}^{\infty} (x_1 - L)^2 \, f(x_1, \cdots, x_n) \, dx_1, \cdots, dx_n$$

is a minimum. It can be shown that for a "best fit" the (regression) coefficients in L are given by

(12.15.3)
$$b_1 = a_1 - \sum_{i,j=2}^{n} a_i C_{1j} C^{ji},$$

$$b_i = \sum_{j=2}^{n} C_{1j} C^{ji}. \qquad (i = 2, \cdots, n)$$

Here $a_i = E[X_i]$, $c_{ij} = E[X_i X_j]$, $C_{ij} = E[(X_i - a_i)(X_j - a_j)] = c_{ij} - a_i a_j$, and $C^{ij} = $ (cofactor of C_{ij} in the determinant $|C_{ij}|$) $\div |C_{ij}| = C^{ii}$,

$$|C_{ij}| = D = \begin{vmatrix} C_{22} & C_{23} & \cdots & C_{2n} \\ \vdots & \ddots & & \vdots \\ C_{n2} & C_{n3} & \cdots & C_{nn} \end{vmatrix} \neq 0.$$

The *minimum value* of S is

(12.15.4) $$\overline{S}^2_{1 \cdot 23 \cdots n} = C_{11} - \sum_{i,j=2}^{n} C_{1i} C_{1j} C^{ji} = C \div D,$$

where

$$C = \begin{vmatrix} C_{11} & \cdots & C_{1n} \\ \vdots & & \vdots \\ C_{n1} & \cdots & C_{nn} \end{vmatrix}.$$

The quantity $\overline{S}^2_{1 \cdot 23 \cdots n}$ is the *variance of X_1 about the (least-square) mean square linear regression function L*, and the *mean square regression plane for X_1 with respect to $X_2, \cdots, X_n$* is

(12.15.5) $$X_1 - a_1 = \sum_{i,j=2}^{n} (x_i - a_i) C_{1j} C^{ji} \equiv L - a_1.$$

The *multiple correlation coefficient between X_1 and $X_2, \cdots, X_n$* is the covariance between X_1 and L divided by the square root of the product of the variance of X_1 and L, that is,

(12.15.6) $$R_{1 \cdot 23 \cdots n} = (H / \sqrt{C_{11}H}) = \sqrt{H/C_{11}},$$

where

$$H = \sum_{i,j=2}^{n} C_{1i} C_{1j} C^{ji}.$$

The minimum value of S is

$$(12.15.7) \qquad \overline{S}^2_{1 \cdot 23 \cdots n} = C_{11}(1 - R^2_{1 \cdot 23 \cdots n}).$$

$R^2_{1 \cdot 23 \cdots n} = 1$ if and only if all of the points of the probability density $f(x_1, \cdots, x_n)$ lie on the regression surface (12.15.5).

When $n = 2$, (12.15.7) reduces to $\overline{S}^2_{1 \cdot 2} = \sigma_1{}^2(1 - \rho^2)$ as in (12.7.3).

The mean square regression plane for any other variable X_j may be defined similarly.

Time series

12.16. Time series. When statistical data are arranged in a sequence in accordance with the time of occurrence the sequence is called a *time sequence*, or *time series*. Time series may be *discrete* or *continuous*.

The analysis of time series consists largely in describing and measuring the variations in the values of the terms as a function of time for selected time intervals. A full discussion of the subject of time series is beyond the scope of this book.

Various methods are used for measuring trends, such as by the use of free hand curve fitting, and the use of semi-averages, moving averages, "least squares", etc. Some of these methods are discussed in §12.1 to §12.9 in connection with curve fitting and the least squares methods.

12.17. Moving averages. Consider the time series

$$(12.17.1) \qquad x_1, x_2, \cdots, x_k, x_{k+1}, \cdots, x_n, x_{n+1}, \cdots$$

Form the sequence

$$(12.17.2) \qquad S(1), S(2), \cdots, S(n), \cdots$$

where

$$(12.17.3) \qquad S(u) = (1/k) \sum_{i=u}^{k+u-1} x_i, \qquad (u = 1, 2, \cdots)$$

k being a positive integer > 1. The sequence (12.17.2) is called a *moving average* based on the sum of k consecutive terms. Such sequences are often useful in analyzing time series.

12.18. Correlation between two sequences. Consider the two time series

$$(12.18.1) \qquad x_1, x_2, \cdots, x_k, x_{k+1}, \cdots, x_n,$$

$$(12.18.2) \qquad y_1, y_2, \cdots, y_k, y_{k+1}, \cdots, y_n,$$

where $x_n = x(t_n)$ and $y_n = y(t_n)$ indicate the values of the terms in the sequences at time t_n. Let $\bar{x}$ and $\bar{y}$ denote the means of the values of x in (12.18.1) and of y in (12.18.2), respectively.

By the *coefficient of correlation* ρ between the two sequences is meant the ratio

$$(12.18.3) \qquad \rho = \frac{\sum\limits_{i=1}^{n} (x_i - \bar{x})(y_i - \bar{y})}{\left[\sum\limits_{i=1}^{n} (x_i - \bar{x})^2 \cdot \sum\limits_{i=1}^{n} (y_i - \bar{y})^2 \right]^{1/2}},$$

where the numerator of (12.18.3) is n times the *covariance* between the x_k in (12.18.1) and the y_k in (12.18.2); and the square of the denominator of (12.18.3) is n^2 times the product of the *variance* of x_k and the *variance* of y_k.

ρ lies between -1 and $+1$. When $|\rho|$ is nearly 1, a strong degree of linear statistical dependence, or *correlation*, is said to exist between the sequences (12.18.1) and (12.18.2); when $|\rho|$ is nearly 0, a low degree of linear statistical dependence, or correlation, is said to be indicated.

Sometimes, for some purposes, merely the numerator of (12.18.3) is used to measure the correlation between the two series. This is advised only when the coordinate system is so selected that the means $\bar{x}$ and $\bar{y}$ are at the origin.

12.19. Serial correlation. The *serial correlation coefficient of order k* for the series (12.18.1) is defined to be

$$(12.19.1) \qquad r_k = \frac{\text{cov } (x_i, x_{i+k})}{[\text{var } x_i \; \text{var } x_{i+k}]^{1/2}}, \qquad (k = 0, 1, 2, \cdots)$$

where

$$(12.19.2) \qquad \text{cov } (x_i, x_{i+k}) = (1/(n-k)) \sum_{i=1}^{n-k} (x_i x_{i+k})$$

$$- (1/(n-k)^2) \left(\sum_{i=1}^{n-k} x_i \right) \left(\sum_{i=1}^{n-k} x_{i+k} \right),$$

$$(12.19.3) \qquad \text{var } x_i = (1/(n-k)) \left(\sum_{i=1}^{n-k} x_i^2 \right) - (1/(n-k)^2) \left(\sum_{i=1}^{n-k} x_i \right)^2,$$

$$(12.19.4) \qquad \text{var } x_{i+k} = (1/(n-k)) \left(\sum_{i=1}^{n-k} x_{i+k}^2 \right) - (1/(n-k)^2) \left(\sum_{i=1}^{n-k} x_{i+k} \right)^2.$$

If r_k be plotted as ordinate against k as abscissa, and the consecutive points so obtained joined by straight line segments, the graph so obtained is called a *correlogram*.

When the sequences (12.18.1) and (12.18.2) become infinite, similar definitions can be made.

12.20. Autocorrelation and cross-correlation. For the real infinite sequence

$$(12.20.1) \qquad \cdots x_{-n}, \cdots, x_{-1}, x_0, x_1, x_2, \cdots, x_i, \cdots, x_{i+k}, \cdots, x_n, \cdots,$$

the correlation between successive terms of the sequence is sometimes measured by the *autocorrelation coefficient* of the sequence (12.20.1), namely, by

$$(12.20.2) \qquad \varphi(k) = \lim_{n \to \infty} [1/(2n + 1)] \sum_{j=-n}^{j=n} x_{i+k}x_j.$$

By the *cross-correlation coefficient* between the sequence (12.20.1) and the real sequence (12.20.3)

$$(12.20.3) \qquad \cdots y_{-n}, \cdots, y_{-1}, y_0, y_1, y_2, \cdots, y_i, \cdots, y_{i+k}, \cdots, y_n, \cdots$$

is meant the function

$$(12.20.4) \qquad \psi(k) = \lim_{n \to \infty} [1/(2n + 1)] \sum_{j=-n}^{j=n} x_{i+k}y_j.$$

$\varphi(k)$ and $\psi(k)$ are functions of the *lag k*.

12.21. Continuous case. When dealing with continuous instead of discrete data, the series (12.20.1) and (12.20.3) are replaced by continuous functions of the time t, namely by $f(t)$ and $g(t)$. In such a case, the *autocorrelation coefficient* of $f(t)$ corresponding to the *time lag τ* is defined as

$$(12.21.1) \qquad \varphi(\tau) = \lim_{T \to \infty} (1/2T) \int_{-T}^{T} f(t + \tau) f(t) \, dt.$$

The *cross-correlation coefficient* between $f(t)$ and $g(t)$ for the lag time τ is defined to be

$$(12.21.2) \qquad \psi(\tau) = \lim_{T \to \infty} (1/2T) \int_{-T}^{T} f(t + \tau) g(t) \, dt.$$

In case the sequences, or functions, involved are complex functions of t, the definitions are made in an analogous manner, as follows:

$$(12.21.3) \qquad \varphi(k) = \lim_{n \to \infty} [1/(2n + 1)] \sum_{j=-n}^{j=n} x_{i+k}\bar{x}_i,$$

$$(12.21.4) \qquad \psi(k) = \lim_{n \to \infty} [1/(2n + 1)] \sum_{i=-n}^{i=n} x_{i+k} \bar{y}_i,$$

and

$$(12.21.5) \qquad \varphi(\tau) = \lim_{T \to \infty} (1/2T) \int_{-T}^{T} f(t + \tau) \overline{f(t)} \, dt,$$

$$(12.21.6) \qquad \psi(\tau) = \lim_{T \to \infty} (1/2T) \int_{-T}^{T} f(t + \tau) \overline{g(t)} \, dt,$$

where $\bar{x}_i$ and $\bar{y}_k$ indicate the conjugate complex values of x_i and y_k, respectively.

Confidence intervals for regression lines when dependent variable is normally distributed

12.22. **Regression line, y distributed normally.** Suppose that for any given independent variable x, y is distributed normally with mean $\alpha + \beta x$ and variance σ^2. α, β and σ^2 are assumed independent of x. For a given value of x the probability density function for y is

$$(12.22.1) \quad f(y) = f_N(y|x) = (1/\sigma\sqrt{2\pi}) \exp [-(y - \alpha - \beta x)^2/2\sigma^2].$$

The expected value of y, given x, is

$$(12.22.2) \qquad E(y|x) = \alpha + \beta x.$$

The joint probability density function of a sample of n pairs (x_1, y_1), $\cdots$, (x_n, y_n) is

$$(12.22.3) \qquad f(y_1) f(y_2) \cdots f(y_n) = \prod_{i=1}^{n} f_N(y_i|x_i).$$

If the method of maximum likelihood is used to estimate α and β, (12.22.3) is maximized, which means that

$$(12.22.4) \qquad \sum_{i=1}^{n} (y_i - \alpha - \beta x_i)^2$$

is minimized. When this is done as in §12.3 the estimates of α and β are, respectively, a and b, where

$$(12.22.5) \quad \begin{cases} a = \bar{y} - b\bar{x} \\ b = \left[\sum_{i=1}^{n} x_i y_i - n\bar{x}\bar{y} \right] \div \left[\sum_{i=1}^{n} x_i^2 - n\bar{x}^2 \right]. \end{cases}$$

Hence the estimate of $E(y|x)$ is

$$(12.22.6) \qquad \tilde{y} = a + bx = \bar{y} + b(x - \bar{x}).$$

Let $\tilde{y}_i$ be the value of $\tilde{y}$ when $x = x_i$. The discrepancy between the observed y_i and $\tilde{y}_i$ is $d_i = y_i - \tilde{y}_i$. a and b are unbiased estimates of α and β, respectively.

12.23. From (12.22.5), b can be written as a linear function of $y_1, \cdots, y_n$

$$(12.23.1) \qquad b = \sum_{i=1}^{n} c_i y_i,$$

where

$$(12.23.2) \qquad c_i = (x_i - \bar{x}) / \sum_{i=1}^{n} (x_i - \bar{x})^2$$

Treating $y_1, \cdots, y_n$ as independent random variables and $x_1, \cdots, x_n$ as constants, with y_i having the normal distribution $N(\alpha + \beta x_i, \sigma^2)$ it follows from (12.22.5) that b is a random variable with normal distribution $N[\beta, \sigma_b^2]$. The *expected value of b* is β, and the *variance of b* is

$$(12.23.3) \qquad \sigma_b^2 = \sigma^2 / \sum_{i=1}^{n} (x_i - \bar{x})^2.$$

Similarly, if in (12.22.6) one picks a specific value x_0 for x, the corresponding value of $\tilde{y}$ is $\tilde{y}_0 = \bar{y} + b(x_0 - \bar{x})$, which is a random variable with normal distribution $N[\alpha, \sigma_a^2]$. Here, the *expected value of a* is α and the *variance of a* is

$$(12.23.4) \qquad \sigma_a^2 = \left[\frac{1}{n} + \frac{\bar{x}^2}{\sum_{i=1}^{n} (x - \bar{x})^2} \right] \sigma^2.$$

The sum of the squares of the deviations is

$$(12.23.5) \qquad D^2 = \sum_{i=1}^{n} (y_i - \tilde{y}_i)^2$$

The expected value of D^2 is $(n - 2)\sigma^2$. The quantity $s_{ey}^2 = D^2/(n - 2)$ is a useful measure of the goodness of fit of the line $y = a + bx$ to the data.

12.24. An *unbiased* estimator of σ^2 in (12.23.4) is

$$(12.24.1) \qquad D^2/(n - 2) = s_{ey}^2 = \left(\frac{n - 1}{n - 2} \right) s_y^2 (1 - r^2)$$

where

$$s_x{}^2 = \sum_{i=1}^{n} (x_i - \bar{x})^2/(n-1)$$

$$s_y{}^2 = \sum_{i=1}^{n} (y_i - \bar{y})^2/(n-1)$$

$$s_{xy} = \sum_{i=1}^{n} (x_i - \bar{x})(y_i - \bar{y})/(n-1)$$

$$r = r_{xy} = s_{xy}/s_x s_y.$$

Also,

(12.24.2) $\qquad a = \bar{y} - b\bar{x}, \qquad b = rs_y/s_x.$

D^2/σ^2 is a random variable having the χ^2 distribution with $n-2$ degrees of freedom. D^2/σ^2 is independent of a, b, and $\tilde{y}_0$.

12.25. Confidence intervals for regression lines $y = \alpha + \beta x$.
The ratio

(12.25.1) $\qquad \dfrac{a - \alpha}{s_{ey}\sqrt{(1/n) + \bar{x}^2/\sum\limits_{i=1}^{n} (x_i - \bar{x})^2}}$

has the t-distribution with $n-2$ degrees of freedom. (See §14.43 to §14.65.)

12.26. Confidence interval for σ^2. Let

(12.26.1) $\qquad \displaystyle\int_{w=\chi_0{}^2}^{w=\infty} f_m(w)\, dw = \epsilon$

be the χ^2 integral given in Table IV. (See §13.26 to §13.30.) Let $\chi^2_{n-2, 1-\gamma/2}$ be the value of $\chi_0{}^2$ for which $\epsilon = 1 - \gamma/2$, and $\chi^2_{n-2, \gamma/2}$ the value of $\chi_0{}^2$ for which $\epsilon = \gamma/2$. A $100(1 - \gamma)$ percent confidence interval for σ^2 is

(12.26.2) $\qquad \dfrac{(n-2)s^2}{\chi^2_{n-2, 1-\gamma/2}} < \sigma^2 < \dfrac{(n-2)s^2}{\chi^2_{n-2, \gamma/2}}$

The $100(1 - \gamma)$ percent confidence intervals for α, β, and $\alpha + \beta x_0$ are shown in Table 12.26.1. Note the confidence interval for $\alpha + \beta x_0$ varies with the value selected for x_0.

Table 12.26.1

Confidence limits for α, β, and $E(y \mid x_0)$

$E(y \mid x_0) = \alpha + \beta x_0$ $100(1 - \gamma)\%$ Confidence limits	For parameter
$a \pm h k_1$	α
$b \pm h k_2$	β
$\tilde{y}_0 \pm h k_3$	$\alpha + \beta x_0$

a, b and $\tilde{y}_0$ are given by Eqs. (12.11.5) and (12.11.6).

$$h = t_{n-2;\gamma/2} \, s_{ey}$$

$$k_1^2 = (1/n) + \bar{x}^2 / \sum_{i=1}^{n} (x_i - \bar{x})^2$$

$$k_2^2 = 1 / \sum_{i=1}^{n} (x_i - \bar{x})^2$$

$$k_3^2 = (1/n) + (x_0 - \bar{x})^2 / \sum_{i=1}^{n} (x_i - \bar{x})^2$$

The quantity $t_{n-2,\gamma/2}$ is a number such that

$$\int_{-\infty}^{t_{n-2;\gamma/2}} g_m(t)dt = 1 - \gamma/2$$

where $g_m(t)$ is the t-distribution function (13.59.1) with m degrees of freedom and m is taken to be $n - 2$. Values of $t_{n-2;\gamma/2}$ may be found in Table XIII.

Orthogonal polynomials

12.27. **Curvilinear regression.** A polynomial

$$(12.27.1) \qquad E(y \mid x) = \alpha + \beta_1 x + \beta_2 x^2 + \cdots + \beta_m x^m$$

is to be fitted to certain data (x_1, y_1), (x_2, y_2), $\cdots$, (x_n, y_n), where the x_i's are considered to be observed without error and for each set of x_i's the values of y_i are normally distributed with common variance σ_y^2. An estimated regression polynomial for (12.27.1) is

$$(12.27.2) \qquad g(x) = a + b_1 x + b_2 x^2 + \cdots + b_m x^m$$

when b_i is an estimate of β_i and a is an estimate of α. The estimates a and b_i can be found as in §12.1 by minimizing the sum

$$(12.27.3) \qquad D = \sum_{j=1}^{n} [y_j - g(x_j)]^2$$

where $g(x_j)$ is the value of the right-hand side of (12.27.2) when $x = x_j$. This is done by solving the equations

$$\frac{\partial D}{\partial a} = 0, \frac{\partial D}{\partial b_i} = 0, \, i = 1, \cdots, m, \text{ for } a, b_1, \cdots, b_m.$$

The variance about the regression curve (12.27.2) is

$$(12.27.4) \qquad s^2(y \,|\, x) = \sum_{j=1}^{n} [y_j - g(x_j)]^2 / (n - m).$$

Here $m + 1$ is the number of parameters to be estimated.

12.28. **Orthogonal polynomials.** Often, rather than use the relation (12.27.1), considerable work can be saved by fitting to the data the regression relation

$$(12.28.1) \qquad E(y \,|\, x) = \alpha' + \beta_1' \, \xi_1 + \cdots + \beta_s' \, \xi_s,$$

where

$$(12.28.2) \qquad\qquad \xi_i = \xi_i(x) \qquad\qquad (i = 1, 2, \cdots, s)$$

are *orthogonal* polynomials in x of the ith degree. The polynomials (12.28.2) satisfy the conditions for orthogonality, namely,

$$(12.28.3) \qquad \begin{cases} \displaystyle\sum_{j=1}^{n} \xi_{ij} = 0, & i = 1, 2, \cdots, s \\[2em] \displaystyle\sum_{j=1}^{n} \xi_{ij} \, \xi_{kj} = 0, & \text{for all } i \neq k. \end{cases}$$

Here $\xi_{ij} = \xi_i(x_j)$ is the value of ξ_i when $x = x_j$. The polynomials ξ_i of (12.28.2) are uniquely determined by (12.28.3) except for a constant factor.

12.29. Let $a', b_1', \cdots, b_s'$ be estimates of $\alpha', \beta_1', \cdots, \beta_s'$, respectively. An estimated regression function for (12.28.1) is

$$(12.29.1) \qquad g'(x) = a' + b_1'\xi_1 + \cdots + b_s'\xi_s.$$

The estimates a' and b_1' are found by minimizing

$$(12.29.2) \qquad S = \sum_{j=1}^{n} [y_j - g'(x_j)]^2$$

with respect to $a', b_1', \cdots, b_s'$, yielding

$$(12.29.3) \qquad \begin{cases} \displaystyle a' = \bar{y} = \sum_{j=1}^{n} y_j / n, \\[2em] \displaystyle b_i' = \sum_{j=1}^{n} y_j \xi_{ij} \Big/ \sum_{j=1}^{n} \xi_{ij}^2, & i = 1, \cdots, s. \end{cases}$$

Equation (12.29.3) gives the value of b_i', irrespective of the order s of the polynomial (12.29.1) being fitted.

12.30. The minimum value of S is

$$(12.30.1) \qquad S_s = \sum_{j=1}^n (y_j - \bar{y})^2 - \sum_{i=1}^n T_i$$

where

$$(12.30.2) \qquad T_i = b_i'^2 \sum_{j=1}^n \xi_{ij}^2 = \left[\sum_{j=1}^n y_i \xi_{ij}\right]^2 / \sum_{j=1}^n \xi_{ij}^2, \qquad i = 1, \cdots, s.$$

T_i represents the contribution of the polynomial ξ_i to the sum of squares $\sum_{j=1}^n (y_j - \bar{y})^2$. The reduction in the residual sum of squares resulting from increasing the order of the regression polynomial from $i - 1$ to i is T_i. Each T_i has a χ^2-distribution of one degree of freedom. The T_i are mutually independent.

An unbiased estimate of the residual variance σ_y^2 is $S_s/(n - s - 1)$.

12.31. **Case when the values of x are equally spaced.** When the x's are equally spaced, the calculation is simpler if a standardized scale is used. Let

$$(12.31.1) \qquad z_j = [x_j - (x_1 - w)]/w, \qquad \text{where } w = x_{j+1} - x_j.$$
$$(j = 1, 2, \cdots, n)$$

Then $z_1 = 1, z_2 = 2, \cdots, z_n = n$.

The orthogonal polynomials of order one, two and three, and higher orders, are found from

$$(12.31.2) \quad \begin{cases} \xi_0 = 1, \ \xi_1 = z - \bar{z}, \ \bar{z} = (n+1)/2, \\ \xi_2 = \xi_1^2 - (n^2 - 1)/12, \ \xi_3 = \xi_1^3 - (3n^2 - 7)\xi_1/20, \\ \xi_{r+1} = \xi_1 \xi_r - r^2(n^2 - r^2)\xi_{r-1}/4(4r^2 - 1). \end{cases}$$

The regression equation (12.29.1) corresponding to (12.28.1) is

$$(12.31.3) \qquad E(y \mid x) = a' + b_1' \xi_1 + \cdots + b_s' \xi_s,$$

with a', b_i' found from (12.29.3).

12.32. **Tables.** To facilitate the computation of (12.31.3), tables of orthogonal polynomials for z equally spread at unit intervals have been computed. Table 12.32.1 is such a table. More extensive tables of ϕ_i, ξ_i, λ_i, D_i are given in E. S. Pearson and H. O. Hartley, *Biometrika Tables for Statisticians*, Vol. I, 3d ed., Cambridge University Press, New York, 1966.

TABLE 12.32.1

Orthogonal polynomials

Notation

$$\phi_0(\xi_1) = 1,\ \phi_1(\xi_1) = \lambda_1\xi_1,\ \phi_2(\xi_1) = \lambda_2\xi_2,$$
$$\phi_3(\xi_1) = \lambda_3\xi_3,\ \cdots,\ \phi_r(\xi_1) = \lambda_r\xi_r,\ \lambda_0 = 1$$
$$\xi_{ij} = \xi_i(x_j) = \text{value of } \xi_i \text{ when } x = x_j, \text{ that is, when } z = j.$$
$$z_j = [x_j - (x_1 - w)]/w,\ w = x_{j+1} - x_j,\ z_j = j \text{ for } j = 1, 2, \cdots, n.$$

(12.32.1)
$$\xi_0 = 1,\ \xi_1 = z - \bar{z},\ \bar{z} = (n+1)/2,$$
$$\xi_2 = \xi_1^2 - (n^2 - 1)/12,\ \xi_3 = \xi_1^3 - (3n^2 - 7)\xi_1/20$$
$$\xi_{r+1} = \xi_1\xi_r - r^2(n^2 - r^2)\xi_{r-1}/4(4r^2 - 1)$$

Entries within blocks are values of $\phi_{ij} = \lambda_i\xi_{ij}$.

The coefficients λ_i are selected so that ϕ_{ij} are positive or negative integers throughout. When n is odd, $\lambda_1 = 1$; when n is even, $\lambda_1 = 2$. The λ_i appear on the bottom line of each block.

$D_i = \sum_{j=1}^{n} [\phi_{ij}]^2$ is the sum of squares for a full range of values of x_j, and z_j, $j = 1, \cdots, n$. D_i appears in the line above the bottom line of each block.

The arguments cover the full range $z = j = 1, 2, \cdots, n$, for $n = 3$ through 7; the upper-half range only for $n > 7$. The lower-half range can be written down by use of the relations

(12.32.2)
$$\begin{cases} \phi_i(\xi_1) = \phi_i(-\xi_1) & \text{when } i \text{ even} \\ \phi_i(\xi_1) = -\phi_i(-\xi_1) & \text{when } i \text{ odd.} \end{cases}$$

If $\sum_{i=1}^{s} A_i \phi_i(x)$ is fitted to $(x_1,y_1), \cdots, (x_n,y_n)$, the estimate of A_i is

(12.32.3)
$$A_i = \sum_{t=1}^{n} y_t \phi_i(x_t) / \sum_{t=1}^{n} \{\phi_i(x_t)\}^2$$

If $a' + \sum_{i=1}^{s} b_i'\xi_i$ is fitted to $(x_1,y_1), \cdots, (x_n,y_n)$, the estimates of a' and b_i' are given by Eq. (12.29.3)

(12.32.4)
$$A_0 = a',\ A_i = b_i'/\lambda_i,\ \phi_i = \lambda_i\xi_i.$$

		$n=3$		$n=4$		
x_j	$z_j = j$	ϕ_{1j}	ϕ_{2j}	ϕ_{1j}	ϕ_{2j}	ϕ_{3j}
x_1	1	-1	1	-3	1	-1
x_2	2	0	-2	-1	-1	3
x_3	3	1	1	1	-1	-3
x_4	4			3	1	1
D_i		2	6	20	4	20
λ_i		1	3	2	1	$\frac{10}{3}$

TABLE 12.32.1 (Continued)

$z_j = j$	ϕ_{1j}	ϕ_{2j}	ϕ_{3j}	ϕ_{4j}	
			$n = 5$		
1	-2	2	-1	1	
2	-1	-1	2	-4	
3	0	-2	0	6	
4	1	-1	-2	-4	
5	2	2	1	1	
D_i	10	14	10	70	
λ_i	1	1	$\dfrac{5}{6}$	$\dfrac{35}{12}$	

$z_j = j$	ϕ_{1j}	ϕ_{2j}	ϕ_{3j}	ϕ_{4j}	ϕ_{5j}
			$n = 6$		
1	-5	5	-5	1	-1
2	-3	-1	7	-3	5
3	-1	-4	4	2	-10
4	1	-4	-4	2	10
5	3	-1	-7	-3	-5
6	5	5	5	1	1
D_i	70	84	180	28	252
λ_i	2	$\dfrac{3}{2}$	$\dfrac{5}{3}$	$\dfrac{7}{12}$	$\dfrac{21}{10}$

$z_j = j$	ϕ_{1j}	ϕ_{2j}	ϕ_{3j}	ϕ_{4j}	ϕ_{5j}	ϕ_{6j}
				$n = 7$		
1	-3	5	-1	3	-1	1
2	-2	0	1	-7	4	-6
3	-1	-3	1	1	-5	15
4	0	-4	0	6	0	-20
5	1	-3	-1	1	5	15
6	2	0	-1	-7	-4	-6
7	3	5	1	3	1	1
D_i	28	84	6	154	84	924
λ_i	1	1	$\dfrac{1}{6}$	$\dfrac{7}{12}$	$\dfrac{7}{20}$	$\dfrac{77}{60}$

TABLE 12.32.1 (*Continued*)

$n = 8$					
$z_j = j$	ϕ_{1j}	ϕ_{2j}	ϕ_{3j}	ϕ_{4j}	ϕ_{5j}	ϕ_{6j}
1	-7	7	-7	7	-7	1
2	-5	1	5	-13	23	-5
3	-3	-3	7	-3	-17	9
4	-1	-5	3	9	-15	-5
5–8	Use relations (12.32.2).					
D_i	168	168	264	616	2,184	264
λ_i	2	1	$\frac{2}{3}$	$\frac{7}{12}$	$\frac{7}{10}$	$\frac{11}{60}$

$n = 9$					
$z_j = j$	ϕ_{1j}	ϕ_{2j}	ϕ_{3j}	ϕ_{4j}	ϕ_{5j}	ϕ_{6j}
1	-4	28	-14	14	-4	4
2	-3	7	7	-21	11	-17
3	-2	-8	13	-11	-4	22
4	-1	-17	9	9	-9	1
5	0	-20	0	18	0	-20
6–9	Use relations (12.32.2).					
D_i	60	2,772	990	2,002	468	1,980
λ_i	1	3	$\frac{5}{6}$	$\frac{7}{12}$	$\frac{3}{20}$	$\frac{11}{60}$

$n = 10$					
$z_j = j$	ϕ_{1j}	ϕ_{2j}	ϕ_{3j}	ϕ_{4j}	ϕ_{5j}	ϕ_{6j}
1	-9	6	-42	18	-6	3
2	-7	2	14	-22	14	-11
3	-5	-1	35	-17	-1	10
4	-3	-3	31	3	-11	6
5	-1	-4	12	18	-6	-8
6–10	Use relations (12.32.2).					
D_i	330	132	8,580	2,860	780	660
λ_i	2	$\frac{1}{2}$	$\frac{5}{3}$	$\frac{5}{12}$	$\frac{1}{10}$	$\frac{11}{240}$

TABLE 12.32.1 (*Continued*)

$z_j = j$	ϕ_{1j}	ϕ_{2j}	ϕ_{3j}	ϕ_{4j}	ϕ_{5j}	ϕ_{6j}
$n = 11$						
1	−5	15	−30	6	−3	15
2	−4	6	6	−6	6	−48
3	−3	−1	22	−6	1	29
4	−2	−6	23	−1	−4	36
5	−1	−9	14	4	−4	−12
6	0	−10	0	6	0	−40
7–11			Use relations (12.32.2).			
D_i	110	858	4,290	286	156	11,220
λ_i	1	1	$\dfrac{5}{6}$	$\dfrac{1}{12}$	$\dfrac{1}{40}$	$\dfrac{11}{120}$

$z_j = j$	ϕ_{1j}	ϕ_{2j}	ϕ_{3j}	ϕ_{4j}	ϕ_{5j}	ϕ_{6j}
$n = 12$						
1	−11	55	−33	33	−33	11
2	−9	25	3	−27	57	−31
3	−7	1	21	−33	21	11
4	−5	−17	25	−13	−29	25
5	−3	−29	19	12	−44	4
6	−1	−35	7	28	−20	−20
7–12			Use relations (12.32.2).			
D_i	572	12,012	5,148	8,008	15,912	4,488
λ_i	2	3	$\dfrac{2}{3}$	$\dfrac{7}{24}$	$\dfrac{3}{20}$	$\dfrac{11}{360}$

$z_j = j$	ϕ_{1j}	ϕ_{2j}	ϕ_{3j}	ϕ_{4j}	ϕ_{5j}	ϕ_{6j}
$n = 13$						
1	−6	22	−11	99	−22	22
2	−5	11	0	−66	33	−55
3	−4	2	6	−96	18	8
4	−3	−5	8	−54	−11	43
5	−2	−10	7	11	−26	22
6	−1	−13	4	64	−20	−20
7	0	−14	0	84	0	−40
8–13			Use relations (12.32.2).			
D_i	182	2,002	572	68,068	6,188	14,212
λ_i	1	1	$\dfrac{1}{6}$	$\dfrac{7}{12}$	$\dfrac{7}{120}$	$\dfrac{11}{360}$

13.1. This chapter describes the sample to sample variation of certain statistical quantities, including sample mean, sample variance, and others, for samples taken from a given population. The distributions of some commonly used functions related to sample parameters, such as the functions conventionally denoted by t, F, z, and χ^2, are also described. The application of these distributions to the problem of making inferences about population parameters from sample data is outlined in this and the next chapter.

The notation of §13.2 and §13.5 below will be used throughout the following chapters.

13.2. **Population, or universe.** A *population*, or *universe*, consists of an infinite or finite collection $\{x\}$ of values of a variable X. The same value of X may occur more than once in $\{x\}$. In the population $\{x\}$ let:

N = number of values in the population $\{x\}$, distinct or not, if the population is finite;

$f(x)$ = frequency distribution function of X, in the collection $\{x\}$, with:

$\bar{x}$ = mean of X;

σ_x = standard deviation of X; $\sigma_x{}^2$ = variance;

α_3 = skewness; α_4 = kurtosis.

It is assumed that the population is such that the last five quantities exist and are finite. Often a population $\{x\}$ is the whole set of values of some quantity under investigation.

13.3. **Sample.** A sample is a set of values x_1, x_2, x_3, $\cdots$ taken from a population $\{x\}$. In most practical problems, samples contain a finite number, n, of values x_1, x_2, x_3, $\cdots$, x_n. n is the *size* of the sample. Samples are assumed to arise from a "random" selection from the population $\{x\}$. "Random" here means that the sample values are obtained by chance in an unbiased manner, and the sequence of values x_1, x_2, $\cdots$, x_n are obtained by n independent unbiased repetitions of the experiment yielding the values. In such a random sample x_1, $\cdots$, x_i, $\cdots$, x_j, $\cdots$, x_n

the value of any observation x_i has no effect on the value of any other observation x_j.

13.4. In the set of all possible random samples $(x_1, x_2, \cdots , x_n)$ of size n, from a given population, the values $x_1, x_2, \cdots , x_n$ may then be considered as statistically independent. In this case the multiple variable $\mathbf{x} = (x_1, x_2, \cdots , x_n)$ has the frequency distribution function

$$(13.4.1) \qquad f^*(\mathbf{x}) \equiv f(x_1)f(x_2) \cdots f(x_n),$$

where the statistically independent variables x_i all have the same population frequency distribution function $f(x)$.

13.5. For the sample $x_1, x_2, \cdots , x_n$ let

$$\bar{x} = \text{mean} = \sum_{i=1}^{n} x_i/n;$$

$(13.5.1)$

$$s^2 = \text{variance} = \sum_{i=1}^{n} (x_i - \bar{x})^2/n; \qquad s = \text{standard deviation};$$

$$\alpha_3{}^* = \text{skewness}; \qquad\qquad \alpha_4{}^* = \text{kurtosis}.$$

13.6. *Sample moments* and other sample parameters are defined as in Chapter II. For example, the rth sample (central) moment m_r about the mean $\bar{x}$ of the sample $x_1, x_2, \cdots , x_n$ is

$$(13.6.1) \qquad m_r = (1/n) \sum_{i=1}^{n} (x_i - \bar{x})^r.$$

13.7. **Remark:** In this book the variance of a sample of size n has been defined as

$$(13.7.1) \qquad s^2 = \sum_{i=1}^{n} (x_i - \bar{x})^2/n.$$

Since an unbiased estimate of the population variance σ_x^2 is $s_1^2 = s^2 n/(n-1)$ some writers use s_1^2 instead of s^2 in calculations involving the variance of a sample and define the variance of a sample of size n as

$$(13.7.2) \qquad \sum_{i=1}^{n} (x_i - \bar{x})^2/(n-1)$$

The two usages must be carefully distinguished.

13.8. **Sampling distributions.** A sampling distribution of a sample statistic is the frequency distribution of the set of values of a sample statistic (such as the mean, $\bar{x}$) obtained from the set of all possible samples of a given size taken from a given population. Sampling distributions can be used to deduce information about the values of population parameters on the basis of values of sample parameters calculated from one or more samples.

13.9. The standard deviation of a sampling distribution of a sample statistic is sometimes called the *standard error* of the statistic.

13.10. **Stochastic convergence.** A random variable X_n which is defined for $n = 1, 2, 3, \cdots$ is said to *converge stochastically* to a value A if, for every fixed positive number ϵ,

$$(13.10.1) \qquad P(|X_n - A| > \epsilon) \to 0 \qquad \text{as} \qquad n \to \infty.$$

Estimators

13.11. **Estimators.** Consider a population with density function $f(x; \tilde{u}_1, \tilde{u}_2, \cdots)$, where x is the variable and $\tilde{u}_1, \tilde{u}_2, \cdots$ are parameters of the distribution. Suppose that from sample observations $x_1, x_2, \cdots, x_n$ one attempts to estimate the parameters $\tilde{u}_1, \tilde{u}_2, \cdots$. This means that one should find functions $u_1, u_2, \cdots$ of the observations such that the sampling distributions of these functions are concentrated as closely as possible to the true values $\tilde{u}_1, \tilde{u}_2, \cdots$ of the population parameters. Such functions are called *point estimators* (or *estimates*). For example, an estimator u of the population mean $\tilde{u} = \bar{x}$ is the sample mean

$$\bar{x} = (1/n) \sum_{i=1}^{n} x_i$$

of a sample $x_1, x_2, \cdots, x_n$.

13.12. **Unbiased estimator.** An estimator u, based on a sample $x_1, x_2, \cdots, x_n$, for a population parameter $\tilde{u}$ is said to be *unbiased* if the expected value of u, in the sampling distribution of u, is $\tilde{u}$; that is,

$$(13.12.1) \qquad E[u] = \int_U u \, f(u) \, du = \tilde{u},$$

where $f(u)$ is the sampling distribution (density) function for the set U of all possible values of u from all possible samples of size n from the population, and where the integral extends over the set U.

EXAMPLE 1. An unbiased estimator of the population mean $\bar{x}$ is the sample mean $\bar{x}$.

EXAMPLE 2. An unbiased estimator (or estimate) of the second (central) moment (i.e. the variance) of a population is $s^2 n/(n-1)$ where $s^2 = \sum_{i=1}^{n} (x_i - \bar{x})^2/n$.

EXAMPLE 3. An unbiased estimator of the third (central) moment of a population is $m_3 n^2/(n - 1)(n - 2)$ where $m_3 = (1/n) \sum_{i=1}^{n} (x_i - \bar{x})^3$ is the third central moment of the sample.

EXAMPLE 4.   An unbiased estimator of the fourth (central) moment of a population is

$$n[(n^2 - 2n + 3)m_4 - 3n(2n - 3)m_2^2] \div [(n - 1)(n - 2)(n - 3)],$$

where m_2 and m_4 are the second and fourth central moments of the sample.

13.13. Consistent estimator.   If an estimator u, based on a sample of size n, of $\tilde{u}$ converges stochastically to $\tilde{u}$ as $n \rightarrow \infty$, u is a *consistent estimator* (or consistent estimate) of $\tilde{u}$. An unbiased estimator is not necessarily a consistent estimator, and a consistent estimator is not necessarily an unbiased estimator.

EXAMPLE 1.   The sample variance $s^2 = \sum_{i=1}^{n} (x_i - \bar{x})^2/n$ is a consistent estimator for the population variance, but is not an unbiased estimator.

13.14. Efficient estimator.   Consider an estimator u such that the sampling distribution of $\sqrt{n}(u - \tilde{u})$ is normal, with mean zero and finite variance v, in the limit as the size n of the sample increases. Suppose that there exist a number of estimators $u^{(1)}, u^{(2)}, \cdots$. In this set of estimators there may exist one or more estimators which have associated with them a limiting variance v smaller than the limiting variances associated with any of the other estimators. Such an estimator is termed an asymptotically *efficient estimator* of $\tilde{u}$. For example, $\bar{x} = (1/n) \sum_{i=1}^{n} x_i$ is an efficient estimator of the population mean $\tilde{x}$.

13.15. An efficient estimator is a consistent estimator. Efficient estimators may, or may not, be unbiased for finite samples.

13.16. If u' is an efficient estimator and u is any other estimator such that in the limit as n increases $\sqrt{n}(u - \tilde{u})$ has a normal distribution with mean zero and finite variance, the *efficiency* of u is $E = v'/v$, where v' and v are the variances of the limiting distributions of $\sqrt{n}(u' - \tilde{u})$ and $\sqrt{n}(u - \tilde{u})$. $0 \leq E \leq 1$. An efficient estimator has efficiency equal to unity, and is a "most efficient estimator."

13.17. The *relative efficiency* of two estimators $u^{(1)}$ and $u^{(2)}$ of $\tilde{u}$, neither of which is necessarily an efficient estimator, is sometimes defined as A_1/A_2, where

$$E[(u^{(1)} - \tilde{u})^2] = A_1, \qquad E[(u^{(2)} - \tilde{u})^2] = A_2,$$

and the expected values A_1 and A_2 are determined from the sampling distributions of $u^{(1)}$ and $u^{(2)}$ for samples of a fixed size n.

13.18. Sufficient estimators. An estimator u for the population parameter $\tilde{u}$ is said to be *sufficient* if u contains all of the information regarding the parameter $\tilde{u}$ which can be obtained from the sample. This means that if u is a sufficient estimator, no further information about $\tilde{u}$ can be obtained from any sample statistic u' which is functionally independent of u; the conditional distribution of u', given u, is functionally independent of $\tilde{u}$. If the population distribution (density) function is $f(x, \tilde{u}_1, \tilde{u}_2, \cdots)$, a necessary and sufficient condition for a quantity u_1 based on a sample $x_1, x_2, \cdots, x_n$ to be a sufficient estimator of $\tilde{u}_1$ is that the joint distribution for the sample be factorable into the form

$$13.18.1) \qquad f(x_1, \tilde{u}_1, \tilde{u}_2, \cdots) \cdots f(x_n, \tilde{u}_1, \tilde{u}_2, \cdots)$$
$$= a(u_1, \tilde{u}_1, \tilde{u}_2, \cdots) \cdot b(x_1, \cdots, x_n, \tilde{u}_2, \tilde{u}_3, \cdots),$$

where b is functionally independent of $\tilde{u}_1$, and a is functionally independent of the sample values $x_1, \cdots, x_n$. Sufficient estimators occur only in special distributions.

EXAMPLE 1. The mean $\bar{x}$ of a sample $x_1, x_2, \cdots, x_n$ from a population having a Poisson distribution is a sufficient estimator of the distribution parameter m, since

$$m^{x_1}e^{-m}/x_1!)(m^{x_2}e^{-m}/x_2!) \cdots (m^{x_n}e^{-m}/x_n!) = [m^{n\bar{x}}e^{-nm}][x_1! \, x_2! \cdots x_n!]^{-1},$$

and the second factor on the right is independent of m.

13.19. Maximum likelihood estimators. Let $f(x_1, x_2, \cdots, x_n; \tilde{u})$ be the density function for a sample of size n taken from a population with parameter u, whose value $u = \tilde{u}$ is to be estimated. A *maximum likelihood estimator* of $\tilde{u}$ is a number $\hat{u}$ such that

$$13.19.1) \qquad f(x_1, x_2, \cdots, x_n; \hat{u}) > f(x_1, x_2, \cdots, x_n; u'),$$

where u' is any possible value of u different than $\hat{u}$. $\hat{u}$ may not exist. If f has a suitable analytic expression, and if $\hat{u}$ does exist, $\hat{u}$ is a solution of the equation $\partial f/\partial u = 0$, or, more conveniently in practice, $\partial \log f/\partial u = 0$. This last is called the *likelihood equation*.

13.20. If there is an efficient estimator of $\tilde{u}$, this estimator is a solution of the likelihood equation, and $\hat{u}$ is also a consistent estimator.

13.21. If there is a sufficient estimator u'' of $\tilde{u}$, any solution of the likelihood equation is functionally dependent on u''.

13.22. Under fairly general conditions, if $\hat{u}$ is a maximum likelihood estimator of $\tilde{u}$ based on a sample of size n, then as n increases, the distribution of $\sqrt{n}(\hat{u} - \tilde{u})$ approaches a normal distribution with mean zero and finite variance. Under the same conditions, $\hat{u}$ is an efficient estimator of $\tilde{u}$.

13.23. When the population depends on several parameters $u_1, u_2, \cdots$ there is a corresponding set of likelihood equations $\partial \log f/\partial u_i = 0$ $i = 1, 2, \cdots$ whose simultaneous solutions are the maximum likelihood estimators of $\tilde{u}_1, \tilde{u}_2, \cdots$.

EXAMPLE 1. Let a sample $x_1, x_2, \cdots, x_n$ be taken from a normally distributed population with mean zero and unknown standard deviation σ_x. The maximum likelihood estimator $\hat{\sigma}_x$ of σ_x is a solution of $\partial \log P/\partial \sigma = 0$, where $P = (2\pi\sigma_x^2)^{-n/2} \exp(-t^2/2\sigma_x^2)$ and $t^2 = x_1^2 + \cdots + x_n^2$ A solution is $\hat{\sigma}_x^2 = t^2/n = (x_1^2 + \cdots + x_n^2)/n$. In this case, $\hat{\sigma}_x^2$ is also a consistent, efficient, and sufficient estimator of σ_x^2.

More generally, if the normally distributed population has known mean $\bar{x}$ and unknown standard deviation σ_x, then the maximum likelihood estimator of σ_x is $\hat{\sigma}_x$, where

$$(13.23.1) \qquad \hat{\sigma}_x^2 = (1/n) \sum_{i=1}^{n} (x_i - \bar{x})^2.$$

EXAMPLE 2. For a sample of n values taken from a normal population with unknown mean and unknown variance, the maximum likelihood estimators of the mean and variance are

$$(13.23.2) \qquad \bar{x} = (1/n) \sum_{i=1}^{n} x_i,$$

$$(13.23.3) \qquad \hat{\sigma}_x^2 = (1/n) \sum_{i=1}^{n} (x_i - \bar{x})^2.$$

EXAMPLE 3. For a sample of n values taken from a population whose distribution is as indicated below:
The maximum likelihood estimator for the parameter p in a binomial distribution (8.1.1) is the observed relative frequency; for m in a Poisson distribution (9.1.1), and for θ in an exponential distribution (9.9.1) is (13.23.2).

EXAMPLE 4. A sample $x_1, \cdots, x_n$ of n independent observations on x is taken from a binomial population whose p.f. is $f(x) = p^x(1-p)^{1-x}$ $x = 0, 1$, and whose mean and variance are $\mu = p$, $\sigma^2 = p(1-p)$. Then $T = x_1 + \cdots + x_n$ has mean and variance

$$E(T) = np, \qquad \text{var}(T) = np(1-p).$$

Let $\hat{p} = T/n$ be the maximum likelihood estimator of p. Then

$$E(\hat{p}) = p, \qquad \text{var}(\hat{p}) = p(1 - p)/n.$$

$\hat{p}$ is an unbiased estimator of p with variance $p(1 - p)/n$.

Remark. In applications to acceptance sampling x might be taken as 1 when the object sampled is defective, and 0 when nondefective.

Sampling distributions

13.24. Gamma distribution. If X is such that

$$(13.24.1) \qquad P(X \leqq x') = F(x') = \int_{-\infty}^{x'} f(x)\, dx = \int_{0}^{x'} f(x)\, dx,$$

where

$$(13.24.2) \quad f(x) = [b^{a+1} \cdot \Gamma(a + 1)]^{-1} x^a e^{-x/b}, \ b > 0, a > -1, \text{ for } x > 0,$$
$$= 0, \qquad\qquad\qquad\qquad\qquad \text{for } x \leqq 0,$$

in which a and b are constants, X is said to have a *Gamma distribution.* $F(x')$ is the *incomplete Gamma function.* Here Γ is the Gamma function.

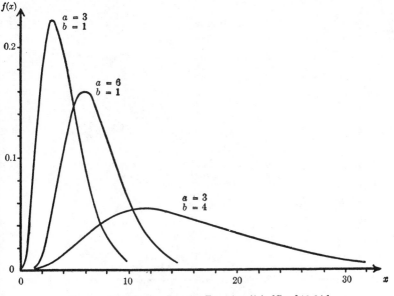

Gamma Distribution Density Function $f(x)$. [See §13.24.]
FIGURE 13.24.1

(See §18.3.) [For extensive tables of this function see Karl Pearson, *Tables of the Incomplete Gamma Function*, Cambridge University Press, 1922.] (See §13.26.)

Properties. The mean and variance of X are $b(a + 1)$ and $b^2(a + 1)$, respectively.

13.25. The Beta distribution. If X is such that

$$(13.25.1) \qquad P(X \leq x') = F(x') = \int_{-\infty}^{x'} f(x)\, dx = \int_{0}^{x'} f(x)\, dx,$$

where

$$f(x) = Ax^a(1 - x)^b, \quad a > -1, \quad b > -1, \quad 0 < x < 1,$$

$$(13.25.2) \qquad = 0, \qquad \text{elsewhere,}$$

$$A = \Gamma(a + b + 2) \div [\Gamma(a + 1) \cdot \Gamma(b + 1)],$$

in which a and b are constants, X is said to have a *Beta distribution*. $F(x')$ is an *incomplete Beta function ratio*. [For extensive tables of this function see Karl Pearson, *Tables of the Incomplete Beta Function*, Cambridge University Press, 1932.] (See §18.7 , §13.68 ; also, Table III.)

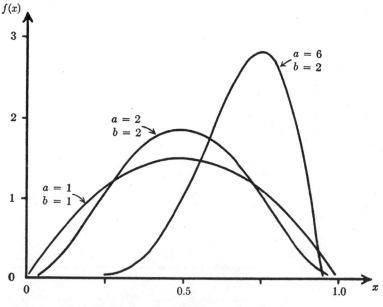

Beta Distribution Density Function $f(x)$. [See §13.25.]
FIGURE 13.25.1

Properties. The mean of X is $\nu_1 = (a + 1)/(a + b + 2)$. The rth moment about the origin is

(13.25.3) $\quad \nu_r = [\Gamma(a + b + 2) \cdot \Gamma(a + r + 1)]$

$$\div [\Gamma(a + b + r + 2) \cdot \Gamma(a + 1)].$$

The variance is

$$\nu_2 - \nu_1{}^2 = (a + 1)(b + 1) \div (a + b + 2)^2(a + b + 3).$$

(13.25.4)

$$F(1) = \int_0^1 f(x) \, dx = 1.$$

The quantity $A = 1 \div B(a + 1, b + 1)$, where $B(a,b)$ is the *Beta function* of a and b. [See §18.5.]

13.26. **The χ^2-distribution ("Chi-squared").** The random variable W is said to possess the χ^2-distribution if the probability that W takes a value in a small interval dw containing a value w is given by the frequency distribution function

(13.26.1) $\quad f_m(w) \, dw = [(w^{(m/2)-1})/(2^{m/2}\Gamma(m/2))] \exp{(-w/2)} \, dw, \, w > 0,$

where m is a positive integer often referred to as the number of *degrees of freedom.* (§§13.31, 13.32.) The probability that $W \leqq w_0$ is

(13.26.2) $\quad P(W \leqq w_0) = \int_{y=0}^{y=w_0} f_m(y) \, dy.$

The density $f_m(w)$ given in (13.26.1) is the Gamma distribution density (13.24.2) with $x = w$, $b = 2$, $a = (m/2) - 1$.

Properties of the χ^2-distribution $f_m(w)$:

(13.26.3)
$$\int_0^\infty f_m(w) \, dw = 1.$$

$$\text{Mean} = \mu = m. \qquad \text{Variance} = \sigma^2 = 2m.$$

Characteristic function for (13.26.1) is $(1 - 2it)^{-m/2}$.

rth moment μ_r about mean m:	rth moment ν_r about $w = 0$:
$\mu_1 = 0.$	$\nu_1 = m.$
$\mu_2 = 2m = \sigma^2.$	$\nu_2 = m(m + 2).$
$\mu_3 = 8m.$	$\nu_3 = m(m + 2)(m + 4).$
$\mu_4 = 12m^2 + 48m.$	$\nu_4 = m(m + 2)(m + 4)(m + 6).$

(13.26·4) . . . .

$$\nu_r = m(m + 2)(m + 4) \cdots (m + 2r - 2).$$

(13.26.5) $$\text{Skewness} = \alpha_3 = \mu_3/\sigma^3 = 2\sqrt{2/m}.$$

$$\text{Kurtosis} = \alpha_4 = \mu_4/\sigma^4 = 3 + (12/m).$$

13.27. As m increases, if W has the distribution $f_m(w)$:

(a) The distribution of W approaches the normal distribution with standard deviation $\sqrt{2m}$ and mean m;

(b) The distribution of $\sqrt{2W}$ approaches the normal distribution with standard deviation 1 and mean $\sqrt{2m-1}$.

If $m > 30$, in practical calculations it is often sufficient to use these normal distributions to approximate the actual distributions.

13.28. **Reproductive property.** If $W_1, W_2, \cdots, W_k$ are independently distributed according to χ^2-distributions with $m_1, m_2, \cdots, m_k$ degrees of freedom, respectively, then $\sum_1^k W_i$ is distributed according to a χ^2-

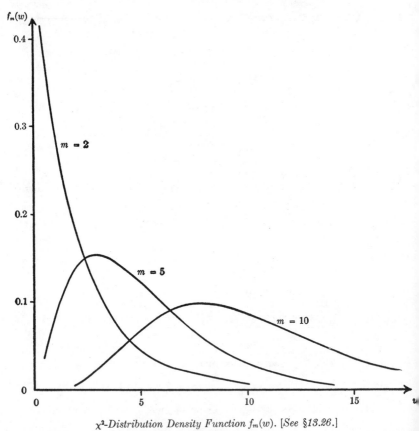

χ^2-*Distribution Density Function* $f_m(w)$. [*See §13.26.*]

FIGURE 13.29.1

distribution (Eq. 13.26.1) with $m = \sum_1^k m_i$ degrees of freedom.

13.29. Examples of the χ^2-distribution are shown in Fig. 13.29.1 for three representative degrees m of freedom.

13.30. Tables of the χ^2-distribution are available showing values of χ_0^2 such that the probability that χ^2 is greater than or equal to χ_0^2, namely

$$(13.30.1) \qquad P(\chi^2 \geqq \chi_0^2) = \int_{y=\chi_0^2}^{y=\infty} f_m(y) \, dy,$$

takes particular values 0.99, 0.98, 0.95, $\cdots$, 0.001. [See Table XIV.]

13.31. **Degrees of freedom.** In selecting a sample of n items from a population, the process may be considered as that of selecting a value for each of n variables, each of the variables ranging over all values in the population. The selection thus involves n degrees of freedom of choice. In examination of quantities derived from samples, the freedom of choice in selecting the samples may be restricted by relations between the sample values. For example, one may be considering samples $x_1, x_2,$ $\cdots$, x_n which have a fixed mean, $\bar{x}$. Then $\sum_{i=1}^{n} x_i = n\bar{x}$, and if $n - 1$ of the values of x_i are selected, the remaining value is determined. In this case, there are $n - 1$ degrees of freedom in selecting the sample values. More generally, if k independent linear relations are imposed on the n values $x_1, x_2, \cdots$, x_n in a sample, then only $n - k$ of the sample values can be selected freely, and there are $n - k$ degrees of freedom.

13.32. In this book, degrees of freedom are associated with quadratic forms. For example, the quadratic expression $A = \sum_{i=1}^{n} (x_i - \bar{x})^2$ has $n - 1$ degrees of freedom if the linear relation $\sum_{i=1}^{n} x_i = n\bar{x}$ is the only restriction on the values x_i. In this example, a non-singular linear transformation of the variables x_i would enable the expression A to be rewritten as the sum of the squares of $n - 1$, and no fewer, new independent variables. In general, if there are k independent linear relations imposed on the n variables in a quadratic form, then by a suitable non-singular linear transformation of variables, the quadratic form can be expressed as the sum of the squares of $n - k$, and no fewer, of the new independent variables. The number of degrees of freedom, $n - k$, is thus the same as the *rank*, $n - k$, of the original quadratic form.

13.33. **Distribution of sample means.** In the sampling distribution of means from samples of size n, let:

$\bar{x}$ = mean of a sample of size n,

$g(\bar{x})$ = frequency distribution function of sample means $\bar{x}$, with:

$\bar{\bar{x}}$ = mean of sample means;

$\sigma_{\bar{x}}$ = standard deviation of sample means;

$\alpha_{3,\bar{x}}$ = skewness of the distribution of sample means;

$\alpha_{4,\bar{x}}$ = kurtosis of the distribution of sample means.

13.34. Infinite population.   In the case of any infinite population, the last four quantities are related to population parameters as follows:

$$(13.34.1) \qquad\qquad \bar{\bar{x}} = \bar{x}.$$

$\bar{x}$ approaches $\bar{\bar{x}}$ stochastically as the size, n, of the sample is increased. An unbiased estimate of $\bar{\bar{x}}$ on the basis of one sample is $\bar{x}$. An unbiased estimate of $\bar{\bar{x}}$ on the basis of k samples with sample means $\bar{x}_1, \bar{x}_2, \cdots, \bar{x}_k$ is $\sum_{i=1}^{k} \bar{x}_i / k$.

$$(13.34.2) \qquad\qquad \sigma_{\bar{x}} = \sigma_x / \sqrt{n}.$$

$$(13.34.3) \qquad\qquad \alpha_{3,\bar{x}} = \alpha_3 / \sqrt{n}.$$

$$(13.34.4) \qquad\qquad \alpha_{4,\bar{x}} = 3 + (\alpha_4 - 3)/n.$$

13.35. Finite population.   In the case of a finite population, containing N values, the analogues of the preceding relations are as follows:

$$(13.35.1) \qquad \bar{\bar{x}} = \bar{x}.$$

$$(13.35.2) \qquad \sigma_{\bar{x}} = [\sigma_x / \sqrt{n}][(N - n)/(N - 1)]^{1/2}.$$

$$(13.35.3) \qquad \alpha_{3,\bar{x}} = \alpha_3(N - 2n) \sqrt{N - 1} \div \sqrt{n}\,(N - 2)\sqrt{N - n}.$$

$$(13.35.4) \qquad \alpha_{4,\bar{x}} = (N - 1)H \div n(N - 2)(N - 3)(N - n),$$

where

$$H = (N^2 - 6Nn + N + 6n^2)\alpha_4 + 3N(N - n - 1)(n - 1)\alpha_2^{\,2}.$$

As $N \to \infty$, these quantities approach the corresponding quantities given above for the infinite population.

Sampling from finite population.   From a population of N items select a random sample $x_1, \cdots, x_n$ of n items without replacement. Np of the population items possess a certain attribute A, and $N(1 - p)$ do not possess A. Suppose $S = np^*$ of the sample items possess A. p and p^* are called the *population proportion* and *sample proportion*, respectively. Let $\bar{x}$ and $\sigma_x^{\,2}$ be the population mean and variance, respectively.

The mean and variance of the sampling distribution of the sample mean $\bar{x}$ are

(13.35.5) $$E(\bar{x}) = \hat{x}$$

(13.35.6) $$\sigma_{\bar{x}}^2 = [\sigma_x^2/n][(N - n)/(N - 1)]$$

The distribution of $\bar{x}$ is approximately normal.

The mean and variance of the sampling distribution of p_s^* are

(13.35.7) $$E(p^*) = p$$

(13.35.8) $$\sigma_{p*}^2 = [p(1 - p)/n][(N - n)/(N - 1)]$$

The distribution of p^* is approximately normal.

In the case of random sampling from a finite population, the sampling being done with replacement (or from an infinite population), (13.35.6) and (13.35.8) become, respectively,

(13.35.9) $$\sigma_{\bar{x}}^2 = \sigma_x^2/n,$$

(13.35.10) $$\sigma_{p*}^2 = p(1 - p)/n.$$

13.36. **Distribution of sample means in normal population.** If an infinite population of values x is normally distributed (see §10.3) with mean $\hat{x}$ and standard deviation σ_x, then the sampling distribution of means $\bar{x}$ is normally distributed with mean $\hat{x}$ and standard deviation $\sigma_x/\sqrt{n}$. In this case if the population values $\hat{x}$ and σ_x are known, the probability p that $\bar{x}$ for any one sample will fall in the interval $\hat{x} - \delta \leqq \bar{x} \leqq \hat{x} + \delta$ is

(13.36.1) $$p = \int_{-\beta}^{\beta} \psi(u)\, du,$$

where

(13.36.2) $$\beta = \delta\sqrt{n}/\sigma_x,$$

and where

(13.36.3) $$u = (\bar{x} - \hat{x})\sqrt{n}/\sigma_x, \qquad \psi(u) = (1/\sqrt{2\pi}) \exp\left[-u^2/2\right].$$

(See §10.3.)

This is true whether the sample is large or small in size. Note that the deviation u is normally distributed with mean zero and variance one. The probability density of u is $\psi(u)$. (See §10.3.) If $y_i = (x_i - \hat{x})/\sigma_x$, $\bar{y} = (\bar{x} - \hat{x})/\sigma_x$, the probability density of the mean $\bar{y}$ is

(13.36.4) $$\sqrt{n/2\pi} \exp\left[-n\bar{y}^2/2\right];$$

the probability density of the mean $\bar{x}$ is

$$(13.36.5) \qquad (\sqrt{n}/\sqrt{2\pi}\,\sigma_x)\exp\left[-n(\bar{x}-\hat{x})^2/2\sigma_x^2\right].$$

If the population values $\hat{x}$ or σ_x are not known, p can be estimated on the basis of sample parameters alone, using methods of the sort given in §13.62.

13.37. Distribution of sample means when population is not normal. If the population is not normally distributed, the distribution of sample means is found in practice often to be approximately normal. Suppose the frequency distribution $g(\bar{x})$ of sample means expressed in terms of $u = (\bar{x}-\hat{x})\sqrt{n}/\sigma_x$ is $g_1(u)$. If the population $\{x\}$ is not normally distributed, and the variance of the population is finite the frequency distribution $g_1(u)$ of sample means $\bar{x}$ approaches the normal distribution $\psi(u)$, given in §10.3, as the sample size n increases indefinitely. (See §13.49.)

13.38. Distribution of sample variances and sample standard deviations. In the sampling distribution of sample variances s^2 of samples of size n, taken from a population of values x, let:

$H(s^2)$ = frequency distribution function of values of s^2;
$\overline{s^2}$ = mean of sample variances s^2;
σ_{s^2} = standard deviation of sample variances.

In the sampling distribution of standard deviations s of samples of size n, let:

$h(s)$ = frequency distribution of values of s;
$\bar{s}$ = mean of sample standard deviations s;
σ_s = standard deviation of sample standard deviations.

For any population,

$$(13.38.1) \qquad \overline{s^2} = \sigma_x^2(n-1)/n.$$

An unbiased estimate of the population variance σ_x^2 is $\overline{s^2}n/(n-1)$. An unbiased estimate of σ_x^2 based on k samples having sizes $n_1, n_2, \cdots, n_k$ and variances $s_1^2, s_2^2, \cdots, s_k^2$ is

$$(13.38.2) \qquad \left[\sum_{i=1}^{k} n_i s_i^2\right] \div \left[\sum_{i=1}^{k} n_i - k\right].$$

13.39. Case when population is normally distributed. If the population $\{x\}$ is normally distributed, with mean $\hat{x}$ and variance σ_x^2, then the frequency (density) distributions for the sample variance s^2 and sample standard deviation s are, respectively:

$$(13.39.1) \qquad H(s^2) = c[s^2]^{(n-3)/2}\exp\left[-ns^2/2\sigma_x^2\right],$$

(13.39.2) $$h(s) = 2cs^{n-2} \exp\left[-ns^2/2\sigma_x^2\right],$$

where

(13.39.3) $$c = [n/2\sigma_x^2]^{(n-1)/2} \div \Gamma[(n-1)/2].$$

Here Γ is the Gamma function. (See §18.3.)

13.40. Both $H(s^2)$ and $h(s)$ approach normal (density) distributions as n increases, and may sometimes be considered approximately normal if $n > 30$. Both $H(s^2)$ and $h(s)$ are independent of the population mean.

The probability that s^2 is less than or equal to x is

(13.40.1) $$P(s^2 \leq x) = \int_{s^2=0}^{s^2=x} H(s^2) \, d(s^2);$$

the probability that s is less than or equal to y is

(13.40.2) $$P(s \leq y) = \int_{s=0}^{s=y} h(s) \, ds.$$

THEOREM 1. If the population is normally distributed the sample mean $\bar{x}$ and sample variance s^2 are statistically independent and $\bar{x}$ is normally distributed with mean $\bar{x}$ and standard deviation $\sigma_x/\sqrt{n}$, while ns^2/σ_x^2 is distributed according to the χ^2-distribution, (Eq. 13.26.1) with $n-1$ degrees of freedom. That is,

(13.40.3) $$P(\bar{x} \leq x) = (\sqrt{n}/\sqrt{2\pi} \, \sigma_x) \int_{t=-\infty}^{t=x} \exp\left[-(t-\bar{x})^2 n/2\sigma_x^2\right] dt,$$

(13.40.4) $$P((ns^2/\sigma_x^2) \leq x) = \int_{w=0}^{w=x} f_{n-1}(w) \, d(w),$$

where $f_{n-1}(w)$ is given by Eq. (13.26.1).

13.41. The density $H(s^2)$ given in (13.39.1) is the Gamma distribution density (13.24.2) with $x = s^2$, $a = (n-3)/2$, $b = 2\sigma_x^2/n$. In this case (13.24.1) reduces to (13.40.1). The density $H(s^2)$ is also identical to the distribution density for $z_{10} = s^2$ given in §13.55 in terms of the χ^2-distribution.

Similarly, the density $h(s)$ given in (13.39.2) is related to the Gamma distribution; Eq. (13.24.1) reduces to (13.40.2) when $x = s^2$, $a = (n-3)/2$, $b = 2\sigma_x^2/n$.

13.42. Examples of the distribution $h(s)$ of sample standard deviations and of the distribution $H(s^2)$ are shown in Figs. 13.42.1 and 13.42.2 for the case of a normally distributed population with $\sigma_x = 1$ and mean $\bar{x}$

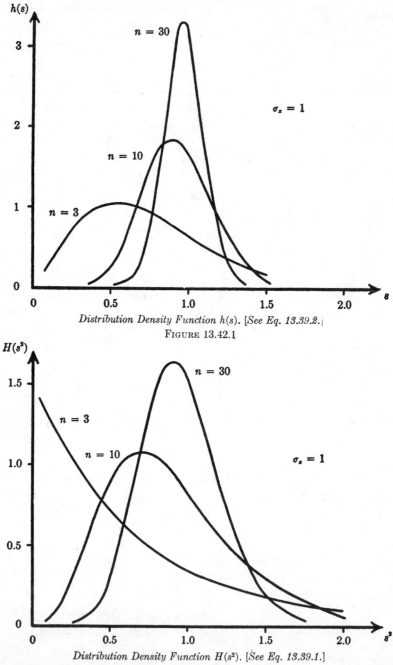

Distribution Density Function h(s). [See Eq. 13.39.2.]
FIGURE 13.42.1

Distribution Density Function H(s²). [See Eq. 13.39.1.]
FIGURE 13.42.2

13.43. Relations between the variances of sample and population.
If $\bar{s^2}$ is the mean of sample variances, and $\bar{s}$ the mean of sample standard deviations s, then

(13.43.1) $$\bar{s^2} = \sigma_x^2(n - 1)/n.$$

(13.43.2) $$\bar{s} = b(n)\,\sigma_x,$$

where

(13.43.3) $$b(n) = (2/n)^{1/2}\Gamma(n/2) \div \Gamma[(n - 1)/2].$$

Some representative values of $b(n)$ are:

n	2	3	4	5	6	7	8	9
$b(n)$	.564	.724	.798	.841	.869	.888	.903	.914

n	10	20	30	50	100
$b(n)$	.923	.962	.975	.985	.992

(13.43.4) $$\sigma_{s^2} = [2(n - 1)]^{1/2}\sigma_x^2/n.$$

(13.43.5) $$\sigma_s = [((n - 1)/n) - [b(n)]^2]^{1/2}\sigma_x.$$

σ_s is sometimes taken as approximately $\sigma_x[1/2n]^{1/2}$.

The modal value of s is $\hat{s} = \sigma_x[(n - 2)/n]^{1/2}$.

The modal value of s^2 is $\sigma_x^2[(n - 3)/n]$.

Modified distributions related to the distribution of sample standard deviations and the distribution of sample means are the Fisher z-distribution, (§13.71), the Student t-distribution, (§13.59), as well as the q-distribution, q^2-distribution, $\cdots$, given below.

13.44. The q-distribution. If in Eq. (13.40.2), which gives the distribution of the standard deviation s of a sample taken from a normal population, one sets $q = \sqrt{n}\,s$, it follows that the probability that q is less than or equal to z is

(13.44.1) $$P(q \leq z) = \int_{q=0}^{q=z} Q(q)\,dq,$$

where

(13.44.2) $$Q(q) = 2cn^{-(n-1)/2}q^{n-2}\exp[-q^2/2\sigma_x^2],$$

(13.44.3) $$c = [n/2\sigma_x^2]^{(n-1)/2} \div \Gamma[(n - 1)/2].$$

The distribution (13.44.1) is known as the *q-distribution*. The mean and variance of q are, respectively:

$$(13.44.4) \qquad \bar{q} = \sqrt{n} \cdot b(n) \cdot \sigma_x \approx \sqrt{(2n - 3)/2}\, \sigma_x,$$

$$(13.44.5) \qquad \sigma_q^2 = [n - 1 - n[b(n)]^2]\sigma_x^2 \approx \sigma_x^2/2.$$

Sometimes (13.44.1) is written with $f = n - 1$. The parameter f is sometimes called the *degree of freedom*.

13.45. Distribution of unbiased estimate of σ_x.

An unbiased estimate of the population standard deviation σ_x is

$$(13.45.1) \qquad u = q/\sqrt{n - 1} = s\sqrt{n/(n - 1)}.$$

The distribution of u for a normal population follows from Eq. (13.40.2). The probability that u is less than or equal to u_0 is

$$(13.45.2) \qquad P(u \leq u_0) = \int_0^{u_0} \zeta(u)\, du,$$

where

$$(13.45.3) \qquad \zeta(u) = 2cn^{-(n-1)/2}(n - 1)^{(n-1)/2}u^{n-2} \exp\left[-(n - 1)u^2/2\sigma_x^2\right].$$

[For definition of c and $b(n)$ see (13.44.3) and (13.43.3).] The mean and variance of u are, respectively:

$$(13.45.4) \qquad \bar{u} = \sqrt{n/(n - 1)} \cdot b(n) \cdot \sigma_x \approx \sigma_x\sqrt{(2n - 3)/2(n - 1)}.$$

$$(13.45.5) \qquad \sigma_u^2 = [1 - [n/(n - 1)][b(n)]^2]\sigma_x^2 \approx \sigma_x^2/2(n - 1).$$

13.46. Distribution of q^2.

The distribution of $q^2 = ns^2$ for a normal population follows from Eq. (13.40.1). The probability that q^2 is less than or equal to q_0^2 is

$$(13.46.1) \qquad P(q^2 \leq q_0^2) = \int_{q^2=0}^{q^2=q_0^2} \xi(q^2) \cdot d(q^2),$$

where

$$(13.46.2) \qquad \xi(q^2) = cn\,[q^2/n]^{(n-3)/2} \exp\left[-q^2/2\sigma_x^2\right].$$

[For definition of c see (13.44.3).] The mean and variance of q^2 are, respectively,

$$(13.46.3) \qquad \overline{q^2} = \sigma_x^2(n - 1), \qquad \sigma_{q^2}^2 = 2(n - 1)\sigma_x^4.$$

The fraction $w = \chi^2 = q^2/\sigma_x^2 = ns^2/\sigma_x^2$ has the χ^2-distribution given in Eq. (13.26.1) with $m = n - 1$.

13.47. Distribution of unbiased estimate of σ_x^2, normal population. Case 1. Population mean $\bar{x}$ unknown. An unbiased estimate of σ_x^2 is

$$(13.47.1) \qquad u^2 = s^2 n/(n-1) = q^2/(n-1).$$

The distribution of u^2 follows from Eq. (13.40.1). The probability that u^2 is less than or equal to u_0^2 is

$$(13.47.2) \qquad P(u^2 \leqq u_0^2) = \int_{u^2=0}^{u^2=u_0^2} z(u^2)\, d(u^2),$$

where

$$(13.47.3) \qquad z(u^2) = c(u^2)^{(n-3)/2}[(n-1)/n]^{(n-1)/2} \exp\left[-(n-1)u^2/2\sigma_x^2\right].$$

For definition of c, see (13.44.3). The mean and variance of u^2 are, respectively:

$$(13.47.4) \qquad \overline{u^2} = \sigma_x^2, \qquad \sigma_{u^2}^2 = 2\sigma_x^4/(n-1).$$

13.48. Case 2. Population mean $\bar{x}$ known. The estimate $\acute{\sigma}_x^2 = (1/n) \sum_{i=1}^{n} (x_i - \bar{x})^2$ for σ_x^2 has the χ^2-distribution density function $(n/\sigma_x^2) \cdot f_n(nz_9/\sigma_x^2)$, with $z_9 = \acute{\sigma}_x^2$, as given in §13.55.

Limit theorems

13.49. Central limit theorem. For an arbitrary population with mean $\bar{x}$ and finite variance σ_x^2, the cumulative distribution function $G_n(z)$ of the sampling distribution of

$$(13.49.1) \qquad z = (\bar{x} - \bar{\bar{x}}) \sqrt{n}/\sigma_x,$$

for samples of size n, approaches the normal distribution (uniformly in z) as $n \rightarrow \infty$, that is,

$$(13.49.2) \qquad \lim_{n \to \infty} G_n(z) = (1/\sqrt{2\pi}) \int_{-\infty}^{z} e^{-u^2/2}\, du.$$

In other words, if a population has a finite variance σ_x^2 and mean $\bar{x}$, then, as the sample size n increases, the distribution of the sample mean $\bar{x}$ approaches the normal distribution with mean $\bar{x}$ and variance σ_x^2/n. The importance of this theorem lies in the fact that the theorem states nothing about the form of the population distribution function.

This theorem is closely related to a similar theorem (§13.62) pertaining to the t-distribution. The latter involves the sample standard deviation s in place of the population standard deviation σ_x.

This theorem is also closely related to a similar theorem (§13.37) pertaining to the normal distribution of sample means.

13.50. The theorem (13.49.2) is a special case of a more general *central limit theorem*, one special form of which may be stated as follows:

THEOREM 1. Let $x_1, x_2, x_3, \cdots, x_n$ be a series of statistically independent random variables having arbitrary distributions for which the means $\mu_1, \cdots, \mu_n$ and variances $\sigma_1^2, \cdots, \sigma_n^2$ exist. If n is sufficiently large, the sum $\sum_{i=1}^n x_i$ will be approximately normally distributed with mean μ and variance σ^2, whether the x_i's are normally distributed or not and where the mean and variance of the sum are given by $\mu = \sum_{i=1}^n \mu_i$ and $\sigma^2 = \sum_{i=1}^n \sigma_i^2$, respectively.

13.51. Tchebycheff's Theorem. If $g(x)$ be any non-negative function of the random variable x, then for every positive number K the probability that $g(x) \geqq K$ does not exceed $E[g(x)] \div K$, where $E[g(x)]$ is the expected value of $g(x)$; that is,

$$(13.51.1) \qquad P(g(x) \geqq K) \leqq E[g(x)] \div K.$$

13.52. Special cases. If $\bar{x}$ and σ_x denote the mean and standard deviation of the variable x in any given distribution, then for every positive number δ,

$$(13.52.1) \qquad P(|x - \bar{x}| \geqq \delta\sigma_x) \leqq 1/\delta^2,$$

or

$$(13.52.2) \qquad P(\bar{x} - \delta\sigma_x < x < \bar{x} + \delta\sigma_x) > 1 - 1/\delta^2.$$

In case the distribution is unimodal and continuous

$$(13.52.3) \qquad P(|x - x_0| \geqq \delta B) \leqq 4/9\delta^2,$$

where x_0 is the mode and $B^2 = \sigma_x^2 + (x_0 - \bar{x})^2$. Also, for every $\delta > |s|$,

$$(13.52.4) \qquad P(|x - \bar{x}| \geqq \delta\sigma_x) \leqq \frac{4(1 + s^2)}{9(\delta - |s|)^2},$$

where $s = (\bar{x} - x_0)/\sigma_x$. In case the distribution is symmetric, $\bar{x} = x_0$, and

$$P(|x - \bar{x}| \geqq \delta\sigma_x) \leqq 4/9\delta^2.$$

13.53. If $\bar{x}$ is the mean of a sample of size n taken from a population whose mean is $\bar{x}$ and whose standard deviation is σ_x, then the probability that the sample mean differs from the population mean by no more than a quantity b is subject to the inequality

$$(13.53.1) \qquad P(-b < \bar{x} - \bar{x} < b) > 1 - \sigma_x^2/nb^2.$$

Relations (13.51.1) to (13.53.1) are known as *Tchebycheff's inequality*.

EXAMPLE. A random sample of size n is drawn from a population having a binomial distribution in which $\mu = p$, $\sigma^2 = p(1 - p)$. The number of defective items observed in the sample is x. From Eq. (13.53.1), with $b > 0$ arbitrarily small, $P(|(x/n) - p| < b) \to 1$. Thus, as n increases one may become more certain that the observed fraction x/n of defectives is a good estimate of the true fraction of defectives in the population.

13.54. **Law of large numbers.** Relation (13.53.1) indicates that by selecting n large enough, the probability that the sample mean will fall within a small interval about the population mean can be made as near to one as desired. This result is known as the *law of large numbers*.

Further sampling distributions

13.55. **Relation of χ^2-distribution to samples from a normal population.** Let $x_1, x_2, \cdots, x_n$ be a random sample from a normally distributed population, with sample mean $\bar{x}$ and sample variance s^2. The population mean is $\bar{x}$ and the population variance is σ_x^2. Let $f_m(w)$ be the χ^2-distribution density function defined in Eq. (13.26.1). If z_j, $j = 1, 2, \cdots$ is a function of the sample as indicated in the first column of the table below, then the probability that z_j is less than or equal to any selected value z_0 is

$$(13.55.1) \qquad P(z_j \leqq z_0) = \int_{z_j=0}^{z_j=z_0} f \, dz_j,$$

where f is the sampling distribution function given in the last column of Table 13.55.1 below corresponding to the value of z_j used. The subscript of f in the last column denotes the number of degrees of freedom.

TABLE 13.55.1

Function z_j of the sample	Population mean, $\bar{x}$	Distribution density function f of z_j
$z_1 = \sum_{i=1}^{n} x_i^2$	0	$(1/\sigma_x^2) \cdot f_n(z_1/\sigma_x^2)$
$z_2 = (1/n) \sum_{i=1}^{n} x_i^2$	0	$(n/\sigma_x^2) \cdot f_n(nz_2/\sigma_x^2)$
$z_3 = (1/\sigma_x^2) \sum_{i=1}^{n} x_i^2$	0	$f_n(z_3)$

TABLE 13.55.1 (*Continued*)

Function z_j of the sample	Population mean, $\bar{x}$	Distribution density function f of z_j
$z_4 = \left[\sum_{i=1}^{n} x_i^2 \right]^{1/2}$	0	$(2z_4/\sigma_x^2) \cdot f_n(z_4^2/\sigma_x^2)$
$z_5 = \left[(1/n) \sum_{i=1}^{n} x_i^2 \right]^{1/2}$	0	$(2nz_5/\sigma_x^2) \cdot f_n(nz_5^2/\sigma_x^2)$
$z_6 = n(\bar{x} - \tilde{x})^2/\sigma_x^2$	$\tilde{x}$	$f_1(z_6)$
$z_7 = (1/\sigma_x^2) \sum_{i=1}^{n} (x_i - \tilde{x})^2$	$\tilde{x}$	$f_n(z_7)$
$z_8 = (ns^2/\sigma_x^2)$ $\quad = (1/\sigma_x^2) \sum_{i=1}^{n} (x_i - \bar{x})^2$	$\tilde{x}$	$f_{n-1}(z_8)$
$z_9 = (1/n) \sum_{i=1}^{n} (x_i - \tilde{x})^2$	$\tilde{x}$	$(n/\sigma_x^2) \cdot f_n(nz_9/\sigma_x^2)$
$z_{10} = s^2 = (1/n) \sum_{i=1}^{n} (x_i - \bar{x})^2$	$\tilde{x}$	$(n/\sigma_x^2) \cdot f_{n-1}(nz_{10}/\sigma_x^2)$

13.56. In the above relations, $x_1, x_2, \cdots, x_n$ were considered as sample values drawn from a normally distributed population. The same relations can be considered as applying to variables $x_1, x_2, \cdots, x_n$ each independently distributed according to the same normal distribution.

13.57. More generally, if variables x_i, $i = 1, 2, \cdots, n$ are independently normally distributed with corresponding means μ_i and variances σ_i^2, not necessarily the same for all i, then the sum

$$y = \sum_{i=1}^{n} [(x_i - \mu_i)/\sigma_i]^2$$

is distributed according to the χ^2-distribution $f_n(y)$, with n degrees of freedom.

13.58. **Fraction of samples for which $s > b\sigma_x$.** The distribution of z_8 can be used to find values of a factor b such that in fraction P of all possible samples of size n, s will exceed $b\sigma_x$. The entries in Table 13.58.1 for given values of n and P, give the value b_i of b such that

$$(13.58.1) \quad P \equiv P(s > b_i \sigma_x) = 1 - \int_{z_8=0}^{z_8=nb_i^2} f_{n-1}(z_8) \, dz_8. \quad \text{(See §13.55.)}$$

Table 13.58.1 shows values of b_i as follows: b_1 for $P = 0.95$, b_2 for $P = 0.05$, b_3 for $P = 0.99$, and b_4 for $P = 0.01$. Thus in 0.90 of all possible samples of size n, s will lie between $b_1 \sigma_x$ and $b_2 \sigma_x$, and in 0.98 of all possible samples of size n, s will lie between $b_3 \sigma_x$ and $b_4 \sigma_x$.

13.59. **The t-distribution.** A random variable T is said to possess the "Student" t-distribution if the probability that T takes a value in a small interval dt containing a value t is given by the frequency distribution function

$$(13.59.1) \qquad g_m(t) \, dt = g^*[1 + t^2/m]^{-(m+1)/2} \, dt, \qquad m = 1, 2, \cdots$$

where

$$(13.59.2) \qquad g^* = \Gamma[(m+1)/2] \div (m\pi)^{1/2} \cdot \Gamma[m/2].$$

The index m is called the number of *degrees of freedom*. (§13.31)

The probability that t is less than or equal to T_0 is

$$(13.59.3) \qquad\qquad P(t \leqq T_0) = \int_{-\infty}^{T_0} g_m(t) \, dt.$$

Properties of the t-distribution:

$$\int_{-\infty}^{\infty} g_m(t) \, dt = 1.$$

(13.59.4) Mean $= \mu = 0$. Variance $= \sigma_t^2 = m/(m-2)$, $m > 2$.

Standard deviation $= \sigma_t = [m/(m-2)]^{1/2}$, $m > 2$.

(13.59.5) Skewness $= \alpha_3 = 0$.

Kurtosis $= \alpha_4 = 3(m-2)/(m-4)$, $m > 4$.

Moments. $\mu_k = \nu_k$ for all k, and $\mu_k = \nu_k = 0$ for odd k,

(13.59.6) i.e., the distribution function $g_m(t)$ is symmetric about $t = 0$.

$$\mu_k = \nu_k = \frac{1 \cdot 3 \, \cdots \, (k-1) m^{k/2}}{(m-2)(m-4) \, \cdots \, (m-k)}, \quad \text{for} \quad k < m, \quad (k \text{ even.})$$

13.60. As m increases indefinitely, the t-distribution approaches the normal (density) distribution, that is,

$$(13.60.1) \qquad \lim_{m \to \infty} g_m(t) = \psi(t) = (1/\sqrt{2\pi}) \exp{(-t^2/2)}.$$

TABLE 13.58.1

Relation of sample standard deviation s and population standard deviation σ_x for normally distributed population

In fraction P of all possible samples of size n, s exceeds $b_i \sigma_x$. (See §13.58)

n	P .95 b_1	.05 b_2	.99 b_3	.01 b_4
5	0.726	2.652	0.614	4.103
6	0.736	2.289	0.631	3.291
7	0.746	2.069	0.645	2.833
8	0.754	1.921	0.658	2.541
9	0.762	1.815	0.669	2.338
10	0.769	1.734	0.679	2.188
11	0.775	1.671	0.688	2.074
12	0.781	1.620	0.697	1.983
13	0.786	1.577	0.704	1.908
14	0.791	1.541	0.711	1.846
15	0.796	1.511	0.717	1.794
16	0.800	1.484	0.723	1.749
17	0.804	1.461	0.729	1.710
18	0.808	1.441	0.734	1.676
19	0.811	1.422	0.739	1.646
20	0.815	1.406	0.743	1.619
21	0.818	1.391	0.748	1.594
22	0.821	1.378	0.752	1.572
23	0.823	1.365	0.756	1.553
24	0.826	1.354	0.759	1.534
25	0.829	1.344	0.763	1.518
26	0.831	1.334	0.766	1.502
27	0.833	1.325	0.769	1.488
28	0.835	1.317	0.772	1.474
29	0.838	1.309	0.775	1.462
30	0.840	1.302	0.778	1.451

In practical applications, $g_m(t)$ may sometimes be considered normal if $m > 30$. (See Table XII.)

13.61. A graph of the distribution $g_m(t)$ for $m = 2$ is shown in Fig. 13.61.1, with the normal density distribution $\psi(t)$ also shown for comparison. It may be noted that for small m, when compared to the normal

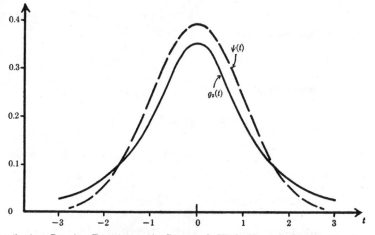

Distribution Density Function $g_2(t)$ Compared With Normal Distribution Density Function $\psi(t)$. [See §§13.59, 13.60.]

FIGURE 13.61.1

distribution, the t-distribution involves greater probabilities of the large deviations from the mean.

13.62. The function $g_m(t)$ is the distribution density of the ratio $t = u\sqrt{m}/v$, where u and v are statistically independent variables, u having a normal distribution about the origin with unit standard deviation, and v^2 having the χ^2-distribution $f_m(w)$, with $w = v^2$ as in Eq. (13.26.1), with m degrees of freedom.

EXAMPLE 1. An important example of the Student t-distribution is the sampling distribution $g_{n-1}(t)$ of the ratio (called the "Student's ratio")

$$(13.62.1) \qquad t = (\bar{x} - \hat{x})\sqrt{n - 1}/s,$$

where $\bar{x}$ is the mean of a sample of size n from a normally distributed population, $\hat{x}$ is the mean of population, and s is the standard deviation of the sample. Both t and $g_{n-1}(t)$ are functionally independent of the population standard deviation σ_x, so that information about sample means and standard deviations can be obtained from $g_{n-1}(t)$ without knowledge of σ_x.

The distribution of t in (13.62.1) according to $g_{n-1}(t)$ is dependent on the fact that the samples come from a normally distributed population. However, in many cases this same distribution of t is approximately applicable even when the population is not normally distributed.

13.63. **Tables.**   The probability ϵ that the variable t differs by more than t_ϵ from its mean, zero, in either direction is

$$(13.63.1) \qquad P(|\,t\,| > t_\epsilon) = \epsilon = 2 \int_{t_\epsilon}^{\infty} g_m(t)\,dt.$$

Tables are available for values of t_ϵ such that

$$(13.63.2) \qquad \int_{-t_\epsilon}^{t_\epsilon} g_m(t)\,dt = 1 - \epsilon,$$

for various values of m, $1 \leq m \leq \infty$, and for various values of ϵ, $.001 \leq \epsilon \leq .90$. (Table XII.)

Table XIII gives values of $\displaystyle\int_{-\infty}^{t_0} g_m(t)\,dt$.

13.64. **Distribution of the difference between means of two samples.** Let $\bar{x}_1$ and $\bar{x}_2$ be means of two samples from the same population, the samples having sizes n_1 and n_2, respectively. Let s_1 and s_2 be the corresponding standard deviations of the samples.

If the population is normally distributed with mean $\bar{x}$ and standard deviation σ_x, then the following statements hold concerning the sampling distribution of the difference $u = \bar{x}_1 - \bar{x}_2$:

(a) $u = \bar{x}_1 - \bar{x}_2$ is normally distributed with mean zero and variance

$$(13.64.1) \qquad \sigma_u^2 = \sigma_x^2[(1/n_1) + (1/n_2)].$$

(b) The quantity $w = u/\sigma_u = (\bar{x}_1 - \bar{x}_2)/\sigma_u$ is normally distributed with unit standard deviation, and mean zero; that is, the probability density for w is $\psi(w)$. (See §10.3.)

(c) If the population standard deviation σ_x is not known, an unbiased estimate of σ_x^2 based on the two samples is

$$(13.64.2) \qquad \acute{\sigma}_x^2 = (n_1 s_1^2 + n_2 s_2^2) \div (n_1 + n_2 - 2). \qquad \text{(See §13.38.)}$$

Let

$$(13.64.3) \qquad \acute{\sigma}_u^2 = \acute{\sigma}_x^2[(1/n_1) + (1/n_2)].$$

The quantity $t = u/\hat{\sigma}_u = (\bar{x}_1 - \bar{x}_2)/\hat{\sigma}_u$ has the t-distribution $g_m(t)$ where $m = n_1 + n_2 - 2$. (See (13.59.1).)

(d) As n_1 and n_2 become large the quantity t approaches $t' = (\bar{x}_1 - \bar{x}_2)/[(s_1{}^2/n_2) + (s_2{}^2/n_1)]^{1/2}$. For large values of n_1 and n_2, the distribution of t' is approximately normal with distribution $\psi(t')$. (See §13.60.)

13.65. **The F-distribution.** A random variable F' is said to possess the Snedecor F-distribution if the probability that F' takes a value in a small interval dF containing a value F is given by the frequency distribution function

(13.65.1) $\quad h_{m_1,m_2}(F)\, dF = B^{-1} r^{m_1/2} F^{(m_1/2)-1} (1 + rF)^{-(m_1+m_2)/2}\, dF,\ F > 0,$

where $B = B(m_1/2,\ m_2/2)$ is the Beta function, (see §18.5), and $r = m_1/m_2$. m_1 and m_2 are parameters called the number of *degrees of freedom*, (§13.31).

The probability that $F \leq F_0$ is

(13.65.2) $\qquad P(F \leq F_0) = \int_{y=0}^{y=F_0} h_{m_1,m_2}(y)\, dy.$

13.66. The distribution $h_{m_1,m_2}(F)$ is the distribution of the ratio $F = (u/m_1) \div (v/m_2)$, where u and v are statistically independent variables having χ^2-distributions (§13.26) with m_1 and m_2 degrees of freedom, respectively.

Properties of the F-distribution:

(13.66.1) $\qquad \int_0^\infty h_{m_1,m_2}(F)\, dF = 1.$

$$\text{Mean} = \mu = m_2/(m_2 - 2), \qquad m_2 > 2.$$

(13.66.2)
$$\text{Variance} = \sigma_F{}^2 = 2m_2{}^2(m_1 + m_2 - 2)$$
$$\div\ m_1(m_2 - 2)^2(m_2 - 4), \qquad m_2 > 4.$$

$$\text{Mode} = \hat{F} = m_2(m_1 - 2) \div m_1(m_2 + 2), \qquad m_1 > 2.$$

The kth moment of the distribution about the origin is

(13.66.3) $\quad \nu_k = \Gamma[(m_1 + 2k)/2]\cdot \Gamma[(m_2 - 2k)/2]\cdot [m_2/m_1]^k$
$$\div\ \Gamma[m_1/2]\cdot \Gamma[m_2/2], \qquad k < m_2/2.$$

13.67. If a variable T is distributed according to the Student t-distribution (§13.59) with m degrees of freedom, then $w = T^2$ is distributed according to the distribution (density) function, $h_{1,m}(w)$.

13.68. The transformation $F = m_2x/m_1(1 - x)$ transforms (13.65.1) into the Beta distribution differential $f(x) dx$ as given in (13.25.2) with $a = (m_1 - 2)/2$, $b = (m_2 - 2)/2$.

13.69. If in (13.65.1), F is replaced by w^2, the formula obtained is known as the w^2-*distribution*.

13.70. Tables are available for values of F_ϵ such that

$$(13.70.1) \qquad \int_0^{F_\epsilon} h_{m_1,m_2}(F) \, dF = 1 - \epsilon,$$

for various values of m_1, m_2, and for $\epsilon = 0.01$ and 0.05. (Table X.) In such tables it is assumed without loss of generality that $F_\epsilon > 1$.

13.71. **The z-distribution.**   Fisher's z-distribution is the distribution obtained from the F-distribution (13.65.1) by the change of variable

$$(13.71.1) \qquad z = (1/2) \log_e F.$$

The frequency distribution (density) function for z is

$(13.71.2)$ $G_{m_1,m_2}(z)$

$$= 2B^{-1}m_1^{m_1/2}m_2^{m_2/2}e^{m_1 z}[m_1e^{2z} + m_2]^{-(m_1+m_2)/2}, \quad -\infty < z < \infty,$$

where $B = B(m_1/2, m_2/2)$ is the Beta function (see §18.5).

The probability that $z \leqq z_0$ is

$$(13.71.3) \qquad P(z \leqq z_0) = \int_{y=-\infty}^{y=z_0} G_{m_1,m_2}(y) \, dy.$$

Properties of the z-distribution:

$$(13.71.4) \qquad \int_{-\infty}^{\infty} G_{m_1,m_2}(z) \, dz = 1.$$

When m_1 and m_2 are both large Eq. (13.71.2) is approximately a normal distribution with mean $[-1/2][(1/m_1) - (1/m_2)]$ and with variance $[1/2][(1/m_1) + (1/m_2)]$.

13.72. Tables are available for values of z_ϵ such that

$$(13.72.1) \qquad \int_{-\infty}^{z_\epsilon} G_{m_1,m_2}(z) \, dz = 1 - \epsilon,$$

for various values of m_1 and m_2, and for $\epsilon = 0.001$, 0.01 and 0.05. (Table XI.)

13.73. Distributions related to the variances or standard deviations from two samples. Let $s_1{}^2$ and $s_2{}^2$ be variances of two samples from the same population, the samples having sizes n_1 and n_2, respectively. Each sample provides an unbiased estimate $\acute{\sigma}_x{}^2$ of the variance $\sigma_x{}^2$ of the population, namely

$$(13.73.1) \qquad \acute{\sigma}_{x1}{}^2 = s_1{}^2 n_1/(n_1 - 1), \qquad (\S13.12,\ \text{Ex. } 2)$$
$$\acute{\sigma}_{x2}{}^2 = s_2{}^2 n_2/(n_2 - 1).$$

If the population is normally distributed, then:

(a) $F = \acute{\sigma}_{x1}{}^2/\acute{\sigma}_{x2}{}^2$ has the F-distribution with distribution density function $h_{n_1-1,n_2-1}(F)$. (13.65.1)

(b) $z = \log_e (\acute{\sigma}_{x1}/\acute{\sigma}_{x2})$ has the z-distribution with distribution density function $G_{n_1-1,n_2-1}(z)$. (13.71.2)

(c) If n_1 and n_2 are both sufficiently large, the quantity $w = (s_1 - s_2) \div [(s_1{}^2/2n_2) + (s_2{}^2/2n_1)]^{1/2}$ is approximately normally distributed about zero with unit standard deviation. If $n_1 > 30$ and $n_2 > 30$, the assumption that the quantity w is normally distributed is often adequate for practical problems.

13.74. w-distribution. Let $q_1 = \sqrt{n_1}\, s_1$ and $q_2 = \sqrt{n_2}\, s_2$ be two independent variables each having the q-distribution (§13.44) with $f_1 = n_1 - 1$, $f_2 = n_2 - 1$ degrees of freedom, respectively.

Suppose s_1 and s_2 are the standard deviations of samples from populations with variances σ_1 and σ_2, respectively. The *variance quotient* is defined as the ratio of the unbiased estimate of σ_1 to the unbiased estimate of σ_2, that is,

$$(13.74.1) \qquad w = [n_1/(n_1 - 1)]^{1/2} s_1 \div [n_2/(n_2 - 1)]^{1/2} s_2$$
$$= (n_2 - 1)^{1/2} q_1 \div (n_1 - 1)^{1/2} q_2.$$

The probability that $w \leqq w_0$ is

$$(13.74.2) \qquad P(w \leqq w_0)$$
$$= \int_{w=0}^{w=w_0} 2B^{-1}[f_1/f_2]^{f_1/2} w^{f_1-1}[1 + f_1 w^2/f_2]^{-(f_1+f_2)/2}\, dw,$$

where $B^{-1} = \Gamma[(f_1 + f_2)/2] \div \Gamma[f_1/2] \cdot \Gamma[f_2/2]$ is the reciprocal of the Beta function $B(f_1/2, f_2/2)$.

Eq. (13.74.1) is called the *w-distribution*. The corresponding distribution of w^2 is precisely the F-distribution, with $F = w^2$, sometimes called the w^2-distribution.

13.75. The r-distribution. If the variable r is such that the probability that r is less than or equal to r_0 is

(13.75.1) $$P(r \leqq r_0) = \int_{r=-\sqrt{m+1}}^{r=r_0} f(r) \, dr,$$

where

(13.75.2) $$f(r) = \Gamma[(m + 1)/2] \cdot [1 - r^2/(m + 1)]^{(m-2)/2}$$
$$\div [\pi(m + 1)]^{1/2} \cdot \Gamma[m/2], \qquad |r| \leqq (m + 1)^{1/2},$$

r is said to have an *r-distribution*. The parameter m is sometimes called the number of degrees of freedom.

13.76. The probability that $|r| \geqq r_1$ is

(13.76.1) $$P = P(|r| \geqq r_1) = 2 \int_{r_1}^{\sqrt{m+1}} f(r) \, dr.$$

Values of r_1 corresponding to $P = 0.10, 0.05, 0.01,$ and 0.001 are shown in the following Table 13.76.1 for various values of m.

TABLE 13.76.1. The *r*-distribution

$$P = P(|r| \geq r_1) = 2 \int_{r_1}^{\sqrt{m+1}} f(u) \, du. \qquad \text{(See §13.76.)}$$

	P			
	0.10	0.05	0.01	0.001
m		r_1		
1	1.397	1.409	1.414	1.414
2	1.559	1.645	1.715	1.730
3	1.611	1.757	1.918	1.982
4	1.631	1.814	2.051	2.178
5	1.640	1.848	2.142	2.329
6	1.644	1.870	2.208	2.447
7	1.647	1.885	2.256	2.540
8	1.648	1.895	2.294	2.616
9	1.649	1.903	2.324	2.678
10	1.649	1.910	2.348	2.730
11	1.649	1.916	2.368	2.774
12	1.649	1.920	2.385	2.812
13	1.649	1.923	2.399	2.845
14	1.649	1.926	2.412	2.874
15	1.649	1.928	2.423	2.899
16	1.649	1.931	2.432	2.921
17	1.649	1.933	2.440	2.941
18	1.649	1.935	2.447	2.959
19	1.649	1.936	2.454	2.975
20	1.649	1.937	2.460	2.990
21	1.649	1.938	2.465	3.003
22	1.648	1.940	2.470	3.015
23	1.648	1.941	2.475	3.026
24	1.648	1.941	2.479	3.037
25	1.648	1.942	2.483	3.047
26	1.648	1.943	2.487	3.056
27	1.648	1.943	2.490	3.064
28	1.648	1.944	2.492	3.071
29	1.648	1.945	2.495	3.078
30	1.648	1.945	2.498	3.085
35	1.648	1.948	2.509	3.113
40	1.648	1.949	2.518	3.134
45	1.647	1.950	2.524	3.152
50	1.647	1.951	2.529	3.166
60	1.646	1.953	2.537	3.186
70	1.646	1.954	2.542	3.201
80	1.646	1.955	2.547	3.211
90	1.646	1.956	2.550	3.220
100	1.646	1.956	2.553	3.227
120	1.646	1.957	2.556	3.237
∞	1.645	1.960	2.576	3.291

This table is based on Arley, N., "On the Distribution of Relative Errors from a Normal Population of Errors," Danske Videnskabernes Selskab., *Mat. fys. Medd.* XVIII, No. 3, 1940, by kind permission of the author.

13.77. As m increases, the r-distribution (13.75.2) approaches the normal distribution with mean zero and variance one.

EXAMPLE 1. An important example of the r-distribution is the distribution of $r_i = (x_i - \bar{x})/s$, where $\bar{x}$ and s are the mean and standard deviation of a sample $x_1, \cdots , x_n$ from a normal population. Here $\sum_{i=1}^{n} r_i = 0$ and $\sum_{i=1}^{n} r_i^2 = n; m = n - 2$. It may be noted that $t = r_i[n - 2]^{1/2} \div [n - 1 - r_i^2]^{1/2}$ has the t-distribution $g_{n-2}(t)$ with $m = n - 2$ degrees of freedom. (13.59.1)

EXAMPLE 2. The quantity $r = \sqrt{n - 1}\, r_{xy}$ has the r-distribution, with $m = n - 2$, where r_{xy} is the correlation coefficient of a sample of n pairs (x_i, y_i) from a bivariate normal population with population correlation coefficient $\rho = 0$. (See §13.78.)

13.78. Distribution of sample correlation coefficient. Suppose the population has a normal bivariate distribution (12.13.1) whose frequency distribution (density) function is

$$(13.78.1) \qquad f(x,y) = [2\pi\sigma_x\sigma_y(1 - \rho^2)^{1/2}]^{-1} \exp H,$$

where

$$(13.78.2) \quad H = \frac{-1}{2(1 - \rho^2)}\left[\frac{(x - \bar{x})^2}{\sigma_x^2} - \frac{2\rho(x - \bar{x})(y - \tilde{y})}{\sigma_x\sigma_y} + \frac{(y - \tilde{y})^2}{\sigma_y^2} \right],$$

and ρ is the population correlation coefficient. The correlation coefficient r_{xy} of a sample of n pairs (x_i, y_i), $i = 1, 2, \cdots , n$, is

$$(13.78.3) \qquad r_{xy} = \sum_{i=1}^{n} (x_i - \bar{x})(y_i - \bar{y})/(ns_x s_y) \equiv \eta,$$

where $\bar{x}$, $\bar{y}$ and s_x, s_y are the means and standard deviations of the x's and y's in the sample.

13.79. The sampling distribution (density) function for η is

$$(13.79.1) \qquad q(\eta) = \frac{[1 - \rho^2]^{(n-1)/2}}{\sqrt{\pi}\cdot\Gamma[(n - 1)/2]\cdot\Gamma[(n - 2)/2]}\, [1 - \eta^2]^{(n-4)/2}$$

$$\sum_{k=0}^{\infty} [(2\rho\eta)^k/k!]\cdot\Gamma^2[(n - 1 + k)/2], \qquad -1 < \eta < 1.$$

The probability that $\eta \leq R$ is

$$(13.79.2) \qquad P(\eta \leq R) = \int_{-1}^{R} q(\eta)\, d\eta.$$

The mean $\bar{\eta}$ of η is approximately $\bar{\eta} = \rho$ when n is large, and the variance $\sigma_\eta{}^2$ of η is approximately $(1 - \rho^2)^2/n$ when n is large. The distribution $q(\eta)$ approaches the normal distribution as n is made large. However, the deviations from normality can be so great for moderate values of n, $n < 50$), that the normal approximation is not useful for many calculations.

13.80. Tables of the distribution $q(\eta)$ are given in *Tables of the Correlation Coefficient* by F. N. David, Biometrika Office, University College, London, 1938.

13.81. Practical calculations may often be based on the following distributions.

(a) *Case when $\rho \neq 0$.* The distribution (density) function $R(v)$ of the quantity

(13.81.1) $v = (1/2) \log_e [(1 + \eta)/(1 - \eta)]$

is approximately normal with mean approximately

(13.81.2) $\bar{v} = (1/2) \log_e [(1 + \rho)/(1 - \rho)] + \rho/2(n - 1)$

and variance $1/(n - 3)$.

If v_1 and v_2 are two values of v calculated from two different samples of sizes n_1 and n_2 from the same population, then $v_1 - v_2$ is approximately normally distributed around mean zero, with variance

(13.81.3) $(n_1 - 3)^{-1} + (n_2 - 3)^{-1}.$

(b) *Case when $\rho = 0$.* In this case, (13.79.1) reduces to

(13.81.4) $q(\eta) = [1 - \eta^2]^{(n-4)/2} \cdot \Gamma[(n - 1)/2] \div \sqrt{\pi} \cdot \Gamma[(n - 2)/2],$

which has mean zero and variance $1/n$.

The quantity $r = \eta(n - 1)^{1/2}$ has the r-distribution (13.75.2) with $n = n - 2$.

When $\rho = 0$, the quantity $t = \eta[(n - 2)/(1 - \eta^2)]^{1/2}$ has the t-distribution $g_{n-2}(t)$ with $n - 2$ degrees of freedom.

13.82. Examples of the distributions $q(\eta)$ and $R(v)$ are shown in Figs. 13.82.1 and 13.82.2.

13.83. The distributions of §13.79 through §13.82 are strictly applicable only if the population is normally distributed. It is found, however, that they often apply approximately if the population has other distributions of a unimodal type.

Sample range

13.84. Distribution of sample range. The *range* R of a sample containing n values $x_1, x_2, \cdots, x_n$ is the difference between the largest value denoted by z, and the smallest value, denoted by y; $R = z - y$. (§2.2.

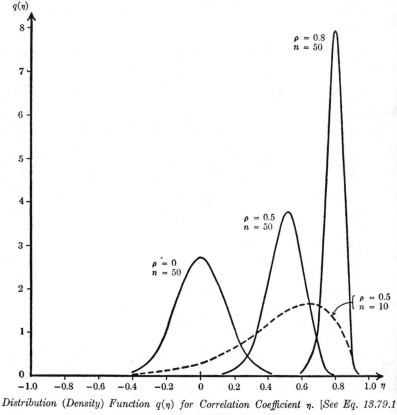

Distribution (Density) Function $q(\eta)$ for Correlation Coefficient η. [See Eq. 13.79.1

FIGURE 13.82.1

Statistical parameters of the sampling distribution of R can be calculated for certain common population distributions. The maximum and minimum values, z and y, can sometimes be used to estimate the mean, $\bar{x}$, or the population, and the standard deviation, σ_x, of the population. Examples for two types of population are as follows.

13.85. Uniform (rectangular) distribution in population. Suppose the population consists of values x uniformly distributed in the interval $a \leqq x \leqq b$. Then the expected value of R in a sample of n values is

$$(13.85.1) \qquad E[R] = \bar{R} = (b - a)(n - 1)/(n + 1).$$

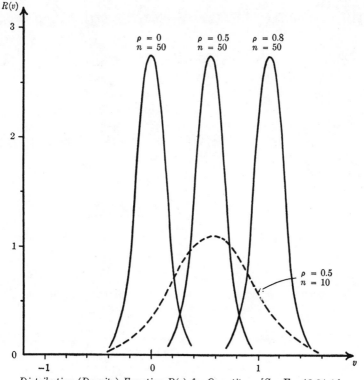

Distribution (Density) Function $R(v)$ for Quantity v. [See Eq. 13.81.1.]

FIGURE 13.82.2

A consistent, unbiased estimate of $b - a$ is $R(n + 1)/(n - 1)$. The standard deviation of values of R, in the sampling distribution of R, is

$$(13.85.2) \qquad \sigma_R = (b - a)[2(n - 1)]^{1/2} \div (n + 1)(n + 2)^{1/2}.$$

TABLE 13.85.1

Mean $\bar{R}$ and standard deviation σ_R of R in units of $(b - a)$

n	$\bar{R}/(b - a)$	$\sigma_R/(b - a)$
2	.33	.24
3	.50	.22
5	.67	.18
10	.82	.11
20	.905	.063
30	.935	.043

A consistent and unbiased estimate of the mean $\bar{x}$ of the population is $(y + z)/2$ since $E[(y + z)/2] = (a + b)/2 = \bar{x}$.

The standard deviation of the sampling distribution of $(y + z)/2$ is $\sigma_m = (b - a)/[2(n + 1)(n + 2)]^{1/2}$.

13.86. Normal distribution in population. If the population is normally distributed, with mean $\bar{x}$ and standard deviation σ_x, the sampling distribution (density) function $k(R)$ for the range R has a complicated expression not readily useful for computations. The expected value of the sample range R is a multiple k_n of σ_x; $\bar{R} = E[R] = k_n \sigma_x$. The standard deviation σ_R of R is also a multiple of σ_x; $\sigma_R = c_n \sigma_x$. Values of k_n and c_n are given in Table 13.86.1 for certain values of n.

The probability that R is less than or equal to R' is given by

$$(13.86.1) \qquad P = P(R \leq R') = \int_0^{R'} k(R)\, dR.$$

Table 13.86.1 gives values of R' for particular values 0.001, 0.005, $\cdots$ of P and n.

TABLE 13.86.1

The distribution of sample ranges, R, when the population is normally distributed with standard deviation σ_x, for samples of size n

(See §13.86.)

n	$k_n = \bar{R}/\sigma_x$	$c_n = \sigma_R/\sigma_x$	\multicolumn{12}{c}{$P = P(R \leq R') = P(R/\sigma_x \leq R'/\sigma_x)$}
			.001	.005	.010	.025	.050	.100	.900	.950	.975	.990	.995	.999
			\multicolumn{12}{c}{R'/σ_x}
2	1.128	.853	.00	.01	.02	.04	.09	.18	2.33	2.77	3.17	3.64	3.97	4.65
3	1.693	.888	.06	.13	.19	.30	.43	.62	2.90	3.31	3.68	4.12	4.42	5.06
4	2.059	.880	.20	.34	.43	.59	.76	.98	3.24	3.63	3.98	4.40	4.69	5.31
5	2.326	.864	.37	.55	.66	.85	1.03	1.26	3.48	3.86	4.20	4.60	4.89	5.48
6	2.534	.848	.54	.75	.87	1.06	1.25	1.49	3.66	4.03	4.36	4.76	5.03	5.62
7	2.704	.833	.69	.92	1.05	1.25	1.44	1.68	3.81	4.17	4.49	4.88	5.15	5.73
8	2.847	.820	.83	1.08	1.20	1.41	1.60	1.83	3.93	4.29	4.61	4.99	5.26	5.82
9	2.970	.808	.96	1.21	1.34	1.55	1.74	1.97	4.04	4.39	4.70	5.08	5.34	5.90
10	3.078	.797	1.08	1.33	1.47	1.67	1.86	2.09	4.13	4.47	4.79	5.16	5.42	5.97
11	3.173	.787	1.20	1.45	1.58	1.78	1.97	2.20	4.21	4.55	4.86	5.23	5.49	6.04
12	3.258	.778	1.30	1.55	1.68	1.88	2.07	2.30	4.29	4.62	4.92	5.29	5.54	6.09
13	3.336	.770	1.38	1.64	1.77	1.97	2.16	2.39	4.35	4.69	4.99	5.35	5.60	6.15
14	3.407	.762	1.47	1.72	1.86	2.06	2.24	2.47	4.41	4.74	5.04	5.40	5.65	6.20
15	3.472	.755	1.55	1.80	1.93	2.14	2.32	2.54	4.47	4.80	5.09	5.45	5.70	6.24
16	3.532	.749	1.63	1.88	2.01	2.21	2.39	2.61	4.52	4.85	5.14	5.49	5.74	6.28
17	3.588	.743	1.69	1.94	2.07	2.27	2.45	2.67	4.57	4.89	5.18	5.54	5.79	6.31
18	3.640	.738	1.75	2.01	2.14	2.34	2.51	2.73	4.61	4.93	5.22	5.58	5.82	6.35
19	3.689	.733	1.82	2.07	2.20	2.39	2.57	2.79	4.65	4.97	5.26	5.61	5.86	6.38
20	3.735	.729	1.88	2.13	2.25	2.45	2.63	2.84	4.69	5.01	5.30	5.65	5.89	6.41
30	4.09													
50	4.50		\multicolumn{12}{l}{P is the probability that the sample range R is less than or}
75	4.81		\multicolumn{12}{l}{equal to R'. The entries in the main body of table are the values of}
100	5.02		\multicolumn{12}{l}{R'/σ_x corresponding to the indicated values of n.}
150	5.3													
			\multicolumn{12}{l}{$\bar{R}$ is the mean of R.}
200	5.5		\multicolumn{12}{l}{σ_R is the standard deviation of R.}

This table is based on Pearson, E. S., "The Percentage Limits for the Distribution of Range in Samples from a Normal Population", *Biometrika*, Vol. 24 (1932); and Pearson, E. S., "The Probability Integral of the Range in Samples of N Observations from a Normal Population", *Biometrika*, Vol. 32 (1941–42), by kind permission of the author and *Biometrika*.

A consistent, unbiased estimate of $\bar{x}$ is $(y + z)/2$. However, the standard deviation of $(y + z)/2$ decreases rather slowly as n is increased. Thus the mean $\bar{x}$ cannot be determined with a high confidence level from the extreme values alone, unless a very large sample is used.

13.87. Probability that at least a fraction h of a population lies between the extreme values y and z of a sample. Let $f(x)$ be the frequency distribution (density) function for any continuous population. The fraction v of the population values lying between y and z is $v = \int_y^z f(x)\ dx$. If y and z are the extreme values in a sample of size n from the population, the sampling distribution of v has the frequency distribution (density) function $n(n - 1)v^{n-2}(1 - v)$, which is independent of the population density $f(x)$. The probability p that at least a fraction h of the population values lies between y and z, that is, lies in the sample range, is given by

$$(13.87.1) \qquad p = 1 - nh^{n-1} + (n - 1)h^n.$$

This equation can be used to determine a sample size n if h and p are given.

13.88. Some values of p for selected values of h and n are shown in Table 13.88.1.

TABLE 13.88.1

Probability p that fraction h of population values lies within the range of a sample of size n

| n | \multicolumn{4}{c}{h} |
	0.99	0.95	0.90	0.80
			p	
5	.00092	.023	.081	.26
10	.0041	.086	.26	.62
15	.0095	.17	.45	.83
20	.017	.26	.61	.93
25	.025	.36	.73	.97
30	.036	.45	.82	.99
40	.060	.60	.92	.999
50	.089	.72	.97	1.000
75	.17	.89	.997	1.000
100	.26	.96	1.000	1.000

13.89. Runs. Suppose elements in a population are of two types, A and B. If a sample be drawn, one element at a time, the sequence of draws

will yield A's and B's in a definite order, such as $AAABBABAABBl$
An unbroken sequence of elements of one kind bounded by elements o
the other kind, or by the end of the whole sequence, is called a *run*
There is a run of 3 A's, then a run of 2 B's, then a run of one A, etc., i
the example just cited. In a sample of size n, let the number of runs o
A's having j elements be denoted by r_{Aj}, and similarly r_{Bj} for the run
of B's. The distribution of *lengths j of runs* in random samples is charac
terized by the joint distribution function $f(r_{Aj})$ of the r_{Aj}'s:

$$(13.89.1) \qquad f(r_{Aj}) = r_A! C_{r_A}{}^{n_B+1} \div r_{A1}! \, r_{A2}! \, \cdots \, r_{An_A}! C_{n_A}{}^n,$$

where

r_A = number of runs of A's, $\qquad\qquad r_B$ = number of runs of B's,

n_A = number of A's in the sample, $\qquad n_B$ = number of B's in the sample

Here, the symbol for the number of combinations of y things taken x a
a time is used, namely, $C_x{}^y = y! \div x!(y - x)!$. $f(r_{Aj})$ gives the prob
ability that, in a random sample of n items for which r_A, r_B, n_A, and n_l
are given, any particular set of values r_{A1}, r_{A2}, $\cdots$, r_{An_A}, with $r_{A1} +$
$\cdots + r_{An_A} = r_A$, will be observed. A similar expression exists for the
joint distribution $f(r_{Bj})$ of the r_{Bj}'s.

The distribution function $g(r_A)$ of the *number r_A of runs of A's* is

$$(13.89.2) \qquad\qquad g(r_A) = C_{r_A-1}{}^{n_A-1} \cdot C_{r_A}{}^{n_B+1} \div C_{n_A}{}^n.$$

$g(r_A)$ gives the probability that, in a random sample of n items for whic
n_A and n_B are given, any particular value of r_A will be observed. A simila
expression exists for the distribution $g(r_B)$ of the r_B's.

The theory can be extended to populations containing three or more
types A, B, C, $\cdots$ of elements.

Order statistics

13.90. In this section consideration is given to probability distributions of statistics of a sample obtained if the n elements of a sample are
ordered from least to greatest. The term "order statistics" is used here
only with reference to values arranged in order of *magnitude*.

Consider the random sample $x_1, x_2, \cdots, x_n$ taken from a large population, where the x_i's are mutually independent, each x_i has the same
probability density function $f(x_i)$, and cumulative distribution function

$$F(x_i) = \int_{-\infty}^{x_i} f(x) \, dx, \qquad\qquad i = 1, \cdots, n.$$

Rearrange $x_1, \cdots, x_n$ in ascending order of magnitude, obtaining $x_1' < x_2' < \cdots < x_n'$. The x's are considered to be continuous random variables. The possibility of any two of the x's being equal is considered negligible.

The principal problem of *order statistics* is that of finding the properties of the x_i's and distributions of statistics based on them.

13.91. Distribution of largest element in sample. The probability that the largest element x_n' in the sample $x_1, \cdots, x_n$ is less than u is

$$(13.91.1) \qquad P(x_n' < u) = \prod_{i=1}^{n} F(x_i)_{x_i=u} = [F(u)]^n \equiv G(u).$$

$G(u)$ is the cumulative distribution function of x_n'.

The probability density function of x_n' is

$$(13.91.2) \qquad g(u) = n[F(u)]^{n-1} f(u).$$

The probability that the largest element x_n' in the sample lies in the interval $(u - \frac{1}{2}\Delta u, u + \frac{1}{2}\Delta u)$ is $g(u)\,\Delta u$.

EXAMPLE. The probability that a mass-produced vacuum tube of a certain type taken at random from production will fail during the time interval $(t, t + \Delta t)$ is $f(t)\,\Delta t$, where $f(t) = ce^{-ct}$, $t > 0$ and $c > 0$ is a constant. $f(t)$ may be considered as the p.d.f. of length of tube life, with

$$(13.91.3) \qquad F(t) = \int_{-\infty}^{t} f(t)\, dt = 1 - e^{-ct}$$

the corresponding c.d.f. Suppose n tubes are taken at random, with lives $t_1, \cdots, t_n$. Let the order statistics be $t_1', \cdots, t_n'$ where t_n' is the life of last tube to fail. The probability that the last one fails during time interval $(u, u + \Delta u)$ is, by (13.91.2),

$$(13.91.4) \qquad g(u)\,\Delta u \cong nc(1 - e^{-cu})^{n-1} e^{-cu}\,\Delta u.$$

13.92. Distribution of smallest element in sample. The probability that the smallest element x_1' is less than v is

$$(13.92.1) \qquad P(x_1' < v) = 1 - [1 - F(v)]^n \equiv G(v)$$

$G(v)$ is the c.d.f. of x_1'.

The p.d.f. of x_1' is

$$(13.92.2) \qquad g(v) = n[1 - F(v)]^{n-1} f(v).$$

The probability that the smallest element x_1' in the sample lies in $(v - \frac{1}{2}\Delta v, v + \frac{1}{2}\Delta v)$ is $g(v)\,\Delta v$.

EXAMPLE. *Case when x's are distributed rectangularly.* If the x_i's are distributed rectangularly on $(0,1)$,

$$(13.92.3) \qquad f(x) = 1, \qquad F(x) = x, \qquad \text{for } 0 \leqq x \leqq 1$$

and

$$(13.92.4) \qquad g(u) = n(u)^{n-1}, \qquad 0 \leqq u \leqq 1$$

the probability that the largest element x_n' in the sample lies in the interval $(u - \frac{1}{2}\Delta u, u + \frac{1}{2}\Delta u)$ is $nu^{n-1}\, \Delta u$. The probability that the smallest element x_1' in the sample lies in $(v - \frac{1}{2}\Delta v, v + \frac{1}{2}\Delta v)$ is

$$n[1 - v]^{n-1} \, \Delta v.$$

The rth moment of x_n' about zero is

$$(13.92.5) \qquad \nu_r'(x_n') = n/(n + r).$$

The mean and variance of x_n' are, respectively,

$$(13.92.6) \qquad \mu = n/(n + 1), \qquad \sigma^2 = n(n + 2)^{-1}(n + 1)^{-2}.$$

13.93. Distribution of kth order statistic. Let x_k' be the kth order statistic of the sample, that is, the kth smallest of $x_1, \cdots, x_n$. The probability that the x_k' lies between $w - \frac{1}{2}\Delta w$ and $w + \frac{1}{2}\Delta w$ is
$$(13.93.1)$$
$$g(w) \, \Delta w = \frac{n!}{(k - 1)! \, (n - k)!} \, F^{k-1}(w)[1 - F(w)]^{n-k} f(w) \, \Delta w$$

$g(w)$ is the p.d.f. of x_k'.

13.94. Distribution of median. If the sample contains $n = 2m + 1$ elements, the probability that the *sample median* x_{m+1}' lies between $w - \frac{1}{2}\Delta w$ and $w + \frac{1}{2}\Delta w$ is

$$(13.94.1) \qquad g(w) \, \Delta w \cong \frac{(2m + 1)!}{(m!)^2} \, F^m(w)[1 - F(w)]^m f(w) \, \Delta w$$

and $g(w)$ is the p.d.f. of x_{m+1}'. (13.94.1) is (13.93.1) with $k = m + 1$, and $n = 2m + 1$.

EXAMPLE. If the x_i's are distributed rectangularly on $(0,1)$, the probability that the sample median x_{m+1}' falls between $w - \frac{1}{2}\Delta w$ and $w + \frac{1}{2}\Delta w$ is

$$(13.94.2) \qquad g(w) \, \Delta w \doteq \frac{(2m + 1)!}{(m!)^2} \, w^m[1 - w]^m \, \Delta w.$$

13.95. Distribution of range. The *sample range* is the difference beween the greatest and smallest values in a sample. If $x_1', \cdots, x_n'$ are he order statistics of a sample $x_1, \cdots, x_n$, $R = (x_n' - x_1')$ is the sample ange. Suppose the x's are drawn from a population with p.d.f. $f(x)$ nd c.d.f. $F(x)$. R is often used as a measure of dispersion, as an alternate o the sample standard deviation, or sample mean deviation.

If the x's are drawn from a population with p.d.f. $f(x)$ and c.d.f. $F(x)$, hen the probability that x_n' falls in interval $(u - \frac{1}{2}\Delta u, u + \frac{1}{2}\Delta u)$ and x_1' falls in $(v - \frac{1}{2}\Delta v, v + \frac{1}{2}\Delta v)$ is

(13.95.1)
$$h(u,v) \, \Delta u \, \Delta v = n(n-1)[F(u) - F(v)]^{n-2} f(u) f(v) \, \Delta u \, \Delta v,$$

where $-\infty < v < u < +\infty$.

Let $R = u - v$ be the sample range. The probability that $R \leq r$ is

$$H(r) = P(R \leq r) = P(u \leq v + r)$$

(13.95.2)
$$= n \int_{-\infty}^{\infty} [F(v+r) - F(v)]^{n-1} f(v) \, dv.$$

$H(r)$ is the c.p.d. of R with p.d.f. $h(r)$ and

(13.95.3)
$$h(r) \, dr = \left\{ n(n-1) \int_{-\infty}^{\infty} [F(v+r) - F(v)]^{n-2} f(v) f(v+r) \, dv \right\} dr$$

The *expected value* of the sample range is $E(x_n') - E(x_1')$. This may be calculated directly, using (13.91.2) and (13.92.2), or from $\int_{-\infty}^{\infty} r \, h(r) \, dr$.

EXAMPLE. If the x_i's are distributed rectangularly on $(0,1)$, the probability that the range R lies in the interval $(r - \frac{1}{2}\Delta r, r + \frac{1}{2}\Delta r)$ is

$$h(r) \, \Delta r = \left\{ n(n-1) \int_0^{1-r} [v + r - v]^{n-2} \, dv \right\} \Delta r$$

(13.95.4)
$$= n(n-1)r^{n-2}(1-r) \, \Delta r.$$

The probability that R lies between a and b, $0 < a < b < 1$, is

$$P(a < R < b) = \int_a^b h(r) \, dr$$

(13.95.5)
$$= n(b^{n-1} - a^{n-1}) - (n-1)(b^n - a^n).$$

The expected value of R over $(0,1)$ is

(13.95.6)
$$\int_0^1 r \, h(r) \, dr = (n-1)/(n+1).$$

The variance of R over $(0,1)$ is $2(n-1)/(n+1)^2(n+2)$.

13.96. **Tables.**   Unless $F(x)$ and $f(x)$ are simple functions, numerical integration may be required to evaluate integrals such as (13.95.2) and (13.95.3). The behavior to be expected of sample range for samples taken from a unit normal distribution may be gained from the extensive tables given in E. S. Pearson and H. O. Hartley, *Biometrika Tables for Statisticians*, Vol. I, 3d ed., Tables 20, 22, 23, Cambridge University Press, New York, 1966, and H. L. Harter, "Tables of Range and Standardized Range," *Annals of Mathematical Statistics*, Vol. 31, 1960.

A brief table for the distribution of sample ranges when the population is normally distributed is given in §13.86.

A short table giving the expected value and the standard deviation of the range when the population is uniformly distributed is given in §13.85.

13.97. **Distribution-free tolerance limits.**   Consider the fraction $F(x_n')$ of the objects in a population having values of x less than or equal to x_n', the largest element in a sample of n elements drawn from a population with c.d.f. $F(x)$. By (13.91.2) the probability that $F(x_n')$ lies in the interval $(t - \frac{1}{2}\Delta t, t + \frac{1}{2}\Delta t)$ is $nt^{n-1}\,\Delta t$ where $t = F(u)$, $\Delta t = f(u)\,\Delta u$.

Upper tolerance limit.   The chance that the fraction of objects in the population having values of x less than x_n' is at least β, $0 < \beta < 1$, is

$$(13.97.1) \qquad P[F(x_n') > \beta] = n \int_\beta^1 t^{n-1}\, dt = 1 - \beta^n = \gamma$$

Relation.   (13.97.1) is true for *any* continuous c.d.f. $F(x)$. The interval $(-\infty, x_n')$ is called a *one-sided statistical tolerance interval*. γ is called a *confidence coefficient*.

If β and γ are given, an integer n may be selected which is closest to the real number solution to $1 - \beta^n = \gamma$. In this case x_n' is called a 100 β percent *upper tolerance limit* with confidence γ.

When n is large and $1 - \beta$ is small,

$$(13.97.2) \qquad\qquad 1 - \beta^n \cong 1 - e^{-(1-\beta)n} \cong \gamma.$$

The required sample size n for given β and γ can be approximated from (13.97.2).

Tables of n calculated for values of β and γ are found in R. B. Murphy "Non-parametric Tolerance Limits," *Annals of Mathematical Statistics* Vol. 19, 1948, and P. N. Somerville, "Tables for Obtaining Nonparametric Tolerance Limits," *Annals of Mathematical Statistics*, Vol. 29, 1958.

EXAMPLE.   How large should the sample size n be in order that the probability γ will be 0.90 that the fraction of elements in the population having values of x less than x_n' is at least $\beta = 0.95$? Here x_n' is the largest x drawn in the sample.

By (13.97.2), $0.90 \cong 1 - e^{-0.05n}$; hence $n = 46$. This result may be stated: In a sample of 46 elements drawn from a large lot, the largest element, x_{46}', is a 95 percent upper statistical tolerance limit with confidence 0.90, and $(-\infty, x_{46}')$ is a one-sided upper tolerance limit with confidence $\gamma = 0.90$.

13.98. Lower tolerance limit. Let x_1' be the smallest x drawn in a sample. $F(x_1')$ is the fraction of the elements in the population having values of x less than or equal to x_1'.

$$(13.98.1) \qquad P[1 - F(x_1') > \beta] = 1 - \beta^n.$$

Thus, the probability is $1 - \beta^n$ that the fraction of elements in the population having values of x greater than x_1' is at least β.

Given β and γ, choose n as the integer nearest to satisfying

$$(13.98.2) \qquad 1 - \beta^n = \gamma.$$

x_1' is a $100\,\beta$ percent lower statistical tolerance limit with confidence γ. The interval $(x_1', +\infty)$ is called a *one-sided lower tolerance interval*.

EXAMPLE. With reference to example of §13.97, in a sample of 46 items the smallest element x_1' drawn from the population is a 95 percent lower statistical tolerance limit with confidence $\gamma = 0.90$.

13.99. Two-sided tolerance limit. The fraction of elements in the population having values of x in (x_1', x_n') is $F(x_n') - F(x_1')$. If $0 < \beta < 1$, the probability that $F(x_n') - F(x_1') > \beta$ is

$$(13.99.1) \qquad 1 - \beta^n - n(1 - \beta)\beta^{n-1} \equiv \gamma.$$

For a given β and γ choose n as the integer most nearly satisfying (13.99.1). The interval (x_1', x_n') is a two-sided $100\,\beta$ percent *statistical tolerance interval* with *confidence* coefficient γ. γ is the probability that (x_1', x_n'), as determined from a sample of size n, covers the values of x of at least $100\,\beta$ percent of the elements in the population (considered to be large).

When n is large and β is close to 1, the relation (13.99.1) can be approximated by

$$(13.99.2) \qquad e^{-n(1-\beta)}[1 + n(1 - \beta)] = 1 - \gamma.$$

which is the sum of the first two terms of the Poisson c.d.f. (See §9.2.)

Further discussion of this subject is given in §13.87. Equation (13.87.1) is Eq. (13.99.1) with $\beta = h$, $\gamma = p$. Some selected values of γ for selected values of β and n are given in Table 13.88.1.

EXAMPLE. How large should a sample be for which (x_1', x_n') is a 95 percent statistical tolerance interval with confidence $\gamma = 0.90$? By (13.99.2), with $\beta = 0.95$, $\gamma = 0.90$,

(13.99.3) $$0.10 \cong e^{-n(0.05)}[1 + n(0.05)].$$

By the use of tables of e^{-x} one finds $n = 78$.

13.100. **Two-order statistics case.** If from a sample of n elements taken from a population with c.d.f. $F(x)$, one takes any two order statistics (x_c', x_{c+k}'), then

13.100.1
$$P[F(x_{c+k}') - F(x_c') > \beta]$$
$$= \frac{n!}{(k-1)!\,(n-k)!} \int_\beta^1 t^{k-1}(1-t)^{n-k}\, dt$$

The integrand on the right does not depend on c or F, only on k and n.

Use of random numbers

13.101. **Selecting a random sample.** From a population having m items a random sample of n items is to be selected. One method of doing this is to assign a number to each item in the population, put a set of numbered marbles corresponding to the items into a box, thoroughly mix the marbles in the box, and draw n marbles from the box. The numbers on these n marbles correspond to the items selected. To shorten the process a table of *random numbers* such as Table XXII may be used.

13.102. **Random number tables.** A *random number* table consists of a sequence of digits designed to represent the result of a simple random sampling from a population consisting of the digits 0, 1, $\cdots$, 9. By using two columns of the table a sequence of numbers formed from the population of the 10^2 numbers 00, 01, $\cdots$, 99 may be obtained; and similarly for three, four, $\cdots$ or any large number of columns.

Table XXII is an example of a random number table. The sequence of figures given in any column of this table may be regarded as the values of independent discrete variables each with the uniform distribution $P[x = i] = 1/10$, where $i = 0, 1, \cdots, 9$. In other words, each of the numbers 0, 1, $\cdots$, 9 appears with approximately the same frequency.

13.103. **Use of Table XXII.** *Purpose:* To select a random sample of n items from a population of m items,

1. Enter table in a random manner.
2. Assign numbers to the m items. For example, if $m = 600$, three place numbers are needed; so three columns of the table are used.

3. Decide on some arbitrary method for selecting the positional digits from each entry selected in keeping with step 1.

4. Begin the selection. Let x be a number formed. If $x \leqq m$ the corresponding item in the population is picked for the random sample. If $x > m$ the number x is passed over, and another selection is made. If the sampling is to be made without replacement, any number which has already been chosen in the process must be passed over. This procedure is continued until the random sample of n items has been selected.

13.104. Other types of random number tables. Table XXII is based on a uniform distribution of $0, 1, \cdots, 9$. Random number tables based on other types of distributions have been constructed. Thus, if the numbers x are distributed normally with population mean μ and variance σ^2, a table of random numbers x constructed to fit this distribution would have the property that numbers x with values near μ would occur more frequently than numbers farther away from μ. Table XXIII is such a table constructed for $\mu = 0$ and $\sigma^2 = 1$.

Several excellent tables of random sampling numbers have been constructed.

Fisher and Yates, *Statistical Tables for Biological, Agricultural and Medical Research*, 6th ed., Hafner Publishing Company, Inc., New York, 1963.

Hald, A., *Statistical Tables and Formulas*, John Wiley & Sons, Inc., New York, 1952.

Kendall, M. G., and Smith, B. Babington, *Tables of Random Sampling Numbers, Tracts for Computers*, No. 24, 1940.

The Rand Corporation, *A Million Random Digits with* 100,000 *Normal Deviates*, The Free Press of Glencoe, Illinois, 1955.

Tippett, L. H. C., *Random Sampling Numbers, Tracts for Computers*, No. 15, 1927.

13.105. Generation of random number tables. It should be noted that when we speak of generating random numbers, we are usually selecting at random a number from the "flat" uniform distribution on $(0,1)$.

Given a table of uniform random numbers, such as Table XXII, a table of random variables may be constructed for any specified cumulative distribution function $F(T)$. Let

$$(13.105.1) \qquad u = F(T) = \int_{-\infty}^{T} f(t)\, dt.$$

The construction is as follows: The quantity u is assumed to be uniformly distributed on the interval $(0,1)$. A value for u is selected a random from a table of uniform random numbers. From (13.105.1) the corresponding value of T is found. T is a random number for the distribution function $F(T)$. This process is repeated as often as may be desired.

Tables of random numbers can be read and stored in a computer This is a slow process, and tables are quickly exhausted in any sizable problem. A number of schemes have been devised to generate random numbers, some of which have been used on digital computers to generate random numbers. For details, see

Hall, T. E., and Dobell, A. R., "Random Number Generators," *Society for Industrial and Applied Mathematics Review*, Vol. 4, pp. 230–254 1962.

fourteen

STATISTICAL INFERENCE.
SIGNIFICANCE TESTS AND
CONFIDENCE INTERVALS

14.1. Sampling distributions can be used as a basis for drawing inferences about a population, on the basis of observations made on one or more samples from the population. The inferences about the population are subject to some uncertainty, unless the observed sample consists of the entire population. The degree of uncertainty can be described quantitatively by various statistical techniques. The methods of *significance tests* and *confidence intervals*, which are among the important techniques for measuring degrees of uncertainty in inferences from sample data to population characteristics, are outlined in this chapter.

The notation used in §13.2 to §13.7 is used throughout this chapter.

14.2. Estimation of population parameters from large samples.
If a sample is very large, the mean of the sample is expected to be very nearly equal to the mean of the population, and similarly for the standard deviation and other parameters of the population distribution. The larger the sample, the better the approximation provided by the sample values, or, in more precise terms, the smaller the probability that the sample value of the parameter will differ from the population value by more than any given quantity d.

14.3. If a large random sample (say, of size greater than 50) is drawn from a *normal* population, certain simple conclusions can be reached about the population mean $\bar{x}$ and standard deviation σ. Let the sample size be n, the sample mean be $\bar{x}$, and the sample standard deviation be s.

Here
$$s^2 = \sum_{i=1}^{n} (x_i - \bar{x})^2/n, \qquad \bar{x} = \sum_{i=1}^{n} x_i/n.$$

(a) An unbiased estimate of $\bar{x}$ is $\bar{x}$. The precision of the estimate can be measured by noting that the sampling distribution of $\bar{x}$ is normal about $\bar{x}$ as a mean with standard deviation $\sigma/\sqrt{n}$. Thus if σ is known, the chance that $\bar{x}$ differs from $\bar{x}$ by as much as any selected quantity d can be estimated from Table 10.8.1 or Table IX. In particular, the chance that $|\bar{x} - \bar{x}|$ exceeds three standard deviations, i.e., $3\sigma/\sqrt{n}$, is so slight (less than 0.003) that in practice the population mean $\bar{x}$ can usually be asserted to lie in the interval $\bar{x} - 3\sigma/\sqrt{n}$ to $\bar{x} + 3\sigma/\sqrt{n}$. If σ is not known, s can be used as an estimate of σ, or $s[n/(n-1)]^{1/2}$ can be used as a slightly more precise (unbiased) estimate as discussed below.

(b) An unbiased estimate of σ is $s[n/(n-1)]^{1/2}$, which is approximately equal to s. The sampling distribution of s is approximately normal about σ as mean with standard deviation approximately $\sigma/\sqrt{2n}$. Thus the probability that s differs from σ by more than any given amount d can be estimated from Table 10.8.1 or Table IX. In particular, s can in practice usually be asserted to lie in the interval $\sigma - 3\sigma/\sqrt{2n}$ to $\sigma + 3\sigma/\sqrt{2n}$. Similar statements can be made for $s[n/(n-1)]^{1/2}$, whose mean is σ and whose standard deviation is $\sigma/\sqrt{2(n-1)}$. If σ is not known, as will often be the case when making these estimates, s (or $s[n/(n-1)]^{1/2}$) can be substituted for σ in the expressions for the mean and standard deviations.

It is emphasized that most of these approximations depend on the population being normal and the sample being large.

EXAMPLE 1. The weight of the fruits in a supply of several thousand must be estimated for a packing process. A sample of 100 items is selected at random. It is assumed, on the basis of past experience, that the weights are normally distributed in the whole supply. The mean weight of the sample of 100 is found to be $\bar{x} = 61$ grams. The standard deviation of the weights of the sample is found to be $s = 11$ grams. The following conclusions can be drawn about the whole supply of fruit.

The mean weight $\bar{x}$ of the fruits in the whole supply can reasonably be estimated as $\bar{x} = \bar{x} = 61$ grams. The standard deviation of means of samples of size $n = 100$ is approximately $s/\sqrt{n} = 11/\sqrt{100} = 1.1$ grams. Thus it is practically certain that the particular sample mean observed, $\bar{x} = 61$ grams, does not differ by more than 3.3 grams from the true mean $\bar{x}$ of the whole supply. According to Table 10.8.1, the chance that $\bar{x}$ differs from the true $\bar{x}$ by more than $2 \times 1.1 = 2.2$ grams is 0.0455. Thus unless the particular sample drawn happens to represent an event whose probability is less than 5%, it follows that the true value of $\bar{x}$ lies between $61 - 2.2$ and $61 + 2.2$ grams. The chance of this conclusion being incorrect is 0.0455, in the sense described. A still smaller interval for $\bar{x}$ can be stated, with corresponding increase in the chance of being in error.

The standard deviation σ of the weights in the whole supply can reasonably be estimated as $\sigma = s = 11$ grams. The standard deviation of standard deviations s for samples of size $n = 100$ is approximately $\sigma_s = s/\sqrt{2n} = 11/\sqrt{200} = 0.78$ grams. Thus, by a procedure similar to that used above, it is concluded that it is practically certain that the true σ lies in the interval $11 - 3 \times 0.78$ to $11 + 3 \times 0.78$ grams, and that there is less than 5% chance that the true σ lies outside the interval $11 - 2 \times 0.78$ to $11 + 2 \times 0.78$ grams.

14.4. The general procedure for making estimates concerning the parameters of a population, and determining the precision of the estimates, on the basis of a sample, is essentially the same as that just described for a particularly simple case. In other cases, however, complications arise from the fact that the sampling distributions may not be normal; this is especially true when the samples are small (say, less than 30), in which cases recourse often must be made to various tabulated special distributions such as the t, F, z, and χ^2-distributions. In the following sections, some procedures applicable to these more complicated situations are outlined. It should be borne in mind that many of these procedures can be materially simplified when the samples are large, or the population is known to be normal, or in other special situations. These simplifications are indicated below in some instances; other possibilities may be derived on the basis of the sampling distributions discussed in the previous chapter.

Statistical inference

14.5. **Theory versus observations.** In order that a mathematical model M of some phenomena W in the physical world be useful there must exist some type of agreement between the theoretical premises and conclusions, and their counterparts in W. If the theory is to be acceptable, when observed relations are possible then observations should (in repeated tests) indicate agreement with the theory.

Whenever a theoretical deduction leads to a definite numerical value for the probability P of a certain observable event E the frequency f of the event, as observed in a long sequence of repeated tests, should be equal to P, approximately.

14.6. **Testing agreement between theory and observations.** Often the problem of testing the agreement between theory and observations is as follows: Given a sample of size n of some variable x. Can this sample be considered as a random sample drawn from a population having a certain cumulative probability distribution function $F(x)$? One scheme for *testing this statistical hypothesis* is as follows:

1. Assume that the hypothesis is true, namely, that the sample has been drawn from population with distribution $F(x)$.
2. Then the cumulative distribution $F^*(x)$ of the sample should approximate $F(x)$ when n is large.
3. Define a non-negative *measure* D of the deviation from F of F^*. This measure has a sampling distribution D^*.

4. From D^* calculate the probability P that the deviation D will exceed any arbitrary given quantity D_0, namely, $P(D > D_0)$.
5. Let $\epsilon > 0$ be any selected small positive number. Select D_0 large enough that $P(D > D_0) = \epsilon$.
6. If ϵ is taken small enough a deviation $D > D_0$ is not likely to occur in one single trial test. For a small value of ϵ assume that the event $D > D_0$ will not occur in a single trial. The number ϵ selected is in effect a significance level ϵ in the sense that the assumption will only be incorrect in a fraction ϵ of all possible trials.
7. Suppose for the sample of size n the quantity D is calculated and found to be D_1.

If $D_1 > D_0$, an event of probability ϵ has occurred which is contrary to assumption (6). The deviation D_1 is then considered *significant*. It is then concluded that hypothesis (1) is *apparently disproved by experience*, at the significance level ϵ. (Logically, D_1 could exceed D_0, yet hypothesis (1) be true.)

If $D_1 \leqq D_0$, the deviation D_1 may be considered as possibly due to random fluctuations, and the sample data are *apparently* consistent with hypothesis (1) at the level ϵ. It is concluded that, in so far as this test is concerned, *hypothesis (1) is a reasonably practical interpretation of the data*, subject to further information that may be obtained later. (Logically, D_1 could be less than or equal to D_0, yet hypothesis (1) be false.)

14.7. A test following the above scheme is called a *test of significance relative to hypothesis* (1). For many purposes the probability ϵ is arbitrarily fixed (as for example, as 0.05, or 0.01, etc.) and is called a *level of significance for the test*. D_0 is the corresponding significance limit. Where the primary interest is with a test concerning the agreement between a distribution of a set of sample values and a theoretical distribution the term *test of goodness of fit* is used.

14.8. **Analysis of causes.** Suppose that a result E may be obtained by two different methods, and that the probability of achieving E by method M_1 is p_1, by method M_2 is p_2. Suppose further that in a series of n_1 trials using method M_1 the result E is obtained in e_1 trials, and that in a series of n_2 trials using method M_2 the result E is obtained in e_2 trials. In general $e_1/n_1 \neq e_2/n_2$. Can it be concluded that $p_1 \neq p_2$, or are the observed results merely indicative of random sample fluctuations which could arise if $p_1 = p_2$? Statistical analyses of such situations are often carried out by making the initial *hypothesis* (sometimes called a *null hypothesis*) that *there is no difference between the effects of the methods* M_1

and M_2. On this hypothesis tests of significance for the observed differences or characteristics may sometimes be devised, such tests leading to statistical criteria for practically determining whether the differences are significant or merely random fluctuations. Applications of this sort are sometimes called *analyses of causes*.

14.9. Theory of estimation. Theory of errors. Let x be a random variable (in any number of dimensions) with a known mathematical distribution involving a number of unknown parameters a_i, $i = 1, \cdots, u$. Suppose a sample of observed values of x are given. The *estimation problem* is that of forming estimates of the parameters a_i, and estimating the precision of these estimates. Many different functions of the sample may be used as estimates for a particular parameter a_i. The properties of these various estimates of a_i are of concern; and of particular interest is the problem of finding (if any) those estimates of *maximum precision*. For such estimates probability statements concerning the deviations of the estimates of a_i from the unknown values of a_i are sought.

14.10. A simple example of this type of theory is the *problem of inverse probability*: Given the frequency of an event E in a sequence of random repetitive tests; what can be stated (and with what precision) concerning the unknown value of the probability p of the event E occurring?

14.11. Prediction. A wide range of statistical problems involve the prediction from given samples and physical (or other) theory of some type of events. A knowledge of associated probability distributions is often of considerable help. However, predictions of this general sort usually require a careful combination of probability statistical theory and an intimate knowledge of the particular physical fields involved.

Tests of significance

14.12. Significance tests. As indicated in §14.6 a significance test of an *hypothesis H* consists in determining whether or not a result observed in a sample could reasonably have been expected if the *hypothesis* about the population were assumed to be true. Some variation in properties of samples is expected due to random variations in the selection of the sample items. This variation is indicated by the sampling distribution of the sample quantity under study. (§14.6) The purpose of the significance test is to determine whether there is a *significant* deviation of some quantity from its expected value, over and above the deviations which might be accounted for by the random sampling variations.

14.13. Suppose that the characteristic being considered is measured by a quantity D, and that, under the assumption that the hypothesis H is true, the probability is ϵ that $D > D_0$ in a sample of size n selected at random. Then if, in a sample which is actually examined, D is observed to be greater than D_0, the hypothesis is rejected at the *level of significance* ϵ; if in the actual sample D is observed to be less than or equal to D_0, the hypothesis is accepted as a reasonable interpretation of the data. D_0 is the corresponding *significance limit*. ϵ must be selected according to the purpose of the test and the nature of the data. Values of ϵ such as 0.05, 0.01, or 0.001 are sometimes used in practice, the smaller values providing the more stringent tests of the hypothesis.

14.14. Two general types of hypotheses H are frequently of interest. One type asserts that the population has some specified frequency distribution, such as normal, Poisson, $\cdots$. This type of hypothesis is often tested by means of the "Chi-squared" test of significance. (§14.15.) Another type of hypothesis H asserts that some parameter associated with the population, such as a mean, or standard deviation, $\cdots$ has a specified magnitude. This type of hypothesis is often tested by consideration of the sampling distribution of the parameter in question, or of some closely related quantity. (§14.37 et seq.) Other types of hypotheses are commonly used. They may concern the relation between two or more populations, or two or more samples, in terms of frequency distributions or selected parameters. (§14.41 et seq.)

14.15. **The agreement of the frequency distribution in a sample with a population distribution. (The Chi-squared test.)** The χ^2-distribution may be used to test how well a sample distribution agrees with a population distribution, the latter being (a) known in advance, or (b) deduced from the sample. The comparison is made on the basis of the observed and the population frequencies for a suitable set of class intervals of the variable of the distribution. Thus the χ^2-procedure examines the whole sample distribution at once in relation to the population distribution, and is in this sense more general than examination of a sample mean, sample variance, etc.

14.16. Let $x_1, x_2, \cdots , x_n$ be a sample of values of x, and let the range of x be divided into r class intervals $X_1 \leqq x \leqq X_2$, $X_2 \leqq x \leqq X_3$, $\cdots , X_r \leqq x \leqq X_{r+1}$. Suppose that the numbers of values x_i from the sample falling in these intervals are $f_1, f_2, \cdots , f_r$, respectively. Suppose that the relative frequencies in these same intervals expected in the population distribution are $g_1, g_2, \cdots , g_r$, so that the numbers of values expected in the class intervals from a sample of n are $f_1' = ng_1$, $f_2' = ng_2$, $\cdots$, $f_r' = ng_r$, respectively. The χ^2-test is concerned with the difference between f_i and f_i' for all class intervals. This difference is measured by

$$(14.16.1) \qquad \chi^2 = \sum_{i=1}^{r} [(f_i - f_i')^2/f_i'] = \sum_{i=1}^{r} (f_i^2/ng_i) - n.$$

14.17. The distribution function for the χ^2 in Eq. (14.16.1) is not the χ^2-distribution given in Eq. 13.26.1, and in Table XIV. However, for n large enough, $\chi^2 = w$ is distributed approximately according to a distribution function $f_m(w)$. (Eq. 13.26.1.) This limiting distribution obtained by increasing n without bound is independent of the population distribution.

14.18. In applying the χ^2-tests outlined below, in order to use the limiting χ^2-distribution $f_m(w)$, tabulated in Table XIV, the restriction to "large enough" n is often taken in practice to require that each f_i must be ≥ 10. This is done in order that the approximation (2) in §14.22 be adequate. It is usually feasible to choose class intervals to assure this, especially by grouping intervals near the tails of distributions. When this can not be done Table XIV should not be used. (Some writers require that each f_i be ≥ 5.)

14.19. The better the fit of the frequencies f_i to the population frequencies f_i', the smaller is χ^2. Table XIV tells how likely χ^2 is to have a value as large as any selected χ_0^2 purely by chance, *if* the sample really is drawn from a universe distributed according to f_i'. It is sometimes arbitrarily assumed that if the chance that an observed χ^2 would occur is only 0.05 or less, then there is significant doubt that the sample really comes from a universe with the distribution defined by f_i'. There is still more significant doubt if the chance is only 0.01 or less, or 0.001 or less; these values are sometimes used in tests.

14.20. It should be noted that Table XIV is applicable only when the number m of degrees of freedom is ≤ 30. When $m > 30$, $\sqrt{2\chi^2}$ is, approximately, normally distributed with mean $\sqrt{2m - 1}$ and unit standard deviation. (See §13.27.)

14.21. In these comparisons between two distributions, the distributions may be given by formulas, distribution functions, or other analytic methods; or the distributions may be given by tables of numbers, sets of observations, or the like. The basic procedure is to calculate χ^2 (as given in Eq. 14.16.1) by whatever means are appropriate and then to refer the value obtained to a distribution of χ^2.

14.22. χ^2-test when population distribution is completely known.
When the parameters of the population distribution $f(x)$ are completely
known, the χ^2-test to determine how well the sample data $x_1, \cdots, x_n$
fit the given population distribution is as follows:

Hypothesis H: The data $x_1, \cdots, x_n$ are a sample of a random variable
x with distribution $f(x)$.

Purpose of test: To determine whether the data $x_1, \cdots, x_n$ may be
considered as consistent with hypothesis H.

Steps in test: (1) The measure of the deviation between the sample and
the population distribution $f(x)$ is $w = \chi^2$ as given in Eq. 14.16.1.

(2) Suppose n is large enough so that the χ^2-distribution function
$f_m(w)$ can be used to approximate the distribution of the χ^2 given in
Eq. (14.16.1). (See §14.18.) The number of degrees of freedom is
$m = r - 1$.

(3) Select a level of significance ϵ, (e.g., $\epsilon = 0.001, 0.01, \cdots$).

(4) From tables of the χ^2-distribution, determine $\chi_0{}^2$ such that
$P(\chi^2 \geqq \chi_0{}^2) \approx \epsilon$.

Interpretation from test: (a) If in an actual sample a value $\chi^2 > \chi_0{}^2$ is
found, the sample $x_1, \cdots, x_n$ shows a significant deviation from H,
and the hypothesis H is rejected at the ϵ level of significance.

(b) If in an actual sample a value $\chi^2 \leqq \chi_0{}^2$ is found, the sample
$x_1, \cdots, x_n$ is considered as consistent with hypothesis H.

14.23. χ^2-test when population distribution is not completely known.
When not all of the parameters of the population distribution $f(x)$ are
known, the χ^2-test outlined above must be modified. In this case some
scheme for estimating the unknown population parameters from the
sample data must be used. The sampling distribution of χ^2 will depend
on the scheme used for estimating the unknown parameters. However,
many of the known methods lead to the same limiting distribution $f_m(w)$
(Eq. 13.26.1) as the sample size n increases; the number of degrees of
freedom being $m = r - 1 - b$, where b is the number of parameters in
the population distribution determined from the sample.

14.24. The χ^2-test outlined in §14.22 is then applicable to the case under
consideration, the only change in the test being that the number of
degrees of freedom is taken to be $m = r - 1 - b$.

14.25. Degrees of freedom. The number of degrees of freedom measures the extent to which the sample frequencies f_i are known in advance
to agree with the population frequencies $f_i{}'$. The extent of this agreement
must be taken into account in order to obtain a meaningful measure of

the chance of occurrence of any given total deviation χ^2 between sample and population distributions. In all cases, the sample size n and the number n of population values used for comparison are the same. Thus if $r - 1$ of the f_i' are given, the remaining value of f_i' can be deduced. It follows that if the f_i' are not otherwise related to the f_i in advance, there are $m = r - 1$ degrees of freedom. This is the value of m to use when the population distribution is prescribed and the χ^2 procedure is used to see if the sample distribution f_i is consistent with it. When the population distribution is determined by fitting some distribution to the sample, then the sample and population distributions are made consistent with respect to mean, variance, moments, etc., and for each such parameter a degree of freedom is lost. Thus if b such parameters are determined from the sample, the total number of degrees of freedom is $m = r - 1 - b$.

14.26. Examples of chi-squared test procedures. The following sections illustrate some uses of the chi-squared test, and indicate how χ^2 is calculated in specific cases.

14.27. Test of estimate of probability of event from one sample. Suppose that in a sequence of n independent trials, event E has occurred v times. It is desired to test whether this is consistent with the hypothesis that the probability of the occurrence of E is $p = 1 - q$ in each trial. There are $r = 2$ class intervals for the observations, namely, E occurred, E did not occur. From Eq. 14.16.1, with $r = 2$, $f_1 = v$, $f_2 = n - v$, $f_1' = np$, $f_2' = nq$, $m = r - 1 = 1$ degree of freedom,

$$(14.27.1) \qquad \chi^2 = (v - np)^2/npq.$$

EXAMPLE 1. If $n = 40$, $v = 12$, $p = 0.25$, then $\chi^2 = [12 - (0.25)(40)]^2 /40(0.25)(0.75) = 0.53$. The χ^2-table (Table XIV) shows that a value of χ^2 as large as 0.53 may be expected in over 40% of samples. Thus the sample value $v = 12$ is not unlikely if the actual probability of E is $p = 0.25$.

14.28. Test of estimate of probability of event from k samples. In k independent sets of observations, with $n_1, n_2, \cdots, n_k$ observations, respectively, event E has occurred $v_1, v_2, \cdots, v_k$ times. It is desired to test whether this is consistent with the hypothesis that the probability of occurrence of E is $p = 1 - q$. This hypothesis may be tested in various ways, e.g.,

(a) *First test*: based on all observations together, treated as in §14.27

with $n = \sum n_i$, $v = \sum v_i$, χ^2 as in (14.27.1), with $m = 1$ degree of freedom.

(b) *Second test*: based on the ith sample, treated as in §14.27, with $n = n_i$, $v = v_i$, with $m = 1$ degree of freedom,

$$(14.28.1) \qquad \chi_i^2 = (v_i - n_i p)^2 / n_i pq.$$

(c) *Third test*: based on a composite distribution obtained from the sample distributions with $m = k$ degrees of freedom and

$$\chi_t^2 = \chi_1^2 + \chi_2^2 + \cdots + \chi_k^2.$$

(d) *Fourth test*: based on sample to sample distribution. Use χ^2-test to determine how the distribution of the quantities χ_1^2, χ_2^2, $\cdots$, χ_k^2 agrees with the χ^2-distribution with $k - 1$ degrees of freedom.

EXAMPLE 1.   If $k = 4$ and $n_1 = 50, v_1 = 22, n_2 = 35, v_2 = 16, n_3 = 40$, $v_3 = 28, n_4 = 46, v_4 = 12$, the hypothesis that $p = 0.40$ may be tested in various ways as follows, in accordance with the above outline.

(a) $n = \sum n_i = 171,$ $v = \sum v_i = 78,$ $m = 1,$

$$\chi^2 = (v - np)^2 / npq = 2.24.$$

From the tables, values of χ^2 this large might be expected in more than 10% of samples, so the hypothesis $p = 0.40$ is not indicated to be unreasonable by this test.

(b) $\chi_1^2 = 0.33,$ $\chi_2^2 = 0.48,$ $\chi_3^2 = 15,$ $\chi_4^2 = 3.7;$

in each case $m = 1$.
From the tables, if the hypothesis that $p = 0.40$ is correct, 0.1% of samples will yield values of χ^2 larger than 10.8, 1% larger than 6.6, 2% larger than 5.4, 5% larger than 3.8. Thus the first two samples lead to little doubt about the hypothesis $p = 0.40$, the third sample has probability less than 0.001 of coming from a population with $p = 0.40$, and the fourth sample is a borderline case whose value of χ^2 is large enough so that it might arise by chance in only about 5% of samples.

(c) $\chi_t^2 = 19.5$, with $m = 4$ degrees of freedom. From the tables, there is less than 0.1% chance that χ_t^2 would be this large in a random sample from a population characterized by $p = 0.40$. It is thus extremely unlikely that the hypothesis $p = 0.40$ is correct.

(d) Since there are only four values of χ_i^2 in this example, the distribution of these values can not be used for a valid χ^2-test.

It might be concluded in the above numerical example that the hypothesis $p = 0.40$ is extremely unlikely (probability less than 0.001 in this example) to be correct. The first test (a) did not show this, since all the available information, i.e., the observations of the separate samples, was not used. The samples are so different, as indicated in (b), that there might be some doubt whether they were obtained in a random manner.

14.29. Test of fit of sample to Poisson distribution. To test the hypothesis that a sample of non-negative integers $x_1, x_2, \cdots, x_n$ has been drawn from a population having some Poisson distribution with unknown parameter λ proceed as follows: Divide all non-negative integers i into r intervals, where the first interval contains the integers $0 \leqq i < c_1$, the second interval contains the integers $c_1 \leqq i < c_2$, and so on with the rth interval containing the integers $c_{r-1} \leqq i < \infty$, where $c_1, c_2, \cdots$ are positive integers. Let the sample frequency of occurrence of values x_j in the kth interval be v_k. For a Poisson distribution, the probability that $x = i$ is $p_i = (\lambda^i/i!)e^{-\lambda}$. (See §9.1.) The population frequency of occurrence of values in the kth interval is $w_k = \sum np_i$, where the sum runs from $i = c_{k-1}$ to $i = c_k - 1$, and where $c_r = \infty$. From (14.16.1),

$$(14.29.1) \qquad \chi^2 = \sum_{k=1}^{r} (v_k - w_k)^2/w_k.$$

A minimum χ^2 is assured if λ is taken as $\lambda = \bar{x} = (1/n) \sum_{j=1}^{n} x_j$. The limiting χ^2-distribution has $r - 2$ degrees of freedom. The test proceeds as in §14.22.

14.30. Test of fit of sample to normal distribution. To test the hypothesis that a sample $x_1, x_2, \cdots, x_n$ has been drawn from a normal population of variable x with unknown parameters $\bar{x}$ and σ proceed as follows: Select r class intervals for x. Let the sample frequency in class interval i be v_i. Let ξ_i be the midpoint of the ith interval, h the length of the intervals. The number of values of x from a sample of size n which are expected to fall in the ith interval I is

$$(14.30.1) \qquad np_i = (n/\sigma\sqrt{2\pi}) \int_I \exp\left(-[x - \bar{x}]^2/2\sigma^2\right) dx,$$

where the integral extends over the ith interval. From (14.16.1)

$$(14.30.2) \qquad \chi^2 = \sum_{i=1}^{r} (v_i - np_i)^2/np_i.$$

Estimates of $\bar{x}$ and σ^2 which minimize χ^2 are

$$(14.30.3) \qquad \bar{x}^* = (1/n) \sum_{i=1}^{r} v_i\xi_i,$$

$$(14.30.4) \qquad \sigma^{*2} = (1/n) \sum_{i=1}^{r} v_i(\xi_i - \bar{x}^*)^2 - (h^2/12).$$

The last term, involving h^2, is Sheppard's correction for using the midpoint of the x interval. (§6.47.) The limiting χ^2-distribution has $r - 3$ degrees of freedom. The test proceeds as in §14.22.

EXAMPLE 1.   Table 3.3.2 gives the following sample of data on measurements of the lengths of rods.

i :	1	2	3	4	5	6	7	8	9
ξ_i:	2.40	2.45	2.50	2.55	2.60	2.65	2.70	2.75	2.80
v_i:	1	4	4	19	6	11	11	3	1

$n = 60$, $h = 0.05$, $\bar{x}^* = 2.60$, $\sigma_x^* = 0.088$.

Can one consider that these data have been drawn from a normal population? From (14.30.2) it is found that $\chi^2 = 12.9$, with $r = 9$, $b = 2$, $m = r - 1 - b = 6$ degrees of freedom. (§14.23.) The χ^2 table indicates that slightly less than 5% of samples from a normal population have a value of χ^2 as large as 12.9. This would not usually be considered sufficient evidence to indicate serious doubt about the normality of the population, although the sample clearly does not itself have a distribution very close to a normal distribution. The actual "fit" of a normal distribution to the sample distribution is shown in Table 10.9.1.

14.31. Test of whether s samples come from the same population. Suppose s different samples have sizes $n_1, n_2, \cdots, n_s$, respectively. Let $\sum_{j=1}^{s} n_j = n$. It is desired to test the hypothesis that these samples can be considered all to come from the same population. Divide the range of the variable into r class intervals. If the samples all come from one population, there is a population relative frequency p_i for the ith class interval which can be compared with results obtained in all the samples. Let v_{ij} be the observed sample frequency for the ith interval from the jth sample. Estimate $p_i = (1/n) \sum_j v_{ij}$. The test is made as outlined in §14.22, with $m = (r - 1)(s - 1)$ degrees of freedom, and

$$(14.31.1) \qquad \chi^2 = \sum_{i,j} (v_{ij} - n_j p_i)^2 / n_j p_i.$$

The summation includes all values $1 \leq i \leq r$, $1 \leq j \leq s$. The larger the value of χ^2, the less likely that the samples all come from the same population.

14.32. Case $r = 2$. A special case is often of interest, where the samples consist of a series of trials, and in any trial an event E may or may not occur. Suppose there are v_j occurrences of E in the n_j trials of the jth sample. There are two "class intervals," namely, E occurred, E did not occur, so $r = 2$. The corresponding population relative frequencies may be denoted by p and $q = 1 - p$. Estimate $p = (1/n) \sum_j v_j$. It is desired to test the hypothesis that E has a constant, but unknown, probability p throughout the series of trials. The test proceeds as in §14.22, with $m = s - 1$ degrees of freedom, and

$$(14.32.1) \qquad \chi^2 = \sum_j (v_j - n_j p)^2 / n_j pq = (1/pq)(\sum_j v_j^2 / n_j) - (np/q).$$

The summation includes all values $1 \leq j \leq s$.

EXAMPLE 1. Suppose that in the four quarters of a particular year the numbers of people in a given community who are affected with a particular disease are $n_1 = 61$, $n_2 = 50$, $n_3 = 23$, and $n_4 = 46$, respectively. Suppose that the corresponding numbers of fatalities from the disease are $v_1 = 12$, $v_2 = 12$, $v_3 = 8$, and $v_4 = 14$, respectively. Is the disease, once contracted, significantly more likely to be fatal at one time of the year than another, or can the variations in fatality rates observed in this particular year be attributed to chance? Select a significance level $\epsilon = 0.05$. Calculate $n = 180$, and then the estimated fatality rate for the whole year, $p = (1/n) \sum_{i=1}^{4} v_i = 0.256$. The quantity χ^2 as given by (14.32.1) is then found to be $\chi^2 = 2.5$. From Table XIV it is found that the chance that χ^2 exceeds $\chi_0^2 = 7.8$ is $\epsilon = 0.05$. Since $\chi^2 = 2.5$ is much less than $\chi_0^2 = 7.8$, the test does not indicate any significant difference in fatality rate at the different periods of the year.

14.33. **Contingency table. A test for statistical independence.** Suppose n objects can be classified according to two parameters x and y, with frequencies v_{ij} in the various class intervals as indicated in the following table, called a *contingency table*. The midpoints of the x class intervals are indicated by the values x_j; of the y class intervals, by y_i.

	x_1	x_2	$\cdots$	x_j	$\cdots$	x_s	Total row frequency
y_1	v_{11}	v_{12}	$\cdots$	v_{1j}	$\cdots$	v_{1s}	h_1
y_2	v_{21}	v_{22}	$\cdots$	v_{2j}	$\cdots$	v_{2s}	h_2
.	.	.		.		.	.
.	.	.		.		.	.
.	.	.		.		.	.
y_i				v_{ij}			h_i
.				.			.
.				.			.
.				.			.
y_r	v_{r1}	v_{r2}	$\cdots$	v_{rj}	$\cdots$	v_{rs}	h_r
Total column frequency	k_1	k_2	$\cdots$	k_j	$\cdots$	k_s	n

In the table,

$$h_i = \sum_{j=1}^{s} v_{ij}, \qquad k_j = \sum_{i=1}^{r} v_{ij}, \qquad n = \sum_{i=1}^{r} h_i = \sum_{j=1}^{s} k_j.$$

14.34. Test. It is desired to test whether the variables x and y affect the frequencies v_{ij} independently in the statistical sense. If so, there is a distribution function $p_{ij} = p_i' p_j''$, where p_i' is the relative frequency for the ith class interval of y, and p_j'' is the relative frequency for the jth class interval of x. Estimate $p_i' = h_i/n$, $p_j'' = k_j/n$. Then from (14.16.1)

$$(14.34.1) \qquad \chi^2 = n \sum_{i,j} \frac{(v_{ij} - h_i k_j/n)^2}{h_i k_j} = n\left(\sum_{i,j} \frac{v_{ij}^2}{h_i k_j} - 1\right),$$

where the sum is carried out over $i = 1$ to $i = r$, $j = 1$ to $j = s$. The limiting χ^2-distribution has $(r-1)(s-1)$ degrees of freedom. The test proceeds as in §14.22.

14.35. The expected value of χ^2 is $n(r-1)(s-1)/(n-1)$. "Large" values of χ^2 indicate that the hypothesis that the population frequency distribution can be factored in the form $p_{ij} = p_i' p_j''$ is doubtful. (See §14.34 for test to determine whether the observed value of χ^2 is large enough to cast doubt on this hypothesis.)

EXAMPLE 1.   A sample of the houses in a certain district which have been inhabited for at least one year might be classified to give the distributions of the sample according to the total income per year, x, of the inhabitants of the houses, and the tax value, y, of the houses. The χ^2-test could then be used to test the hypothesis that the quantities x and y are statistically independent.

EXAMPLE 2.   A range test of 150 shells gave the data of Table 14.35.1. The entries v_{ij} give the number of shells fired which fell in the indicated range and deflection intervals. It is desired to test the hypothesis that at the 5 percent level of significance the range and deflection are (statistically) independent characteristics.

TABLE 14.35.1

Contingency table for test with 150 shells

Range, yd	Deflection, mils			Total
	-200 to -51	-50 to 49	50 to 249	
0 to 1,000	10	18	14	42
1,001 to 2,000	14	6	18	38
2,001 to 3,000	16	42	12	70
Total	40	66	44	150

In the notation of §14.33, $v_{11} = 10$, $v_{12} = 18$, $\cdots$, $h_1 = 42$, $\cdots$, $k_1 = 40$, $\cdots$, $k_3 = 44$, $n = 150$, $r = 3$, $s = 3$, and $(r-1)(s-1) = 4$ is the degrees of freedom. From Table XIV the chance that x^2 exceeds $x_0^2 = 9.488$ is $\epsilon = 0.05$. From (14.34.1), calculation gives $x^2 = 20.9$. Since $20.9 > 9.488$, the hypothesis that the lateral deflection and range are independent characteristics is rejected at the 5 percent level.

14.36. **Two-by-two case.** The particular case of a two-by-two table is of common interest, for which the classification according to x is "succeed" or "fail", and according to y is "succeed" or "fail"; or a similar two-way classification. In this case, $r = s = 2$ and there is one degree of freedom, with

$$(14.36.1) \qquad \chi^2 = n\, \frac{(v_{11}v_{22} - v_{12}v_{21})^2}{h_1 h_2 k_1 k_2}.$$

EXAMPLE 1. A lot of 132 fuses is tested under water and in air. If the fuses perform properly in a test the result is scored a success, otherwise a failure. The data obtained are shown in Table 14.36.1. The entries v_{ij} give the number of fuses which fell in the indicated categories. It is desired to test the hypothesis that there is no significant difference at the 5 percent level in the proportion of fuse failures under the two environments.

TABLE 14.36.1

Contingency table for fuses tested

Score	Fuses tested		Total
	Under water	In air	
Success	63 (63.5)	54 (53.5)	117
Failure	9 (8.5)	6 (6.5)	15
Total	72	60	132

In the notation of §14.36, $n = 132$, $r = s = 2$, $v_{11} = 63$, $v_{12} = 54$, $v_{21} = 9$, $v_{22} = 6$, $h_1 = 117$, $h_2 = 15$, $k_1 = 72$, $k_2 = 60$.

The expected number of failures in air is $(60)(15)/132 = 6.8$, which exceeds $v_{22} = 6$. If a continuity correction (see §8.7) is applied entry 6 is replaced by 6.5, and the other entries in the table are adjusted to leave the marginal totals unchanged. The entries obtained in this way are indicated in the table by the figures in parentheses.

By Eq. (14.36.1) the x^2 as calculated from the "corrected" entries is $x^2 = 0.031$. From Table XIV, for one degree of freedom, the chance that x^2 exceeds $x_0^2 = 3.841$ is $\epsilon = 0.05$. Since $0.031 < 3.841$ the hypothesis that the proportion defective is the same under water as in air cannot be rejected at the 5 percent level.

14.37. Examples of significance tests for parameter values. In accordance with §14.6, a significance test can be based upon any quantities for which a sampling distribution is known. A few examples illustrating the procedures are given below.

14.38. Significance test for population proportion. Suppose that an event E may or may not occur in each trial in a population of trials, and that the probability of the occurrence of E in a single trial is p. Here p is the *population proportion*. Suppose that p is unknown, but that in a sample of n trials the event E has occurred nh_1 times. Here h_1 is a particular value of the *sample proportion* h. It is desired to test whether this sample result is consistent, at the significance level ϵ, with the hypothesis that p has the particular value p_0.

14.39. It can usually be assumed, on the basis of additional knowledge of the problem, that the sample proportion h is distributed according to some sampling frequency distribution function $f_n(h)$, such as a binomial or a normal distribution. It is then possible to compute a quantity $k(\epsilon)$ such that

$$(14.39.1) \qquad \int_{p_0 - k(\epsilon)}^{p_0 + k(\epsilon)} f_n(h) \, dh = 1 - \epsilon.$$

Here ϵ is the probability that the sample proportion h will have a value outside the interval $p_0 - k(\epsilon) < h < p_0 + k(\epsilon)$. If the observed value h_1 is such that $| h_1 - p_0 | > k(\epsilon)$, then the hypothesis that the population proportion is p_0 is rejected at the significance level ϵ.

EXAMPLE 1.   A die is thrown 90 times and a three-spot is observed 18 times. Is the die true? Adopt the significance level $\epsilon = 0.01$. If the die is true, then the probability of occurrence of a three-spot in a single trial is $p_0 = 1/6 = 0.167$. The test is applied to determine whether the observed frequency $18/90 = 0.200$ is consistent with this assumed value of p_0. The sampling distribution of the proportion h of three-spots obtained in groups of 90 throws is assumed to be the binomial distribution. That is, the probability of obtaining exactly proportion $h = m/90$ three-spots in 90 throws is $f_{90}(h) = C_m^{90}(1/6)^m(5/6)^{90-m}$. (§8.3.) Here m is the number of observed three-spots in 90 throws. The standard deviation σ_m of m is $(90p_0q_0)^{1/2} = 3.54$, (§8.2), and the standard deviation σ_h of h is

3.54/90 = 0.039. The function $f_{90}(h)$ is closely approximated (§8.5, Ex. 5) by the normal density distribution $\psi(t)$ when $t = (h - p_0)/\sigma_h$, with mean $p_0 = 0.167$ and standard deviation $\sigma_h = 0.039$. From the normal probability Table IX,

$$\int_{p_0-k(\epsilon)}^{p_0+k(\epsilon)} f_{90}(h) \, dh \cong \int_{-t_1}^{t_1} \psi(t) \, dt = 1 - \epsilon = 0.99$$

where $t_1 = 2.576$. Hence $k(\epsilon) = 2.576\sigma_h = 0.100$. Since the discrepancy between the assumed p_0 and the observed proportion is $0.200 - 0.167 = 0.033$, which is much less than $k(\epsilon) = 0.100$, it is concluded that the sample observed does not itself give reason to suspect that the die is not true. In fact, by reference to normal distribution tables again it is seen that a deviation of h from the mean value p_0 of h would be expected to be as large as 0.033 in about 40% of all possible samples of size 90. [See §14.27 for a similar example worked out using the χ^2-test.]

EXAMPLE 2. If in 90 throws of a die a three-spot is observed 2 times, a calculation similar to that in Example 1 above shows that the discrepancy between the assumed p_0 and the observed proportion $(2/90 = 0.022)$ is $0.167 - 0.022 = 0.145$. Since this is larger than $k(\epsilon) = 0.100$ it is concluded that the hypothesis that the die is true (i.e., $p_0 = 0.167$) must be rejected at the significance level $\epsilon = 0.01$.

14.40. Significance test for mean of normal population. Let the hypothesis be that the mean of a normal population takes the particular value $\bar{x}$. If this hypothesis is true, then the likelihood that a sample selected at random from the population will have mean $\bar{x}$ and standard deviation s can be determined from the t-distribution of $t = (\bar{x} - \bar{x}) \cdot \sqrt{n - 1}/s$, with $n - 1$ degrees of freedom. Select a level of significance ϵ. Table XII indicates the value of t_ϵ such that the chance is ϵ that $|t| > t_\epsilon$. If the value of t observed in a sample is such that $|t| > t_\epsilon$, the hypothesis that the population mean is $\bar{x}$ would be rejected at the level of significance ϵ. [Here s is as in Eq. (13.7.1).]

EXAMPLE 1. Suppose a population is known to be normal, and the hypothesis to be tested is that $\bar{x} = 3.5$. Suppose that $\epsilon = 0.01$ is selected. In a sample of $n = 20$, suppose values $\bar{x} = 4.5$, $s = 1.3$ are observed. $t_\epsilon = 2.861$ from Table XII. From the sample, $t = (4.5 - 3.5)\sqrt{19}/1.3 = 3.35$. Thus $|t| > t_\epsilon$, i.e., the probability is less than $\epsilon = 0.01$ that such a sample would be drawn from a normal population whose mean is $\bar{x} = 3.5$. It is concluded, at the 0.01 level of significance, that the population mean is not $\bar{x} = 3.5$. If such a test were applied in a large number of separate instances, in about 0.01 of the instances when the assumed value of $\bar{x}$

was rejected by the test, the test would not in fact justify rejecting the assumed value of $\bar{x}$.

14.41. Significance test for difference between means of two samples.
Let $\bar{x}_1$ and $\bar{x}_2$ be the means of two independent samples of sizes n_1 and n_2 from normal populations, with corresponding sample variances $s_1{}^2$ and $s_2{}^2$. It is desired to test whether the samples come from the same normal population. If this hypothesis is correct, the quantity

$$(14.41.1) \qquad t = (\bar{x}_1 - \bar{x}_2)[(n_1 + n_2 - 2)n_1n_2]^{1/2}$$
$$\div \; [(n_1 + n_2)(n_1s_1{}^2 + n_2s_2{}^2)]^{1/2}$$

is distributed according to the t-distribution $g_m(t)$ (§13.64c) with $m = n_1 + n_2 - 2$ degrees of freedom. Select a level of significance ϵ. From Table XII determine t_ϵ such that the chance that $|t| > t_\epsilon$ is ϵ. If the value of t calculated according to (14.41.1) from the samples exceeds t_ϵ, the hypothesis that the two samples come from the same normal population would be rejected at the level ϵ. (See §13.7.)

If there is reason to assume that $\sigma_1 = \sigma_2$, the rejection of the hypothesis implies that the population means $\bar{x}_1$ and $\bar{x}_2$ are different.

EXAMPLE 1. Suppose sample 1 from a normal population has size $n_1 = 15$, mean $\bar{x}_1 = 4$, and variance $s_1{}^2 = 3$, and sample 2 from a possibly different normal population has size $n_2 = 22$, mean $\bar{x}_2 = 3$, and variance $s_2{}^2 = 8$. It is to be estimated whether the two samples come from the same normal population. Select significance level $\epsilon = 0.05$. From Table XII, $t_\epsilon = 2.03$, for $m = n_1 + n_2 - 2 = 35$. The value of t calculated from the samples according to formula (14.41.1) is $t = 1.2$. Since t is less than t_ϵ, the test does not give reason to doubt that the two samples come from the same population, at significance level 0.05.

14.42. Significance test for difference between variances of two samples. Let $s_1{}^2$ and $s_2{}^2$ be the variances of two samples, of sizes n_1 and n_2. It is desired to test whether the two samples come from the same normal population. On the basis of §13.38, unbiased estimates of the population variance are

$$(14.42.1) \qquad \acute{\sigma}_1{}^2 = n_1s_1{}^2/(n_1 - 1), \qquad \acute{\sigma}_2{}^2 = n_2s_2{}^2/(n_2 - 1).$$

Suppose $\acute{\sigma}_1{}^2 > \acute{\sigma}_2{}^2$. The quantity $F = \acute{\sigma}_1{}^2/\acute{\sigma}_2{}^2$ has the F-distribution (§13.65). Choose a level of significance ϵ. In Table X find F_ϵ, corresponding to $n_1 - 1$, $n_2 - 1$ degrees of freedom, such that the chance that $F > F_\epsilon$ is ϵ. If the value of F calculated from the samples exceeds F_ϵ, then the hypothesis that the samples come from the same normal population is rejected at the level ϵ.

EXAMPLE 1. Suppose two samples are observed with $n_1 = 40$, $s_1{}^2 = 6$, $n_2 = 70$, $s_2{}^2 = 5.2$. Then $F = 1.17$. For $\epsilon = 0.01$, $F_\epsilon \approx 1.89$. Since $F < F_\epsilon$, the test does not itself give reason to doubt the hypothesis that the samples come from the same normal population.

Confidence intervals

14.43. Confidence intervals. Consider a sample $x_1, x_2, \cdots, x_n$ from a population with distribution density function $f(x;\tilde{a})$, where $\tilde{a}$ is a fixed but unknown value of a parameter a. By an operation G (described below) on the particular sample, an interval I, $c_1 \leqq a \leqq c_2$, can sometimes be determined in which the true value $\tilde{a}$ of a can be asserted to lie *with confidence β*, in the following sense. Suppose a sequence of m independent samples is drawn from the same population. By means of the operation G, determine the corresponding intervals $I_1, I_2, \cdots, I_m$ for all of the m samples. The value $\tilde{a}$ will be contained within some number k of these intervals. Then k/m converges stochastically to β as m increases indefinitely. Consequently, the probability that the interval I determined from any one sample will contain the true value $\tilde{a}$ of a is β.

14.44. The operation G for determining the bounds c_1 and c_2 of the interval I from a sample is as follows. Let a^* be an estimate of a based on a sample. For any particular value of a, let the sampling distribution of a^* be $A(a^*;a)$. By some scheme such as illustrated in §14.49, determine two functions $g_1(a,\beta)$ and $g_2(a,\beta)$ such that the probability that $g_1(a,\beta) \leqq a^* \leqq g_2(a,\beta)$, for any particular values of a and β, is β,

$$(14.44.1) \qquad P(g_1(a,\beta) \leqq a^* \leqq g_2(a,\beta)) = \int_{g_1(a,\beta)}^{g_2(a,\beta)} A(a^*;a) \, da^* = \beta.$$

For a fixed β, there are many ways of selecting $g_1 = g_1(a,\beta)$ and $g_2 = g_2(a,\beta)$ so that (14.44.1) holds. Thus, if $\epsilon_1 + \epsilon_2 = \epsilon = 1 - \beta$, g_1 and g_2 may be found from

$$(14.44.2) \qquad \int_{-\infty}^{g_1} A(a^*;a) \, da^* = \epsilon_1, \qquad \int_{g_2}^{\infty} A(a^*;a) \, da^* = \epsilon_2.$$

As a is varied, with β fixed, the points (a,g_1) and (a,g_2) describe two curves as shown in Fig. 14.44.1. It is assumed that any parallel to the a-axis cuts each of these curves exactly once. Let a_1^* be the value of a^* obtained from a particular sample. The line $a^* = a_1^*$ cuts the curves g_2 and g_1 in points whose abscissae are $c_1 = c_1(a_1^*,\beta)$, $c_2 = c_2(a_1^*,\beta)$, respectively. Thus the interval I is obtained by taking c_1 as the value of a such that $g_2(a,\beta) = a_1^*$, and c_2 as the value of a such that $g_1(a,\beta) = a_1^*$.

14.45. The interval I is a *confidence interval*. The quantities c_1 and c_2 are *confidence limits*, sometimes called *fiducial limits*. The quantity β is the *confidence coefficient*, and $\epsilon = 1 - \beta$ is the *confidence level*. The random interval (c_1,c_2) is an *interval estimator* for a.

Since the functions $g_1(a,\beta)$ and $g_2(a,\beta)$ may be selected in many ways in (14.44.1), confidence intervals are not unique. It is usually desirable to use schemes for calculating c_1 and c_2 which yield confidence intervals as short as possible.

14.46. When $\epsilon_1 = \epsilon_2$ the confidence intervals are *two-sided*. If one does not care how much the parameter $\tilde{a}$ being estimated may err in one direction, provided it is not too far off in the other, the interest may be in *one-sided* confidence intervals, e.g., confidence intervals calculated for $\epsilon_1 = 0$ and $\epsilon_2 = \epsilon$; or for $\epsilon_2 = 0$ and $\epsilon_1 = \epsilon$. Thus, if the concern is for the upper limit, the interest is in an interval for which $P(\tilde{a} < c_2) = \beta = 1 - \epsilon$; if the concern is for the lower limit, the interest is in an interval for which $P(\tilde{a} > c_1) = \beta$.

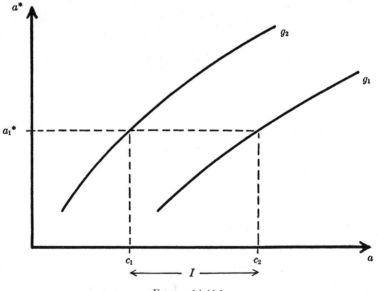

FIGURE 14.44.1

14.47. When the theory outlined above is applied to *discrete* distributions involving one unknown parameter a, the only changes are that (1) Eq. (14.44.1) should read $P(g_1(a,\beta) \leq a^* \leq g_2(a,\beta)) \geq \beta$, and (2) the probability that the interval I determined from any one sample will contain the value $\tilde{a}$ is *at least equal to* β.

14.48. Case of several parameters in the population distribution. The case when the population contains several unknown parameters a_1, a_2, $\cdots$, a_r follows in a similar manner, except that the discussion leads to a definition of a *confidence region* for the "point" $(a_1, a_2, \cdots, a_r)$ corresponding to the confidence level ϵ.

14.49. Confidence interval for mean, $\bar{x}$, of a normal population with known σ. Suppose a sample contains n items, and that the mean of the particular sample is $\bar{x}_1$. The mean of a sample is taken as an approximation to the mean of the population. The sampling distribution of sample means $\bar{x}$ of samples of size n is normal (§13.36), with the mean value of $\bar{x}$ equal to $\bar{x}$, and the standard deviation of values of $\bar{x}$ equal to $\sigma/\sqrt{n}$. Thus if $\int_0^{\lambda_\beta} \psi(t)\, dt = \beta/2$, where $\psi(t)$ is the normal density function (§10.3), the probability is β that $\bar{x}$, in the sampling distribution of $\bar{x}$, falls between $g_1(\bar{x},\beta) = \bar{x} - \lambda_\beta\sigma/\sqrt{n}$ and $g_2(\bar{x},\beta) = \bar{x} + \lambda_\beta\sigma/\sqrt{n}$. The calculation of λ_β is facilitated by Table IX and Table 10.8.1. In accordance with §14.44, the limits c_1 and c_2 of a confidence interval for $\bar{x}$, with confidence coefficient β, are obtained by taking c_1 as the value of $\bar{x}$ such that $g_2(\bar{x},\beta) = \bar{x}_1$, and c_2 as the value of $\bar{x}$ such that $g_1(\bar{x},\beta) = \bar{x}_1$. This yields the confidence limits $c_1 = \bar{x}_1 - \lambda_\beta\sigma/\sqrt{n}$ and $c_2 = \bar{x}_1 + \lambda_\beta\sigma/\sqrt{n}$. If it is asserted, whenever a sample is drawn, that the unknown population mean $\bar{x}$ lies between the limits c_1 and c_2 calculated for that particular sample, then the probability that the assertion is correct is β.

EXAMPLE 1. Suppose that a sample of $n = 34$ values is known to come from a normal population with population standard deviation $\sigma = 8.2$ and unknown population mean $\bar{x}$. If the sample mean is $\bar{x}_1 = 12.1$, a confidence interval for the population mean $\bar{x}$ is obtained as follows, corresponding to confidence coefficient $\beta = 0.95$. From Table IX, $\lambda_\beta = 1.96$. Then confidence limits for $\bar{x}$, with $\beta = 0.95$, are $c_1 = 12.1 - 1.96(8.2/\sqrt{34}) = 9.3$ and $c_2 = 12.1 + 1.96(8.2/\sqrt{34}) = 14.9$. It can be asserted with confidence 0.95 that the population mean $\bar{x}$ lies in the interval $9.3 \leq \bar{x} \leq 14.9$.

It may be noted that if the same type of calculation is carried through for $\beta = 0.99$, the resulting confidence interval for $\bar{x}$ is $8.5 \leq \bar{x} \leq 15.7$. On the basis of a given sample, the larger the confidence coefficient β, the larger is the confidence interval for $\bar{x}$.

14.50. Confidence interval for mean, $\bar{x}$, of a normal population with unknown σ. Suppose that a sample of size n has mean $\bar{x}$ and standard deviation s. The quantity $t = \sqrt{n - 1}\,(\bar{x} - \bar{x})/s$ has the t-distribution $g_m(t)$ with $m = n - 1$ degrees of freedom (Eq. 13.59.1). (See §13.7.)

(a) *In case $n < 30$:* From Table XII find t_ϵ such that $2\int_0^{t_\epsilon} g_m(t)\, dt =$

$\beta = 1 - \epsilon$. Then in fraction β of samples of size n, $-t_\epsilon \leq \sqrt{n-1} \cdot (\bar{x} - \hat{x})/s \leq t_\epsilon$. This relation is independent of σ. Rearranging yields a confidence interval $\bar{x} - t_\epsilon s/\sqrt{n-1} \leq \hat{x} \leq \bar{x} + t_\epsilon s/\sqrt{n-1}$ for the mean $\hat{x}$, corresponding to the confidence level ϵ.

(b) *In case* $n \geq 30$: The tables of $g_m(t)$ do not extend to all values $m \geq 30$. In this case, $t = \sqrt{n-1}\,(\bar{x} - \hat{x})/s$ is approximately normally distributed about zero with unit standard deviation. (§13.60.) Find t_ϵ such that $2 \int_0^{t_\epsilon} \psi(t)\, dt = \beta$. (See Table XII.) The confidence interval for $\hat{x}$ is then found as in (a) above, using the value t_ϵ found from the normal distribution.

EXAMPLE 1.   Suppose that a sample of 18 values from a normal population has sample mean $\bar{x} = 12.1$ and sample standard deviation $s = 7.5$, and that the population mean $\hat{x}$ and standard deviation σ are unknown. In order to find a confidence interval for $\hat{x}$ with confidence coefficient $\beta = 0.95$, t_ϵ is found from Table XII to be $t_\epsilon = 2.11$, using $m = 17$. Then the confidence limits for $\hat{x}$ are $c_1 = 12.1 - 2.11(7.5/\sqrt{17}) = 8.3$ and $c_2 = 12.1 + 2.11(7.5/\sqrt{17}) = 15.9$, for confidence coefficient 0.95.

14.51. **Confidence interval for the difference in means, $\hat{x}_1 - \hat{x}_2$, of two normal populations having the same but unknown σ.**   Suppose there are n_1 items in a sample from one population, and n_2 items in a sample from the other population. Let $\bar{x}_1, \bar{x}_2$ be the means, and s_1, s_2 the standard deviations for the respective samples. The quantity

(14.51.1) $t = (\bar{x}_1 - \bar{x}_2 - \hat{x}_1 + \hat{x}_2) \cdot J$,

where

(14.51.2) $J = [n_1 n_2 (n_1 + n_2 - 2)]^{1/2} \div [(n_1 + n_2)(n_1 s_1^2 + n_2 s_2^2)]^{1/2}$,

has the t-distribution function $g_m(t)$ with $m = n_1 + n_2 - 2$. (§13.59.) From Table XII find t_ϵ such that $\int_{-t_\epsilon}^{t_\epsilon} g_m(t)\, dt = \beta = 1 - \epsilon$. It follows according to §14.44 that a confidence interval for the difference $\hat{x}_1 - \hat{x}_2$ is bounded by $\bar{x}_1 - \bar{x}_2 \mp t_\epsilon/J$, corresponding to a confidence level ϵ.

14.52. **Confidence interval for the variance σ^2 of a normal population.** Suppose a sample of size n has standard deviation s. The quantity $z_8 = ns^2/\sigma^2$ has the χ^2-distribution $f_m(w)$ with $w = z_8$ and $m = n - 1$ degrees of freedom. (Table 13.55.1.) From Table XIV find χ_2^2 such that the probability that $\chi^2 > \chi_2^2$ is $\epsilon/2$. Likewise, find χ_1^2 such that the probability that $\chi^2 > \chi_1^2$ is $1 - \epsilon/2$. Then the probability is $\beta = 1 - \epsilon$ that $\chi_1^2 \leq ns^2/\sigma^2 \leq \chi_2^2$. Rearrangement yields a confidence interval $ns^2/\chi_2^2 \leq \sigma^2 \leq ns^2/\chi_1^2$ for the variance σ^2, corresponding to a confidence level ϵ. (s as in Eq. (13.7.1).)

EXAMPLE 1.   Suppose a sample of $n = 24$ values from a normal population has sample variance $s^2 = 60$. In order to find a confidence interval for the population variance σ^2, with confidence level $\epsilon = 0.02$, enter Table XIV for $m = 23$, and find $\chi_2^2 = 41.64$, $\chi_1^2 = 10.20$. Confidence limits for the population variance σ^2, for confidence level $\epsilon = 0.02$, are then $c_1 = 24(60/41.64) = 35$ and $c_2 = 24(60/10.20) = 141$. It may be noted that a sample of size 24 is not sufficient to locate the population variance σ^2 very closely with such a high degree of confidence as $\beta = 1 - \epsilon = 0.98$.

Confidence interval for proportion of successes

14.53. Consider a population of trials such that each trial may or may not result in the occurrence of an event E. The occurrence of E will be called a "success." It is desired to estimate the *population proportion p* of all trials which yield successes, on the basis of the number S of successes observed in a sample of n trials. The quantity $p^* = S/n$ is called the *sample proportion*. If the chance of success in one trial has a fixed value p, and $q = 1 - p$, then it is ordinarily assumed that the relative chances of exactly $0, 1, \cdots, n$ successes in n trials are distributed according to the magnitudes of the successive terms of the expansion of $(q + p)^n$. In order to determine a confidence interval, it is sometimes convenient to consider this binomial distribution to be approximated by a normal or a Poisson distribution, depending on whether p, or q, is very small or not. It is also sometimes convenient to use the χ^2 technique. The appropriateness of each of these approximations depends upon the sample size, n, and the general magnitude of p. There are situations in which the normal or Poisson distributions are appropriate in themselves, not merely as approximations to the binomial distribution.

14.54. **Use of binomial distribution to find a confidence interval for proportion of successes.** If the probability of success in the population of trials is p, then the chances of obtaining $0, 1, 2, \cdots, n$ successes out of n trials are given by

$$q^n, \quad pq^{n-1}n!/1!(n-1)!, \quad p^2q^{n-2}n!/2!(n-2)!, \cdots . \qquad [\S 8.3.]$$

The expected number of successes is np. Suppose the observed number of successes is S, where $S \neq 0$. In order to find a confidence interval $p_1 \leq p \leq p_2$ with confidence level ϵ, p_1 may be selected so that the last $n + 1 - S$ terms of the expansion of $(q_1 + p_1)^n$ has the sum $\epsilon/2$. Likewise, p_2 is then selected so that the first $S + 1$ terms of the expansion of $(q_2 + p_2)^n$ has the sum $\epsilon/2$.

14.55. Two convenient procedures, using tables, for obtaining numerical values p_1 and p_2 by determining sums of terms on one end or the other of binomial expansions may be noted.

(a) The incomplete Beta function ratio $I_p(x, n - x + 1)$ [see Eq. (8.5.3)] may be used. In order to obtain p_1 by use of this ratio, set $I_p = \epsilon/2$ in Table III, and find the value p_1 of p in this table for which $x = S$. Likewise, set $I_p = \epsilon/2$, find the value p' of p from this table for which $x = n - S$, and then take $p_2 = 1 - p'$.

(b) The F-distribution [see §13.65] may be used, if $\epsilon/2$ is taken to have a value 0.05 or 0.01 which is a level of significance used in the F-table [Table X]. In order to determine p_1, find $F = F_1$ in this table corresponding to m_1 and m_2 degrees of freedom, where $m_1 = 2S$ and $m_2 = 2(n - S + 1)$. Then p_1 can be found from $(m_1/m_2)F_1 = p_1/(1 - p_1)$. In order to determine p_2, find $F = F_2$ in the table corresponding to m_1 and m_2 degrees of freedom, where $m_1 = 2(S + 1)$ and $m_2 = 2(n - S)$. Then $(m_1/m_2)F_2 = (1 - p_2)/p_2$ can be used to find p_2.

14.56. **Case when observed number of successes is zero (or n).** In the special case when $S = 0$ it does not follow that p is zero. The interval $p_1 \leqq p \leqq p_2$ for a confidence level ϵ takes the form $0 \leqq p \leqq p_2$. The largest value p_2 of p which yields a probability of at least ϵ that $p^* = S/n$ could be as small as zero is found from $(1 - p_2)^n = \epsilon$.

In case $S = n$ the interval $p_1 \leqq p \leqq p_2$ takes the form $p_1 \leqq p \leqq 1$. The smallest value p_1 of p which yields a probability of at least ϵ that $p^* = S/n$ could be as large as 1 is found from $p_1{}^n = \epsilon$.

14.57. **Tabulation of confidence intervals.** Representative confidence intervals calculated as described in §14.54 to §14.56 are tabulated in Tables 14.57.1 and 14.57.2, for confidence coefficients $\beta = 1 - \epsilon = 0.95$ and $\beta = 0.99$, respectively, (i.e., confidence levels $\epsilon = 0.05$ and 0.01, respectively.)

EXAMPLE 1. If a sample of 50 small metal clips in a large batch are inspected and 6 are found defective, what is a 95% confidence interval for the fraction of defective clips in the whole batch? Table 14.57.1 shows the confidence limits to be 0.05 and 0.24. It may be noted that if 12 defects were found in a sample of 100 clips, which is the same proportion as before, the 95% confidence interval is then reduced to 0.06 to 0.20. This narrower confidence interval can be given because of the larger sample on which it is based. On the other hand, the 99% confidence interval based on 6 defects observed in a sample of 50 is given by Table 14.57.2 as 0.03 to 0.29. This interval fixes the estimated population proportion of defects less closely than the 95% confidence intervals, illus-

TABLE 14.57.1

Confidence interval for proportion of successes

Confidence Coefficient $\beta = 0.95$,
(i.e. Confidence Level $\epsilon = 1 - \beta = 0.05$).
[See §14.54 to §14.56.]

The two entries for each sample size, n, and number of successes observed, S, are the confidence limits for proportion of successes in the population.

Number of Successes Observed, S	Sample size, n											
	10		15		20		30		50		100	
0	0	.31	0	.22	0	.17	0	.12	0	.07	0	.04
1	0	.45	0	.32	0	.25	0	.17	0	.11	0	.05
2	.03	.56	.02	.40	.01	.31	.01	.22	0	.14	0	.07
3	.07	.65	.04	.48	.03	.38	.02	.27	.01	.17	.01	.08
4	.12	.74	.08	.55	.06	.44	.04	.31	.02	.19	.01	.10
5	.19	.81	.12	.62	.09	.49	.06	.35	.03	.22	.02	.11
6	.26	.88	.16	.68	.12	.54	.08	.39	.05	.24	.02	.12
7	.35	.93	.21	.73	.15	.59	.10	.43	.06	.27	.03	.14
8	.44	.97	.27	.79	.19	.64	.12	.46	.07	.29	.04	.15
9	.55	1.00	.32	.84	.23	.68	.15	.50	.09	.31	.04	.16
10	.69	1.00	.38	.88	.27	.73	.17	.53	.10	.34	.05	.18
11			.45	.92	.32	.77	.20	.56	.12	.36	.05	.19
12			.52	.96	.36	.81	.23	.60	.13	.38	.06	.20
13			.60	.98	.41	.85	.25	.63	.15	.41	.07	.21
14			.68	1.00	.46	.88	.28	.66	.16	.43	.08	.22
15			.78	1.00	.51	.91	.31	.69	.18	.44	.09	.24
16					.56	.94	.34	.72	.20	.46	.09	.25
17					.62	.97	.37	.75	.21	.48	.10	.26
18					.69	.99	.40	.77	.23	.50	.11	.27
19					.75	1.00	.44	.80	.25	.53	.12	.28
20					.83	1.00	.47	.83	.27	.55	.13	.29
21							.50	.85	.28	.57	.14	.30
22							.54	.88	.30	.59	.14	.31
23							.57	.90	.32	.61	.15	.32
24							.61	.92	.34	.63	.16	.33

Table 14.57.1 (*Continued*)

Number of Successes Observed, S	Sample size, n							
	10	15	20	30		50		100
25				.65 .94		.36 .64		.17 .35
26				.69 .96		.37 .66		.18 .36
27				.73 .98		.39 .68		.19 .37
28				.78 .99		.41 .70		.19 .38
29				.83 1.00		.43 .72		.20 .39
30				.88 1.00		.45 .73		.21 .40
31						.47 .75		.22 .41
32						.50 .77		.23 .42
33						.52 .79		.24 .43
34						.54 .80		.25 .44
35						.56 .82		.26 .45
36						.57 .84		.27 .46
37						.59 .85		.28 .47
38						.62 .87		.28 .48
39						.64 .88		.29 .49
40						.66 .90		.30 .50
41						.69 .91		.31 .51
42						.71 .93		.32 .52
43						.73 .94		.33 .53
44						.76 .95		.34 .54
45						.78 .97		.35 .55
46						.81 .98		.36 .56
47						.83 .99		.37 .57
48						.86 1.00		.38 .58
49						.89 1.00		.39 .59
50						.93 1.00		.40 .60
								*

*If S exceeds 50, read $100 - S$ in place of S and subtract each tabulated confidence limit from 1.00.

This table is taken from Snedecor, G. W., *Statistical Methods*, Iowa State College Press, Ames, Iowa, (4th ed.) 1946, by kind permission of the author and publishers.

TABLE 14.57.2

Confidence interval for proportion of successes

Confidence Coefficient $\beta = 0.99$,
(i.e. Confidence Level $\epsilon = 1 - \beta = 0.01$).

[See §14.54 to §14.56.]

The two entries for each sample size, n, and number of successes observed, S, are the confidence limits for proportion of successes in the population.

Number of Successes Observed, S	Sample size, n										
	10		15		20		30		50		100
0	0	.41	0	.30	0	.23	0	.16	0	.10	0	.05
1	0	.54	0	.40	0	.32	0	.22	0	.14	0	.07
2	.01	.65	.01	.49	.01	.39	0	.28	0	.17	0	.09
3	.04	.74	.02	.56	.02	.45	.01	.32	.01	.20	0	.10
4	.08	.81	.05	.63	.04	.51	.03	.36	.01	.23	.01	.12
5	.13	.87	.08	.69	.06	.56	.04	.40	.02	.26	.01	.13
6	.19	.92	.12	.74	.08	.61	.06	.44	.03	.29	.02	.14
7	.26	.96	.16	.79	.11	.66	.08	.48	.04	.31	.02	.16
8	.35	.99	.21	.84	.15	.70	.10	.52	.06	.33	.03	.17
9	.46	1.00	.26	.88	.18	.74	.12	.55	.07	.36	.03	.18
10	.59	1.00	.31	.92	.22	.78	.14	.58	.08	.38	.04	.19
11			.37	.95	.26	.82	.16	.62	.10	.40	.04	.20
12			.44	.98	.30	.85	.18	.65	.11	.43	.05	.21
13			.51	.99	.34	.89	.21	.68	.12	.45	.06	.23
14			.60	1.00	.39	.92	.24	.71	.14	.47	.06	.24
15			.70	1.00	.44	.94	.26	.74	.15	.49	.07	.26
16					.49	.96	.29	.76	.17	.51	.08	.27
17					.55	.98	.32	.79	.18	.53	.09	.29
18					.61	.99	.35	.82	.20	.55	.09	.30
19					.68	1.00	.38	.84	.21	.57	.10	.31
20					.77	1.00	.42	.86	.23	.59	.11	.32
21							.45	.88	.24	.61	.12	.33
22							.48	.90	.26	.63	.12	.34
23							.52	.92	.28	.65	.13	.35
24							.56	.94	.29	.67	.14	.36

TABLE 14.57.2 (*Continued*)

Number of Successes Observed, S	Sample size, n					
	10	15	20	30	50	100
25				.60 .96	.31 .69	.15 .38
26				.64 .97	.33 .71	.16 .39
27				.68 .99	.35 .72	.16 .40
28				.72 1.00	.37 .74	.17 .41
29				.78 1.00	.39 .76	.18 .42
30				.84 1.00	.41 .77	.19 .43
31					.43 .79	.20 .44
32					.45 .80	.21 .45
33					.47 .82	.21 .46
34					.49 .83	.22 .47
35					.51 .85	.23 .48
36					.53 .86	.24 .49
37					.55 .88	.25 .50
38					.57 .89	.26 .51
39					.60 .90	.27 .52
40					.62 .92	.28 .53
41					.64 .93	.29 .54
42					.67 .94	.29 .55
43					.69 .96	.30 .56
44					.71 .97	.31 .57
45					.74 .98	.32 .58
46					.77 .99	.33 .59
47					.80 .99	34 .60
48					.83 1.00	.35 .61
49					.86 1.00	.36 .62
50					.90 1.00	.37 .63
						*

*If S exceeds 50, read $100 - S$ in place of S and subtract each tabulated confidence limit from 1.00.

This table is taken from Snedecor, G. W., *Statistical Methods*, Iowa State College Press, Ames, Iowa, (4th ed.) 1946, by kind permission of the author and publishers.

trating the general rule that the more closely it is attempted to estimate a population parameter on the basis of given sample data, the less certain it is that the estimate is correct.

EXAMPLE 2. Suppose it were desired to have 95% confidence that *at least* 80% of the boxes coming off of a packing machine weighed more than one pound. Table 14.57.1 indicates that such a conclusion could not be based on a sample of 10 boxes or of 15 boxes, even if every one of the boxes in the sample were found to weigh more than one pound. If a sample of 20 boxes were examined, every box in the sample would have to weigh over one pound in order to justify the desired conclusion. The table shows that the conclusion could be reached on the basis of 29 or 30 boxes weighing more than one pound out of a sample of 30, 46 or more out of a sample of 50, or 88 or more out of a sample of 100. It is thus clear that in order to conclude (with confidence 95%) that at least 80% of a population have some property, it is necessary that considerably more than 80% of an observed sample have the property, even for samples as large as 100.

14.58. Use of normal distribution to find confidence interval for proportion of successes. Observations in terms of proportion of successes.

The binomial frequency distribution of sample proportions $p^* = S/n$ can sometimes be approximated by a normal distribution. [§8.7.] This approximation is often justified if $n > 30$ and $0.03 \leq p \leq 0.97$, and it can sometimes be used for smaller values of n. The standard deviation of values of p^*, as obtained from the binomial distribution, is $\sigma_{p^*} = \sqrt{pq/n}$. The mean of p^*, as obtained from the binomial distribution, is the population proportion p. These values of the mean and standard deviation can be used for approximating the binomial distribution of sample proportions p^* by means of the normal distribution.

14.59. Suppose a particular value p_1^* of p^* is observed for the proportion of successes in a sample of n trials. To find a confidence interval $p_1 \leq p \leq p_2$ for the probability of success p, let $t = (p - p^*) \div (pq/n)^{1/2}$. From a normal probability table [Table IX], and for confidence level ϵ, find G such that

$$(14.59.1) \qquad (1/\sqrt{2\pi}) \int_G^\infty e^{-t^2/2} \, dt = \epsilon/2.$$

Set

$$(14.59.2) \qquad G = (p - p^*) \div (pq/n)^{1/2}.$$

Since p^* and n are known, this yields two solutions p_1 and p_2 for p. These solutions are the bounds of a confidence interval desired corresponding to confidence level ϵ.

14.60. Use of normal distribution to find confidence interval for proportion of successes. Observations in terms of number of successes. Suppose the number S of times a success occurs in n trials is distributed binomially with mean $\overline{S} = np$ and variance $\sigma_S{}^2 = npq$, $q = 1 - p$, p being the chance that a success is obtained in one trial. For n large enough, z_β can be found so that the probability that S lies between $np - z_\beta\sigma$ and $np + z_\beta\sigma$ is approximately β. For any selected confidence coefficient $\beta = 1 - \epsilon$, the corresponding value of z_β can be obtained from the normal probability table. [Table IX.] Some representative values are as follows.

β:	0.80	0.85	0.90	0.95	0.9545	0.98	0.99	0.9973
z_β:	1.282	1.440	1.645	1.960	2.000	2.326	2.576	3.000

The probability that $(S - np)/\sqrt{npq}$ lies between $-z_\beta$ and $+z_\beta$ is approximately β. Suppose that a particular value S_1 is observed for the number of successes S in a sample of n trials. Confidence limits p_1 and p_2 for the probability p of success in one trial are found by solving the equation

$$(14.60.1) \qquad (S_1 - np)/\sqrt{npq} = \pm z_\beta$$

for p.

14.61. Graphical presentation of confidence limits. Suppose that in any sample of size n the proportion of successes observed is $p^* = S/n$. For any specified value of p^* the confidence interval $p_1 \leqq p \leqq p_2$, for confidence coefficient $\beta = 0.90$, as found from (14.59.2) or (14.60.1), is the interval between the ordinates on Fig. 14.61.1 for the two curves corresponding to the sample size n. For a given p^*, the lower curve of a pair gives values of p_1, and the upper curve of a pair gives values of p_2.

This graph illustrates the reduction in length of confidence interval which can be achieved by increasing the sample size n.

Similar graphs can be constructed for other values of β, and by various other types of calculation, not necessarily based on the normal distribution.

Figures 14.61.1, 14.61.2, and 14.61.3 give *two-sided* confidence intervals for confidence coefficients $\beta = 1 - \epsilon = 0.90$, 0.95, and 0.99, for population proportion p as a function of the sample proportion $p^* = S/n$.

Figures 14.61.1, 14.61.2, and 14.61.3 can be used for *one-sided* confidence intervals with confidence coefficients $\beta = 0.95$, 0.975, and 0.995, respectively. This is done by referring to only one of a symmetrical pair of curves.

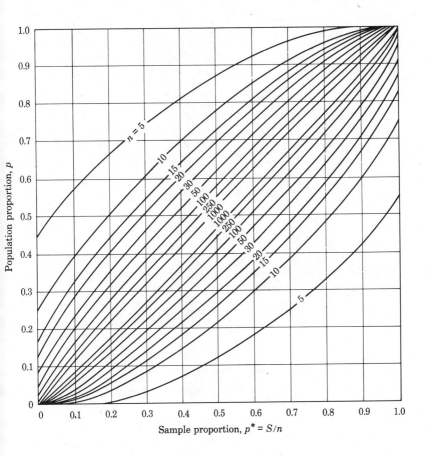

Confidence Intervals for Proportions
(for confidence coefficient $\beta = 1 - \epsilon = 0.90$)

FIGURE 14.61.1

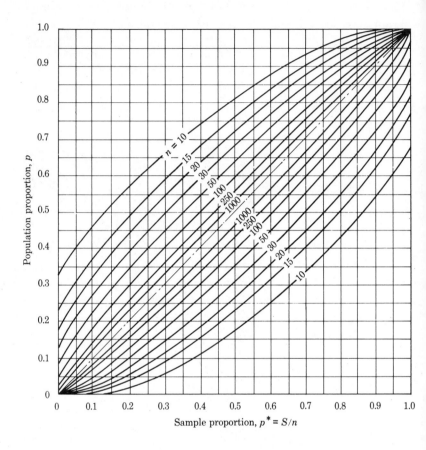

Confidence Intervals for Proportions
(for confidence coefficient $\beta = 1 - \epsilon = 0.95$)

This chart is reproduced with the permission of Professor E. S. Pearson from Clopper, C. J., and E. S. Pearson, "The Use of Confidence or Fiducial Limits Illustrated in the Case of the Binomial," *Biometrika*, Vol. 26 (1934), p. 404.

FIGURE 14.61.2

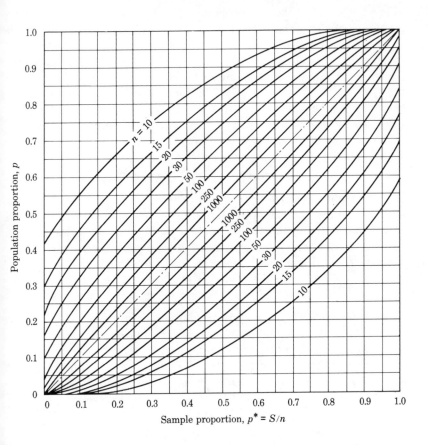

Confidence Intervals for Proportions
(for confidence coefficient $\beta = 1 - \epsilon = 0.99$)

This chart is reproduced with the permission of Professor E. S. Pearson from Clopper, C. J., and E. S. Pearson, "The Use of Confidence or Fiducial Limits Illustrated in the Case of the Binomial," *Biometrika*, Vol. 26 (1934), p. 404.

FIGURE 14.61.3

14.62. Use of Poisson distribution to find confidence interval for proportion of successes. The binomial frequency distribution of sample proportions $p^* = S/n$, where S is the observed number of occurrences of event E (successes) in n trials, can be approximated by a Poisson distribution [§9.5] under some conditions. This approximation is sometimes justified if $n > 30$ and $0 \leq p \leq 0.03$ with the expected number of successes np between 0 and 10; or $0 \leq q \leq 0.03$ with nq between 0 and 10. Let $\lambda = np$ be the parameter of the Poisson distribution. The probability of 0, 1, 2, $\cdots$ occurrences of event E (successes) out of n trials is given by the successive terms of the series

$$e^{-\lambda}(1 + \lambda + (\lambda^2/2!) + (\lambda^3/3!) + \cdots).$$

The probability of not more than S occurrences out of n trials is the sum D of the first $S + 1$ terms of this series. The probability of not less than S occurrences out of n trials is the sum T of the terms of the series after the Sth term. To find a confidence interval $p_1 \leq p \leq p_2$ for the probability of success p in a population of trials, with confidence level ϵ, select p_1 to be the smallest value of p so that $T = \epsilon/2$, and p_2 the largest value of p so that $D = \epsilon/2$.

14.63. Table VIII may be used to find D and T. Table VIII gives the sum T of all of the terms of the series after and including the $x' + 1$ term, for $\lambda = m$, the first term of the Poisson series being the one for which $x' = 0$. The sum of the first x' terms is $1 - T$. With $x' = S$, p_1 is the smallest value of $p = \lambda/n = m/n$ for which $T = T_1 = \epsilon/2$. With $x' = S + 1$, and corresponding value T_2 of T, p_2 is the largest value of $p = m/n$ for which $D = 1 - T_2 = \epsilon/2$.

14.64. Use of χ^2-distribution to find a confidence interval for proportion of successes. Suppose proportion p of an infinite set of trials would result in "successes." Let S successes be observed in n trials. In §14.27, the χ^2-test was used to determine whether the observed proportion S/n was consistent with a particular assumed value of p. The quantity

$$(14.64.1) \qquad \chi^2 = (S - np)^2/npq, \qquad q = 1 - p,$$

has the χ^2-distribution (Eq. 13.26.1) with $m = 1$ degrees of freedom. This distribution also may be used to calculate a confidence interval for values of p, based on the observed value of S, with confidence level ϵ. The use of the χ^2-distribution is often justified if both S and $n - S$ are as large as 10. It can sometimes be used when one of these class frequencies is smaller than 10.

14.65. In order to calculate confidence limits for a confidence level ϵ, first find χ^2 from Table XIV corresponding to the level $\epsilon/2$ and $m = 1$.

Then insert this value χ_1^2 of χ^2 and the observed values of S and n in Eq. (14.64.1). The resulting equation can be put in the form

$$(14.65.1) \qquad (n + \chi_1^2)p^2 - (2S + \chi_1^2)p + S^2/n = 0.$$

The two solutions for p are the bounds p_1 and p_2 of a confidence interval for p, with the confidence level ϵ.

Case of finite binomial population. Suppose a random sample of n elements is drawn without replacement from a population of N elements, in which there are only two kinds of elements: A and B. The proportion of A's in the population is p, the proportion of B's is q. Suppose the sample contains S elements of type A, $S = np^*$, p^* being the sample proportion. The distribution of S has mean np and variance $npq\ [(N - n)/(N - 1)]$. If n is large and $N > n$, S is approximately normally distributed with this mean and variance. In such a case confidence intervals for the population proportion p may be obtained as in §14.58 to §14.60 by replacing npq in (14.59.2) and (14.60.1) by $npq\ [(N - n)/(N - 1)]$.

If the sampling is made with replacement the distribution of S has mean np and variance npq. In such a case confidence intervals are obtained as in §14.58 to §14.60.

Type I and Type II errors in statistical tests

14.66. In testing a statistical hypothesis H_0 versus an alternate hypothesis H_1, two types of errors may be made.

Type I error: H_0 is rejected (H_1 is accepted) when H_0 is true.

Type II error: H_0 is accepted (H_1 is rejected) when H_1 is true.

Let the probability of making a Type I error be α. α is called *size*, or the *significance* level of the test.

Let the probability of making a Type II error be β. When testing a specific hypothesis by a given experiment, the value of $1 - \beta$ is called the *power of the test*.

The selection of values for α and β should hinge on the consequences of making Types I and II errors, respectively. For example, in the purchase of a lot of gyros, one might assume that the lot is of satisfactory quality. If the hypothesis is true and it is rejected, no particular harm to the purchaser may result (unless he is short of gyros); so α need not necessarily be very small. To avoid an unacceptable number of "duds," β should be quite small. However, the manufacturer might feel differently about the selection of β and α.

In practice, α and β are commonly chosen to be 0.01, 0.05, 0.10, or 0.20.

14.67. Power function. Suppose H_0 is expressible in terms of a single parameter μ_0, and H_1 in terms of μ_1. α is a function of μ_0 and β a function of μ_1, say $\beta(\mu_1)$. The actual (unknown) value of the parameter is μ. The function $\beta(\mu_1)$ is the *operating characteristic* (OC) *function* and the graph of $\beta(\mu_1)$ as a function of μ_1 is the *operating characteristic curve* of the test of H_0 against H_1.

The function $\gamma(\mu_1) \equiv 1 - \beta(\mu_1)$ is called the *power function* of the test. $\gamma(\mu_1)$ is the probability of rejecting H_0 if H_1 is true. *Note:* $\gamma(\mu_0) = 1 - \beta(\mu_0) = \alpha$.

The quantity μ_0 is called a *test statistic*. The set of all possible values of the test statistic is the *sample space*. The region of the sample space for which the hypothesis is rejected is the *rejection* (or *critical*) *region*. The region for which the hypothesis is accepted is the *acceptance region*.

14.68. Examples of Type I and Type II errors. Let the probability density function for the statistic μ be $f_0 = p(\mu|\mu_0)$ if H_0 is true, and $f_1 = p(\mu|\mu_1)$ if H_1 is true. Suppose the critical region is $\mu > c$.

The probability of rejecting H_0 if H_0 is true is

$$\alpha = \int_c^\infty f_0 \, d\mu$$

and the probability of accepting H_0 when H_0 is false is

$$\beta = \int_{-\infty}^c f_1 \, d\mu$$

14.69. Illustrations of Type I and Type II errors are given in the following examples involving the testing of statistical hypotheses.

EXAMPLE 1. *Test of mean of a normal population with known variance.* Suppose a sample of size n of sample mean $\bar{x}$ is drawn from a population having a normal distribution $N(\mu,\sigma^2)$, μ being unknown and σ^2 known.

Left-sided test. The hypothesis $H_0 : \mu = \mu_0$ versus alternates $H_1 : \mu = \mu_1 < \mu_0$ is to be tested.

To control the Type I error so that its probability of occurrence is α, a value k is sought such that

(14.69.1) $\alpha = P(\bar{x} < k \,|\, \mu = \mu_0)$

Let k be selected to satisfy (14.69.1). The distribution of $\bar{x}$ is $N(\mu_0, \sigma^2/n)$. Hence

(14.69.2) $\alpha = \Psi[z_{1-\alpha}]$

$$(14.69.3) \qquad \beta = P(\bar{x} > k \,|\, \mu = \mu_1) = 1 - \Psi(z_\beta)$$

where Ψ is as defined in (10.3.1) and Table 10.8.3, and

$$(14.69.4) \qquad z_{1-\alpha} = (k - \mu_0)\sqrt{n}/\sigma$$

$$(14.69.5) \qquad z_\beta = (k - \mu_1)\sqrt{n}/\sigma$$

From (14.69.4)

$$(14.69.6) \qquad k = \mu_0 + \sigma z_{1-\alpha}/\sqrt{n}.$$

With the value of k given in (14.69.6), β is given by (14.69.3). The critical region for $\bar{x}$ is the set of values of $\bar{x}$ such that $\bar{x} < k$.

Any observed value of $\bar{x}$ falling in the critical region is *significantly smaller than* μ_0 at the level of significance α.

The test described above is called a *left-sided test*.

Right-sided test. The hypothesis $H_0 : \mu = \mu_0$ versus alternates $H_1 : \mu = \mu_1 > \mu_0$ is to be tested.

The critical region for $\bar{x}$ is the set of values of $\bar{x}$ such that $\bar{x} > k$, and where k is determined by

$$(14.69.7) \qquad \alpha = P(\bar{x} > k \,|\, \mu = \mu_0).$$

Then

$$(14.69.8) \qquad \beta = P(\bar{x} \leq k \,|\, \mu = \mu_1).$$

The critical region for $\bar{x}$ is

$$(14.69.9) \qquad \bar{x} > \mu_0 + \sigma z_\alpha/\sqrt{n} \equiv k.$$

Any observed value of $\bar{x}$ falling in the critical region is *significantly larger than* μ_0 at the level α.

The test thus described is a *right-sided test*.

Two-sided test. The hypothesis $H_0 : \mu = \mu_0$ versus $H_1 : \mu = \mu_1 \neq \mu_0$ is to be tested.

The critical region for $\bar{x}$ is the set of values of $\bar{x}$ such that $|\bar{x} - \mu_0| > k$, where k is determined by

$$(14.69.10) \qquad \alpha = P(|\bar{x} - \mu_0| > k \,|\, \mu = \mu_0)$$

From (14.69.10)

$$(14.69.11) \qquad \alpha/2 = 1 - \Psi(k\sqrt{n}/\sigma)$$

The critical region for $\bar{x}$ is given by

$$(14.69.12) \qquad |\bar{x} - \mu_0| > \sigma z_{\alpha/2}/\sqrt{n} \equiv k$$

Any observed value of $\bar{x}$ falling in the critical region is *significantly different from* μ_0 at the level α.

The power of the two-sided test for H_0 versus H_1 is

$$(14.69.13) \qquad \gamma(\mu_1) = 1 - \{ \Psi[v_1 + z_{\alpha/2}] - \Psi[v_1 - z_{\alpha/2}] \}$$

where $v_1 = (\mu_0 - \mu_1)\sqrt{n}/\sigma$. The operating characteristic (OC) curve of the test of H_0 versus H_1 is $\beta(\mu_1) = 1 - \gamma(\mu_1)$. A $100(1 - \alpha)$ percent confidence interval for μ_0 is $[\bar{x} \pm \sigma z_{\alpha/2}/\sqrt{n}]$.

EXAMPLE 2.   *Test of mean of a normal population with unknown variance.*   Consider Example 1 but suppose σ^2 is unknown.

The procedures and interpretations relative to the test parallel those of Example 1.

Left-sided test.   To test hypothesis $H_0:\mu = \mu_0$ versus $H_1:\mu = \mu_1 < \mu_0$ proceed as follows.

Let the sample variance be s^2, defined as in (13.7.1) and $t = (\bar{x} - \mu_0)\sqrt{n - 1}/s$. The critical region in the $(\bar{x},s)$ plane is the set of values of $(\bar{x},s)$ for which $t < t_{1-\alpha}$, where $t_{1-\alpha}$ is the value of t in the t-distribution with $n - 1$ degrees of freedom (see §13.59 to §13.63) such that

$$P(t > t_{1-\alpha}) = 1 - \alpha$$

The probability of Type I error is

$$(14.69.14) \qquad\qquad \alpha = P[t < t_{1-\alpha} | \mu = \mu_0]$$

The power of the test of H_0 versis H_1 is

$$(14.69.15) \qquad\qquad \gamma(\mu_1) = P[t < t_{1-\alpha} | \mu = \mu_1]$$

Right-sided test.   To test the hypothesis $H_0:\mu = \mu_0$ versus $H_1:\mu = \mu_1 > \mu_0$, the critical region is the set of values of $(\bar{x},s)$ such that $t > t_\alpha$. In this case,

$$(14.69.16) \qquad\qquad \alpha = P[t > t_\alpha | \mu = \mu_0]$$

Two-sided test.   To test hypothesis $H_0:\mu = \mu_0$ versus $H_1:\mu = \mu_1 \neq \mu_0$, the critical region is the set of values $(\bar{x},s)$ such that $|t| > t_{\alpha/2}$, and

$$(14.69.17) \qquad\qquad \alpha = P[|t| > t_{\alpha/2} | \mu = \mu_0].$$

14.70. Tolerance intervals.   A tolerance interval T, $l_1 \leq x \leq l_2$, is an interval which contains a given fraction α of the values x in a (population) distribution. The quantities l_1 and l_2 are called *tolerance limits*. In the case of a normal distribution, certain tolerance intervals have been defined in §10.8 assuming the distribution of x is completely known. In the case of any population, not necessarily normal, the use of the maximum and minimum values of x in a sample drawn from the population as the bounds of a tolerance interval for population values is discussed in §13.87.

14.71. Tolerance intervals apply to the random variable x of a distribution. Confidence intervals apply generally to a parameter such as the mean, standard deviation, $\cdots$, of a distribution. Tolerance intervals correspond to statements of the type: "The probability is α that the variable x in a distribution falls between l_1 and l_2." Confidence intervals correspond to statements of the type: "The probability is β that the interval $c_1 \leqq a \leqq c_2$ contains the true value $\tilde{a}$ of a population parameter a, where c_1 and c_2 are calculated on the basis of an observed sample." The two types of intervals should be carefully distinguished.

14.72. Notation $A \pm B$ for intervals. Expressions of the form $A \pm B$ (for example, 2.11 ± 0.03) are often used to describe magnitudes of a quantity x. Such expressions may mean any of several things, such as:

(a) x has mean value A, and no values of x differ from A by more than B.

(b) x has mean A and standard deviation B.

(c) All values of x are somewhere in the interval $A - B$ to $A + B$.

(d) It is intended that x should be equal to A, but values differing from A by as much as B are acceptable.

(e) $A - B$ and $A + B$ are 95% (or 99%, $\cdots$) confidence limits for x; $\cdots$.

It is important that the meaning of $A \pm B$ be made clear in each discussion; failure to do so results in ambiguity and misunderstanding. It is unsafe to assume that there is a standard convention.

14.73. Tolerance limits for normal populations. Suppose the parent population is described by a normal distribution with mean μ and standard deviation σ. A sample of size n is drawn having mean $\bar{x}$ and unbiased sample standard deviation s.

The interval $(\mu - 1.645\sigma, \ \mu + 1.645\sigma)$ will include 0.90 of the population. Generally μ and σ are not known. If $\bar{x}$ and s are used to estimate μ and σ, respectively, it cannot be said with certainty that $(\bar{x} - 1.645s, \ \bar{x} + 1.645s)$ will include 0.90 of the population.

If many random samples are drawn from a normal population it is possible to find a number K so that a fraction γ of the intervals $(\bar{x} - Ks, \ \bar{x} + Ks)$ will cover at least a specified proportion P of the population. γ is called the *confidence coefficient*. $(\bar{x} - Ks, \ \bar{x} + Ks)$ is called a *two-sided tolerance range.*

Tables for two-sided tolerance intervals have been calculated for various values of P, γ and n. Table XX is such a table. More extensive tables may be found in

Bowker, A. H., and Lieberman, G. J., *Engineering Statistics*, Prentice-Hall, Inc., Englewood Cliffs, N.J., 1959.

National Bureau of Standards, Handbook 91, *Experimental Statistics*, U.S. Department of Commerce, Springfield, Va., 1963.

Owen, D. B., *Tables of Factors for One-sided Tolerance Limits for a Normal Distribution*, Sandia Corporation Monograph SCR-13, 1958.

Statistical Research Group, Columbia University, Eisenhart, C., Hastay, M. W., and Wallis, W. A., (eds.), *Selected Techniques of Statistical Analysis*, McGraw-Hill Book Company, New York, 1947.

Weissberg, A., and Beatty, G. H., *Tables of Tolerance Limit Factors for Normal Distributions*, Battelle Memorial Institute, 1959.

In some cases it is desirable to specify a single limit (a *one-sided tolerance limit*) which has the property that a given proportion P of the population will be less than (or greater than) this limit.

Table XXI is typical of an abbreviated table calculated for one-sided tolerance factors.

14.74. If the sampled population is not normal, and is considerably different from normal, then tolerance limits and factors K as determined from Tables XX and XXI may be considerably in error. In such cases some other approach should be used.

For more extensive tables, and for further information on this topic the following references may be consulted:

Lieberman, G. J., "Tables of One-sided Statistical Tolerance Limits," *Industrial Quality Control*, Vol. 14, No. 10, April, 1958.

Ostle, B., *Statistics in Research*, 2d ed., The Iowa State University Press, Ames, Iowa, 1963.

Owen, D. B., *Factors for One-sided Tolerance Limits and for Variables Sampling Plans*, Sandia Corporation Monographs SCR-607 and SCR-13, Clearing House for Federal Scientific and Technical Information, U.S. Department of Commerce, Springfield, Va., 22151.

Owen, D. B., *Distribution-free Tolerance Limits*, Sandia Corporation Technical Memorandum SCTM 66A-57-51, 1957.

Non-parametric tests

14.75. This section is concerned with a class of tests called *non-parametric tests* which may be used for samples from continuous population distributions of any shape. Such tests are useful when little is known concerning the form of the population distribution.

14.76. The sign test. Suppose n independent pairs of sample values (x_1, x_1'), (x_2, x_2'), $\cdots$, (x_n, x_n') are drawn from two c.d.f.'s $F_1(x)$, and $F_2(x)$, respectively. It is desired to test the hypothesis H_0 that $F_1(x) \equiv F_2(x)$ against alternates $H_1 : F_1(x) \neq F_2(x)$, $H_2 : F_1(x) > F_2(x)$, and $H_3 : F_1(x) < F_2(x)$.

Let $v_i = 1$ if $x_i - x_i' > 0$, and $v_i = 0$ otherwise, $i = 1, 2, \cdots, n$. Let $r = \sum_{i=1}^{n} v_i$.

It can be shown: If H_0 is true, then $P(v_i = 1) = P(x_i > x_i') = \frac{1}{2}$, r is a random variable with binomial distribution

$$(14.76.1) \qquad b(r) = \binom{n}{r}(1/2)^n, \qquad r = 0, 1, \cdots, n,$$

and the expected value of r is $E(r) = n/2$.

It may be said that, at the α level of significance:

(a) r differs significantly from $n/2$ if r falls outside the interval $[r_\alpha, n - r_\alpha]$, where r_n is the largest integer for which $P(r_\alpha \leq r \leq n - r_\alpha) > 1 - \alpha$, and $r_\alpha < n/2$ when H_0 is true.

(b) r is significantly large if $r \geq r_{l\alpha}$, where $r_{l\alpha}$ is the smallest integer such that $P(r \geq r_{l\alpha}) \leq \alpha$ when H_0 is true.

(c) r is significantly small if $r \leq r_{s\alpha}$, where $r_{s\alpha}$ is the largest integer such that $P(r \leq r_{s\alpha}) \leq \alpha$ when H_0 is true.

14.77. Table 14.77.1 gives values of r_α, $r_{l\alpha}$, and $r_{s\alpha}$ for $\alpha = 0.01$ and $\alpha = 0.05$ and for $n = 5, 6, \cdots, 30$. (This table is derivable from the cumulative (summed) binomial distribution table.)

TABLE 14.77.1

Values of r_α, $r_{l\alpha}$, $r_{s\alpha}$ for use in the sign test

n	r_α $\alpha = 0.01$	0.05	$r_{l\alpha}$ 0.01	0.05	$r_{s\alpha}$ 0.01	0.05
5				5		0
6				6		0
7			7	7	0	0
8	1	1	8	7	0	1
9	1	2	9	8	0	1
10	1	2	10	9	0	1
11	1	2	10	9	1	2
12	2	3	11	10	1	2
13	2	3	12	10	1	3
14	2	3	12	11	2	3

TABLE 14.77.1 (Continued)

Values of r_α, $r_{l\alpha}$, $r_{s\alpha}$ for use in the sign test

	r_α		$r_{l\alpha}$		$r_{s\alpha}$	
n	$\alpha = 0.01$	0.05	0.01	0.05	0.01	0.05
15	3	4	13	12	2	3
16	3	4	14	12	2	4
17	3	5	14	13	3	4
18	4	5	15	13	3	5
19	4	5	15	14	4	5
20	4	6	16	15	4	5
21	5	6	17	15	4	6
22	5	6	17	16	5	6
23	5	7	18	16	5	7
24	6	7	19	17	5	7
25	6	8	19	18	6	7
26	7	8	20	18	6	8
27	7	8	20	19	7	8
28	7	9	21	19	7	9
29	8	9	22	20	7	9
30	8	10	22	20	8	10

14.78. Approximate values of r_α, $r_{l\alpha}$, $r_{s\alpha}$ may be found by using a normal approximation to the binomial $b(r)$. The necessary formulas are as follows. (See Eq. (10.3.1), Table 10.8.1, and Table IX.)

$$(14.78.1) \qquad P(r \leqq r_0) = \Psi(t_0)$$

where

$$(14.78.2) \qquad t_0 = \left(r_0 + \frac{1}{2} - \frac{n}{2}\right) \Big/ \frac{1}{2}\sqrt{n},$$

If t_α is the value of t_0 for which

$$(14.78.3) \qquad \Psi(t_\alpha) = 1 - \alpha,$$
$$(14.78.4) \qquad 2r_\alpha \cong n + 1 - \sqrt{n}\, t_{\alpha/2}$$
$$(14.78.5) \qquad 2r_{l\alpha} \cong n + 1 + \sqrt{n}\, t_\alpha$$
$$(14.78.6) \qquad 2r_{s\alpha} \cong n - 1 - \sqrt{n}\, t_\alpha.$$

EXAMPLE. Forty-nine samples of steel are tested for their nickel content by method C and method D. The nickel contents found by C and D, respectively, were x and x'. Let $d_i = x_i - x_i'$. $v_i = 1$ if $d_i > 0$, $v_i = 0$ if $d_i \leqq 0$. $r = \sum_{i=1}^{49} v_i$. Select level $\alpha = 0.05$. Calculate r_α for $\alpha = 0.05$, $n = 49$.

From Table 10.8.1 or Table IX, $t_{\alpha/2} = t_{0.025} = 1.96$, and from (14.78.4)

$$r_\alpha = r_{0.05} \cong [50 - 7(1.96)]/2 = 18.14.$$

The *hypothesis* that methods C and D are equivalent at the $\alpha = 0.05$ level of significance is accepted if the value r_1 of r resulting from the tests falls in the interval $[r_\alpha, n - r_\alpha]$, that is, in the interval $18 < r < 31$. If the value r_1 obtained from the tests falls outside this interval, the hypothesis is rejected.

14.79. Mann-Whitney (Wilcoxon) w test for two samples. Suppose $(x_1, \cdots, x_m)$ and $(y_1, \cdots, y_n)$ are independent samples from populations having continuous distribution functions $F_1(x)$ and $F_2(y)$, respectively. Pool the two samples together into a single sample having $m + n = v$ observations. Order this combined sample, the order statistics being $z_{(1)}, z_{(2)}, \cdots, z_{(v)}$. Consider the subscripts (ranks) of all the z's which represent the elements of $(x_1, \cdots, x_m)$. Let T be the *sum* of these ranks. Define:

$$(14.79.1) \qquad w = mn - T + m(m + 1)/2.$$

w is a random variable called the *Mann-Whitney statistic*, T the *Wilcoxon statistic*.

THEOREM. If the hypothesis $H_0 : F_1(x) \equiv F_2(x)$ is true (e.g., if both samples come from populations with common c.d.f.'s), then if $m \geq 8$, $n \geq 8$, w is distributed (approximately) normally with mean $\bar{w} = mn/2$ and variance $\sigma_w^2 = mn(v + 1)/12$.

THEOREM. As $m \to +\infty$ and $n \to +\infty$, the random variable

$$(14.79.2) \qquad U \equiv (w - \bar{w})/\sigma_w$$

has the normal distribution $N(0,1)$ as its limiting distribution.

EXAMPLE. Individuals A and B make $m = 15$ and $n = 17$ measurements, respectively, of a certain material. Determine whether A and B are obtaining significantly different results. The measurements of A are $x_1, \cdots, x_m$, of B, $y_1, \cdots, y_n$. The pooled sample gives the order statistics $z_{(1)}, z_{(2)}, \cdots, z_{(v)}$. The sum of the ranks of all the z's of the x observations is found to be $T = 247$. Hence by (14.79.1) $w = 128$, $\bar{w} = 127.5$, $\sigma_w^2 = 701$, $\sigma_w \cong 26.5$. The observed value of (14.79.2) is $U = 0.019$. This is an insignificant difference. The hypothesis that the results of the analyses of A and B do not differ significantly is accepted.

14.80. Tests of runs. Let $x_1, x_2, \cdots, x_{2n}$ be a sample drawn from a population having a continuous c.d.f. $F(x)$. A test to give evidence of non-randomness in the fluctuations in the sequence of drawings is as follows.

Order the statistics of the sample obtaining the sequence $x_{(1)}$, $x_{(2)}$, $\cdots$, $x_{(2n)}$. Suppose $x_{(n)} \neq x_{(n+1)}$. Graph $x_{(i)}$ as a function of i, $i = 1, \cdots, 2n$. Draw a line M between $x_{(n)}$ and $x_{(n+1)}$ parallel to the i-axis.

If $x_{(i)}$ is above M write A below the i, if $x_{(i)}$ is below M write B below i, etc. There results a sequence such as $AA\,BB\,AB \cdots A$. In any such sequence (arrangement) there will be *clusters (runs)* of one or more A's separated by clusters of one or more B's. Let u be the total number of clusters of A's and B's. The *number of runs* above and below line M is u. The number of distinct arrangements of A's and B's possible is

$$\binom{2n}{n} = (2n)!/(n!)^2.$$

Remark. If the number of elements in the sample is odd, say $2n + 1$, the line M could be drawn through the median of the sample.

14.81. The probability function for u is

$$(14.81.1) \qquad \begin{cases} p(u) = 2b^2/a, & u = 2, 4, \cdots, 2n \\ p(u) = 2cd/a, & u = 3, 5, \cdots, 2n - 1 \end{cases}$$

where

$$a = \binom{2n}{n}, \qquad b = \binom{n-1}{\dfrac{u}{2}-1}, \qquad c = \binom{n-1}{\dfrac{u}{2}-\dfrac{1}{2}}, \qquad d = \binom{n-1}{\dfrac{u}{2}-\dfrac{3}{2}}$$

The mean and variance of u are

$$(14.81.2) \qquad \bar{u} = n + 1, \qquad \sigma_u^2 = n(n-1)/(2n-1).$$

Where n and u are sufficiently large, u is distributed approximately normal with mean and variance as in (14.81.2).

The procedure for testing whether an observed value differs significantly from its mean value one way or the other is similar to the test procedure used in the sign test §14.76, though (14.81.1) is used rather than (14.76.1).

If the measurements drawn as in Figure 14.81.1 in the order x_1, x_2, $\cdots$, x_{2n} tend to alternate too much u tends to be significantly large. If the measurements tend to drift up (or down) from the expected value of u, u may be significantly small. If u is too small (or too large) non-randomness in the sequence of sample drawings may be present.

EXAMPLE. A chart such as Figure 14.81.1 is constructed for measure-

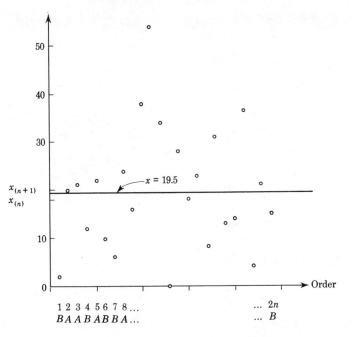

Chart Showing Runs Above and Below Median

FIGURE 14.81.1

ments $x_1, x_2, \cdots, x_{24}$, namely, for 02, 20, 21, 12, 22, 10, 06, 24, 16, 38, 54, 34, 00, 28, 18, 23, 08, 31, 13, 14, 36, 04, 21, 15. The order statistics is $x_{(1)}, x_{(2)}, \cdots, x_{(24)}$, that is, 00, 02, 04, 06, 08, 10, 12, 13, 14, 15, 16, 18, 20, 21, 21, 22, 23, 24, 28, 31, 34, 36, 38, 54.

To test the hypothesis that the sequence $x_1, x_2, \cdots, x_{24}$ is random at the 1 percent level of significance proceed as follows: $x_{(n)} = x_{(12)} = 18$, $x_{(n+1)} = x_{(13)} = 20$. Select a central line between $x_{(12)}$ and $x_{(13)}$, say $x = 19.5$. If a value $x_{(i)}$ exceeds 19.5, write A; if not write B. The sequence $BAABABBBABAAABABABABBABAB$ of $u = 19$ clusters results. By (14.81.2), $\bar{u} = 13$, $\sigma_u{}^2 = 5.74$, $\sigma_u = 2.4$.

If $|u - \bar{u}| \geqq \sigma_u(2.576)$, at the $\alpha = 0.01$ level of significance the hypothesis is rejected; otherwise the hypothesis is accepted. Since $|u - \bar{u}| = |u - 13| = 6$ and $\sigma_u(2.576) = 2.4(2.576) = 6.18$, the hypothesis that the sequence $x_1, x_2, \cdots, x_{24}$ is random is accepted at the 1 percent level. Here $t_{0.005}$ is the value of t for which $\Psi(t) = 1 - \alpha/2$. [See Eq. (10.3.1), Table 10.8.1, or Table IX.]

14.82. Wald-Wolfowitz run test. Let $x_1, \cdots, x_{n_1}$ and $x_1', \cdots, x_{n_2}'$ be two samples drawn from two populations having continuous c.d.f.'s $F_1(x)$ and $F_2(x)$, respectively.

To test the hypotheses H_0 that $F_1(x) \equiv F_2(x)$ against alternative hypotheses $H_1 : F_1(x) \not\equiv F_2(x)$, proceed as follows: Pool the samples. Order the observations in the pool. The resulting sequence is $y_1, y_2, \cdots, y_i, \cdots$. If y_i is an x write 0; if y_i is an x', write 1. Do this for $i = 1, \cdots, i = n_1 + n_2 \equiv n$. The resulting sequence is a sequence of zero's and one's. The number of distinct arrangements of the n_1 0's and n_2 1's is

$$\binom{n}{n_1}.$$

By a *run* is meant a cluster of one or more zero's, or a cluster of one or more one's. Let u be the total number of runs in $y_1, y_2, \cdots, y_n$.

14.83. The probability function for u is

$$(14.83.1) \qquad p(u) = \begin{cases} 2b_1b_2/a, & \text{if } u \text{ is even} \\ (c_1d_2 + c_2d_1)/a, & \text{if } u \text{ is odd} \end{cases}$$

where for $j = 1, 2$

$$a = \binom{n}{n_1}, \quad b_j = \binom{n_j - 1}{\frac{u}{2} - 1}, \quad c_j = \binom{n_j - 1}{\frac{u}{2} - \frac{1}{2}}, \quad d_j = \binom{n_j - 1}{\frac{u}{2} - \frac{3}{2}}$$

The mean and variance of u are

$$(14.83.2) \qquad \begin{cases} \bar{u} = (2n_1n_2 + n)/n \\ \sigma_u^2 = 2n_1n_2(2n_1n_2 - n)/n^2(n - 1) \end{cases}$$

When n_1 and n_2 are large enough, u is approximately normally distributed with mean and variance given by (14.83.2). When $n_1 > 10$, $n_2 > 10$, the normal approximation is usually adequate.

14.84. The test. Let α be the level of significance. If the total number u of runs is such that $|u - \bar{u}| \geqq \sigma_u t_{\alpha/2}$, the hypothesis that the two samples $x_1, \cdots, x_{n_1}$ and $x_1', \cdots, x_{n_2}'$ come from populations having identical c.d.f.'s is rejected at the α level; otherwise the hypothesis is accepted. Here $t_{\alpha/2}$ is the value of t for which $\Psi(t) = 1 - \alpha/2$. [See Eq. (10.3.1), Table 10.8.1, or Table IX.]

14.85. References. For further information on non-parametric statistics the following works may be consulted:

Frazer, D. A. S., *Non-parametric Methods in Statistics*, John Wiley & Sons, Inc., New York, 1957.

Siegel, S., *Nonparametric Statistics for the Behavioral Sciences*, McGraw-Hill Book Company, New York, 1956.

Walsh, J. E., *Handbook of Non-parametric Statistics*, Vols. I, II, D. Van Nostrand Company, Inc., Princeton, N.J., 1962, 1965.

fifteen

DESIGN OF EXPERIMENTS AND ANALYSIS OF VARIANCE

Statistical design of experiments

15.1. The design of an experiment is a complete sequence of steps taken before the experiment to insure that appropriate data will be obtained, data which permit an objective analysis leading to valid inferences with respect to the problem at hand. The individual who formulates the design should clearly understand the objective of the investigation.

15.2. The purpose of an experimental design is to provide a maximum amount of information relevant to the problem, at a minimum cost. The design should provide for (a) the best possible precision that can be achieved with a given experimental procedure, and (b) means of detecting sources of constant error in the procedure. The design should be kept as simple as possible. The investigation should be conducted as efficiently as possible. The need for the conservation of time, effort, money, personnel, and material must be kept in mind.

There are three basic principles of concern: *replication, randomization,* and *local control.*

15.3. **Replication.** The *repetition* of an experiment is called a *replication.* Replication is desirable because: (a) it provides an estimate of experimental error which may be used as a "unit of measurement" for assessing the significance of observed differences or for determining the length of a confidence interval; (b) under certain conditions, where an estimate of experimental error is available from other sources, replication may furnish a more accurate estimate; (c) it may provide a more precise estimate of the mean effect of any factor. Multiple readings do not necessarily represent true replication.

15.4. The term *experimental unit* refers to that unit to which a single treatment (investigation) is applied in one replication of the basic experiment.

15.5. The term *experimental error* is used to describe the failure of two identically treated experimental units to yield identical results. To the statistician an experimental error reflects: errors of experimentation, errors of observation, errors of measurement, the variation of experi-

nental material, and the combined effects of extraneous factors which
re not singled out for attention in the investigation.

15.6. Experimental error may often be reduced by the use of more
homogeneous experimental material, careful stratification of available
material, greater care in conducting the experiment, more efficient experimental design, related information provided by other sources.

15.7. Two or more effects are *confounded* ("mixed together") in an
experiment if it proves to be impossible to separate the effects when
statistical analysis is performed. While a good design should avoid un-
warranted confounding, confounding is sometimes deliberately used to
reduce the size of an experiment.

15.8. **Randomization.** Every test procedure has associated with it
certain underlying assumptions which must be satisfied if the test is to
be considered valid. A common assumption is that the observations
(or, the associated errors) are independently distributed.

15.9. While one cannot be sure that this assumption holds true, one
can insist that the samples selected from a population are random, or
that a random assignment of treatments to the experimental units is
used. With this approach one can proceed as though the assumption is
true. Such *randomization* permits one to proceed as though the observations are independently distributed. In practice, true and complete independence of errors can never be fully achieved. For example, errors associated with experimental units which are close to each other in time or
position may tend to be correlated; randomization tends to minimize
the effects of such correlations.

Randomization is also used as a device for "eliminating" bias.

15.10. There may be reasons why a non-random, systematic design
of an experiment has merit. It is a matter of debate whether it is always
best to use a random design rather than a systematic design. Most designs presently in use involve both the random and the systematic approaches. The statistician favors the random approach since he wishes
to make inferences of various sorts from the observed data and likes to
be able to attach a measure of reliability to such inferences.

15.11. **Local control.** The term *local control* refers to the amount of
balancing, blocking, and grouping of experimental units used in the
experimental statistical design adopted. The purpose of local control is
to improve the efficiency of the experimental design. Local control tends
to improve the sensitivity of tests for significance and to reduce the

magnitude of experimental error estimates. Common methods of local control are listed in §15.6.

15.12. Grouping, blocking, balancing. The placing of a set of homogeneous experimental units into groups for the purpose of subjecting the different groups to different treatments is called *grouping*.

The allocation of experimental units to blocks in such a way that units within a block are relatively homogeneous, while the bulk of the predictable variation among the units has been confounded with the effect of the blocks, is called *blocking*. In blocking the investigator's prior knowledge of the nature of the experimental units is used to design the experiment so that much of the expected variation will not be imbedded in the experimental error. The term "block" arose from its use in agricultural problems.

The statistician attempts to *balance* the selection of experimental units, the grouping, blocking and assignment of the treatments to the experimental units. The term is often used in a loose sense.

15.13. Treatments. The term *treatment* (or *treatment combination*) is used to refer to a particular set of experimental conditions which are imposed on an experimental unit, subject to the chosen experimental design. The term "treatment" stems from its use in agronomic experimentation (e.g., treatment of a soil by a fertilizer).

15.14. Factors, factorials, factor levels. The term *factorial* is an abbreviation for lengthy phrases describing the nature of experiments. In speaking of experimental design the term *factor* refers to an independent variable. The various values of classification of the factors are called the *levels* of the factors (e.g., the three levels of weights used in an experiment might be written w_1, w_2, w_3). *Level* is a general term used in a variety of ways.

15.15. Many experiments involve factorial treatment arrangements. Several systems of notation are used in the literature. Table 15.15.1 illustrates five types of *methods* (*notations*) used to represent factorial treatment combinations. In the table six treatment combinations are involved formed from one level of factor a, two levels of factor b, and three levels of factor c. In the case of Method I, $a_i b_j c_k$ represents the treatment combination formed by using the ith level of factor a, the jth level of factor b, the kth level of factor c. Table 15.15.1 is called a $1 \times 2 \times 3$ *factorial*.

TABLE 15.15.1

Notations for representing factorial treatment combinations

($1 \times 2 \times 3$ factorial case)

Treatment combination	Method				
	I*	II	III	IV	V
1	$a_1b_1c_1$	111	$a_0b_0c_0$	000	(1)
2	$a_1b_1c_2$	112	$a_0b_0c_1$	001	c
3	$a_1b_1c_3$	113	$a_0b_0c_2$	002	c^2
4	$a_1b_2c_1$	121	$a_0b_1c_0$	010	b
5	$a_1b_2c_2$	122	$a_0b_1c_1$	011	bc
6	$a_1b_2c_3$	123	$a_0b_1c_2$	012	bc^2

* Here a, b, c represent factors; the subscripts represent the level of the factor; e.g., a_i indicates the ith level of factor a.

15.16. Effects and interactions. Factors that do not act independently are said to exhibit an *interaction*. *Interaction* is an effect due to the combined influence of two or more factors.

An *effect* of a factor is a measure of the change in response variable to changes in the level of the factor under some stated conditions.

The *main effect* of a factor is a measure of the change in response variable to changes in the level of the factor averaged over all levels of all the other factors. (See §15.30.)

EXAMPLE 1. The yield y of a certain (e.g., chemical) reaction depends on two factors a and b (e.g., temperature and pressure). Perhaps the effects of changes in a and b on y are not independent. An experiment is designed which utilizes treatment combinations formed by combining various levels of the two factors. The four treatments are a_0b_0, a_0b_1, a_1b_0, a_1b_1.

Let $\alpha_i\beta_j$ ($i = 0, 1$; $j = 0, 1$) represent the average yields from all experimental units, which have been subjected to the similarly designed treatment combinations a_ib_j. Each average yield is obtained from the same number of experimental units. The problem is a 2^2 factorial case.

(1) The effect of a at level b_0 of b is $\alpha_1\beta_0 - \alpha_0\beta_0$.

(2) The effect of a at level b_1 of b is $\alpha_1\beta_1 - \alpha_0\beta_1$.

    The *main effect of a* is the average of (1) and (2) and is equal to
$$(\alpha_1 - \alpha_0)(\beta_1 + \beta_0)/2 = A.$$

(3) The effect of b at level a_0 of a is $\alpha_0\beta_1 - \alpha_0\beta_0$.

(4)   The effect of b at level a_1 of a is $\alpha_1\beta_1 - \alpha_1\beta_0$.
   The *main effect of b* is the average of (3) and (4) and is equal to
$$(\alpha_1 + \alpha_0)(\beta_1 - \beta_0)/2 = B.$$

Relations (1) and (2) would be equal if a and b were acting independently. [Similarly, for (3) and (4)].

The *interaction* between a and b is defined as one-half (2) minus (1); that is,

$$AB = (\alpha_1 - \alpha_0)(\beta_1 - \beta_0)/2.$$

The *mean yield* of all experimental units (the *mean effect*) is

$$M = (\alpha_0 + \alpha_1)(\beta_0 + \beta_1)/4$$

The process illustrated in Example 1 may be extended to calculating effects in 2^n factorials.

EXAMPLE 2.   Let a_ib_j and $\alpha_i\beta_j$ $(i = 0, 1, 2; j = 0, 1, 2)$ represent the treatment combinations and the yields from the treatment combinations, respectively.

The (main) linear effect of a is $A_L = (\alpha_2 - \alpha_0)(\beta_0 + \beta_1 + \beta_2)/3$.
The quadratic effect of a is $A_Q = (\alpha_2 - 2\alpha_1 + \alpha_0)(\beta_0 + \beta_1 + \beta_2)/6$.
Linear $\times$ quadratic interaction is $A_LB_Q = (\alpha_2 - \alpha_0)(\beta_2 - 2\beta_1 + \beta_0)/4$
Similarly, for other effects and interactions.

15.17. Contrast. Treatment comparisons.   As an aid in making comparisons among treatment means, "contrasts" are often used.

Let T_i be the sum of n_i observations. A *contrast* (or *comparison*) C_j among $T_1, \cdots, T_k$ is defined as

$$(15.17.1) \qquad C_j = c_{1j}T_1 + c_{2j}T_2 + \cdots + c_{kj}T_k,$$

where

$$(15.17.2) \qquad \sum_{i=1}^{k} n_i c_{ij} = 0.$$

Several such contrasts $C_1, C_2, \cdots$ may be used in a given case.

If each $n_i = n$, $\sum_{i=1}^{k} c_{ij} = 0$ is a necessary condition for a contrast.

The coefficients c_{ij} are selected so that the comparison involved will not be affected by differing numbers of observations associated with the various treatments.

15.18.   Let $C_p = \sum_{i=1}^{k} c_{ip}T_i$ and $C_q = \sum_{i=1}^{k} c_{iq}T_i$ be two contrasts. The contrast C_p is *orthogonal* to contrast C_q if the coefficients are such that

$$(15.18.1) \qquad \sum_{i=1}^{k} n_i c_{ip} c_{iq} = 0, \qquad\qquad p \neq q.$$

When $n_i = n$ for all i, the orthogonality condition becomes

15.18.2) $$\sum_{i=1}^{k} c_{ip}\, c_{iq} = 0, \qquad\qquad p \neq q.$$

15.19. **Steps in designing an experiment.** A statistically designed xperiment typically requires steps such as:

(a) Formulate a clear statement of the problem.
(b) Collect and organize background information.
(c) Formulate hypotheses relating to the problem.
(d) Design the test program.
(e) Plan and carry out the experimental work.
(f) Analyze the data of the experiment.
(g) Interpret the results.
(h) Write a report.

Analysis of variance

15.20. A collection of values of a quantity y can sometimes be classified nto subgroups corresponding to different conditions in experiments ielding the different values. The values may be simply divided into mutually exclusive subgroups, or they may be classified according to wo or more methods of classification at once, for example, by an arrangement corresponding to "rows and columns." It is then often of interest o determine how quantities such as mean or variance of y change from ne subgroup to another, and to determine some of the features of the lata which can be ascribed to the various conditions in the experiment. An important type of application, for which many of the statistical echniques were first developed, is in agricultural experiments. For example, the yield of corn depends upon seed quality and soil fertility, mong other variables. A suitable experiment and statistical analysis can eparate the effects of these two variables, and can estimate how much of he observed variability in yield must be ascribed to other causes.

15.21. In the *analysis and design of experiment* it is often useful to esolve the variance of a variable relative to its sample mean into separate components of variance, that is, into the sum of several distinct ums of squares, each corresponding to a source, real or suspect, of ariation. This process of resolution into separate components of variance s sometimes called *analysis of variance*.

15.22. The following sections outline some of the elementary techniques f the analysis of variance. These techniques are used in the analysis of

subgroups within a collection of data for the purpose of studying th
contributions to variance arising from different experimental causes.

15.23. Analysis of variance for one-way classification. Let $S_1, S_2, \cdots$
S_m be samples, of sizes $l_1, l_2, \cdots, l_m$, of values y drawn from m differen
normal populations, the populations having the same variance σ^2, an
means $\tilde{y}_1, \tilde{y}_2, \cdots, \tilde{y}_m$, respectively, which are not known in advance a
to be the same. Denote the elements in sample S_i by $y_{ij}, j = 1, 2, \cdots, l$
The whole collection of values y_{ij} is said to have a *one-way classificatio*
into the sets S_i, (*one-factor experiment.*)

15.24. Suppose that the mean of sample S_i is $\bar{y}_i$. An analysis of var
ance, based on the observed samples, can be carried out as outlined belo
to test the hypothesis that all of the population means $\tilde{y}_i$ are equal.

15.25. The mean of the whole collection of values y_{ij} in all of the san
ples together is

$$(15.25.1) \qquad \bar{y} = (1/n) \sum_{i=1}^{m} \sum_{j=1}^{l_i} y_{ij},$$

where $n = \sum_{i=1}^{m} l_i$. If the means $\tilde{y}_i$ are in fact all equal, then all of th
different samples S_i come from the same normal population, and a
unbiased estimate of σ^2 is

$$(15.25.2) \qquad \hat{\sigma}^2 = [1/(n-1)] \sum_{i=1}^{m} \sum_{j=1}^{l_i} (y_{ij} - \bar{y})^2.$$

The appropriate number of degrees of freedom is $n-1$ since the y_{ij} ar
subject to the one linear restriction $\sum_{i=1}^{m} \sum_{j=1}^{l_i} y_{ij} = n\bar{y}$.

15.26. The means $\bar{y}_i$ of the samples S_i would be expected to diff
among themselves somewhat due to chance alone even if all the observa
tions were from the same population. The analysis of variance procee
by breaking up the estimated total variance $\hat{\sigma}^2$ into components due t
the variations of sample values within each sample, and due to the varia
tions from one sample to another sample. By comparing these types c
variations (using the z or F distributions) it is possible to estimat
whether the variations between samples is greater than would be expecte
if all the samples were obtained from the same population.

15.27. The following is an identity:

$$(15.27.1) \qquad \sum_{i=1}^{m} \sum_{j=1}^{l_i} (y_{ij} - \bar{y})^2 = \sum_{i=1}^{m} l_i(\bar{y}_i - \bar{y})^2 + \sum_{i=1}^{m} \sum_{j=1}^{l_i} (y_{ij} - \bar{y}_i)$$

or

$$A \qquad = \qquad B \qquad + \qquad C.$$

A is the total sum of squares.

B is the sum of squares "between samples."

C is the sum of squares "within samples."

The notation may be summarized as follows:

	Samples					
	S_1	S_2	$\cdots$	S_i	$\cdots$	S_m
Items in sample	y_{11}	y_{21}	$\cdots$	y_{i1}	$\cdots$	y_{m1}
	y_{12}	y_{22}	$\cdots$	y_{i2}	$\cdots$	y_{m2}
	·	·		·		·
	·	·		·		·
	·	·		·		·
				y_{ij}		
				·		
				·		
				·		
				y_{il_i}		
Number of items in sample	l_1	l_2	$\cdots$	l_i	$\cdots$	l_m
Average of sample	$\bar{y}_1$	$\bar{y}_2$	$\cdots$	$\bar{y}_i$	$\cdots$	$\bar{y}_m$

15.28. *A* involves $n - 1$ degrees of freedom, since $\sum_{i=1}^{m} \sum_{j=1}^{l_i} y_{ij} = n\bar{y}$ is one linear restriction on the y_{ij}'s. *B* involves $m - 1$ degrees of freedom, since $\sum_{i=1}^{m} l_i\bar{y}_i = n\bar{y}$ is one linear restriction on the $\bar{y}_i$'s. *C* involves

Analysis of variance for one-way classification

Variation	Degrees of freedom	Sum of squares	Mean square
Between samples	$m - 1$	$B = \sum_{i=1}^{m} l_i(\bar{y}_i - \bar{y})^2$	$\hat{\sigma}_1^2 = B/(m - 1)$
Within samples	$n - m$	$C = \sum_{i=1}^{m} \sum_{j=1}^{l_i} (y_{ij} - \bar{y}_i)^2$	$\hat{\sigma}_2^2 = C/(n - m)$
Total	$n - 1$	$A = \sum_{i=1}^{m} \sum_{j=1}^{l_i} (y_{ij} - \bar{y})^2$	$\hat{\sigma}^2 = A/(n - 1)$

$n - m$ degrees of freedom, since the y_{ij}'s are subject to the m linear restrictions $\sum_{j=1}^{l_i} y_{ij}/l_i = \bar{y}_i$, $i = 1, 2, \cdots, m$. Thus, if the hypothesis that all of the population means $\tilde{y}_i$ are equal is correct, unbiased estimates of the population variance σ^2 are given by each of the three separate quantities $\hat{\sigma}^2 = A/(n - 1)$, $\hat{\sigma}_1^2 = B/(m - 1)$, and $\hat{\sigma}_2^2 = C/(n - m)$. Numerical tabulations of these quantities are commonly arranged in the form shown in the table on page 277.

Computational scheme. The sums of squares in the above table may be calculated using the following scheme. Let

$$(15.28.1) \qquad U_i = \sum_{j=1}^{l_i} y_{ij}, \qquad V_i = \sum_{j=1}^{l_i} y_{ij}^2$$

Then

$$B = \sum_{i=1}^{m} (U_i^2/l_i) - \left(\sum_{i=1}^{m} U_i\right)^2/n$$

$$(15.28.2) \qquad C = \sum_{i=1}^{m} V_i - \sum_{i=1}^{m} (U_i^2/l_i)$$

$$A = \sum_{i=1}^{m} V_i - \left(\sum_{i=1}^{m} U_i\right)^2/n$$

15.29. If the hypothesis under test is true so that all the observations y_{ij} are from the *same* normal population, then $z = (1/2) \log_e (\hat{\sigma}_1^2/\hat{\sigma}_2^2)$ has the z-distribution $G_{m_1, m_2}(z)$, (§13.71) with $m_1 = m - 1$ and $m_2 = n - m$; and $F = \hat{\sigma}_1^2/\hat{\sigma}_2^2$ has the F-distribution $h_{m_1, m_2}(F)$, (§13.65), with $m_1 = m - 1$ and $m_2 = n - m$. If the value of z (or F) calculated from a set of observations does not exceed z_ϵ (Table XI), (or F_ϵ (Table X)), for, say, $\epsilon = 0.05$, then there is not more than probability 0.05 that the difference between $\hat{\sigma}_1^2$ and $\hat{\sigma}_2^2$ would be as large as observed due solely to chance, if the hypothesis that the various samples come from the same population is true. If z (or F) calculated from the sample exceeds z_ϵ (or F_ϵ), it would be concluded that the various samples probably are not all from the same population, i.e., that there probably are real differences between the means $\tilde{y}_i$ of the populations from which the samples are drawn. [Here ϵ is the (confidence) level of significance used.]

15.30. **Models I and II.** Suppose the elements in the samples are representable by the linear statistical model

$$(15.30.1) \qquad y_{ij} = u_i + \epsilon_{ij} = \mu + \alpha_i + \epsilon_{ij},$$

where

$$u_i = \mu + \alpha_i$$

$\mu = \sum_{i=1}^{m} l_i u_i / n$ is the true *overall mean* (or true mean effect)

(15.30.2) α_i = *effects* or *(explained) deviations* due to treatments (conditions of experiment) of group i (or, the true effect of the ith treatment).

ϵ_{ij} = *effects* of *unexplained deviations* (or, the true effect of the jth experimental unit subject to the ith treatment).

Note: μ is constant and the ϵ_{ij} are independent and normally distributed with mean zero and variance σ^2. The selection of α_i is made in terms of the interest of the investigator.

15.31. Model I (fixed effects model). If the interest is in comparing the effectiveness of the m treatments used in the investigations, certain parameters are used. In this case one may take $\sum_{i=1}^{m} l_i \alpha_i = 0$, which means that μ is a weighted mean of the u_i's.

(15.31.1) $$\mu = n^{-1} \sum_{i=1}^{m} l_i u_i, \qquad n = \sum_{i=1}^{m} l_i$$

The statistical model (15.30.1) with μ given by (15.31.1) is called the *analysis of variance (fixed effects) model* (or *parametric*, or *systematic Model I*). In Model I parameters are used to denote factor level effects. The parameters α_i are also called *differential effects*.

15.32. Model II (random effects model). If the interest is in a population of treatments of which only a random sample of size m is present in the experiments, then one may assume that the α_i are normally distributed with mean zero and variance σ_a^2. In this case the α_i are assumed to be independent of the ϵ_{ij}.

The statistical model (15.30.1) used with these assumptions is called the *component of variance (random effects) model* (or *random*, or *Model II*).

From (15.27.1), with $\hat{\sigma}_1^2 = B/(m-1)$ and $\hat{\sigma}_2^2 = C/(n-m)$, it can be shown that the expected value of the "between samples" mean square is

(15.32.1) $$E(\hat{\sigma}_1^2) = \sigma^2 + \sum_{i=1}^{m} l_i(u_i - \mu)^2/(m-1),$$

in the case of the fixed effects Model I, and

(15.32.2) $$E(\hat{\sigma}_2^2) = \sigma^2 + (m-1)^{-1}\left[n - \left(\sum_{i=1}^{m} l_i^2\right)/n\right]\sigma_a^2,$$

in the random effects Model II case.

The expected value of the "within samples" mean square is

$$(15.32.3) \qquad E(\hat{\sigma}_2{}^2) = \sigma^2$$

in both the Model I and Model II cases.

15.33. Test for effects in Model I case. To test the hypothesis that the differential effects due to samples $S_1, S_2, \cdots, S_m$ are essentially zero, that is, to test the hypothesis

$$H_0 : \alpha_1 = \cdots = \alpha_m = 0$$

against alternate hypothesis H_1: that the α_i are not all zero, the F-test of §15.29 may be used, since the ratio $F = \hat{\sigma}_1{}^2/\hat{\sigma}_2{}^2$ has the F-distribution $h_{m_1, m_2}(F)$ with $m_1 = m - 1$ and $m_2 = n - m$ degrees of freedom under H_0. If for the samples the value of $\hat{\sigma}_1{}^2/\hat{\sigma}_2{}^2 > F_\epsilon$, where F_ϵ is the appropriate value given in Table X, the hypothesis H_0 is rejected at the $100\,\epsilon$ percent level of significance; otherwise H_0 may be accepted.

If H_0 is rejected the effects $\alpha_1, \cdots, \alpha_m$ may be estimated from the fact that the expected values of $\bar{y}$ and $\bar{y}_j$ are, respectively,

$$E(\bar{y}) = \mu, \qquad E(\bar{y}_j) = \mu + \alpha_j$$

so that $\bar{y}$ and $\bar{y}_j - \bar{y}$ are unbiased estimators of μ and α_j, respectively.

$\hat{\sigma}_2{}^2$ is an unbiased estimator for σ^2.

The results given above may be listed as in Table 15.33.1 (sometimes called an ANOVA table).

TABLE 15.33.1

Analysis of variance table
For one-way classification

Variation	Degrees of freedom	Mean square	Expected mean square for	
			Fixed model	Random model
Between samples	$m - 1$	$\hat{\sigma}_1{}^2$	$\sigma^2 + \sum_{i=1}^{m} l_i \alpha_i{}^2/(m - 1)$	$\sigma^2 + n_0 \sigma_\alpha{}^2$
Within samples	$n - m$	$\hat{\sigma}_2{}^2$	σ^2	σ^2
Total	$n - 1$			

Model: $y_{ij} = \mu + \alpha_i + \epsilon_{ij}$; $i = 1, 2, \cdots, m, j = 1, 2, \cdots, l_i$

Note:
$$n_0 = (m - 1)^{-1} \left[n - \left(\sum_{i=1}^{m} l_i{}^2 \right)/n \right]$$

EXAMPLE 1. Three different methods of teaching algebra are used in three groups S_1, S_2, S_3 of students. Random samples of size 5 are drawn from each group. The results by grades (based on a 10-point maximum grade) are as listed in Table 15.33.2. A test to estimate whether the different methods had any appreciable effect on the grades is desired.

The collection of grades y_{ij} has a one-way classification into the groups S_1, S_2, S_3. The analysis of variance follows the pattern given above. The results of the calculations are shown in Tables 15.33.2 and 15.33.3.

Consider the hypothesis that the various grades come from the same population is true. The fraction $F = \hat{\sigma}_1^2/\hat{\sigma}_2^2$ has the F-distribution $\phi_{m_1,m_2}(F)$, with $m_1 = m - 1 = 2$ and $m_2 = n - m = 12$. But $F = 5/2.17 = 2.3$. From Table X for the $\epsilon = 0.05$ level of significance, $F_\epsilon = 3.88$. Since $F < F_\epsilon$ there is not more than probability 0.05 that the difference between $\hat{\sigma}_1^2$ and $\hat{\sigma}_2^2$ would be as large as observed owing solely to chance, if the hypothesis that the various samples of grades come from the same population (and the three means are equal) is true. This means that the various teaching methods do not appear (at level $\epsilon = 0.05$) to affect the average grade.

TABLE 15.33.2

	Groups		
	S_1	S_2	S_3
Grades in group	8	8	7
	4	7	3
	7	6	6
	7	7	5
	4	7	4
Number of grades in group	5	5	5
Average of group	6	7	5

Summary of calculation $m = 3$, $n = 15$. $l_1 = l_2 = l_3 = 5$. $\bar{y}_1 = 6$. $\bar{y}_2 = 7$. $\bar{y}_3 = 5$.
A = total sum of squares = 36.
B = sum of squares "between groups" = 10.
C = sum of squares "within groups" = 26. $A = B + C$.
Note: In computing A, B, C use the identity $\sum(x - \bar{x})^2 = \sum x^2 - n\bar{x}^2$ to organize and simplify the calculations.

TABLE 15.33.3

Analysis of variance for S_1, S_2, S_3

Variation	Degrees of freedom	Sum of squares	Mean square
Between groups	$3 - 1 = 2$	$B = 10$	$\hat{\sigma}_1^2 = 10/2 = 5$
Within groups	$15 - 3 = 12$	$C = 26$	$\hat{\sigma}_2^2 = 26/12 = 2.17$
Total	$15 - 1 = 14$	$A = 36$	$\hat{\sigma}^2 = 36/14 = 2.57$

EXAMPLE 2. Three different machines U, V, W produce an objec that weighs w pounds. The weights of the outputs of machine i ar normally distributed with mean u_i and variance σ_i^2, $i = 1, 2, 3$. Th variances of the outputs (populations) of the machines are equal, tha is, $\sigma_1^2 = \sigma_2^2 = \sigma_3^2 = \sigma^2$. The actual measured weights (outputs) ar listed in the form used in Table 15.33.2.

A test to estimate whether the means u_i of the outputs of the machine are equal to the true overall mean μ of all the outputs is desired, that is a test of the hypothesis that $u_1 = u_2 = u_3 = \mu$.

Suppose the jth output of the ith machine is y_{ij}, $(j = 1, 2, 3)$. Since th interest is in the differences of the means u_i, the *fixed effects* Model I i used.

Since the outputs of the machines are normally distributed, the ϵ are distributed normally with $E(\epsilon_{ij}) = 0$ and $\text{var}(\epsilon_{ij}) = \sigma^2$.

Suppose a random sample of size $l_i = 5$, $i = 1, 2, 3$, is taken for eacl machine. Calculate the values of A, B, C, $\cdots$, listing the results as ii Tables 15.33.2 and 15.33.3.

The variance ratio $F = \hat{\sigma}_1^2/\hat{\sigma}_2^2$ has the F-distribution with $m_1 = 3 - 1 = 2$, $m_2 = 15 - 3 = 12$ degrees of freedom. For the $\epsilon = 0.0$ level of significance, if $F > 3.88$, we reject the hypothesis that $u_1 = u_2 = u_3 = \mu$; if $F < 3.88$, we accept the hypothesis.

EXAMPLE 3. *Another interpretation of Example 2.* The three ma chines U, V, W are selected from different sources. Is there a differenc in their performances? We test the hypothesis that there is no differenc in the outputs (performance) of the machines, that is, test the hypothe sis that the means are equal, $u_1 = u_2 = u_3$. Rejection of the hypothesi implies that the performance outputs of the machines are different.

EXAMPLE 4. The analysis of variance has found much use in agricul tural investigations. To a plot of land a fertilizer is added. The fertilize is the *treatment* applied to the *plot*. As a result the *yield* of corn may var significantly from an overall average. This variation is the *effect* due to the *treatments*.

Suppose the plot is divided into equal area subplots in each one of which one of the fertilizers U, V, W is applied to treat the ground. Suppose a random sample of size $l_i = 5$ is taken for each subplot treated with fertilizer i, $i = 1$, 2, 3. Is there a difference in the yields? To test the hypothesis that there is no difference in yields, test the hypothesis that the mean yields are equal for all cases.

15.34. Random effects model. In the type of problems illustrated above the values of α_i in (15.30.1) were fixed. Thus, in Example 2 of §15.33 the three machines had fixed means u_1, u_2, u_3 which deviated by α_1, α_2, α_3, respectively, from the overall mean μ.

A somewhat different point of view is often of interest. For example, in Example 2 of §15.33 suppose machine U is selected from a large group of similar machines produced by a given manufacturer. The value of α_1 will vary depending on the particular machine selected. Suppose that it is assumed that α_1 is a random variable normally distributed with mean $E(\alpha_1) = 0$, and variance $\text{var}(\alpha_1) = \sigma_{\alpha_1}^2$. Machine V is selected in a similar random manner from a different manufacturer; likewise for machine W. The variables α_2 and α_3 are considered random, normally distributed with means 0 and equal variances $\sigma_{\alpha_1}^2 = \sigma_{\alpha_2}^2 = \sigma_{\alpha_3}^2 = \sigma_\alpha^2$.

If it is desired to check whether there is a difference in the populations from which machines U, V, and W are selected, the interest is not just in the single machines U, V, and W. The test procedures and related computation are the same as illustrated above for the fixed effects model case.

Where the viewpoint is that illustrated in the two preceding paragraphs (i.e., where the interest is in differences in the populations from which the machines came), the model (15.30.1) is a *random effects model*.

15.35. Analysis of variance for two-way classification. One observation per cell. Suppose an array of values y_{ij} is given by an experiment, or otherwise, arranged in r rows and s columns, $i = 1, 2, \cdots, r; j = 1, 2, \cdots, s$. Each y_{ij} may be considered to apply to a cell at the (i,j)-position in the array. Let each y_{ij} represent a random sample value from a normal population, a different normally distributed population being associated with each (i,j)-cell of the array. The classification of the values $y = y_{ij}$ according to two factors, one called factor A characterized by the subdivision into rows and the other called factor B characterized by the subdivision into columns, is called a *two-way classification*, (*two-factor experiment*.) Here, only one observation appears in each cell (i,j).

15.36. Suppose that the standard deviations of all of the normal populations from which the y_{ij} are drawn are equal to the same value σ. Suppose that the means of these population distributions are of the form $m_{ij} = m + R_i + C_j$, where m is a constant, R_i is fixed for all cells in the ith row, C_j is fixed for all cells in the jth column, and $\sum_{i=1}^{r} R_i = 0$, $\sum_{j=1}^{s} C_j = 0$. The last two relations involve no loss of generality, since if the mean of the R_i's (or C_j's) were not zero, the mean could be included in the constant m and R_i (or C_j) could be replaced by the difference between R_i and the mean of the R_i's (or similarly for C_j).

15.37. All the cells in any one column are sometimes called a *block*. In describing the variation of the y_{ij}'s from cell to cell, the C_j's may be called the "column effects" due to factor B or "block effects," and the R_i's may be called the "row effects" due to factor A. The columns are called *randomized blocks* since the row effects R_i are distributed at random among the cells in a column or block, the distribution of R_i's in a block not depending on the block.

15.38. In the interpretation of experiments, it is often important to estimate, on the basis of the observed values y_{ij}, whether the underlying population distributions vary from row to row. If these distributions do not vary from row to row, it is possible to assume $R_i = 0$ for all values of i. Similarly it may be desirable to determine whether the underlying distributions vary from column to column. If these distributions do not vary from column to column, it is possible to assume $C_j = 0$ for all values of j. If the underlying distributions are the same for all cells in the array, then $m_{ij} = m$ and $R_i = 0$, $C_j = 0$ for all i and j.

15.39. The dependence of the means m_{ij} on row and column is tested by an analysis of variance, much as in §15.28.

Let

$$\bar{y} = (1/rs) \sum_{i=1}^{r} \sum_{j=1}^{s} y_{ij}$$ be the general mean of the whole sample array,

$$\bar{y}_{i\cdot} = (1/s) \sum_{j=1}^{s} y_{ij}$$ be the mean of the ith row,

$$\bar{y}_{\cdot j} = (1/r) \sum_{i=1}^{r} y_{ij}$$ be the mean of the jth column.

The total sum of squares S can be subdivided into

$$S = S_R + S_C + S_E$$

as follows:

Analysis of variance for two-way classification

Source of Variation	Sum of squares	Degrees of freedom	Mean square
Row effects	$S_R = s \sum_{i=1}^{r} (\bar{y}_{i\bullet} - \bar{y})^2$	$r - 1$	$\hat{\sigma}_R^2 = S_R/(r - 1)$
Column effects	$S_C = r \sum_{j=1}^{s} (\bar{y}_{\bullet j} - \bar{y})^2$	$s - 1$	$\hat{\sigma}_C^2 = S_C/(s - 1)$
Error	$S_E = \sum_{i=1}^{r} \sum_{j=1}^{s} H_{ij}$	G	$\hat{\sigma}_E^2 = S_E/G$
Total	$S = \sum_{i=1}^{r} \sum_{j=1}^{c} (y_{ij} - \bar{y})^2$	$rs - 1$	$S/(rs - 1)$

$$H_{ij} = (y_{ij} - \bar{y}_{i\bullet} - \bar{y}_{\bullet j} + \bar{y})^2. \qquad G = (r - 1)(s - 1).$$

S_R is the sum of squares due to variation "between rows;" S_C is the sum due to variation "between columns." S_E is the portion of the sum of squares not accounted for by the variations of y_{ij} between rows and between columns, and thus may be considered as the scatter in values y_{ij} due to random experimental errors (i.e., S_E is the portion not accounted for after removing the effects of factors A and B). S_E is sometimes called "interaction" or "discrepance."

15.40. In order to determine whether the observed sum of squares between rows (or columns) represents a significant difference between the values of R_i for the different rows (or the values of C_j for the different columns), the row (or column) sum S_R (or S_C) must be compared with the sum of squares S_E inherent in the accuracy of the data, as described below.

If $R_i = 0$ for all i, that is if there is no variation of population means m_{ij} with row, then (as in §15.29)

(15.40.1) $$F = (s - 1)S_R/S_E$$

has the F distribution $h_{(r-1),(r-1)(s-1)}(F)$, (§13.65). Thus in order to test the hypothesis that $R_i = 0$ for all i, calculate F and compare with values F_ϵ from Table X for $r - 1$ and $(r - 1)(s - 1)$ degrees of freedom. If $\epsilon = 0.05$, say, and all $R_i = 0$, there is only probability 0.05 that the observed value (15.19.1) of F will exceed F_ϵ. Thus if $F > F_\epsilon$, the hypothesis that all $R_i = 0$ is rejected at level of significance ϵ, and $\bar{y}_{i\cdot} - \bar{y}$ is an estimator for R_i.

If $C_j = 0$ for all j, that is if there is no variation of population mean m_{ij} with column, then

$$(15.40.2) \qquad F = (r - 1)S_C/S_E$$

has the F distribution $h_{(s-1),(r-1)(s-1)}(F)$, (§13.65). The hypothesis that $C_j = 0$ for all j can be tested in a manner similar to the test just described. If this hypothesis is rejected, an estimator for C_j is $\bar{y}_{\cdot j} - \bar{y}$.

15.41. If all R_i are zero then S_R/σ^2 is distributed according to the χ^2-distribution with $r - 1$ degrees of freedom.

If all C_j are zero then S_C/σ^2 is distributed according to the χ^2-distribution with $s - 1$ degrees of freedom.

Regardless of the values of R_i and C_j, S_E/σ^2 is distributed according to the χ^2-distribution with $(r - 1)(s - 1)$ degrees of freedom, and $\hat{\sigma}_E^2$ is an unbiased estimate of the population variance σ^2.

If all the R_i's and the C_j's are zero, S/σ^2 has the χ^2-distribution with $rs - 1$ degrees of freedom.

15.42. **Two-way model.** Suppose the elements in the samples are representable by

$$(15.42.1) \qquad y_{ij} = m + R_i + C_j + \epsilon_{ij}$$

where the ϵ_{ij} are assumed to have the normal distribution $N(0,\sigma^2)$.

The expected value of the "between rows" mean square is

$$(15.42.2) \qquad E[S_R/(r - 1)] = \sigma^2 + S \sum_{i=1}^{r} R_i^2/(r - 1)$$

The expected value of the "between columns" mean square is

$$(15.42.3) \qquad E[S_C/(s - 1)] = \sigma^2 + r \sum_{j=1}^{s} C_j^2/(s - 1)$$

The expected value of the "error sum of squares" mean square is

$$(15.42.4) \qquad E(S_E/G) = E(\hat{\sigma}_E^2) = \sigma^2.$$

Hence, $\hat{\sigma}_E^2$ is an estimator of σ^2. An estimator of m is $\bar{y}$.

EXAMPLE 1.   The classical applications of the randomized block analysis are in agricultural experiments of the following type, although many other applications can also be made.

Suppose each of three types of corn is planted in four different plots having four different fertilizer treatments. Let the yield per acre be y_{ij} where $i = 1, 2, 3, 4$ represent the type of fertilizer and $j = 1, 2, 3$ represent the type of corn. On the basis of the observed values y_{ij} it is desired to test whether there is a significant difference in yield of the three types of corn.

Suppose the yields y_{ij} are as follows, in suitable units. It is assumed that the standard deviation of the population of values y_{ij} (for fixed i and j) which would be obtained by an infinite number of repetitions of the whole experiment is the same for all values i,j.

	j 1	2	3	← (Corn)
i				
(Fertilizer) 1	100	160	150	
↓ 2	195	180	210	Entries in the table are the yields y_{ij}
3	135	125	175	
4	100	110	110	

Then in accordance with §15.39:

$$r = 4, \qquad s = 3$$

$$\bar{y} = 146$$

$$\bar{y}_{1\cdot} = 137, \qquad \bar{y}_{2\cdot} = 195, \qquad \bar{y}_{3\cdot} = 145, \qquad \bar{y}_{4\cdot} = 107$$

$$\bar{y}_{\cdot 1} = 132.5, \qquad \bar{y}_{\cdot 2} = 144, \qquad \bar{y}_{\cdot 3} = 161$$

$$S_R = 12,012, \qquad S_C = 1,645, \qquad S_E = 2,435, \qquad S = 16,092.$$

In order to test the hypothesis that there is no significant difference in the yields which can be attributed to the type of corn (i.e., that all $C_j = 0$), calculate $F = (r - 1)S_C/S_E = 2.03$. If the significance level $\epsilon = 0.05$ be adopted, then Table X shows that $F_\epsilon = 5.14$ for $m_1 = s - 1 = 2$ and $m_2 = (r - 1)(s - 1) = 6$. Since the observed value 2.03 is much less than 5.14, it is concluded that the test data do not indicate a significant difference in yields which can be attributed to the type of corn.

On the other hand, in order to test the hypothesis that there is no significant difference in the yields which can be attributed to the different fertilizers (i.e., that all $R_i = 0$), calculate $F = (s - 1)S_R/S_E = 9.9$. For $\epsilon = 0.05$, $F_\epsilon = 4.76$ for $m_1 = 3$ and $m_2 = 6$. Therefore, the hypothesis that there is no dependence of yield on the effect of the different fertilizers is rejected. It appears that the yield is significantly affected by the type of fertilizer.

For the level $\epsilon = 0.05$, it is concluded that the estimators for C_i are $C_1 = C_2 = C_3 = 0$, and the estimators for m and σ^2 are, respectively, $\bar{y} = 146$ and $S_E/G = 2,435/6 = 40.6$. The estimators for R_i are $R_1 = 137 - 146 = -9$, $R_2 = 49$, $R_3 = -1$.

15.43. **Replicated two-way classification.** Suppose the elements in the samples are representable by

$$(15.43.1) \qquad y_{ijk} = m + R_i + C_j + \lambda_{ij} + \epsilon_{ijk},$$

where the ϵ_{ijk} are rst independent variables each having the normal distribution $N(0,\sigma^2)$. Suppose

$$\sum_{i=1}^{r} \lambda_{ij} = 0 \text{ for each } j, \qquad \sum_{j=1}^{s} \lambda_{ij} = 0 \text{ for each } i,$$

$$(15.43.2)$$

$$\sum_{i=1}^{r} R_i = \sum_{j=1}^{s} C_j = 0.$$

λ_{ij} is called the (differential) *interaction effect* between the row i and the column j.

For each cell (i,j) there are t observations y_{ijk}, where $k = 1, \cdots, t$. This repetition of an experiment for a fixed (i,j) is called a *replication* of the ij^{th} cell.

Let

$$\bar{y}_{ij\bullet} = \frac{1}{t} \sum_{k=1}^{t} y_{ijk} \equiv S_{ij\bullet}/t, \qquad \text{(cell means)}$$

$$\bar{y}_{i\bullet\bullet} = \frac{1}{st} \sum_{j=1}^{s} \sum_{k=1}^{t} y_{ijk} \equiv S_{i\bullet\bullet}/st, \qquad \text{(row means)}$$

$$(15.43.3)$$

$$\bar{y}_{\bullet j\bullet} = \frac{1}{rt} \sum_{j=1}^{r} \sum_{k=1}^{t} y_{ijk} \equiv S_{\bullet j\bullet}/rt, \qquad \text{(column means)}$$

$$\bar{y} = \bar{y} \ldots = \frac{1}{rst} \sum_{i=1}^{r} \sum_{j=1}^{s} \sum_{k=1}^{t} y_{ijk} \equiv S/rst, \qquad \text{(grand mean)}$$

The total sum of the squares of the rst observations is

$$(15.43.4) \qquad S_T = S_A + S_B + S_I + S_E$$

where

$$S_A = \sum \sum \sum (\bar{y}_{i\cdot\cdot} - \bar{y})^2$$
$$S_B = \sum \sum \sum (\bar{y}_{\cdot j\cdot} - \bar{y})^2$$
$$S_I = \sum \sum \sum (\bar{y}_{ij\cdot} - \bar{y}_{i\cdot\cdot} - \bar{y}_{\cdot j\cdot} + \bar{y})^2$$
$$S_E = \sum \sum \sum (y_{ijk} - \bar{y}_{ij\cdot})^2$$
$$S_T = \sum \sum \sum (y_{ijk} - \bar{y})^2$$

and $\sum \sum \sum$ denotes $\displaystyle\sum_{i=1}^{r} \sum_{j=1}^{s} \sum_{k=1}^{t} \equiv \sum_{i,j,k}$

S_E is the *error sum of squares* representing the variation of y_{ijk} after removing the row effects, column effects and interaction effects.

S_I is the *interaction sum of squares* for the variation in y_{ijk} due to interaction effects.

S_A is the *row sum of squares* for the variation in y_{ijk} due to row effects.

S_B is the *column sum of squares* for the variation in y_{ijk} due to column effects.

15.44. Consider the hypotheses

$$\begin{cases} H_0: & R_1 = \cdots = R_r = 0 \\ H_1: & R_1, \cdots, R_r \quad \text{are not all zero} \end{cases} \quad \text{versus alternates}$$

$$\begin{cases} H_0': & C_1 = \cdots = C_s = 0 \\ H_1': & C_1, \cdots, C_s \quad \text{are not all zero} \end{cases} \quad \text{versus alternates}$$

$$\begin{cases} H_0'': \lambda_{11} = \cdots = \lambda_{rs} = 0 \\ H_1'': \lambda_{11}, \lambda_{12}, \cdots, \lambda_{rs} \quad \text{are not all zero} \end{cases} \quad \text{versus alternates}$$

Under hypothesis H_0'', S_I/σ^2 and S_E/σ^2 are independent and distributed as chi-square variables with $(r-1)(s-1)$ and $rs(t-1)$ degrees of freedom, respectively, and the ratio

(15.44.1) $$F'' = rs(t-1) S_I \div (r-1)(s-1) S_E$$

has the F-distribution with $(r-1)(s-1)$ and $rs(t-1)$ degrees of freedom. If $F'' > F_\epsilon$ the hypothesis H_0'' is rejected at the $100\,\epsilon$ percent level of significance. If H_0'' is rejected an unbiased estimator of λ_{ij} is $(\bar{y}_{ij\cdot} - \bar{y}_{i\cdot\cdot} - \bar{y}_{\cdot j\cdot} + \bar{y})$.

Under hypothesis H_0, the ratio

(15.44.2) $$F = rs(t-1) S_A/(r-1) S_E$$

has the F-distribution with $(r-1)$ and $rs(t-1)$ degrees of freedom. If by F-test H_0 is rejected, an unbiased estimate of R_i is $(\bar{y}_{r\cdots} - \bar{y})$.

Under hypothesis H_0', the ratio

(15.44.3) $$F' = rs(t-1)S_B/(s-1)\,S_E$$

has the F-distribution with $(s-1)$ and $rs(t-1)$ degrees of freedom. If hypothesis H_0' is rejected, an unbiased estimate of C_j is $(\bar{y}_{\bullet\bullet\bullet} - \bar{y})$.

The expected values of m and σ^2 are, respectively, $\bar{y}$ and $S_E/rs(t-1)$. The expected value of the "between rows" mean square is

(15.44.4) $$E[S_A/(r-1)] = \sigma^2 + st \sum_{i=1}^{r} R_i^2/(r-1).$$

Similar expressions for the expected mean square for the various sources of variation are shown in Table 15.44.1.

15.45. Computational scheme. In connection with Table 15.44.1 the following scheme of calculation may be used. From (15.43.3) calculate

$$S_{ij\bullet}, \ S_{i\bullet\bullet}, \ S_{\bullet j\bullet}, \ S$$

$$V = \sum_{i=1}^{r} \sum_{j=1}^{s} \sum_{k=1}^{t} y_{ijk}^{2}$$

$$S_T = V - (S^2/rst)$$

$$S_A = \left[\left(\sum_{i=1}^{r} S_{i\bullet\bullet}^{2}\right)\bigg/st\right] - (S^2/rst)$$

$$S_B = \left[\left(\sum_{j=1}^{s} S_{\bullet j\bullet}^{2}\right)\bigg/rt\right] - (S^2/rst)$$

$$S_E = V - \sum_{i=1}^{r} \sum_{j=1}^{s} S_{ij\bullet}^{2}/t$$

$$S_I = S_T - (S_A + S_B + S_E)$$

(15.45.1)

$$S_I = \left[\sum_{i=1}^{r} \sum_{j=1}^{s} S_{ij\bullet}^{2}/t\right] - \left[\left(\sum_{i=1}^{r} S_{i\bullet\bullet}^{2}\right)\bigg/st\right]$$

$$- \left[\left(\sum_{j=1}^{s} S_{\bullet j\bullet}^{2}\right)\bigg/rt\right] + (S^2/rst)$$

15.46. Completely randomized design. A completely randomized design is one in which the treatments are assigned completely at random to the experimental units, or vice versa. Such designs are widely used. The method is applicable to cases in which homogeneous experimental units are available.

TABLE 15.44.1

Analysis of variance table

For a replicated two-way experimental layout

Source of variation	Degrees of freedom	Sum of squares	Mean squares	Expected mean square
Row effects (A factor)	$r-1$	S_A	$S_A/(r-1)$	$\sigma^2 + st\sum_{i=1}^{r} R_i^2/(r-1)$
Column effects (B factor)	$s-1$	S_B	$S_B/(s-1)$	$\sigma^2 + rt\sum_{j=1}^{s} C_j^2/(s-1)$
Interaction effects	$(r-1)(s-1)$	S_I	$S_I/(r-1)(s-1)$	$\sigma^2 + t\sum_{i=1}^{r}\sum_{j=1}^{s} \lambda_{ij}^2/(r-1)(s-1)$
Error	$rs(t-1)$	S_E	$S_E/rs(t-1)$	σ^2
Total	$rst-1$	S_T		

Model: $y_{ijk} = m + R_i + C_j + \lambda_{ij} + \epsilon_{ijk}$.

15.47. A *randomized complete block design* is one in which: (a) the experimental units are allocated to *blocks* in such a way that the units within a block are relatively homogeneous and the number of units is equal to the number of treatments being investigated, (b) the treatments *within each block* are assigned at *random* to the units.

A typical statistical model associated with a completely randomized design involving t treatments and n experimental units per treatment is one in which the sample is considered to be of the form

$$y_{ij} = m + R_i + \epsilon_{ij}, \qquad i = 1, \cdots, t, \qquad j = 1, \cdots, n.$$

15.48. **Example involving factorial treatment combinations.** If the t' treatments are combinations of r levels of factor A, s levels of factor B, $t' = rs$, the model is

$$y_{ijk} = m + R_i + C_j + \lambda_{ij} + \epsilon_{ijk} \qquad i = 1, \cdots, r;$$

(15.48.1) $$\qquad j = 1, \cdots, s; k = 1, \cdots, t.$$

Here m is the true mean effect, R_i is the true effect of the ith level of factor A, C_j is the true effect of the jth level of factor B, λ_{ij} is the true effect of the interaction of the ith level of A with the jth level of B, and ϵ_{ijk} is the true effect of the kth experimental unit subject to the ij^{th} treatment combination. m is taken to be a constant and ϵ_{ijk} to have distribution $N(0,\sigma)$.

The data of interest are commonly studied by using identities expressing the sum of the squares of all the observations from the mean of all the y_{ijk} in terms of the sums of squares associated with the various levels of A and B and their interactions.

Model I. Fixed effects model. If the concern is centered on the r levels of factor A and the s levels of factor B present in the experiment, the model is assumed to be such that

(15.48.2) $$\sum_{i=1}^{r} R_i = \sum_{j=1}^{s} C_j = \sum_{i=1}^{r} \lambda_{ij} = \sum_{j=1}^{s} \lambda_{ij} = 0,$$

the model is called a *fixed effects model*.

Model II. Random effects model. If the concern is with: (a) a population of levels of factor A of which only a random sample of r levels are present in the experiment, (b) a population of levels of factor B of which only a random sample of s levels are present in the experiment; and R_i, C_j, and λ_{ij} have distributions $N(0,\sigma_R{}^2)$, $N(0,\sigma_C{}^2)$, and $N(0,\sigma_\lambda{}^2)$, respectively, the model is a *random effects model*.

Model III. Mixed model (A fixed, B random). If the concern is with:

(a) only the r levels of factor A present in the experiment, (b) a population of levels of factor B of which only a random sample of s levels are present; and $\sum_{i=1}^{r} R_i = \sum_{i=1}^{r} \lambda_{ij} = 0$; C_j have distributions $N(0,\sigma_C^2)$, the model is called a *mixed model*.

Model III. Mixed model (A random, B fixed). A model in which the concern is with: (a) only the s levels of factor B present, (b) a population of levels of factor A of which only a random sample of r levels are present; and $\sum_{j=1}^{s} C_j = \sum_{j=1}^{s} \lambda_{ij} = 0$; the R_i having distributions $N(0,\sigma_R^2)$ is also a *mixed model*.

The expected mean squares for a two-factor factorial in a completely randomized design are shown in Table 15.48.1. This result may be compared with that given in Table 15.44.1. (See page 294.)

15.49. **More general cases.** The system of analysis illustrated above is used for many sorts of m-factor classifications. Other designs of experiments are given in the literature (e.g., Latin squares, Greco-Latin squares, Youden squares, hierarchical designs). Such schemes of analysis are designed to fill the needs of experiments in which the effects of many factors are studied simultaneously. To list fully the more elementary cases is beyond the scope of this book.

15.50. **Analysis of covariance.** This is a technique employed in analyzing factorial experiments when the subject matter is related to certain observable parameters by means of regression functions. In such cases the variances may be related to such regression functions. The methods involve those of regression analysis and the analysis of variation. The theory is known as the *analysis of covariance*.

15.51. **Latin squares.** Suppose an array of values y_{ij} is given, arranged in a square array of r rows and r columns, $i = 1, 2, \cdots, r, j = 1, 2, \cdots, r$. Let each y_{ij} represent a random sample value from a normal distribution associated with the (i,j)-cell of the array. Suppose that the standard deviations of all of these distributions have the same value σ. Suppose that the means of these distributions are of the form $m_{ij} = m + R_i + C_j + T_t$, where $t = 1, 2, \cdots, r$, with $y_{ij} = m_{ij} + \epsilon_{ij}$, ϵ_{ij} being r^2 independent variables with normal distribution $N(0,\sigma^2)$ and

$$\sum_{i=1}^{r} R_i = 0, \qquad \sum_{j=1}^{r} C_j = 0, \qquad \sum_{t=1}^{r} T_t = 0.$$

Here m is a constant, each R_i is fixed for all cells in the ith row, each C_j is fixed for all cells in the jth column, and T_t is a fixed factor called a "treatment" associated with an experiment.

TABLE 15.48.1

Expected mean squares in a completely randomized design for a two-factor classification

Expected mean square

Model	Source of variation (treatment)		
	A	B	Interaction of A, B
Model I (fixed effects)	$\sigma^2 + ts \sum_{i=1}^{r} R_i^2/(r-1)$	$\sigma^2 + tr \sum_{j=1}^{s} C_j^2/(s-1)$	$\sigma^2 + t \sum_{i=1}^{r}\sum_{j=1}^{s} \lambda_{ij}^2/(r-1)(s-1)$
Model II (random effects)	$\sigma^2 + t\sigma_\lambda^2 + ts\sigma_R^2$	$\sigma^2 + t\sigma_\lambda^2 + tr\sigma_C^2$	$\sigma^2 + t\sigma_\lambda^2$
Model III (mixed) A fixed, B random	$\sigma^2 + t\sigma_\lambda^2 + ts \sum_{i=1}^{r} R_i^2/(r-1)$	$\sigma^2 + tr\sigma_C^2$	$\sigma^2 + t\sigma_\lambda^2$
Model III (mixed) A random, B fixed	$\sigma^2 + ts\sigma_R^2$	$\sigma^2 + t\sigma_\lambda^2 + tr \sum_{j=1}^{s} C_j^2/(s-1)$	$\sigma^2 + t\sigma_\lambda^2$

Experimental error $= \sigma^2$ in all cases.

Suppose further that each value of T_t occurs exactly once in each row and exactly once in each column of the square array. There are many possible arrangements of the set of T_t's which satisfy this condition. Such an arrangement is called a *latin square*.

A test layout leading to a latin square array of the type described above is used, for example, where an area of ground is laid out in r^2 plots, and r different fertilizer treatments are being tested. It is of interest to determine whether the different treatments result in different crop yields. It is assumed that there may be different yields from row to row, and from column to column, due to differences in the initial soil conditions, regardless of the treatments used. The problem is to determine whether there is a significant additional variation in yield due to the different treatments.

15.52. An analysis of variance can be used, much as in §15.16. Let

$$\bar{y} = (1/r^2) \sum_{i=1}^{r} \sum_{j=1}^{r} y_{ij}, \qquad \bar{y}_{i\cdot} = (1/r) \sum_{j=1}^{r} y_{ij},$$

$$\bar{y}_{\cdot j} = (1/r) \sum_{i=1}^{r} y_{ij}, \qquad \bar{y}_{(t)} = (1/r) \sum_{(t)} y_{ij},$$

where the last quantity is summed over all cells associated with T_t, for the indicated t.

The sum S can be subdivided as follows into

$$S = S_R + S_C + S_T + S_E ,$$

each treatment occurring once in each row and once in each column.

S_R is the sum of squares due to variations "between rows"; S_C is the sum due to variations "between columns"; S_T is the sum due to variation in "treatments." S_E is the sum due to random experimental errors not accounted for by the variations of y_{ij} between rows, or between columns, or between treatments. (See Table 15.52.1.)

15.53. In order to determine whether the observed sum of squares between rows (or columns, or treatments) represents a significant difference between the values of y_{ij} in the different rows (or columns, or associated with the different treatments), the sum of squares between rows (or columns, or treatments) must be compared with the sum of squares S_E inherent in the accuracy of the test or observations, as described below.

If $R_i = 0$ for all i, that is if there is no significant variation in values

TABLE 15.52.1

Analysis of variance for latin square

Variation	Sum of squares	Degrees of freedom	Mean square
Rows	$S_R = r \sum_{i=1}^{r} (\bar{y}_{i\bullet} - \bar{y})^2$	$r - 1$	$S_R/(r - 1)$
Columns	$S_C = r \sum_{j=1}^{r} (\bar{y}_{\bullet j} - \bar{y})^2$	$r - 1$	$S_C/(r - 1)$
Treatments	$S_T = r \sum_{(t)} (\bar{y}_{(t)} - \bar{y})^2$	$r - 1$	$S_T/(r - 1)$
Errors	$S_E = \sum_{i=1}^{r} \sum_{j=1}^{r} J_{ij}$	K	S_E/K
Total	$S = \sum_{i=1}^{r} \sum_{j=1}^{r} (y_{ij} - \bar{y})^2$	$r^2 - 1$	$S/(r^2 - 1)$

$$J_{ij} = (y_{ij} - \bar{y}_{i\bullet} - \bar{y}_{\bullet j} - \bar{y}_{(t)} + 2\bar{y})^2. \quad K = (r - 1)(r - 2).$$

y_{ij} with row, then

$$(15.53.1) \qquad\qquad F = (r - 2)S_R/S_E$$

has the F-distribution $h_{(r-1),(r-1)(r-2)}(F)$, (§13.65). Thus in order to test the hypothesis that $R_i = 0$ for all i, calculate F and compare with values F_ϵ from Table X for $(r - 1)$, $(r - 1)(r - 2)$ degrees of freedom. If $\epsilon = 0.05$, say, and all $R_i = 0$, there is a probability 0.05 that the observed value (15.34.1) of F will exceed F_ϵ. Thus, if $F > F_\epsilon$, the hypothesis that all $R_i = 0$ is rejected at the significance level ϵ. If this hypothesis is rejected an unbiased estimator for R_i is $\bar{y}_{i\bullet} - \bar{y}$.

In a similar manner, if $C_j = 0$ for all j, that is, if there is no significant variation in values y_{ij} with column, then

$$(15.53.2) \qquad\qquad F = (r - 2)S_C/S_E$$

has the F-distribution $h_{(r-1),(r-1)(r-2)}(F)$, and a test of the hypothesis that all $C_j = 0$ may be carried out in a similar manner to the test outlined above for the case of $R_i = 0$. If the hypothesis that all $C_j = 0$ is rejected an unbiased estimator for C_j is $\bar{y}_{\cdot j} - \bar{y}$.

Likewise, if $T_t = 0$ for all t, that is, if there is no significant variation in values y_{ij} with treatment, then

$$(15.53.3) \qquad F = (r - 2)S_T/S_E$$

has the F-distribution $h_{(r-1),(r-1)(r-2)}(F)$, and a test of the hypothesis that all $T_t = 0$ may be carried out in a similar manner to the test outlined above for the case of $R_i = 0$. If the hypothesis that all $T_t = 0$ is rejected an unbiased estimator for T_t is $\bar{y}_t - \bar{y}$.

If all R_i, C_j, and T_t are zero, then

(a) S_R/σ^2, S_C/σ^2, S_T/σ^2, S_E/σ^2 are independently distributed according to χ^2-distributions with $r - 1$, $r - 1$, $r - 1$, $(r - 1)(r - 2)$ degrees of freedom, respectively;

(b) S/σ^2 has the χ^2-distribution with $r^2 - 1$ degrees of freedom, and $S/(r^2 - 1)$ is an unbiased estimate of σ^2.

No matter what values m, R_i, C_j and T_t may have, S_E/σ^2 has the χ^2-distribution with $(r - 1)(r - 2)$ degrees of freedom, and $S_E/(r - 1)(r - 2)$ is an unbiased estimate of the population variance σ^2 and $\bar{y}$ is an unbiased estimate of m.

15.54. Restrictions on usefulness of latin squares. The effect of the grouping in a latin square is to reduce or eliminate from the errors all differences among rows and all differences among columns. The usefulness of latin squares is restricted somewhat by the fact that the array of quantities y_{ij} must be a square, the number of "treatments" must equal the number r of rows, and when r becomes large the number of replications becomes impractical. In practice squares larger than 12×12 are seldom used.

15.55. Tables. Examples of latin squares are given in Table 15.55.1. For a more complete representation of squares up to 8×8 see W. G. Cochran and G. M. Cox, *Experimental Designs*, 2d ed., John Wiley & Sons, Inc., New York, 1957.

TABLE 15.55.1

Selected latin squares

3 × 3

A	B	C
B	C	A
C	A	B

4 × 4

1				2				3				4			
A	B	C	D	A	B	C	D	A	B	C	D	A	B	C	D
B	A	D	C	B	C	D	A	B	D	A	C	B	A	D	C
C	D	B	A	C	D	A	B	C	A	D	B	C	D	A	B
D	C	A	B	D	A	B	C	D	C	B	A	D	C	B	A

5 × 5

A	B	C	D	E
B	A	E	C	D
C	D	A	E	B
D	E	B	A	C
E	C	D	B	A

6 × 6

A	B	C	D	E	F
B	F	D	C	A	E
C	D	E	F	B	A
D	A	F	E	C	B
E	C	A	B	F	D
F	E	B	A	D	C

7 × 7

A	B	C	D	E	F	G
B	C	D	E	F	G	A
C	D	E	F	G	A	B
D	E	F	G	A	B	C
E	F	G	A	B	C	D
F	G	A	B	C	D	E
G	A	B	C	D	E	F

8 × 8

A	B	C	D	E	F	G	H
B	C	D	E	F	G	H	A
C	D	E	F	G	H	A	B
D	E	F	G	H	A	B	C
E	F	G	H	A	B	C	D
F	G	H	A	B	C	D	E
G	H	A	B	C	D	E	F
H	A	B	C	D	E	F	G

sixteen

FINITE DIFFERENCES.
INTERPOLATION

16.1. In some branches of statistics, in numerical analysis and computation, considerable use is made of various forms of interpolation theory, factorial polynomials, and differences. A brief outline of some of these topics is given in this chapter. These schemes also furnish methods for approximating a given function by means of polynomials.

16.2. Factorial polynomials.

The polynomial

$$(16.2.1) \qquad x^{[n]} = x(x - 1)(x - 2) \cdots (x - n + 1)$$

of degree n in x is called a *factorial polynomial*. If x is an integer and $x > n - 1$, then

$$(16.2.2) \qquad x^{[n]} = x!/(x - n)!.$$

Eq. (16.2.1) can be written

$$(16.2.3) \qquad x^{[n]} = S_0^n x^n + S_1^n x^{n-1} + \cdots + S_{n-1}^n x,$$

where S_i^n is a *Stirling number of the first kind*. (Table 16.2.1)
Properties.

$$(16.2.4) \qquad (x + 1)^{[n]} - x^{[n]} = nx^{[n-1]}, \qquad x^{[0]} = 1, \qquad x^{[1]} = x.$$

$$(16.2.5) \quad x^{[k]}/k! = x(x - 1) \cdots (x - k + 1) \div k! = C_k^x$$

$$= \text{binomial coefficient, } (k \text{ a positive integer, see §8.1.)}$$

$$(16.2.6) \qquad S_i^{n+1} = S_i^n - nS_{i-1}^n.$$

TABLE 16.2.1

Values of S_i^n, Stirling number of the first kind

n	S_0^n	S_1^n	S_2^n	S_3^n	S_4^n	S_5^n	S_6^n	S_7^n	S_8^n	S_9^n
1	1									
2	1	−1								
3	1	−3	2							
4	1	−6	11	−6						
5	1	−10	35	−50	24					
6	1	−15	85	−225	274	−120				
7	1	−21	175	−735	1,624	−1,764	720			
8	1	−28	322	−1,960	6,769	−13,132	13,068	−5,040		
9	1	−36	546	−4,536	22,449	−67,284	118,124	−109,584	40,320	
10	1	−45	870	−9,450	63,273	−269,325	723,680	−1,172,700	1,026,576	−362,880

16.3. **Binomial theorem.** The binomial theorem states that $(a + b)^x = \sum_{i=0}^{L} C_i^x a^{x-i} b^i$. When x is a positive integer, or zero, $L = x$; when x is any other real number the series indicated is infinite, and converges when $b^2 < a^2$. Thus, the binomial coefficients C_i^x, given in (16.2.5), and factorial polynomials $x^{[j]}$ are of interest for any real x, j a non-negative integer. For example,

$$C_3^x = x^{[3]} \div 3! = x(x - 1)(x - 2) \div 3!,$$

$$C_3^{-2.1} = (-2.1)^{[3]} \div 3! = (-2.1)(-3.1)(-4.1) \div 3!.$$

16.4. **Remark about factorial moments** (see §§ 2.14, 6.13). The relation (16.2.3) implies the relation between factorial moments, $\nu_{[r]}$, and ordinary moments about the origin, ν_r, namely,

$$(16.4.1) \qquad \nu_{[r]} = S_0^r \nu_r + S_1^r \nu_{r-1} + \cdots + S_{r-1}^r \nu_1.$$

16.5. **Scheme for calculating factorial moments.** Suppose the frequency of occurrence of x_i is $f(x_i) = f_i$, with $f_1 + f_2 + \cdots + f_m = n$, as shown below in Table 16.5.1, with $x_0 = 0$, $x_1 = 1$, $\cdots$, $x_m = m$. (For simplicity, the table is constructed for $m = 4$.)

TABLE 16.5.1

x_i	f_i	Col. 1	Col. 2	Col. 3	Col. 4	Col.5
0	f_0	$f_0 + f_1 + f_2 + f_3 + f_4$				
1	f_1	$f_1 + f_2 + f_3 + f_4$	$f_1 + 2f_2 + 3f_3 + 4f_4$			
2	f_2	$f_2 + f_3 + f_4$	$f_2 + 2f_3 + 3f_4$	$f_2 + 3f_3 + 6f_4$		
3	f_3	$f_3 + f_4$	$f_3 + 2f_4$	$f_3 + 3f_4$	$f_3 + 4f_4$	
4	f_4	f_4	f_4	f_4	f_4	f_4

The entry in row i, column j is the sum of the quantities in column $j - 1$ from the bottom up to and including the term in the ith row. The sums Σ_i in the top rows of the successive columns are the "reduced factorial moments"

$$(16.5.1) \qquad \Sigma_1 = n\nu_{[0]}, \qquad \Sigma_2 = n\nu_{[1]},$$
$$\Sigma_3 = n\nu_{[2]} \div 2!, \cdots, \Sigma_{r+1} = n\nu_{[r]} \div r!, \cdots.$$

This furnishes a scheme for calculating systematically the factorial moments $\nu_{[r]}$. From this scheme other schemes applicable to special situations are derivable.

Finite differences

16.6. Notation. Consider the sequence of numbers

$$(16.6.1) \qquad \cdots F_{-n}, F_{-n+1}, \cdots, F_{-2}, F_{-1}, F_0, F_1, F_2, \cdots, F_n, \cdots$$

The following notation is in common use, where i denotes any integer (including zero).

$$
\begin{aligned}
& \Delta^0 F_i = F_i, \qquad \Delta^1 F_i = \Delta F_i = F_{i+1} - F_i, \\
& \Delta^2 F_i = \Delta(\Delta F_i) = \Delta F_{i+1} - \Delta F_i, \cdots,
\end{aligned}
$$

$$(16.6.2)
\begin{aligned}
& \Delta^n F_i = \Delta(\Delta^{n-1} F_i) = \Delta(\Delta^{n-2} F_{i+1} - \Delta^{n-2} F_i) \\
& \qquad\quad = \Delta^{n-1} F_{i+1} - \Delta^{n-1} F_i, \cdots,
\end{aligned}
$$

$$
\begin{aligned}
& \Delta^n F_i = \Delta_n F_{i+n}, \qquad (n \text{ a non-negative integer}) \\
& \Delta_n F_m = \Delta^n F_{m-n}. \qquad (m - n \text{ any integer})
\end{aligned}
$$

16.7. Diagonal differences. Consider the function $y = f(x)$. Let $y_0, y_1, \cdots, y_n$ denote a set of values of this function corresponding to the values $x_0, x_1, \cdots, x_n$ of x. The *first (diagonal) differences* of y are

$$(16.7.1) \qquad \Delta y_0 = y_1 - y_0, \ \Delta y_1 = y_2 - y_1, \cdots;$$

the *second (diagonal) differences*

$$(16.7.2) \qquad \Delta^2 y_0 = \Delta y_1 - \Delta y_0, \ \Delta^2 y_1 = \Delta y_2 - \Delta y_1, \cdots;$$

TABLE 16.7.1

Diagonal differences

x	y	Δy	$\Delta^2 y$	$\Delta^3 y$	$\Delta^4 y$	$\Delta^5 y$	$\Delta^6 y$
x_0	y_0						
		Δy_0					
x_1	y_1		$\Delta^2 y_0$				
		Δy_1		$\Delta^3 y_0$			
x_2	y_2		$\Delta^2 y_1$		$\Delta^4 y_0$		
		Δy_2		$\Delta^3 y_1$		$\Delta^5 y_0$	
x_3	y_3		$\Delta^2 y_2$		$\Delta^4 y_1$		$\Delta^6 y_0$
		Δy_3		$\Delta^3 y_2$		$\Delta^5 y_1$	
x_4	y_4		$\Delta^2 y_3$		$\Delta^4 y_2$		
		Δy_4		$\Delta^3 y_3$			
x_5	y_5		$\Delta^2 y_4$				
		Δy_5					
x_6	y_6						

the *third (diagonal) differences*

$$(16.7.3) \qquad \Delta^3 y_0 = \Delta^2 y_1 - \Delta^2 y_0, \; \Delta^3 y_1 = \Delta^2 y_2 - \Delta^2 y_1, \; \cdots ;$$

and so on. Table 16.7.1 shows how such differences are formed. Also,

$$(16.7.4) \qquad \Delta^n y_0 = \sum_{j=0}^{n} (-1)^j C_j{}^n y_{n-j}.$$

16.8. Horizontal differences. The *first (horizontal) differences* of y are

$$(16.8.1) \qquad \Delta_1 y_1 = y_1 - y_0, \; \Delta_1 y_2 = y_2 - y_1, \; \cdots ;$$

the *second (horizontal) differences*,

$$(16.8.2) \qquad \Delta_2 y_2 = \Delta_1 y_2 - \Delta_1 y_1, \; \Delta_2 y_3 = \Delta_1 y_3 - \Delta_1 y_2, \; \cdots ;$$

the *third (horizontal) differences*,

$$(16.8.3) \qquad \Delta_3 y_3 = \Delta_2 y_3 - \Delta_2 y_2, \; \Delta_3 y_4 = \Delta_2 y_4 - \Delta_2 y_3, \; \cdots ;$$
$$\vdots$$

the pth *(horizontal) differences*,

$$(16.8.4) \qquad \Delta_p y_p = \Delta_{p-1} y_p - \Delta_{p-1} y_{p-1}, \; \Delta_p y_{p+1} = \Delta_{p-1} y_{p+1} - \Delta_{p-1} y_p, \; \cdots$$

TABLE 16.8.1

Horizontal differences

x	y	$\Delta_1 y$	$\Delta_2 y$	$\Delta_3 y$	$\Delta_4 y$	$\Delta_5 y$	$\Delta_6 y$
x_0	y_0						
x_1	y_1	$\Delta_1 y_1$					
x_2	y_2	$\Delta_1 y_2$	$\Delta_2 y_2$				
x_3	y_3	$\Delta_1 y_3$	$\Delta_2 y_3$	$\Delta_3 y_3$			
x_4	y_4	$\Delta_1 y_4$	$\Delta_2 y_4$	$\Delta_3 y_4$	$\Delta_4 y_4$		
x_5	y_5	$\Delta_1 y_5$	$\Delta_2 y_5$	$\Delta_3 y_5$	$\Delta_4 y_5$	$\Delta_5 y_5$	
x_6	y_6	$\Delta_1 y_6$	$\Delta_2 y_6$	$\Delta_3 y_6$	$\Delta_4 y_6$	$\Delta_5 y_6$	$\Delta_6 y_6$

The relations between the entries of Tables 16.7.1 and 16.8.1 are:

$$(16.8.5) \qquad \Delta^m y_k = \Delta_m y_{k+m}, \qquad \text{(interpolation forward from } y_k)$$

$$(16.8.6) \qquad \Delta_m y_n = \Delta^m y_{n-m}. \qquad \text{(interpolation backward from } y_n)$$

Newton's formulas, Lagrangian interpolation

16.9. Interpolation forward. Suppose $x_0, x_1, x_2, \cdots, x_n$ are spaced equidistant, and $x_{i+1} > x_i$; and $x_{i+1} - x_i = h$, for $i = 0, 1, \cdots, n - 1$.

Let $f(x)$ have values y_i for x_i, $i = 0, 1, \cdots, n$. Then, if $x = x_0 + hu$, Newton's formula for interpolating forward is

$$(16.9.1) \qquad \varphi(x) = \varphi(x_0 + hu)$$

$$= y_0 + C_1{}^u \, \Delta y_0 + C_2{}^u \, \Delta^2 y_0 + \cdots + C_n{}^u \, \Delta^n y_0,$$

where the $C_i{}^u$ are binomial coefficients. (See §8.1.)

Eq. (16.9.1) can be written in various ways, e.g.,

$$(16.9.2) \qquad \varphi(x) = \sum_{j=0}^{n} (u^{[j]}/j!) \, \Delta^j y_0 = (1 + \Delta)^u y_0 = \sum_{j=0}^{n} C_j{}^u \, \Delta^j y_0,$$

$$\text{(where } \Delta^0 y_0 = y_0 \text{).}$$

The polynomial $\varphi(x)$ is such that $\varphi(x_i) = y_i$ for $i = 1, \cdots, n$. For other values of x, $\varphi(x)$ approximates $f(x)$.

Formula (16.9.1) is used to interpolate values of y to the right of (forward from) y_0 near the beginning of a set of tabular values, in which case u is positive, and for extrapolating (backward) a short distance to the left from y_0, in which case u is negative.

When only the first two terms of (16.9.1) are used for interpolation, formula (16.9.1) reduces to the usual formula for *linear interpolation*; namely,

$$(16.9.3) \qquad \varphi(x_0 + hu) = y_0 + u(y_1 - y_0).$$

When the first n terms of (16.9.1) are used, Eq. (16.9.1) reduces to a form F, linear in $y_0, y_1, \cdots, y_n$. F is known as a "n-point interpolation" formula, sometimes called a *Lagrangian formula*. Extensive tables for the coefficients of this form are available, but would require too much space for inclusion here.

16.10. **Interpolation backward.** Newton's formula for interpolating backward is (when $x = x_n + hu$)

$$(16.10.1) \qquad \varphi(x) = \varphi(x_n + hu) = y_n + u \, \Delta_1 y_n + \frac{u(u + 1)}{2!} \, \Delta_2 y_n$$

$$+ \cdots + \frac{u(u + 1) \cdots (u + n - 1)}{n!} \, \Delta_n y_n.$$

Eq. (16.10.1) can be written in various ways; e.g.,

$$(16.10.2) \qquad \varphi(x) = \sum_{j=0}^{n} (-1)^i C_j^{-u} \, \Delta_i y_n$$

$$= \sum_{j=0}^{n} C_j^{u+i-1} \, \Delta_i y_n, \qquad (\text{where } \Delta_0 y_n = y_n)$$

$$= \sum_{j=0}^{n} (-1)^i C_j^{-u} \, \Delta^i y_{n-i} = \sum_{j=0}^{n} C_j^{u+i-1} \, \Delta^i y_{n-j}.$$

The polynomial $\varphi(x)$ is such that $\varphi(x_i) = y_i$ for $i = 1, \cdots, n$. For other values of x, $\varphi(x)$ approximates $f(x)$.

Formula (16.10.1) is used for interpolating a value of y to the left of (backward from) y_n near the end of a set of tabular values, in which case u is negative, and for extrapolating values of y a short distance (forward) to the right of y_n, in which case u is positive.

Central difference formulas

16.11. Another class of interpolation formulas in general use are those known as *central difference formulas*. Such formulas often converge more rapidly than Newton's formulas (16.9.1) and (16.10.1). Two such formulas are given below. They are well adapted for interpolating values of a function near the middle of a tabulated set.

Notation. The central difference formulas make use of the following notation, starting with a set of numbers $\cdots, y_{-i}, y_{-i+1}, \cdots, y_{-1}, y_0, y_1, \cdots, y_i, \cdots$. The central differences are represented by expressions of the form $\delta^n y_j$. The subscript j, which is not necessarily an integer, takes the value midway between the subscripts on the two quantities whose difference is being represented. By definition,

$$(16.11.1) \qquad \delta y_i = y_{i+(1/2)} - y_{i-(1/2)}, \qquad \delta^n y_i = \delta^{n-1} y_{i+(1/2)} - \delta^{n-1} y_{i-(1/2)},$$

$$n > 1.$$

The differences may be arranged in an array such as illustrated in Table 16.11.1.

TABLE 16.11.1.

y_{-2}					
	$\delta y_{-3/2}$				
y_{-1}		$\delta^2 y_{-1}$			
	$\delta y_{-1/2}$		$\delta^3 y_{-1/2}$		
y_0		$\delta^2 y_0$		$\delta^4 y_0$	$\delta^5 y_{1/2}$
	$\delta y_{1/2}$		$\delta^3 y_{1/2}$		
y_1		$\delta^2 y_1$		$\delta^4 y_1$	
	$\delta y_{3/2}$		$\delta^3 y_{3/2}$		
y_2		$\delta^2 y_2$			
	$\delta y_{5/2}$				
y_3					

It is convenient to define the mean

(16.11.2) $\mu \, \delta^n y_i = (1/2)(\delta^n y_{i-(1/2)} + \delta^n y_{i+(1/2)}).$

The quantities $\mu \, \delta^n y_i$ may be thought of as falling between the corresponding pairs $\delta^n y_{i-(1/2)}$, $\delta^n y_{i+(1/2)}$ in the columns of an array such as shown in Table 16.11.1. Stirling's and Bessel's interpolation formulas given below are based on horizontal rows in such a supplemented array, the successive quantities in a given row alternating between differences and means.

The central differences represent the same quantities as the diagonal differences, but with a changed notation. The relation between the two notations is

(16.11.3)
$$\delta^n y_i = \Delta^n y_{i-(n/2)},$$
$$\mu \, \delta^n y_i = (1/2)(\Delta^n y_{i-(n+1)/2} + \Delta^n y_{i-(n-1)/2}).$$

16.12. Examples of this notation are given as follows:

$\Delta y_0 = y_1 - y_0, \, \Delta y_{-1} = y_0 - y_{-1},$

$$\Delta y_{-2} = y_{-1} - y_{-2}, \, \Delta y_{-3} = y_{-2} - y_{-3}, \, \cdots ,$$

$\delta y_{1/2} = \Delta y_0, \, \delta^2 y_1 = \Delta^2 y_0, \, \delta^3 y_{3/2} = \Delta^3 y_0, \, \cdots , \, \delta^n y_{n/2} = \Delta^n y_0.$

(16.12.1)
$\mu y_i = (y_{i+(1/2)} + y_{i-(1/2)}) \div 2, \quad \delta y_i = y_{i+(1/2)} - y_{i-(1/2)},$

$\mu \, \delta y_0 = (\Delta y_{-1} + \Delta y_0) \div 2, \qquad \delta^2 y_0 = \Delta^2 y_{-1},$

$\mu \, \delta^3 y_0 = (\Delta^3 y_{-2} + \Delta^3 y_{-1}) \div 2, \qquad \delta^4 y_0 = \Delta^4 y_{-2},$

$\delta^m y_{n/2} = \Delta^m y_{(n-m)/2}, \qquad$ if the integers m and n, both are odd,

or both are even.

If $y = f(x)$ denotes a function of x;

$$\delta f(x) = f(x + (1/2)h) - f(x - (1/2)h),$$
$$\mu f(x) = [f(x + (1/2)h) + f(x - (1/2)h)] \div 2,$$
$$f(x_i + (1/2)h) \equiv y_{i+(1/2)}.$$

16.13. Stirling's central difference formula.
Suppose the values $\cdots$, $y_{-n}, y_{-n+1}, \cdots , y_{-1}, y_0, y_1, y_2, \cdots , y_n, \cdots$ are a set of values corresponding to the values $\cdots , x_{-n}, x_{-n+1}, \cdots , x_{-1}, x_0, x_1, x_2, \cdots , x_n, \cdots$ of a

function $y = f(x)$. Suppose the x's are equally spaced with $x_{i+1} - x_i = h$; and $f(x)$ has the value y_i when $x = x_i$, where i can be any integer. Let $x = x_0 + hu$. *Newton-Stirling's* formula for interpolating is

$$(16.13.1) \qquad \varphi(x) = \varphi(x_0 + hu) = y_0 + u(\mu\ \delta y_0) + \frac{u^2}{2!}(\delta^2 y_0)$$

$$+ \frac{u(u^2 - 1^2)}{3!}(\mu\ \delta^3 y_0) + \frac{u^2(u^2 - 1^2)}{4!}(\delta^4 y_0)$$

$$+ \frac{u(u^2 - 1^2)(u^2 - 2^2)}{5!}(\mu\ \delta^5 y_0)$$

$$+ \frac{u^2(u^2 - 1^2)(u^2 - 2^2)}{6!}(\delta^6 y_0) + \cdots$$

$$+ \frac{u(u^2 - 1^2)\cdots[u^2 - (n-1)^2]}{(2n-1)!}(\mu\ \delta^{2n-1} y_0)$$

$$+ \frac{u^2(u^2 - 1^2)\cdots[u^2 - (n-1)^2]}{(2n)!}(\delta^{2n} y_0) + \cdots$$

If in (16.13.1) the terms beyond the one containing $\delta^{2n} y_0$ are ignored the resulting polynomial $\varphi(x)$ is such that $\varphi(x_i) = y_i$ for $i = -n, \cdots, -1, 0, 1, \cdots, n$; for other values of x, $\varphi(x)$ approximates $f(x)$.

16.14. **Bessel's central difference interpolation formula.** The *Newton-Bessel* formula for interpolating is

$$(16.14.1) \quad \varphi(x) = \varphi(x_0 + hu) = \varphi(x_0 + h(v + 1/2)) = \mu\ y_{1/2} + v(\delta y_{1/2})$$

$$+ \frac{(v^2 - 1/4)}{2!}(\mu\ \delta^2 y_{1/2}) + \frac{v(v^2 - 1/4)}{3!}(\delta^3 y_{1/2})$$

$$+ \frac{(v^2 - 1/4)(v^2 - 9/4)}{4!}(\mu\ \delta^4 y_{1/2})$$

$$+ \frac{v(v^2 - 1/4)(v^2 - 9/4)}{5!}(\delta^5 y_{1/2}) + \cdots$$

$$+ \frac{(v^2 - 1/4)(v^2 - 9/4)\cdots[v^2 - (2n-1)^2/4]}{(2n)!}(\mu\ \delta^{2n} y_{1/2})$$

$$+ \frac{v(v^2 - 1/4)(v^2 - 9/4)\cdots[v^2 - (2n-1)^2/4]}{(2n+1)!}(\delta^{2n+1} y_{1/2}) + \cdots$$

where

$$u = v + (1/2), \qquad\qquad \delta y_{r+(1/2)} = \Delta y_r,$$

$$\mu\, y_{1/2} = (y_0 + y_1) \div 2, \qquad \delta y_{1/2} = \Delta y_0,$$

(16.14.2)
$$\mu\, \delta^2 y_{1/2} = (\Delta^2 y_{-1} + \Delta^2 y_0) \div 2, \qquad \delta^3 y_{1/2} = \Delta^3 y_{-1},$$

$$\mu\, \delta^4 y_{1/2} = (\Delta^4 y_{-2} + \Delta^4 y_{-1}) \div 2, \qquad \delta^5 y_{1/2} = \Delta^5 y_{-2},$$

$$\vdots \qquad\qquad\qquad \vdots$$

$$\mu\, \delta^{2n} y_{1/2} = (\Delta^{2n} y_{-n} + \Delta^{2n} y_{-n+1}) \div 2, \quad \delta^{2n+1} y_{1/2} = \Delta^{2n+1} y_{-n}.$$

If in (16.14.1) the terms beyond the one containing $\delta^{2n+1} y_{1/2}$ are ignored the resulting polynomial $\varphi(x)$ is such that $\varphi(x_i) = y_i$ for $i = -n, \cdots, -1, 0, 1, \cdots, n + 1$; for other values of x, $\varphi(x)$ approximates $f(x)$.

In general, Bessel's formula gives a more accurate result when interpolating near the middle of an interval; Stirling's formula gives a more accurate interpolation near the ends of an interval.

16.15. **Other interpolation formulas.** There are many other notations, formulas and schemes used for interpolation. They are too extensive for inclusion in this book.

For further information on this subject the user is referred to

Guest, P. G., *Numerical Methods of Curve Fitting*, Cambridge University Press, New York, 1961.

Milne, W. E., *Numerical Calculus*, Princeton University Press, Princeton, N.J., 1949.

Milne-Thomson, L. M., *The Calculus of Finite Differences*, The Macmillan Company, New York, 1933.

Scarborough, J. B., *Numerical Mathematical Analysis*, The Johns Hopkins Press, Baltimore, 1930.

Thompson, A. J., *Tables of Coefficients of Everett's Central Difference Interpolation Formula*, Cambridge University Press, New York, 1943.

seventeen

SEQUENTIAL ANALYSIS, SAMPLING INSPECTION, QUALITY CONTROL, RELIABILITY THEORY

17.1. This chapter discusses briefly several applications of statistical theory to practical sampling test problems which are commonly used for sampling inspection of manufactured products, surveillance of stores of products, monitoring of a continuous process, reliability studies, and in other ways in making quantitative estimates concerning large collections of items without having information on all of the collection. These applications each represent large fields. Detailed discussions and associated tables for each of these fields are given in numerous other books devoted to these specialties. The discussion here is intended only to give an indication or example of the procedures; a full exposition would be too lengthy for inclusion in this book.

Sequential analysis

17.2. In a sequential test of a statistical hypothesis, certain calculations are made after each observation (or group of observations); on the basis of the calculated results a decision is made as to whether: (1) to accept the hypothesis under test, (2) to accept a selected alternative hypothesis, or (3) to withhold judgment until more data have been examined or have become available. The decisions are so governed that the hypothesis is accepted or rejected as soon as it appears that the available data are adequate for making a decision with preselected degrees of reliability.

17.3. Three quantities play important roles in a sequential process; namely, the number of observations N, the probability α of erroneously rejecting the hypothesis under test, the probability β of erroneously accepting this hypothesis. In the theory, the number of trials considered necessary to reach a decision is minimized while fixing the risks α and β. This procedure is somewhat in contrast with that used in non-sequential tests where N is likely to be fixed, and one of the risks minimized while fixing the other. Sequential tests are frequently more economical than non-sequential tests, especially when N is readily changed; whereas non-sequential tests are apt to be more economical when N is not easily changed. Oftentimes the number of tests N called for in a sequential test of a given reliability is considerably less than the number of tests required by the corresponding non-sequential test.

17.4. The hypothesis under test may be denoted by H_1. The alternate hypothesis may be denoted by H_2. It is assumed in advance that one of these two hypotheses is correct, and the sequential test is used to choose between them.

17.5. The acceptable risk of error is decided in advance and is defined by α and β, where:

α is the largest acceptable fraction of cases in which H_2 will be selected by the test procedure when in fact H_1 is true; β is the largest acceptable fraction of cases in which H_1 will be selected by the procedure when in fact H_2 is true.

It is required that $\alpha + \beta < 1$. α is a limit on the occurrence of the "α-error" or "Type I error," which is the risk of rejecting a true hypothesis H_1. β is a limit on the occurrence of the "β-error" or "Type II error," which is the risk of accepting a false hypothesis H_1.

17.6. In a sequential test it is customary to make use of the "likelihood ratio," the ratio being calculated after each observation (or group of observations), each computation making use of all available observations. The *likelihood ratio L* is defined as

$$(17.6.1) \qquad\qquad L = P_2/P_1,$$

where P_1 is the probability of obtaining the observations at hand on the basis that the hypothesis H_1 under test is true, and P_2 is the probability of obtaining the observations at hand assuming the selected alternative hypothesis H_2 is true. According to the theory, if L ever reaches (exceeds) a certain value A, $(A > 1)$, the collection of data from observations is stopped and the hypothesis H_2 accepted; if L ever falls as low as (or below) a certain level $1/B$, $(B > 1)$, the collection of data is stopped and hypothesis H_1 is accepted. As long as L is between A and $1/B$, no decision is made and the collection of data proceeds, since the evidence at hand is then inadequate to reach either decision subject to the selected reliabilities α and β.

17.7. As long as successive observations can be considered as independent statistically, the ratio L for any set of observations is the ratio for the last observation multiplied by the ratio calculated for the set of preceding observations.

17.8. Sequential procedure to choose between hypotheses H_1 and H_2
It is found that the defined risks will not be exceeded if

$$H_1 \text{ is accepted if } L \leq \beta/(1 - \alpha) = 1/B,$$

$$H_2 \text{ is accepted if } L \geq (1 - \beta)/\alpha = A,$$

where $P_2/P_1 = L$. The condition $\alpha + \beta < 1$ ensures that $1/B < A$.

17.9. The sequential procedure consists in computing L after each observation. If $L \leq 1/B$, accept H_1; if $1/B < L < A$, make another observation and compute a new L; if $L \geq A$, accept H_2. If a new L has been computed, repeat the same comparisons with the new L. Ultimately either H_1 or H_2 will be accepted.

17.10. Since the value of L after $n + 1$ observations is obtained from the value of L based on n observations by multiplying the latter by the ratio of probabilities p_2/p_1 of the $n + 1^{\text{th}}$ observation according to the two hypotheses, the successive values of L are obtained by a continuing product. In practice, it is often convenient to take logarithms so that each new observation yields an increment to a sum.

17.11. If it is more practical to take m observations before applying the test procedure, and then (if more observations are required) to take another m observations, etc., the same procedures are applicable as discussed above, except that $n = m, 2m, \cdots$. The procedure is just as before except the test is not applied at the intermediate observations. Such procedures may lead to use of more observations than would have been required with single observations between test calculations, since it might have been possible to reach a decision before a group of m observations was completed.

17.12. It is convenient to express the limits, and the procedure for computation to be made after each observation, in terms of the results of the observations in a simple form. This can be done for each type of hypothesis which may be tested sequentially. Two examples are outlined below.

17.13. Sequential test for mean value of a normal population. Assume the values of x are normally distributed, with standard deviation σ and mean m. Suppose m is unknown. The hypothesis H_1 to be tested is that $m = m_1$; the alternative hypothesis H_2 is that $m = m_2$, $m_1 < m_2$. If n observations $x_1, x_2, \cdots, x_n$ are made, the probability P_1 that this sequence is obtained is

$$(17.13.1) \quad P_1 = (\sigma \sqrt{2\pi})^{-n} \exp \left\{ (-1/2\sigma^2) \sum_{i=1}^{n} (x_i - m_1)^2 \right\},$$

under hypothesis H_1; the probability P_2 that this sequence is obtained under hypothesis H_2 is given by the same formula with m_1 replaced by m_2. The natural logarithm of the likelihood ratio $L_n = P_2/P_1$ is

$$(17.13.2) \quad J_n = \log_e L_n = [(m_2 - m_1)/\sigma^2] \sum_{i=1}^{n} x_i - (m_2^2 - m_1^2)n/2\sigma^2.$$

Let

$$(17.13.3) \quad A = (1 - \beta)/\alpha, \ B = (1 - \alpha)/\beta, \ a = \log_e A, \ b = \log_e B.$$

The quantities m_1, m_2, a, b are fixed. If $J_n \leqq -b$, then

$$(17.13.4) \quad \sum_{i=1}^{n} x_i \leqq [-b\sigma^2/(m_2 - m_1)] + (m_2 + m_1)n/2 \equiv M_1,$$

and hypothesis H_1 is accepted. If $J_n \geqq a$, it follows that

$$(17.13.5) \quad \sum_{i=1}^{n} x_i \geqq [a\sigma^2/(m_2 - m_1)] + (m_2 + m_1)n/2 \equiv M_2,$$

and hypothesis H_2 is accepted. The limits M_1 and M_2 ,when considered as functions of n, are parallel lines in the plane of coordinates $M \equiv \sum_{i=1}^{n} x_i$ and n.

17.14. The testing procedure consists in calculating $\sum_{i=1}^{n} x_i = M$ after each observation. The number pair (M,n) may be represented by a point on the graph. If the point is below the lower line, accept H_1; if it is above the upper line, accept H_2; if it is in between, withhold judgment, and make another observation. This procedure can be put in tabular form if desired.

EXAMPLE 1. Suppose that a certain type of casting is designed to weigh 1.2 pounds, but that due to a change in process it is suspected that the weights will run somewhat high. It is desired to check the average weight, with a minimum of measurements of the weight x. The hypothesis to test is $H_1: m = m_1 = 1.2$ lb. where m is the mean weight. An alternate hypothesis, determined by a knowledge of the new process, is H_2: $m = m_2 = 1.3$ lb. Long experience with these castings indicates that the weights of the individual castings are approximately normally distributed with standard deviation around the mean approximately $\sigma = 0.3$ lb; this value of σ is considered applicable to both old and new processes.

Let $\alpha = 0.10$, $\beta = 0.10$. It is found from (17.13.3) that $a = 2.197$, $b = 2.197$. Then in accordance with (17.13.4) and (17.13.5), $M_1 =$

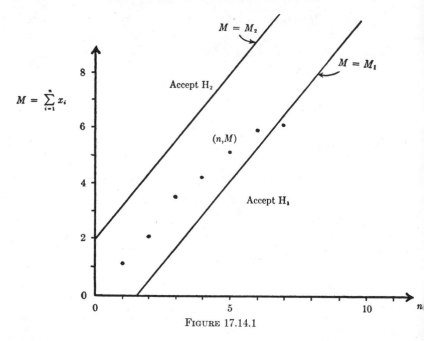

$$M = \sum_{i=1}^{n} x_i$$

FIGURE 17.14.1

$1.25n - 1.98$, $M_2 = 1.25n + 1.98$. The lines $M = M_1$ and $M = M_2$ are shown on Fig. 17.14.1.

17.15. It can be seen that in this example hypothesis H_1 could not possibly be accepted on the basis of the first observation alone. If the successive values of M observed in an actual sequence of weight measurements were as shown by the points plotted on Fig. 17.14.1, the hypothesis H_1 would be accepted after the seventh measurement. The risks of error in this procedure are given by the particular values of α and β assumed, α and β having the meanings defined in §17.5.

17.16. The general scheme of fixing two parallel lines as limits for the regions of decision in favor of either hypothesis or of taking a further observation, can be worked out for other types of hypotheses also.

17.17. **Sequential test for population proportion of items having a property** E. Suppose the proportion p of items in a lot which have property E is unknown. The number of items in a sample which have property E may be assumed to follow the binomial distribution, but with unknown fraction or proportion p. Suppose that by some means (e.g., experience) two values of p are selected for consideration. Suppose that the hypothesis H_1 is that $p = p_1$; the alternate hypothesis H_2 being that $p = p_2$. Let the observed number of E's in a sample of n be h. The probability that h successes would have occurred after n operations is $p_1^h(1 - p_1)^{n-h}$ under hypothesis H_1; $p_2^h(1 - p_2)^{n-h}$ under hypothesis H_2. The likelihood ratio is

$$(17.17.1) \qquad L_n = p_2^h(1 - p_2)^{n-h} \div p_1^h(1 - p_1)^{n-h}.$$

Take logarithms to base 10.

$$(17.17.2) \qquad K_n = \log_{10} L_n = h \log_{10} (p_2/p_1) - (n - h)c,$$

$a = \log_{10} [(1 - \beta)/\alpha]$, $b = \log_{10} [(1 - \alpha)/\beta]$, $c = \log_{10} [(1 - p_1)/(1 - p_2)]$.

Limits for the sequential test are given by parallel lines derived from $K_n = b$ and $K_n = a$, on a (h,n) graph. Let

$$(17.17.3) \qquad d = \log_{10} [p_2(1 - p_1)/p_1(1 - p_2)].$$

If $K_n \leqq -b$, then

$$(17.17.4) \qquad h \leqq (cn - b)/d \equiv S_1,$$

and hypothesis H_1 is accepted.

If $K_n \geqq a$, then

$$(17.17.5) \qquad h \geqq (cn + a)/d \equiv S_2,$$

and hypothesis H_2 is accepted.

17.18. The limits S_1 and S_2, when considered as functions of n, are parallel lines in the plane of coordinates h and n. The testing procedure consists in calculating h after each observation. The number pair (h,n) may be represented by a point on the graph. If the point is below the lower line $h = S_1$, accept H_1; if above the upper line $h = S_2$, accept H_2; if between, withhold judgment and make another observation.

17.19. References. The methods of sequential analysis are discussed at greater length in the following references among others.

Wald, A.: *Sequential Analysis*, John Wiley & Sons, Inc., New York, 1947.

Wald, A.: *Sequential Tests of Statistical Hypotheses*, Annals of Math. Statistics, Vol. 16 (1945), pp. 117–186.

Statistical Research Group, Columbia University: *Sampling Inspection*, McGraw-Hill Book Co., New York, 1947, edited by Freeman, Friedman, Mosteller, and Wallis.

Statistical Research Group, Columbia University: *Techniques of Statistical Analysis*, McGraw-Hill Book Co., New York, 1947, edited by M. W. Hastay, G. C. Eisenhart, and W. A. Wallis.

Sampling inspection by attributes

17.20. A group of items forming a lot may have to be judged as to whether the lot contains more than some given proportion of defects, or whether the variation of some quantity is too great within the lot, or whether the average of some measurement is within an acceptable limit, etc. It is often impractical or expensive or destructive to inspect the whole lot, so inspection is confined to one or more samples. A judgment (accept or reject) concerning the whole lot is then made on the basis of the sample, with a calculated risk of error.

17.21. Operating characteristic curve. The performance of lot or sampling inspection procedures, and the risk of error in accepting or rejecting the lot, may be described in terms of an *operating characteristic curve*. Such a curve shows the likelihood of accepting a lot which contains any given proportion of defective items. An operating characteristic curve (o.c.c.) for a typical sampling plan is often plotted as shown in Fig. 17.21.1. The method of constructing the chart depends on the sampling plan used.

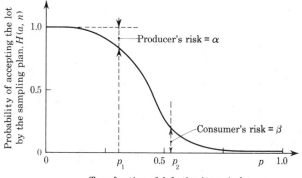

Operating Characteristic Curve

FIGURE 17.21.1

17.22. It may be decided in advance that some fraction p_2 of defects in the lot is acceptable. The quantity $100p_2$ is sometimes called the "tolerance percent defective." It may also be decided that it is uneconomical to reject lots having less than some fraction p_1 of defects. A sampling plan is selected so that the curve drops steeply from near 100% acceptance of lots having fraction of defects near p_1 to near 0% acceptance for lots having fraction of defects near p_2. In general, in order to make the curve steeper, a larger sample from the lot has to be inspected. The amount of inspection required must be balanced against the amount of risk involved in accepting a lot which has more than fraction p_2 of defects, or in rejecting a lot which has fewer than fraction p_1 of defects. The probability of accepting a lot having fraction p_2 or more of defects is sometimes called the *consumer's risk*. The probability of rejecting a lot having fraction p_1 or less of defects is sometimes called the *producer's risk*.

17.23. Sampling plans. A sampling plan is a schedule of sample sizes and *acceptance numbers, a,* and *rejection numbers, b.* Each sample is inspected and the number, d, of defects is compared with the acceptance and rejection numbers. If $d \leq a$, the lot is accepted. If $d \geq b$, the lot is rejected, and if $a < d < b$ (if this is possible) the next sample on the schedule is drawn and inspected, the accumulated sample being then judged according to a new set of acceptance and rejection numbers given in the schedule.

17.24. A given sampling plan has a particular operating characteristic curve. Extensive compilations of sampling plans have been made available corresponding to a wide variety of operating characteristic curves, and making use of samples of various sizes.

17.25. The sampling plan is sometimes selected so as to yield a given consumer's risk (or producer's risk). It is sometimes selected so as to yield a given average number of defects in lots after inspection ("average outgoing quality"). In the latter case, the "rejected" lots are sometimes reinspected completely; i.e., every item is inspected instead of a sample, and the acceptable items are sent out with the other lots which were accepted by the sampling plan. In all sampling plans, the intent is to achieve the desired type of control with the smallest possible samples.

17.26. One sampling procedure is to inspect a sample containing a specified number n of items, and base the decision on the observed features of the sample. ("Single sampling.") In an attempt to reduce the number of inspections required, another procedure involves inspecting first a sample of n' items ($n' < n$). If the sample evidence is sufficiently convincing statistically, a decision about the lot is made. If the sample evidence is not sufficiently one-sided, a second sample of n'' items is inspected, and a decision is based on the combined sample of $n' + n''$

items. ("Double sampling.") In an attempt to reduce still further the amount of inspection required, the method of sequential analysis may be used which consists in inspecting one item at a time, or one group of items at a time, and after each inspection determining whether the accumulated evidence justifies a decision concerning the lot or whether another item or group must be inspected. ("Sequential sampling.") Examples of these different types of sampling plans are outlined below; there are also other possible ways of setting up sampling plans of these general types.

17.27. Single sampling plan. Inspect a sample of n items. The acceptance number is a and the rejection number is $b = a + 1$. The observed number of defects is d. If $d \geq b$, reject the lot. If $d \leq a$, accept the lot.

17.28. The numbers a and b can be found as follows if one point on the operating characteristic curve is prescribed. Suppose the proportion defective in the whole lot is p. Then the probability that r defects will be observed in a sample of n is given by the $(r + 1)^{\text{st}}$ term of the expansion of $[(1 - p) + p]^n \equiv w$, namely,

$$(17.28.1) \qquad P(r,n) = [n!/r!(n - r)!](1 - p)^{n-r}p^r.$$

Hence the probability of acceptance of the lot on the basis of a random sample of size n from the lot is the sum $H(a,n)$ of the first $(a + 1)$ terms of the binomial expansion w. Methods of calculating or approximating these sums are indicated in §8.5, and Tables II and III can be used. Thus a and $b = a + 1$ can be found such that the chance F of accepting the lot will take approximately any desired fixed value F_1 for a selected value of p and n. The operating characteristic curve can be made to pass through a selected point $(p,H(a,n))$, when a is determined by making $H(a,n)$ take the selected value F_1 of the probability of acceptance for a particular p. The particular point $(p,H(a,n))$ is often taken to correspond to the tolerance fraction defective $p = p_2$, and the consumer's risk $H(a,n)$.

17.29. Double sampling plan. Inspect a sample of n_1 items. The first acceptance number is a_1. The first rejection number is $b_1 = a_2 + 1$. The observed number of defects is d_1. If $d_1 \geq b_1$, reject the lot. If $d_1 \leq a_1$, accept the lot. If $a_1 < d_1 < b_1$, draw another sample of $n_2 = 2n_1$ items. The number of defects in the second sample is d_2. The acceptance number for the combined sample of $n_1 + n_2$ is a_2. The rejection number for the combined sample is $b_2 = a_2 + 1$. If $d_1 + d_2 \leq a_2$, accept the lot. If $d_1 + d_2 \geq b_2$, reject the lot.

17.30. The probability A of accepting a lot on the basis of a given set of numbers a_1, b_1, a_2, b_2 is found as follows. It is noted that when a_1 and a_2 are determined, the values of b_1 and b_2 are automatically fixed. Suppose the proportion defective in the whole lot is p. The probability A of accepting the lot is equal to the sum of the probability of there being a_1 or less defects in a sample of n_1 items, *plus* the probability of there being more than a_1 defects in n_1 items and less than or equal to a_2 defects in $n_1 + n_2$ items;

$$(17.30.1) \qquad A = H(a_1,n_1) + \sum_{k=a_1+1}^{a_2} P(k,n_1)\, H(a_2 - k,\, 2n_1),$$

where

(17.30.2) $P(k,n)$ is the $(k + 1)^{\text{th}}$ term of $[(1 - p) + p]^n$, as in the case of single sampling;

(17.30.3) $H(a,n)$ is the sum of the first $(a + 1)$ terms of $[(1 - p) + p]^n$, as in the case of single sampling.

17.31. The setting up of a particular sampling plan is usually done in practice by computing the operating characteristic curves (A as a function of p) for families of values of a_1, a_2, n_1, and selecting a suitable sampling plan for the intended purpose. A systematic trial and error method is thus evolved for building various sampling plans of interest.

17.32. **Sequential sampling plan.**   Inspect one item at a time. At any stage, n items have been inspected, and d_n items are found defective. The acceptance number is a_n and the rejection number is b_n. If $d_n \leqq a_n$, accept the lot. If $d_n \geqq b_n$, reject the lot. If $a_n < d_n < b_n$, inspect another item and apply the test with the sample of $n + 1$ items.

17.33. In order to find a_n and b_n, proceed as follows. Decide on $\alpha =$ acceptable probability of rejecting the lot if the proportion defective is p_1, where it is desired that the test accept the lot if the proportion defective is not greater than p_1. Decide on $\beta =$ acceptable probability of accepting the lot if the proportion defective is p_2, where it is desired that the test reject the lot if the proportion defective is greater than p_2. Thus $(1 - \alpha,\, p_1)$ and (β,p_2) are two points on the operating characteristic curve of the test. It is found that $a_n = S_1$, $b_n = S_2$, where S_1 and S_2 are defined in (17.17.4), (17.17.5).

17.34. **Grouped sequential sampling plan.**   It may be impractical to inspect item by item. It is then convenient to inspect a group of n_1 items, apply a test, and either accept, reject, or decide to inspect a further group of n_1 items. In the latter case, a new test is applied to the $2n_1$ items thus far inspected, leading either to acceptance, rejection, or decision to inspect still another group of n_1 items.

17.35. The acceptance and rejection numbers for an accumulated sample of size n are approximately the same as those computed for the item by item sequential procedure. Minor adjustments can be made to allow for the effects of grouping, and in order to yield essentially the same operating characteristic curve as for the item by item scheme.

17.36. **Examples of sampling plans.** Specific numerical examples of the sampling plans outlined above are given in Table 17.36.1. This one particular set of plans was selected for illustration only. Large catalogs of such plans are available in various books, but are too extensive for inclusion here. The examples of sampling plans given in Table 17.36.1 all accomplish approximately the same degree of control over the lots, in the sense that they all have very nearly the same operating characteristic curves. Other similar plans have been devised for a great variety

TABLE 17.36.1

Examples of sampling plans for lot inspection

Type of sampling	Individual samples to be examined		Accumulated sample		
	Order of sample in sequence	Sample size	Size	Accept- ance number	Rejec- tion number
Single	1	75	75	10	11
Double	1	50	50	6	17
	2	100	150	16	17
Item by item sequential	n	1	n	$0.134n$ -2.09	$0.134n$ $+2.68$
Grouped sequential	1	20	20	1	5
	2	20	40	3	8
	3	20	60	6	11
	4	20	80	9	13
	5	20	100	13	16
	6	20	120	16	18
	7	20	140	18	19

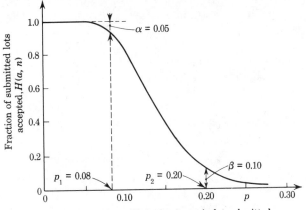

Operating Characteristic Curve Approximately Applicable to All Sampling Plans in the Examples of Table 17.36.1.

FIGURE 17.36.1

of operating characteristics, and for various sample sizes and groups for inspection.

17.37. In the particular sampling plans tabulated, the consumer's risk is no more than $\beta = 0.10$ that a lot with fraction $p_2 = 0.20$ or more of defects would be accepted, and the producer's risk is no more than $\alpha = 0.05$ that a lot with fraction $p_1 = 0.08$ or less of defects would be rejected.

17.38. **Certain widely used sampling plans.** A widely used set of sampling plans is one known as Military Standard 105 D (MIL-STD-105 D). This standard establishes sampling plans and procedures for inspection by attributes for use in the determination of the acceptability of products procured by the United States government. In this plan defects are classified as critical, major, and minor.

An *acceptable quality level* (AQL) is a *nominal* value expressed in terms of fraction or percent defective. The AQL's in MIL-STD are not tied to a particular producer's risk. The AQL values to be used for a given product are specified by the buyer (e.g., the government). The acceptability of a lot (found in accordance with MIL-STD-105 D) submitted to the buyer is determined by use of designated sampling plans associated with a specified value of AQL.

The probability of accepting a lot having a fraction of defectives θ is denoted by $P(A|\theta)$. The *operating characteristic curve* (OC curve) *of the sampling plan* is the graph of $P(A|\theta)$ plotted as a function of θ. MIL-STD-105 D gives certain OC curves which indicate the fraction P_a of lots of a given quality which, when submitted, may be expected to be accepted under the various sampling plans given. P_a is called the *probability of acceptance*. The OC curves are given for single, double, and multiple sampling.

By *process average* is meant the average fraction defective of product submitted for original inspection.

The *average outgoing quality* (AOQ) is the average quality of outgoing product including all accepted lots plus all rejected lots after the rejected lots have been effectively 100 percent inspected and all defectives have been replaced by non-defectives (i.e., *rectified*). AOQ is the fraction of defectives left in a large pool of inspected lots in which rejected lots have been rectified, each lot having fraction θ defective before inspection.

The quantity $\theta \cdot P(A|\theta)$ is the average outgoing quality $AOQ(\theta)$ for incoming fraction defective θ.

The maximum value of $AOQ(\theta)$ when treated as a function of θ is called the *average outgoing quality limit* (AOQL). AOQL is the upper limit of the fraction of defectives in a large pool of rectifiable lots which have been screened of defectives by the sampling plan.

Each *sampling plan* indicates the sample size from each lot which are to be inspected and gives criteria (e.g., acceptance and rejection numbers) for determining the acceptability of a lot.

The buyer specifies the *inspection level* to be used for any particular requirement. Generally, three inspection levels are used, but when the risks are high or the samples small, four additional levels are provided.

Three types of sampling plans are given: single, double, and multiple. In general, the average sample size of a multiple plan is less than the corresponding single sample size.

MIL-STD-105 D gives procedures to be followed when switching from one level of inspection to another (e.g., from normal to reduced, or to high level inspection), and when terminating the inspection procedure.

For further information, tables, and charts on this subject the reader should consult:

Bowker, A. H., and Goode, H. P., *Sampling Inspection by Variables*, McGraw-Hill Book Company, New York, 1952.

Dodge, H. F., and Romig, H. G., *Sampling Inspection Tables, Single and Double Sampling*, John Wiley & Sons, Inc., New York, 1959.

Freeman, Friedman, Mosteller, and Wallis (eds.), Statistical Research

Group, Columbia University, *Sampling Inspection*, McGraw-Hill Book Company, New York,1948.

Military Standard 105 D, *Sampling Procedures and Tables for Inspection by Attributes*, Government Printing Office, Washington, D.C., Apr. 29, 1963.

Quality control

17.39. Quality control is used to obtain control of quality in a production process, and to monitor the output of a repetitive production process. It is recognized that the product will occasionally not meet specifications, and that any measured characteristic will show some variation from one item to the next, even though all the items are supposedly made by the same process. The likelihood of failure in the process under test, or the variation of measured characteristics of a product, may be examined for a while under conditions believed to be normal or "in control" in order to establish a basis of comparison for judging future items. In some cases the basis of comparison may be prescribed by a specification. The quality control procedure is then put into operation in order to detect any significant change in the likelihood of failure or the variation of measured characteristics of the product, with the purpose of having early warning that some feature of the production process has changed. When a significant change in the output is found, the production is said to be "out of control."

17.40. Quality is determined either by quantitative measurements, or by a classification into "accept" or "reject", "good" or "bad", etc. The latter is known as *inspection by attributes*.

17.41. **Quality control on the basis of quantitative measurements.** In the case of quantitative measurements, quality control is often carried out by finding the mean $\bar{x}$ and the standard deviation s or the range R of the measurements in each of a sequence of small samples drawn periodically from the continuous flow of items to be judged. Any large changes in the mean would disclose a systematic trend in the general magnitude of the measurements. Any large change in the standard deviation or the range would disclose a major change in the scatter of the measurements.

17.42. **Control procedure.** Let the observations on the individual items be quantities x. Let u be any of the quantities (e.g., average $\bar{x}$, standard deviation s, range R) which might be used to characterize the distribution of observations x, in a sample. The control procedure for u is as follows, assuming that the population of values of x is normal or nearly normal.

(a) Determine a mean value $u = u_0$ which is accepted as representing values of x for items which are "in control". This value might be known from a physical specification, or from an appropriate theory, or from past experiences, or by authority, or it might be estimated by observing a sequence I_0 of items which are considered to be "in control" and finding the average value of u for this sequence. In the latter case, the sequence I_0 should be large enough so that u_0 can be used with confidence. When plotted, $u = u_0$ is called the *central line*.

(b) Estimate the standard deviation σ_u of values of u calculated from samples of size n. This step is based on sampling distributions of u as described in §17.43 below.

(c) Set *control limits* equal to $u_0 \pm 3\sigma_u$. The use of $3\sigma_u$ is arbitrary but has been found in practice to yield a reasonable balance between failing to take note of an out of control situation and investigating occasional chance wild data. About 0.003 of the observations of u would be expected to fall outside of the $u_0 \pm 3\sigma_u$ limits by chance if u is normally distributed. Other limits can be set if the circumstances warrant. *Note*: If a limit $u_0 + 3\sigma_u$ or $u_0 - 3\sigma_u$ would have an impossible value (e.g., a negative value for a quantity known to be positive), the nearest allowable value should be substituted for the limit.

(d) Decide on a sampling schedule, which involves choosing a sample size n and how often the samples are to be drawn. This will have to be decided on the basis of the practical considerations of cost of testing, importance of early detection of changes in the process, technical knowledge, familiarity with conditions of production and sampling, and so on; but the individual samples need not always be large and may be taken with n less than 10 in many applications.

(e) Compute u from each of a succession of samples of size n. If any sample yields a value of u outside of the control limits, the process should be considered "out of control" and should be investigated and corrected. It is a reasonable presumption that a significant change in the process being examined has taken place. The nature of the particular parameter u sometimes aids in identifying the cause of the change in the process.

17.43. Methods for estimating central line and control limits. If the population is normal, with mean $\bar{x}$ and standard deviation σ_x, the central line u_0 and the control limits $u_1 = u_0 - 3\sigma_u$, $u_2 = u_0 + 3\sigma_u$, can be estimated as indicated in Table 17.43.1. Numerical coefficients for use with Table 17.43.1 are given in Table 17.43.2. The estimates of central

TABLE 17.43.1

Estimates of central line and control limits

Note: $b(n)$ is given in §13.43. k_n and c_n are given in Table 13.86.1.

$$\sigma_s = G\sigma_x \qquad G = [((n-1)/n) - [b(n)]^2]^{1/2}.$$

u	Estimate of central line: u_0	Estimate of control limits: u_1, u_2
(a) Case when $\bar{x}$ and σ_x are known.		
Sample mean, $\bar{x}$	$\bar{\bar{x}} = \bar{x}$	$a_1 = \bar{\bar{x}} - M, \qquad a_2 = \bar{\bar{x}} + M$ $\qquad M = 3\sigma_x/\sqrt{n}$
Sample standard deviation, s	$\bar{s} = b(n)\cdot\sigma_x$	$s_1 = \bar{s} - 3G\sigma_x, \qquad s_2 = \bar{s} + 3G\sigma_x$
Sample range, R	$\overline{R} = k_n\sigma_x$	$R_1 = \overline{R} - 3c_n\sigma_x, \qquad R_2 = \overline{R} + 3c_n\sigma_x$

(b) Case when $\bar{x}$ and σ_x are unknown, and sample values $\bar{x}_0$, s_0, and R_0 are available from a preliminary sample "in control."

Sample mean, $\bar{x}$	$\bar{\bar{x}} = \bar{x}_0$	$a_1' = \bar{\bar{x}} - M', \qquad a_2' = \bar{\bar{x}} + M'$ $\qquad M' = 3s_0/(b(n)\cdot\sqrt{n})$
Sample standard deviation, s	$\bar{s} = s_0$	$s_1' = \bar{s} - 3s_0G/b(n), \quad s_2' = \bar{s} + 3s_0G/b(n)$
Sample range, R	$\overline{R} = R_0$	$R_1' = \overline{R} - 3c_ns_0/b(n), \quad R_2' = \overline{R} + 3c_ns_0/\bar{o}(n)$

In Tables 17.43.1 and 17.43.2, $s^2 = \sum_{i=1}^{n}(x_i - \bar{x})^2/n$.

TABLE 17.43.2

Numerical coefficients for estimating central lines and control limits

Sample size	Factors for averages	Factors for standard deviations			Factors for ranges		

(a) Case when $\bar{x}$ and σ_x are known.

n	M/σ_x	$b(n)$	s_1/σ_x	s_2/σ_x	k_n	R_1/σ_x	R_2/σ_x
2	2.121	0.5642	0	1.843	1.128	0	3.686
3	1.732	0.7236	0	1.858	1.693	0	4.358
4	1.500	0.7979	0	1.808	2.059	0	4.698
5	1.342	0.8407	0	1.756	2.326	0	4.918
6	1.225	0.8686	0.026	1.711	2.534	0	5.078
7	1.134	0.8882	0.105	1.672	2.704	0.205	5.203
8	1.061	0.9027	0.167	1.638	2.847	0.387	5.307
9	1.000	0.9139	0.219	1.609	2.970	0.546	5.394
10	0.949	0.9227	0.262	1.584	3.078	0.687	5.469
12	0.866	0.9359	0.331	1.541	3.258	0.924	5.592
14	0.802	0.9453	0.384	1.507	3.407	1.121	5.693
16	0.750	0.9523	0.427	1.478			
18	0.707	0.9576	0.461	1.454			
20	0.671	0.9619	0.491	1.433			

(b) Case when $\bar{x}$ and σ_x are unknown, and sample values $\bar{x}_0$, s_0, and R_0 are available from a preliminary sample "in control."

n	M'/s_0	M'/R_0	s_1'/s_0	s_2'/s_0	R_1'/R_0	R_2'/R_0
2	3.760	1.880	0	3.267	0	3.267
3	2.394	1.023	0	2.568	0	2.575
4	1.880	0.729	0	2.266	0	2.282
5	1.596	0.577	0	2.089	0	2.115
6	1.410	0.483	0.030	1.970	0	2.004
7	1.277	0.419	0.118	1.882	0.076	1.924
8	1.175	0.373	0.185	1.815	0.136	1.864
9	1.094	0.337	0.239	1.761	0.184	1.816
10	1.028	0.308	0.284	1.716	0.223	1.777
12	0.925	0.266	0.354	1.646	0.284	1.716
14	0.848	0.235	0.406	1.594	0.329	1.671
16	0.788		0.448	1.552		
18	0.738		0.482	1.518		
20	0.697		0.510	1.490		

This table is based on American Society for Testing Materials *Manual on Quality Control of Materials*, Philadelphia, Pa., 1951, by permission of the publishers.

line and control limits are made in different ways, depending on whether the population mean $\bar{x}$ and standard deviation σ_x are known, or whether, on the other hand, the estimates must be based on samples only.

17.44. Use of range. It is generally preferable, from the standpoint of most effective control with a given sample size, to use the standard deviation s rather than the range R as a measure of scatter in the samples. The range should not be used if $n > 10$. For smaller values of n, the range is sometimes used because it is so simple to compute in a practical sampling procedure. When $n \leq 10$, and $\bar{x}$ and σ_x are not known, it is sometimes convenient to estimate σ_x as having the value R_0/k_n, where

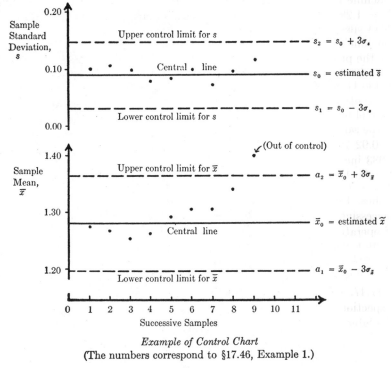

Example of Control Chart
(The numbers correspond to §17.46, Example 1.)

FIGURE 17.45.1

R_0 is the range for the preliminary sample I_0 in control, and k_n is obtained from Table 17.43.2.

17.45. Control charts. The quality control process can conveniently be followed on a "control chart." In the case of quantitative measurements, the mean $\bar{x}$ and the standard deviation s or range R or other parameters of the sample are plotted as ordinates, along with their respective means and control limits. As each sample is inspected, the corresponding values of mean $\bar{x}$, standard deviation s, range R or other quantity are plotted. The trend of the samples can easily be discerned, and if any point falls outside the limiting lines, the process is considered out of control. The control chart often assists in indicating when and where trouble has occurred. The identification and elimination of the trouble should then be pursued as an engineering problem.

17.46. It is customary to keep track of both the mean and the standard deviation (or other measure of dispersion in a sample) at the same time.

A control chart can be constructed for inspection by attributes, as described below, using the same general layout.

EXAMPLE 1. The length of screws being produced by an automatic machine has been measured for 400 screws, with resulting mean length $\bar{x}_0 = 1.28$ inches and standard deviation $s_0 = 0.09$ inches. These values meet specifications, and they are considered to be representative of the machine's capabilities, so these values of $\bar{x}_0$ and s_0 are taken as a standard for the process when "in control." It is desired to set up a control chart in order to detect any significant change in the process. According to Table 17.43.1, the central line for the mean is $\bar{\bar{x}} = \bar{x}_0 = 1.28$ inches, and the central line for the standard deviation is $\bar{s} = s_0 = 0.09$ inches. The control limits a_1' and a_2' for the mean are found by using Table 17.43.2. If the sample size be selected as $n = 12$, it is found that $M' = (M'/s_0)s_0 = 0.92 \times 0.09 = 0.083$, so $a_1' = 1.28 - 0.083$ inches and $a_2' = 1.28 + 0.083$ inches. Likewise for $n = 12$, the control limits s_1' and s_2' for the standard deviation are found from Table 17.43.2 to be $s_1' = (s_1'/s_0)s_0 = 0.35 \times 0.09 = 0.032$ inches, and $s_2' = (s_2'/s_0)s_0 = 1.65 \times 0.09 = 0.15$ inches. The values obtained in this example are used in Fig. 17.45.1. If successive samples of size 12, taken, say, from successive 4-hour periods of operation, yield the values plotted on Fig. 17.45.1, it is seen that the mean is out of control in the 9th period. Since the standard deviation is still well within control limits, it is possible that a setting on the machine has slipped.

17.47. Quality control on the basis of attributes. In the case of inspection by attributes, the mean, standard deviation, or range are not available for control purposes. The inspection can be summarized in terms of fraction p of defective items in a sample of n, called "fraction defective." The quality control procedure, in terms of p, is similar to that for $\bar{x}$ and s. First determine the average, $\bar{p}$, for an adequate sample considered to be "in control" or determine $\bar{p}$ by a specification or otherwise. Set control limits at $\bar{p} \pm 3\sigma_p$, central line at $\bar{p}$. It may usually be assumed that p has a binomial distribution. Then by §14.58, $\sigma_p = [\bar{p}(1 - \bar{p})/n]^{1/2}$. Control limits are set at $\bar{p} \pm 3[\bar{p}(1 - \bar{p})/n]^{1/2}$. The lower control limit is taken to be zero when $\bar{p} - 3\sigma_p$ is negative.

17.48. The samples used in inspection by attributes must generally be larger than for inspection by quantitative measurements. In general, the sample size n should be large enough to have a good chance of detecting at least one defective item; i.e., n at least as large as $1/\bar{p}$.

17.49. When all the samples have the same size n, the control chart may be constructed for the *number of "defects,"* pn. For values of pn, the central line is at the average $\bar{p}n$, and the control limits at $\bar{p}n \pm 3\sigma_{pn}$, where $\sigma_{pn} = [\bar{p}n(1 - \bar{p})]^{1/2}$. The lower control limit is taken as zero when $\bar{p}n - 3\sigma_{pn} < 0$.

17.50. References. The following books are representative of the many expositions of quality control now available.

American Standards Association: *Guide for Quality Control and Control Chart Method of Analyzing Data*, New York, 1941.

ASTM Manual on Quality Control of Materials: American Society for Testing and Materials, Philadelphia, 1951.

Burr, I. W.: *Engineering Statistics and Quality Control*, McGraw-Hill Book Company, New York, 1953.

Duncan, A. J.: *Quality Control and Industrial Statistics*, Richard D. Irwin, Inc., Homewood, Ill., 1959.

Grant, Eugene L.: *Statistical Quality Control*, 3d ed., McGraw-Hill Book Company, New York, 1964.

Shewhart, W. A.: *Economic Control of Quality of Manufactured Product*, Van Nostrand, New York, 1931.

Simon, L. E.: *An Engineers' Manual of Statistical Methods*, John Wiley and Sons, Inc., New York, 1941.

Statistical Research Group, Columbia University: *Techniques of Statistical Analysis*, McGraw-Hill Book Co., New York, 1947, edited by M. W. Hastay, G. C. Eisenhart, and W. A. Wallis.

Acceptance sampling plans based on parameters

17.51. In many physical situations a parameter associated with an object is measured. By using measured values of this parameter a sensitive plan can be developed. Here it is supposed that experience with many lots indicates that the measurements are distributed approximately normal with a known variance σ^2.

17.52. One-sided plans based on sample means. Consider the following plan.

(a) A sample of n objects is taken from a lot and the parameter x measured. Let μ be the lot mean and $\bar{x}$ the sample mean of measurements.

(b) Let k and n be two integers determined as indicated below.

(c) If $\bar{x} > k$, the lot is accepted.

(d) If $\bar{x} \leqq k$, the lot is rejected.

The numbers k and n are determined subject to the following conditions:

If $\mu = \mu_1$, the consumer's risk is β.

If $\mu = \mu_2$, the producer's risk is α. ($\mu_1 < \mu_2$.) When the sample is drawn from a lot with mean μ_1 and variance σ^2,

$$(17.52.1) \qquad P(\bar{x} > k \,|\, \mu = \mu_1) = P(v_1 > w_1) = \beta;$$

when drawn from a lot with mean μ_2, and variance σ^2,

$$(17.52.2) \qquad P(\bar{x} > k \,|\, \mu = \mu_2) = P(v_2 > w_2) = 1 - \alpha,$$

where

$$(17.52.3) \qquad (\bar{x} - \mu_i)\sqrt{n}/\sigma \equiv v_i, \qquad (k - \mu_i)\sqrt{n}/\sigma \equiv w_i, \qquad i = 1, 2.$$

In a given problem, values of α, β, μ_1, μ_2, σ are given or agreed upon. The quantities v_i, $i = 1$, 2, are distributed approximately normal $N(0,1)$. To find n and k, solve

$$(17.52.4) \qquad \begin{aligned} 1 - \Psi(w_1) &= \beta \\ 1 - \Psi(w_2) &= 1 - \alpha \end{aligned}$$

Table 10.8.3 or Table IX can be used to find w_1 and w_2. k and n can be found from w_1 and w_2, and (17.52.3). The *acceptance region* is the set of values of $\bar{x}$ in the interval $(k, + \infty)$; the left-sided *rejection region*, the interval $(- \infty, k)$.

17.53. Two-sided plans based on sample means. It is often desirable to use a sampling plan which (with a high probability) rejects lots with high means as well as lots with low means (e.g., to avoid underweights and overweights, overlength and underlength).

Case when variance is known. One such plan is as follows: Let n be the number of items in the sample, μ the mean of the lot, and k a certain integer. n and k are determined as indicated below.

(a) n items are drawn from a lot. A measurement of the parameter x is made for each item and the sample mean $\bar{x}$ calculated.

(b) If $-k + \mu < \bar{x} < \mu + k$, accept the lot.

(c) If (b) is not so, reject the lot.

Suppose it is known from experience that x is normally distributed and that the variance of lots of the type of interest is σ^2 for any lot having mean near μ. Let δ be a positive number. Determine n and k so that the probability of acceptance P is β if the mean μ' of the lot is near $\mu_1 = \mu - 2\delta$, and near $\mu_4 = \mu + 2\delta$; and P is $1 - \alpha$ if the mean μ' of the lot is near $\mu_2 = \mu - \delta$, and near $\mu_3 = \mu + \delta$.

These conditions are

$$(17.53.1) \quad P(\mu - k < \bar{x} < \mu + k) = \begin{cases} \beta, & \text{if } \mu' = \mu_1 \text{ or } \mu_4. \\ 1 - \alpha, & \text{if } \mu' = \mu_2 \text{ or } \mu_3. \end{cases}$$

Each of the quantities $z_i \equiv (\bar{x} - \mu_i)\sqrt{n}/\sigma$, $i = 1, \cdots, 4$ is normally distributed $N(0,1)$.

The quantities (17.53.1) may be written

$$(17.53.2) \quad \begin{aligned} P[(2\delta - k)\sqrt{n}/\sigma < z_1 < (2\delta + k)\sqrt{n}/\sigma] &= \beta \\ P[(\delta - k)\sqrt{n}/\sigma < z_2 < (\delta + k)\sqrt{n}/\sigma &= 1 - \alpha. \end{aligned}$$

In view of the symmetry of the normal distribution one need only consider the cases where $i = 1$ and $i = 2$.

If $(2\delta + k)\sqrt{n}/\sigma$ is sufficiently large, (17.53.2) can be approximated by

$$(17.53.3) \quad \begin{aligned} P[(2\delta - k)\sqrt{n}/\sigma < z] &\cong \beta \\ P[(\delta - k)\sqrt{n}/\sigma < z] &\cong 1 - \alpha. \end{aligned}$$

Once α and β are selected, say $\alpha = \beta = 0.05$, the normal probability Table IX can be used to give solutions for n and k in terms of δ and σ.

Suppose for selected values of α, β, δ, and σ, the solutions are $n = n_1$ and $k = k_1$. Then the sampling plan is essentially complete. If the mean $\bar{x}$ of a sample lies in the interval $(\mu - k, \mu + k)$, accept the lot, and reject the lot otherwise.

In case the variance σ^2 is unknown, a similar procedure may be used involving the t-distribution. (See §14.40.)

17.54. Plans based on other parameters. Many acceptance sampling plans based on sample parameters (e.g., sample variance) may be found in the literature. In general such plans use the principles of statistical testing treated in this chapter and the related sections of Chapters XIII and XIV.

Reliability theory

17.55. The term reliability. The term *reliability* is used in a variety of ways to indicate the capability of a system (or equipment, or device) to operate properly without breakdowns. In reliability engineering the concern is commonly centered on the elimination and prevention of early life failures, wearout failures, and chance failures.

Many definitions of "reliability" have been given. A typical one is as follows: The *reliability* of a device is the probability that, when called upon to act, the device will perform its function (purpose, or mission) adequately for the period of time intended under specified operating conditions. In order to use this definition in a given situation the device must be described properly, its function, mission, and purpose clearly delineated, and fully defined criteria covering what is meant by "the device will perform its purpose adequately . . ., under the operating conditions" given.

What is meant by "adequate performance," "malfunction," "success," "failure," "frequency of failures," "period of time," "environment," "operating conditions," "states of stress and strain," etc., must be spelled out for each case in advance. In some cases, instead of requiring that the device or system operate properly for a stated period of time, the requirement may be worded in terms of a specified number of *cycles*, or *switchings*, or some other unit which may characterize the operational life of the device. All important performance parameters should be measured.

Failure rate is commonly measured in terms of the *number of failures per unit of operating hour* (or *per cycle*). The *reciprocal* of the *failure* rate is called the *mean time between failures* (MTBF).

In order to evaluate statistically a given equipment (device or system), the equipment must be operated and its performance properly observed for a specified time period under actual operating conditions. In some cases the system may be operated in a laboratory under simulated conditions. Each part of the system may work satisfactorily yet the system may not operate properly. A part in the system may fail and the system may or may not operate satisfactorily.

17.56. Reliability function. Suppose tests are applied to a lot of new devices of the same type all purported to be produced according to some sort of a controlled production process. The tests are begun at time $t = 0$. Suppose N_0 of these devices are tested and operated repeatedly according to some well-defined process. N_S are found to *survive* the tests after an operating time t (i.e., are still operable after operating for time t). N_F are found to fail the tests some time prior to time t (i.e., cease to be operable some time prior to time t).

The events of device *survival* and device *failure* are mutually exclusive events and hence are *complementary events*.

Let $R(t)$ be the probability that the device will survive the tests after an operating time t, the device being new at time $t = 0$, at which instant the operating tests are begun. $R(t)$ is called the *reliability* of the device. $R(t)$ describes the hazard of a device to fail as a function of its age.

Let $Q(t) = 1 - R(t)$ be the probability that the device will not survive the tests after an operating time t.

An estimate of $R(t)$ is given by N_S/N_0, where $N_0 = N_S + N_F$. The expected number of devices operable at time t is $N_S = N_0 R(t)$; the expected number of devices not operable at time t is $N_F = N_0(1 - R(t)) = N_0 Q(t)$.

17.57. The probability density function $f(t)$ for "time to failure" is

$$(17.57.1) \qquad f(t) \equiv \frac{dQ(t)}{dt} = -\frac{dR(t)}{dt}.$$

$f(t)$ is called the *failure density function.*

The rate of change of the number of non-operable devices with respect to time per number of operable devices is

$$(17.57.2) \quad \lambda(t) = \frac{1}{N_S}\frac{dN_F}{dt} = \frac{1}{R(t)}\frac{dQ(t)}{dt} = \frac{f(t)}{R(t)} = -\frac{1}{R(t)}\frac{dR(t)}{dt}$$

$\lambda(t)$ is called the *hazard rate* or *instantaneous failure rate* of the device, or *failure rate.*

From (17.57.2)

$$(17.57.3) \qquad \log_e R(t) = -\int_0^t \lambda(\tau)\, d\tau.$$

If $R = 1$ at $t = 0$,

$$(17.57.4) \qquad R(t) = \exp\left[-\int_0^t \lambda(\tau)\, d\tau\right].$$

In general, $\lambda(t)$ may be a complicated function which should be determined from the data of a careful series of experiments. For some purposes it has been found that $\lambda(t)$ may be approximated as a positive constant. Variations of the definition of $R(t)$ are found in the literature, as, for example, in the treatment of renewable systems (e.g., overhauls and part replacement theory).

Remark. There are two common cases of interest in testing devices: (a) testing to failure and (b) testing to a specified time T. Some devices may fail prior to T; others may still be operable at time T.

Each individual item being tested has its critical failure time, and when this is passed the part fails. It is of interest to estimate the distribution of such failure times for the population of items being tested. It is with this in mind that engineers often speak of failures *attributable to chance.*

17.58. Case when λ = constant. If λ is a constant (say positive), then from (17.57.1), (17.57.2), and (17.57.4)

$$R(t) = e^{-\lambda t}, \qquad f(t) = \lambda e^{-\lambda t}$$

(17.58.1)
$$Q(t) = \int_0^t f(\tau)\, d\tau = 1 - e^{-\lambda t},$$

$$R(t) = 1 - Q(t) = \int_t^\infty f(\tau)\, d\tau = e^{-\lambda t}$$

The quantity $m = 1/\lambda$ is the *mean time between failures* (MTBF). The reliability of the device when $t = m$ is $R(m) = e^{-1} = 0.368$. (For further properties of the exponential function used here, see §9.9.) [Some writers define the *mean time between failures* (MTBF) for a device (or system) having reliability R to be $m = \int_0^\infty R\, dt$.]

17.59. Instead of t being interpreted as a time interval t may be taken as the number c of cycles during which the device is operated. In this case the reliability function $R(t)$ may be written $R(c) = e^{-\lambda c} = e^{-c/m}$ where λ is the failure rate in terms of failures per cycle, and m is the mean number of cycles between failures of the device. The reliability of the device for one cycle is $e^{-\lambda} = (0.368)^\lambda$.

17.60. Probability of survival for period t_1 to t_2. The a priori probability that the device will fail during the period from time t_1 to time t_2 within the operating time (e.g., within the mission time t) is

$$Q_{t_1, t_2} = \int_{t_1}^{t_2} f(\tau)\, d\tau = Q(t_2) - Q(t_1)$$
(17.60.1)
$$= R(t_1) - R(t_2)$$

The probability that the device will not fail during the period t_1 to t_2 is

$$R_{t_1, t_2} = 1 - Q_{t_1, t_2}.$$

Note: R_{t_1, t_2} is not the reliability of the device for functioning from t_1 to t_2, since the device may fail during the period $t = 0$ to $t = t_1$. The probability that the device will survive up to t_2 is the product.

$$(17.60.2) \qquad R(t_2) = R(t_1) \cdot R_{t_1, t_2}.$$

The operation of the device may be such that the failure rate λ_1 during the interval $t = 0$ to t_1 is different from the failure rate λ_2 for the interval t_1 to t_2. Thus, for the constant failure rate case, if

$$R(t_1) = e^{-\lambda_1 t_1} \qquad \text{and} \qquad R_{t_1, t_2} = e^{-\lambda_2 (t_2 - t_1)},$$

then

$$R(t_2) = e^{-\lambda_1 t_1} \, e^{-\lambda_2 (t_2 - t_1)}$$

This illustrates the fact that a device may exhibit a failure rate $\lambda_1(t)$ when operated under one state of stress, another rate $\lambda_2(t)$ under a different state of stress, etc.

17.61. Failure rate as function of age. The failure rate of a device (or system, or component) is generally a variable depending on the age of the device, the manner in which it is used and maintained, etc. A typical form of failure rate λ plotted as a function of operating life is shown in Figure 17.61.1. When the device is first operated a *debugging* or *burn-in period* occurs during which the failure rate changes considerably as the "bugs" are found and removed. After the debugging stage the failure rate may level out to a constant value and the failures may be considered as chance events. As the device approaches the *wearout stage* the failure rate may rise appreciably with age. In order to prolong the life of a device it is good practice to replace parts as they fail and in

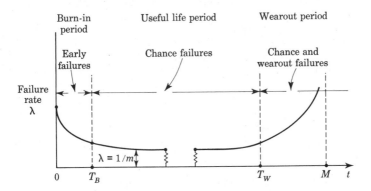

Component Failure Rate as a Function of Operating Life (Age) t

FIGURE 17.61.1

addition to replace each component of the device at such times as experience indicates that the component has reached the end of its expected useful life. The latter form of action is called *preventive maintenance*.

17.62. The form of the density function $f(t)$, where t is the operating age of the device, may be Weibull, exponential, normal, log normal, or some other form. For example, the form of $f(t)$ during the wearout phase may be normal $N(M, \sigma^2)$ where M is the mean wearout life and σ^2 is the variance of the lifetime of the device from the mean M, and t is the age or accumulated operating time since the device was new.

17.63. The reliability $R(t)$ of a device is calculated from Eqs. (17.57.4), (17.57.1), and (17.57.2) and the appropriate density function $f(t)$ and time interval. If the cumulated time t of operation of the device has reached a point where t is in excess of, say, $M - 3\sigma$, and the device is still operable, wearout may be expected to show itself.

17.64. The conditional or a posteriori probability of a failure during the time interval T_1 to T_2, given no failures up to time T_1, is given by

$$(17.64.1) \qquad F_{T_1, T_2} = \int_{T_1}^{T_2} f(\tau)\, d\tau / R(T_1)$$

where $f(\tau)$ is the density function for time to failure which corresponds to the situation, stress and device under consideration, and $R(T_1)$ is the probability that the device (new at $T = 0$) survives from $T = 0$ to T_1. Here the a priori probability of failure during the interval T_1 to T_2 is

$$(17.64.2) \qquad Q_{T_1, T_2} = \int_{T_1}^{T_2} f(\tau)\, d\tau.$$

F_{T_1, T_2} is called the *conditional unreliability* for the time interval T_1 to T_2.

17.65. The cumulative probability of failure from $T = 0$ to time T, called the *cumulative unreliability*, is given by

$$(17.65.1) \quad Q(T) = 1 - R(T) = \int_0^T f(\tau)\, d\tau = 1 - \int_T^\infty f(\tau)\, d\tau.$$

Note: The definition of unreliability in a given interval has been defined throughout this section as the number of devices failing in that interval divided by the total number of devices living at the beginning of the interval. This is the reason why the conditional probability approach is used and the term $R(T_1)$ appears in the definition of F_{T_1, T_2}.

17.66. **An example. Combined effects of chance and wearout.** Suppose:

(a) A device has a mission (operating) time of t hours, and at the beginning of the mission the device is operable and of age T hours.

(b) Two types of failure may occur, one attributable to chance, the other to wearout.

(c) The probability $Q_c(t)$ of failure due to chance in the time interval T to $T + t$ is independent of age T and $Q_c(t) = 1 - e^{-\lambda t}$, where λ is a constant.

(d) The probability $Q_w(t)$ of failure due to wearout in the interval T to $T + t$ is

(17.66.1) $$Q_w(t) = \int_T^{T+t} f_N(\tau)\, d\tau \Big/ \int_T^\infty f_N(\tau)\, d\tau$$

where

$$f_N(\tau) = (1/\sigma\sqrt{2\pi}) \exp\left[-(\tau - M)^2/2\sigma^2\right]$$

M is the mean wearout life and σ^2 is the variance of the lifetime of the device relative to M.

Then, subject to premises (a) to (d):

(α) The probability that the device will not fail because of wearout during the mission of t hours covering the time interval T to $T + t$ is $R_w(t) = 1 - Q_w(t)$. $R_w(t)$ is the *reliability* of the device for the time interval T to $T + t$.

(β) The probability that the device will fail in interval T to $T + t$ because of chance or wearout, or both, is

(17.66.2) $$\begin{aligned} Q(t) &= 1 - [1 - Q_c(t)][1 - Q_w(t)] \\ &= Q_c(t) + Q_w(t) - Q_c(t)\, Q_w(t) \end{aligned}$$

(γ) The overall probability that the device will not fail because of chance or wearout, or both, in interval T to $T + t$ is $R(t) = 1 - Q(t)$.

(δ) From (c) and (d)

(17.66.3) $Q(t) = 1 - e^{-\lambda t} + e^{-\lambda t} Q_w(t)$ $R(t) = e^{-\lambda t} R_w(t)$

(ϵ) From (17.66.1)

(17.66.4)
$$R_w(t) = R_w(T + t)/R_w(T) \qquad \text{and} \qquad R(t) = e^{-\lambda t} R_w(T + t)/R_w(T).$$

$R(t)$ is the reliability of the device for the interval T to $T + t$. Here

(17.66.5) $$R_w(T) = \int_T^\infty f_N(\tau)\, d\tau, \qquad R_w(T + t) = \int_{T+t}^\infty f_N(\tau)\, d\tau$$

The overall failure rate λ_0 may be found from

(17.66.6) $$R(t) = \exp\left[-\int_T^{T+t} \lambda_0\, dt\right].$$

The per hour failure rate due to wearout is

(17.66.7) $$\lambda_w(T) = f_N(T)/R_w(T).$$

An approximation to (17.66.6) is given by

(17.66.8) $$R(t) = \exp\left[-(\lambda_c + \lambda_{wm})t\right],$$

where λ_c is the constant chance rate, and

$$\lambda_{wm} = [\lambda_w(T) + \lambda_w(T + t)]/2.$$

17.67. Reliability of series systems. A set of n devices having reliabilities $R_1(t), \cdots, R_n(t)$ are coupled into a series to form a system. If any one of the devices fails the system fails. $R_i(t)$ is the probability that the ith device will remain operable during time interval $t = 0$ to t. If the devices are independent the reliability of the system is the product

(17.67.1) $$R(t) = R_1(t)\, R_2(t) \cdots R_n(t) = \prod_{j=1}^{n} R_j(t)$$

If each R_i is exponential in form, with constant failure rate λ_i, the reliability of the system is

(17.67.2) $$R(t) = \exp\left[-\sum_{i=1}^{n} \lambda_i t\right].$$

EXAMPLE. If 3 transistors and 20 resistors are wired in series to form a system, and the failure rates per hour of the transistors are each $\lambda_t = 0.00001$, and of the resistors each $\lambda_r = 0.000001$, the expected hourly failure rate λ of the system is $\lambda = \sum \lambda_i = 3\lambda_t + 20\lambda_r = 0.00005$ and the instantaneous failure rate is $\lambda = -(dR/dt)/R = 0.00005$. The estimated reliability of the system is

$$R(t) = \exp\left[-\sum_{i=1}^{23} \lambda_i t\right] = \exp\left[-0.00005t\right].$$

For a 100-hour operation the probability that the system will not fail in the 100 hours, that is, the reliability, is

$$R(100) = \exp\left[-0.005\right] = 0.995$$

The expected mean time between failures of the system is

$$m = 1/\lambda = 1/0.00005 = 20,000 \text{ hours.}$$

17.68. Reliability of parallel systems. A set of n devices are coupled in parallel to form a system. The system will operate as long as at least one device remains operable. If all the devices fail, the system fails. Let

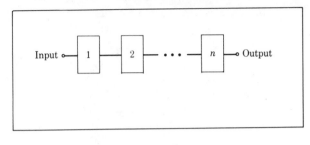

n Devices in Series

FIGURE 17.67.1

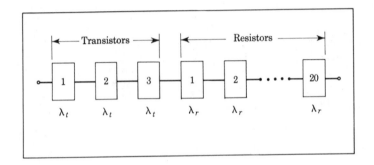

FIGURE 17.67.2

$Q_i(t)$ be the chance that the ith device fails during the time interval $t = 0$ to t. If the failures are independent, the chance that the system fails prior to time t is the product

$$(17.68.1) \qquad Q(t) = Q_1(t)\, Q_2(t) \cdots Q_n(t).$$

The reliability of the system is

$$(17.68.2) \qquad R(t) = 1 - Q(t).$$

The instantaneous failure rate of the system is

$$(17.68.3) \qquad \lambda = -(dR/dt)/R.$$

EXAMPLE. Two devices of reliabilities $R_1 = e^{-\lambda_1 t}$ and $R_2 = e^{-\lambda_2 t}$, respectively, are wired in parallel to form a system. The chance that

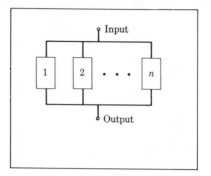

n Devices in Parallel

FIGURE 17.68.1

device i fails during time $t = 0$ to t is $Q_i = 1 - R_i$, $i = 1, 2$. The chance that the system fails is

$$Q = Q_1 Q_2 = (1 - e^{-\lambda_1 t})(1 - e^{-\lambda_2 t}).$$

The reliability of the system is

$$R = 1 - Q = e^{-\lambda_1 t} + e^{-\lambda_2 t} - e^{-(\lambda_1 + \lambda_2)t}.$$

The instantaneous failure rate of the system is

$$\lambda = -(dR/dt)/R.$$

The mean time between failures is

$$m = \int_0^\infty R \, dt = \frac{1}{\lambda_1} + \frac{1}{\lambda_2} - \frac{1}{\lambda_1 + \lambda_2}.$$

17.69. Standby systems. In some cases it is not practical to operate devices in parallel and a "standby" arrangement is used (i.e., when a device is operating, one or more like devices are standing by idle ready on signal to take over the operation when the first device fails). Expressions for the reliability, failure rate, mean time between failures, etc., have been derived for standby systems.

17.70. Reliability of complex systems. If several devices are coupled together in more involved ways than merely series or parallel couplings a similar probability argument based on the manner in which the devices are interrelated leads to an expression for the reliability of the system in terms of the reliabilities of the component devices.

In cases where the components of a system are replaced as they fail, or according to a prearranged schedule of replacements designed to minimize the effects of wearout, various expressions for the reliability of the system have been derived.

17.71. Availability, maintainability, safety, and dependability of systems. The principles sketched above have been applied to the study of optimum methods for the scheduled, preventive, and corrective maintenance of systems. Various definitions have been used to estimate the availability and dependability of systems.

One estimate of the availability of a system is taken to be the ratio of the interval of time a system is available for use to the total time during which it is desired that the system be available for use. Many other definitions of availability are in common use.

The term *system dependability* is commonly used to mean the probability that a system will be available for operation between any two scheduled maintenances, or to mean the probability that a system is available at a moment's notice, or to operate without interruptions.

All these topics involving reliability, availability, maintenance, and dependability bear in most important ways on the *safe* operation of systems (e.g., aircraft).

17.72. References. For further details on this extensive subject see

Barlow, R. E., and Proschan, F., *Mathematical Theory of Reliability*, John Wiley & Sons, Inc., New York, 1965.

Bazovsky, I., *Reliability Theory and Practice*, Prentice-Hall, Inc., Englewood Cliffs, N.J., 1961.

Calabro, S. R., *Reliability Principles and Practices*, McGraw-Hill Book Company, New York, 1962.

Cox, D. R., *Renewal Theory*, John Wiley & Sons, Inc., New York, 1963.

Epstein, B., *Sampling Procedures and Tables for Life and Reliability Testing, Quality Control and Reliability Handbook* H-108, Government Printing Office, Washington, D.C., 1961.

Goldman, A. S., and Slattery, T. B., *Maintainability*, John Wiley & Sons, Inc., New York, 1964.

Lloyd, D. K., and Lipow, M., *Reliability: Management, Methods, and Mathematics*, Prentice-Hall, Inc., Englewood Cliffs, N.J., 1962.

Product Engineering, *A Manual of Reliability*, McGraw-Hill Publications Division, New York, 1960.

eighteen

**SHORT TABLE OF INTEGRALS.
SOME MATHEMATICAL
RELATIONSHIPS**

In this table u and v denote functions of x; a, b, and C constants. Interpret symbols for inverse relations to be principal values.

Where an integration gives a term of the form $A \log N$, where A is a constant and $N < 0$, use $A \log |N|$; if $N > 0$, use $A \log N$.

In formulas 23 to 29 use $a > 0$.

18.1 Indefinite integrals.*

1. $\displaystyle \int df(x) = f(x) + C.$

2. $\displaystyle d \int f(x)\, dx = f(x)\, dx.$

3. $\displaystyle \int 0 \cdot dx = C.$

4. $\displaystyle \int a\, f(x)\, dx = a \int f(x)\, dx.$

5. $\displaystyle \int (u \pm v)\, dx = \int u\, dx \pm \int v\, dx.$

6. $\displaystyle \int u\, dv = uv - \int v\, du.$

7. $\displaystyle \int u^n\, du = \frac{u^{n+1}}{n+1} + C, \qquad n \neq -1.$

8. $\displaystyle \int \frac{du}{u} = \log_e u + C.$

9. $\displaystyle \int e^u\, du = e^u + C.$

10. $\displaystyle \int b^u\, du = \frac{b^u}{\log_e b} + C, \qquad b > 0, b \neq 1.$

---

* For a more complete table of integrals see R. S. Burington: *Handbook of Mathematical Tables and Formulas*, 4th ed., McGraw-Hill Book Company, New York, 1965.

11. $\int \sin u \, du = -\cos u + C.$

12. $\int \cos u \, du = \sin u + C.$

13. $\int \tan u \, du = \log_e \sec u + C = -\log_e \cos u + C.$

14. $\int \operatorname{ctn} u \, du = \log_e \sin u + C = -\log_e \csc u + C.$

15. $\int \sec u \, du = \log_e (\sec u + \tan u) + C = \log_e \tan \left(\dfrac{u}{2} + \dfrac{\pi}{4} \right) + C.$

16. $\int \csc u \, du = \log_e (\csc u - \operatorname{ctn} u) + C = \log_e \tan \dfrac{u}{2} + C.$

17. $\int \sin^2 u \, du = \dfrac{1}{2} u - \dfrac{1}{2} \sin u \cos u + C.$

18. $\int \cos^2 u \, du = \dfrac{1}{2} u + \dfrac{1}{2} \sin u \cos u + C.$

19. $\int \sec^2 u \, du = \tan u + C.$

20. $\int \csc^2 u \, du = -\operatorname{ctn} u + C.$

21. $\int \tan^2 u \, du = \tan u - u + C.$

22. $\int \operatorname{ctn}^2 u \, du = -\operatorname{ctn} u - u + C.$

23. $\int \dfrac{du}{u^2 + a^2} = \dfrac{1}{a} \tan^{-1} \left(\dfrac{u}{a} \right) + C.$

24. $\int \dfrac{du}{u^2 - a^2} = \dfrac{1}{2a} \log_e \left(\dfrac{u - a}{u + a} \right) + C = -\dfrac{1}{a} \operatorname{ctnh}^{-1} \left(\dfrac{u}{a} \right) + C,$

$$\text{if } u^2 > a^2,$$

$$= \dfrac{1}{2a} \log_e \left(\dfrac{a - u}{a + u} \right) + C = -\dfrac{1}{a} \tanh^{-1} \left(\dfrac{u}{a} \right) + C, \quad \text{if } u^2 < a^2.$$

25. $\int \dfrac{du}{\sqrt{a^2 - u^2}} = \sin^{-1} \left(\dfrac{u}{a} \right) + C.$

26. $\displaystyle\int \frac{du}{\sqrt{u^2 \pm a^2}} = \log_e (u + \sqrt{u^2 \pm a^2}) + C.$

27. $\displaystyle\int \frac{du}{u\sqrt{u^2 - a^2}} = \frac{1}{a}\sec^{-1}\left(\frac{u}{a}\right) + C = \frac{1}{a}\cos^{-1}\left(\frac{a}{u}\right) + C.$

28. $\displaystyle\int \frac{du}{u\sqrt{a^2 \pm u^2}} = -\frac{1}{a}\log_e \left(\frac{a + \sqrt{a^2 \pm u^2}}{u}\right) + C.$

29. $\displaystyle\int \sqrt{a^2 - u^2}\, du = \frac{1}{2}\left[u\sqrt{a^2 - u^2} + a^2 \sin^{-1}\left(\frac{u}{a}\right)\right] + C.$

30. $\displaystyle\int \sqrt{u^2 \pm a^2}\, du = \frac{1}{2}[u\sqrt{u^2 \pm a^2} \pm a^2 \log_e (u + \sqrt{u^2 \pm a^2})] + C.$

18.2 Definite integrals.

31. $\displaystyle\int_0^\infty x^{n-1}e^{-x}\, dx = \Gamma(n).$ (Gamma function, see §18.3.)

32. $\displaystyle\int_0^\infty e^{-zx}\cdot z^n \cdot x^{n-1}\, dx = \Gamma(n), \qquad z > 0.$

33. $\displaystyle\int_0^1 x^{m-1}(1 - x)^{n-1}\, dx = \int_0^\infty \frac{x^{m-1}\, dx}{(1 + x)^{m+n}} = \frac{\Gamma(m)\Gamma(n)}{\Gamma(m + n)}.$

34. $\displaystyle\int_0^{\pi/2} \sin^n x\, dx = \int_0^{\pi/2} \cos^n x\, dx$

$$= (1/2)\sqrt{\pi}\; \frac{\Gamma[(n + 1)/2]}{\Gamma[(n + 2)/2]}, \qquad \text{if} \quad n > -1;$$

$$= \frac{1\cdot 3\cdot 5\, \cdots\, (n - 1)}{2\cdot 4\cdot 6\, \cdots\, (n)}\cdot\frac{\pi}{2}, \qquad \text{if } n \text{ is an even integer;}$$

$$= \frac{2\cdot 4\cdot 6\, \cdots\, (n - 1)}{1\cdot 3\cdot 5\cdot 7\, \cdots\, n}, \qquad \text{if } n \text{ is an odd integer.}$$

35. $\displaystyle\int_0^\infty e^{-x^2}\, dx = \sqrt{\pi}/2.$

36. $\displaystyle(1/\sqrt{2\pi})\int_{-\infty}^\infty e^{-x^2/2}\, dx = 1.$

37. $\displaystyle\int_0^\infty xe^{-x^2/2}\, dx = 1.$

38. $\displaystyle\int_0^\infty e^{-a^2x^2}\, dx = \frac{\sqrt{\pi}}{2a} = \frac{1}{2a}\Gamma\left(\frac{1}{2}\right), \qquad \text{if} \quad a > 0.$

39. $\displaystyle\int_0^\infty x^n \cdot e^{-ax}\, dx = \frac{\Gamma(n+1)}{a^{n+1}}$

$\displaystyle = \frac{n!}{a^{n+1}},$ if n is a positive integer, $a > 0$.

40. $\displaystyle\int_0^\infty x^{2n} e^{-ax^2}\, dx = \frac{1 \cdot 3 \cdot 5 \cdots (2n-1)}{2^{n+1} a^n}\sqrt{\frac{\pi}{a}}.$

41. $\displaystyle\int_0^\infty \sqrt{x}\, e^{-ax}\, dx = \frac{1}{2a}\sqrt{\frac{\pi}{a}}.$

42. $\displaystyle\int_0^\infty \frac{e^{-ax}}{\sqrt{x}}\, dx = \sqrt{\frac{\pi}{a}}.$

43. $\displaystyle\int_0^1 x^m\left(\log\frac{1}{x}\right)^n dx = \frac{\Gamma(n+1)}{(m+1)^{n+1}},$ if $m + 1 > 0$, $n + 1 > 0$.

44. $\displaystyle\int_{-1}^{+1} (1 - x^2)^{(n-4)/2}\, dx = \sqrt{\pi} \cdot \Gamma[(n-2)/2] \div \Gamma[(n-1)/2].$

Some mathematical functions and theorems

18.3. The Gamma function $\Gamma(n)$.

(18.3.1) $\displaystyle\Gamma(n) = \int_0^\infty x^{n-1} e^{-x}\, dx = \int_0^1 [\log_e (1/x)]^{n-1}\, dx,$ if $n > 0$.

$\displaystyle\Gamma(n) = 2\int_0^\infty y^{2n-1} e^{-y^2}\, dy.$ $\displaystyle\int_0^\infty e^{-y^2}\, dy = \frac{1}{2}\sqrt{\pi}.$

(18.3.2) $\Gamma(n+1) = n \cdot \Gamma(n),$ if $n > 0$.

$\sqrt{\pi} \cdot \Gamma(2n) = 2^{2n-1} \cdot \Gamma(n) \cdot \Gamma[n + (1/2)].$

$\Gamma(n+1) = n!,$ if n is a positive integer. $\Gamma(0) = \infty$.

$\Gamma(1) = 1.$ $\Gamma(2) = 1.$ $\Gamma(3) = 2.$ $\Gamma(4) = 6.$ $\Gamma(1/2) = \sqrt{\pi}.$

$n! = n(n-1)(n-2) \cdots (1).$ $1! = 1.$ $0! = 1.$

$\Gamma(n)$ can be defined by means of the recursion formula $\Gamma(n) = \Gamma(n+1)/n$, for all negative values of n, except for $n = 0, -1, -2, \cdots$, where $\Gamma(n)$ becomes infinite. For example,

$\Gamma(-1/2) = \Gamma(1/2)/(-1/2) = -2\sqrt{\pi}.$

(18.3.3) $\displaystyle\frac{1}{\sqrt{\pi} \cdot \Gamma\left(\dfrac{n-1}{2}\right) \cdot \Gamma\left(\dfrac{n-2}{2}\right)} = \frac{2^{n-3}}{\pi \cdot \Gamma(n-2)}.$

Table XVII gives values of $\Gamma(n)$ for $n = 1.0$ to $n = 4.99$. Table XVIII gives values of $\log_{10} \Gamma(n)$ for $n = 1.01$ to $n = 2.00$. For other values of $\Gamma(n)$ use the recursion formula $\Gamma(n + 1) = n \cdot \Gamma(n)$.

18.4. Incomplete Gamma function $\Gamma_x(n)$.

$$(18.4.1) \qquad \Gamma_x(n) = \int_0^x x^{n-1} e^{-x}\, dx.$$

18.5. The Beta function $B(m,n)$.

$$(18.5.1) \qquad B(m,n) = \int_0^1 x^{m-1}(1 - x)^{n-1}\, dx, \quad \text{where} \quad m > 0, \quad n > 0.$$

$$B(m,n) = 2 \int_0^{\pi/2} \sin^{2m-1} \theta \, \cos^{2n-1} \theta \, d\theta, \quad m > 0, \quad n > 0.$$

$$B(m,n) = B(n,m).$$

$$(18.5.2) \qquad B(m,n) = \frac{\Gamma(m)\,\Gamma(n)}{\Gamma(m + n)}.$$

18.6. Incomplete Beta function $B_x(m,n)$.

$$B_x(m,n) = \int_0^x x^{m-1}(1 - x)^{n-1}\, dx, \quad \text{where} \quad m > 0, \quad n > 0.$$

18.7. Incomplete Beta function ratio $I_x(m,n)$.

$$(18.7.1) \qquad I_x(m,n) = B_x(m,n) \div B(m,n).$$

$$I_x(m,n) = 1 - I_y(n,m), \quad \text{where} \quad x + y = 1.$$

Table III gives values of $I_x(m,n)$.

18.8 Stirling's formula. For values of $n > 0$,

$$\sqrt{2n\pi} \, (n/e)^n < \Gamma(n + 1) < \sqrt{2n\pi} \, (n/e)^n[1 + 1/(12n - 1)],$$

$$\sqrt{2n\pi} \, (n/e)^n < n! < \sqrt{2n\pi} \, (n/e)^n[1 + 1/(12n - 1)],$$

where $\qquad \pi = 3.14159 \cdots, \qquad e = 2.71828 \cdots.$

$$\log n! \cong (n + (1/2)) \log n - n \log e + \log \sqrt{2\pi}.$$

18.9. Change of variable in definite double integral. If $x = g(u,v)$ and $y = h(u, v)$,

$$\iint\limits_{(R)} f(x,y)\, dx\, dy = \iint\limits_{(R)} f(g,h)\, dA,$$

where

$$dA = \left| J\!\left(\frac{x,y}{u,v}\right) \right| du\, dv,$$

and

$$J\!\left(\frac{x,y}{u,v}\right) = \begin{vmatrix} \dfrac{\partial x}{\partial u} & \dfrac{\partial x}{\partial v} \\[2ex] \dfrac{\partial y}{\partial u} & \dfrac{\partial y}{\partial v} \end{vmatrix} = \frac{\partial x}{\partial u}\frac{\partial y}{\partial v} - \frac{\partial x}{\partial v}\frac{\partial y}{\partial u}.$$

J is called the *Jacobian* of x and y with respect to u and v. The integrals are to be carried out over the region R in the x,y-plane. For the more general case, see Burington and Torrance, *Higher Mathematics*, McGraw-Hill Book Company, New York, 1939.

18.10. Functional independence and statistical independence. The notion of functional independence should not be confused with that of statistical independence. (See §5.35.)

Functional independence. Suppose

$$u = f(x,y), \qquad v = g(x,y),$$

where f and g have continuous first partial derivatives. u and v are *functionally related* (or *functionally dependent*) if, for each point (x,y) of some region of the xy-plane, u and v always form a solution of an equation of the form $\psi(u,v) = 0$, where ψ is differentiable and ψ_u and ψ_v are never simultaneously zero at any point (u,v).

If u and v are not functionally dependent they are said to be *functionally independent*.

THEOREM 1. u and v are functionally dependent when and only when

$$J\!\left(\frac{u,v}{x,y}\right) = 0$$

over some region of the xy-plane.

The notion of functional dependence is readily extended to functions of several variables.

18.11. Volume V of n-dimensional ellipsoid whose equation is

$$Q = c^2, \qquad \text{where} \qquad Q = \sum_{i,k=1}^{n} a_{ik}x_i x_k \geqq 0,$$

is

(18.11.1) $$V = \iint \cdots \int_{Q < c^2} dx_1 \, dx_2 \, \cdots \, dx_n = \frac{\pi^{n/2}}{\Gamma[(n/2) + 1]} \cdot \frac{c^n}{\sqrt{A}},$$

where

$$A = |a_{ik}| \text{ is the determinant of } Q.$$

Also,

(18.11.2) $$\int_{-\infty}^{\infty} \int_{-\infty}^{\infty} \cdots \int_{-\infty}^{\infty} e^{-Q/2} \, dx_1 \, dx_2 \, \cdots \, dx_n = \frac{(2\pi)^{n/2}}{\sqrt{A}}.$$

Table I

BINOMIAL DISTRIBUTION FUNCTION $C_x{}^n p^x (1-p)^{n-x}$

(See Chapter VIII)

Entries in the table are values of $C_x{}^n p^x (1-p)^{n-x}$ for the indicated values of n, x and p. When $p > 0.5$, the value of $C_x{}^n p^x (1-p)^{n-x}$ for a given n, x and p is obtained by finding the tabular entry for the given n, with $n - x$ in place of the given x, and $1 - p$ in place of the given p.

n	x	.05	.10	.15	.20	.25	.30	.35	.40	.45	.50
1	0	.9500	.9000	.8500	.8000	.7500	.7000	.6500	.6000	.5500	.5000
	1	.0500	.1000	.1500	.2000	.2500	.3000	.3500	.4000	.4500	.5000
2	0	.9025	.8100	.7225	.6400	.5625	.4900	.4225	.3600	.3025	.2500
	1	.0950	.1800	.2550	.3200	.3750	.4200	.4550	.4800	.4950	.5000
	2	.0025	.0100	.0225	.0400	.0625	.0900	.1225	.1600	.2025	.2500
3	0	.8574	.7290	.6141	.5120	.4219	.3430	.2746	.2160	.1664	.1250
	1	.1354	.2430	.3251	.3840	.4219	.4410	.4436	.4320	.4084	.3750
	2	.0071	.0270	.0574	.0960	.1406	.1890	.2389	.2880	.3341	.3750
	3	.0001	.0010	.0034	.0080	.0156	.0270	.0429	.0640	.0911	.1250
4	0	.8145	.6561	.5220	.4096	.3164	.2401	.1785	.1296	.0915	.0625
	1	.1715	.2916	.3685	.4096	.4219	.4116	.3845	.3456	.2995	.2500
	2	.0135	.0486	.0975	.1536	.2109	.2646	.3105	.3456	.3675	.3750
	3	.0005	.0036	.0115	.0256	.0469	.0756	.1115	.1536	.2005	.2500
	4	.0000	.0001	.0005	.0016	.0039	.0081	.0150	.0256	.0410	.0625
5	0	.7738	.5905	.4437	.3277	.2373	.1681	.1160	.0778	.0503	.0312
	1	.2036	.3280	.3915	.4096	.3955	.3602	.3124	.2592	.2059	.1562
	2	.0214	.0729	.1382	.2048	.2637	.3087	.3364	.3456	.3369	.3125
	3	.0011	.0081	.0244	.0512	.0879	.1323	.1811	.2304	.2757	.3125
	4	.0000	.0004	.0022	.0064	.0146	.0284	.0488	.0768	.1128	.1562
	5	.0000	.0000	.0001	.0003	.0010	.0024	.0053	.0102	.0185	.0312
6	0	.7351	.5314	.3771	.2621	.1780	.1176	.0754	.0467	.0277	.0156
	1	.2321	.3543	.3993	.3932	.3560	.3025	.2437	.1866	.1359	.0938
	2	.0305	.0984	.1762	.2458	.2966	.3241	.3280	.3110	.2780	.2344
	3	.0021	.0146	.0415	.0819	.1318	.1852	.2355	.2765	.3032	.3125
	4	.0001	.0012	.0055	.0154	.0330	.0595	.0951	.1382	.1861	.2344
	5	.0000	.0001	.0004	.0015	.0044	.0102	.0205	.0369	.0609	.0938
	6	.0000	.0000	.0000	.0001	.0002	.0007	.0018	.0041	.0083	.0156
7	0	.6983	.4783	.3206	.2097	.1335	.0824	.0490	.0280	.0152	.0078
	1	.2573	.3720	.3960	.3670	.3115	.2471	.1848	.1306	.0872	.0547
	2	.0406	.1240	.2097	.2753	.3115	.3177	.2985	.2613	.2140	.1641
	3	.0036	.0230	.0617	.1147	.1730	.2269	.2679	.2903	.2918	.2734
	4	.0002	.0026	.0109	.0287	.0577	.0972	.1442	.1935	.2388	.2734
	5	.0000	.0002	.0012	.0043	.0115	.0250	.0466	.0774	.1172	.1641
	6	.0000	.0000	.0001	.0004	.0013	.0036	.0084	.0172	.0320	.0547
	7	.0000	.0000	.0000	.0000	.0001	.0002	.0006	.0016	.0037	.0078
8	0	.6634	.4305	.2725	.1678	.1001	.0576	.0319	.0168	.0084	.0039
	1	.2793	.3826	.3847	.3355	.2670	.1977	.1373	.0896	.0548	.0312
	2	.0515	.1488	.2376	.2936	.3115	.2965	.2587	.2090	.1569	.1094
	3	.0054	.0331	.0839	.1468	.2076	.2541	.2786	.2787	.2568	.2188
	4	.0004	.0046	.0185	.0459	.0865	.1361	.1875	.2322	.2627	.2734
	5	.0000	.0004	.0026	.0092	.0231	.0467	.0808	.1239	.1719	.2188
	6	.0000	.0000	.0002	.0011	.0038	.0100	.0217	.0413	.0703	.1094
	7	.0000	.0000	.0000	.0001	.0004	.0012	.0033	.0079	.0164	.0312
	8	.0000	.0000	.0000	.0000	.0000	.0001	.0002	.0007	.0017	.0039

Linear interpolation with respect to p will generally not be accurate to more than two decimal places, and sometimes less.

For extensive tables of $C_x{}^n p^x (1-p)^{n-x}$ see *Tables of the Binomial Probability Distribution*, National Bureau of Standards, Applied Mathematics Series 6, Washington, D. C., 1950.

Table I continued

BINOMIAL DISTRIBUTION FUNCTION $C_x^n p^x (1-p)^{n-x}$

n	x	.05	.10	.15	.20	.25	.30	.35	.40	.45	.50
9	0	.6302	.3874	.2316	.1342	.0751	.0404	.0207	.0101	.0046	.0020
	1	.2985	.3874	.3679	.3020	.2253	.1556	.1004	.0605	.0339	.0176
	2	.0629	.1722	.2597	.3020	.3003	.2668	.2162	.1612	.1110	.0703
	3	.0077	.0446	.1069	.1762	.2336	.2668	.2716	.2508	.2119	.1641
	4	.0006	.0074	.0283	.0661	.1168	.1715	.2194	.2508	.2600	.2461
	5	.0000	.0008	.0050	.0165	.0389	.0735	.1181	.1672	.2128	.2461
	6	.0000	.0001	.0006	.0028	.0087	.0210	.0424	.0743	.1160	.1641
	7	.0000	.0000	.0000	.0003	.0012	.0039	.0098	.0212	.0407	.0703
	8	.0000	.0000	.0000	.0000	.0001	.0004	.0013	.0035	.0083	.0176
	9	.0000	.0000	.0000	.0000	.0000	.0000	.0001	.0003	.0008	.0020
10	0	.5987	.3487	.1969	.1074	.0563	.0282	.0135	.0060	.0025	.0010
	1	.3151	.3874	.3474	.2684	.1877	.1211	.0725	.0403	.0207	.0098
	2	.0746	.1937	.2759	.3020	.2816	.2335	.1757	.1209	.0763	.0439
	3	.0105	.0574	.1298	.2013	.2503	.2668	.2522	.2150	.1665	.1172
	4	.0010	.0112	.0401	.0881	.1460	.2001	.2377	.2508	.2384	.2051
	5	.0001	.0015	.0085	.0264	.0584	.1029	.1536	.2007	.2340	.2461
	6	.0000	.0001	.0012	.0055	.0162	.0368	.0689	.1115	.1596	.2051
	7	.0000	.0000	.0001	.0008	.0031	.0090	.0212	.0425	.0746	.1172
	8	.0000	.0000	.0000	.0001	.0004	.0014	.0043	.0106	.0229	.0439
	9	.0000	.0000	.0000	.0000	.0000	.0001	.0005	.0016	.0042	.0098
	10	.0000	.0000	.0000	.0000	.0000	.0000	.0000	.0001	.0003	.0010
11	0	.5688	.3138	.1673	.0859	.0422	.0198	.0088	.0036	.0014	.0005
	1	.3293	.3835	.3248	.2362	.1549	.0932	.0518	.0266	.0125	.0054
	2	.0867	.2131	.2866	.2953	.2581	.1998	.1395	.0887	.0513	.0269
	3	.0137	.0710	.1517	.2215	.2581	.2568	.2254	.1774	.1259	.0806
	4	.0014	.0158	.0536	.1107	.1721	.2201	.2428	.2365	.2060	.1611
	5	.0001	.0025	.0132	.0388	.0803	.1321	.1830	.2207	.2360	.2256
	6	.0000	.0003	.0023	.0097	.0268	.0566	.0985	.1471	.1931	.2256
	7	.0000	.0000	.0003	.0017	.0064	.0173	.0379	.0701	.1128	.1611
	8	.0000	.0000	.0000	.0002	.0011	.0037	.0102	.0234	.0462	.0806
	9	.0000	.0000	.0000	.0000	.0001	.0005	.0018	.0052	.0126	.0269
	10	.0000	.0000	.0000	.0000	.0000	.0000	.0002	.0007	.0021	.0054
	11	.0000	.0000	.0000	.0000	.0000	.0000	.0000	.0000	.0002	.0005
12	0	.5404	.2824	.1422	.0687	.0317	.0138	.0057	.0022	.0008	.0002
	1	.3413	.3766	.3012	.2062	.1267	.0712	.0368	.0174	.0075	.0029
	2	.0988	.2301	.2924	.2835	.2323	.1678	.1088	.0639	.0339	.0161
	3	.0173	.0852	.1720	.2362	.2581	.2397	.1954	.1419	.0923	.0537
	4	.0021	.0213	.0683	.1329	.1936	.2311	.2367	.2128	.1700	.1208
	5	.0002	.0038	.0193	.0532	.1032	.1585	.2039	.2270	.2225	.1934
	6	.0000	.0005	.0040	.0155	.0401	.0792	.1281	.1766	.2124	.2256
	7	.0000	.0000	.0006	.0033	.0115	.0291	.0591	.1009	.1489	.1934
	8	.0000	.0000	.0001	.0005	.0024	.0078	.0199	.0420	.0762	.1208
	9	.0000	.0000	.0000	.0001	.0004	.0015	.0048	.0125	.0277	.0537
	10	.0000	.0000	.0000	.0000	.0000	.0002	.0008	.0025	.0068	.0161
	11	.0000	.0000	.0000	.0000	.0000	.0000	.0001	.0003	.0010	.0029
	12	.0000	.0000	.0000	.0000	.0000	.0000	.0000	.0000	.0001	.0002
13	0	.5133	.2542	.1209	.0550	.0238	.0097	.0037	.0013	.0004	.0001
	1	.3512	.3672	.2774	.1787	.1029	.0540	.0259	.0113	.0045	.0016
	2	.1109	.2448	.2937	.2680	.2059	.1388	.0836	.0453	.0220	.0095
	3	.0214	.0997	.1900	.2457	.2517	.2181	.1651	.1107	.0660	.0349
	4	.0028	.0277	.0838	.1535	.2097	.2337	.2222	.1845	.1350	.0873
	5	.0003	.0055	.0266	.0691	.1258	.1803	.2154	.2214	.1989	.1571
	6	.0000	.0008	.0063	.0230	.0559	.1030	.1546	.1968	.2169	.2095
	7	.0000	.0001	.0011	.0058	.0186	.0442	.0833	.1312	.1775	.2095
	8	.0000	.0000	.0001	.0011	.0047	.0142	.0336	.0656	.1089	.1571
	9	.0000	.0000	.0000	.0001	.0009	.0034	.0101	.0243	.0495	.0873
	10	.0000	.0000	.0000	.0000	.0001	.0006	.0022	.0065	.0162	.0349
	11	.0000	.0000	.0000	.0000	.0000	.0001	.0003	.0012	.0036	.0095
	12	.0000	.0000	.0000	.0000	.0000	.0000	.0000	.0001	.0005	.0016
	13	.0000	.0000	.0000	.0000	.0000	.0000	.0000	.0000	.0000	.0001

BINOMIAL DISTRIBUTION FUNCTION $C_x^n p^x (1-p)^{n-x}$

n	x	.05	.10	.15	.20	.25	.30	.35	.40	.45	.50
14	0	.4877	.2288	.1028	.0440	.0178	.0068	.0024	.0008	.0002	.0001
	1	.3593	.3559	.2539	.1539	.0832	.0407	.0181	.0073	.0027	.0009
	2	.1229	.2570	.2912	.2501	.1802	.1134	.0634	.0317	.0141	.0056
	3	.0259	.1142	.2056	.2501	.2402	.1943	.1366	.0845	.0462	.0222
	4	.0037	.0349	.0998	.1720	.2202	.2290	.2022	.1549	.1040	.0611
	5	.0004	.0078	.0352	.0860	.1468	.1963	.2178	.2066	.1701	.1222
	6	.0000	.0013	.0093	.0322	.0734	.1262	.1759	.2066	.2088	.1833
	7	.0000	.0002	.0019	.0092	.0280	.0618	.1082	.1574	.1952	.2095
	8	.0000	.0000	.0003	.0020	.0082	.0232	.0510	.0918	.1398	.1833
	9	.0000	.0000	.0000	.0003	.0018	.0066	.0183	.0408	.0762	.1222
	10	.0000	.0000	.0000	.0000	.0003	.0014	.0049	.0136	.0312	.0611
	11	.0000	.0000	.0000	.0000	.0000	.0002	.0010	.0033	.0093	.0222
	12	.0000	.0000	.0000	.0000	.0000	.0000	.0001	.0005	.0019	.0056
	13	.0000	.0000	.0000	.0000	.0000	.0000	.0000	.0001	.0002	.0009
	14	.0000	.0000	.0000	.0000	.0000	.0000	.0000	.0000	.0000	.0001
15	0	.4633	.2059	.0874	.0352	.0134	.0047	.0016	.0005	.0001	.0000
	1	.3658	.3432	.2312	.1319	.0668	.0305	.0126	.0047	.0016	.0005
	2	.1348	.2669	.2856	.2309	.1559	.0916	.0476	.0219	.0090	.0032
	3	.0307	.1285	.2184	.2501	.2252	.1700	.1110	.0634	.0318	.0139
	4	.0049	.0428	.1156	.1876	.2252	.2186	.1792	.1268	.0780	.0417
	5	.0006	.0105	.0449	.1032	.1651	.2061	.2123	.1859	.1404	.0916
	6	.0000	.0019	.0132	.0430	.0917	.1472	.1906	.2066	.1914	.1527
	7	.0000	.0003	.0030	.0138	.0393	.0811	.1319	.1771	.2013	.1964
	8	.0000	.0000	.0005	.0035	.0131	.0348	.0710	.1181	.1647	.1964
	9	.0000	.0000	.0001	.0007	.0034	.0116	.0298	.0612	.1048	.1527
	10	.0000	.0000	.0000	.0001	.0007	.0030	.0096	.0245	.0515	.0916
	11	.0000	.0000	.0000	.0000	.0001	.0006	.0024	.0074	.0191	.0417
	12	.0000	.0000	.0000	.0000	.0000	.0001	.0004	.0016	.0052	.0139
	13	.0000	.0000	.0000	.0000	.0000	.0000	.0001	.0003	.0010	.0032
	14	.0000	.0000	.0000	.0000	.0000	.0000	.0000	.0000	.0001	.0005
	15	.0000	.0000	.0000	.0000	.0000	.0000	.0000	.0000	.0000	.0000
16	0	.4401	.1853	.0743	.0281	.0100	.0033	.0010	.0003	.0001	.0000
	1	.3706	.3294	.2097	.1126	.0535	.0228	.0087	.0030	.0009	.0002
	2	.1463	.2745	.2775	.2111	.1336	.0732	.0353	.0150	.0056	.0018
	3	.0359	.1423	.2285	.2463	.2079	.1465	.0888	.0468	.0215	.0085
	4	.0061	.0514	.1311	.2001	.2252	.2040	.1553	.1014	.0572	.0278
	5	.0008	.0137	.0555	.1201	.1802	.2099	.2008	.1623	.1123	.0667
	6	.0001	.0028	.0180	.0550	.1101	.1649	.1982	.1983	.1684	.1222
	7	.0000	.0004	.0045	.0197	.0524	.1010	.1524	.1889	.1969	.1746
	8	.0000	.0001	.0009	.0055	.0197	.0487	.0923	.1417	.1812	.1964
	9	.0000	.0000	.0001	.0012	.0058	.0185	.0442	.0840	.1318	.1746
	10	.0000	.0000	.0000	.0002	.0014	.0056	.0167	.0392	.0755	.1222
	11	.0000	.0000	.0000	.0000	.0002	.0013	.0049	.0142	.0337	.0667
	12	.0000	.0000	.0000	.0000	.0000	.0002	.0011	.0040	.0115	.0278
	13	.0000	.0000	.0000	.0000	.0000	.0000	.0002	.0008	.0029	.0085
	14	.0000	.0000	.0000	.0000	.0000	.0000	.0000	.0001	.0005	.0018
	15	.0000	.0000	.0000	.0000	.0000	.0000	.0000	.0000	.0001	.0002
	16	.0000	.0000	.0000	.0000	.0000	.0000	.0000	.0000	.0000	.0000
17	0	.4181	.1668	.0631	.0225	.0075	.0023	.0007	.0002	.0000	.0000
	1	.3741	.3150	.1893	.0957	.0426	.0169	.0060	.0019	.0005	.0001
	2	.1575	.2800	.2673	.1914	.1136	.0581	.0260	.0102	.0035	.0010
	3	.0415	.1556	.2359	.2393	.1893	.1245	.0701	.0341	.0144	.0052
	4	.0076	.0605	.1457	.2093	.2209	.1868	.1320	.0796	.0411	.0182
	5	.0010	.0175	.0668	.1361	.1914	.2081	.1849	.1379	.0875	.0472
	6	.0001	.0039	.0236	.0680	.1276	.1784	.1991	.1839	.1432	.0944
	7	.0000	.0007	.0065	.0267	.0668	.1201	.1685	.1927	.1841	.1484
	8	.0000	.0001	.0014	.0084	.0279	.0644	.1134	.1606	.1883	.1855
	9	.0000	.0000	.0003	.0021	.0093	.0276	.0611	.1070	.1540	.1855
	10	.0000	.0000	.0000	.0004	.0025	.0095	.0263	.0571	.1008	.1484
	11	.0000	.0000	.0000	.0001	.0005	.0026	.0090	.0242	.0525	.0944
	12	.0000	.0000	.0000	.0000	.0001	.0006	.0024	.0081	.0215	.0472
	13	.0000	.0000	.0000	.0000	.0000	.0001	.0005	.0021	.0068	.0182
	14	.0000	.0000	.0000	.0000	.0000	.0000	.0001	.0004	.0016	.0052

BINOMIAL DISTRIBUTION FUNCTION $C_x^n p^x (1-p)^{n-x}$

n	x	.05	.10	.15	.20	.25	.30	.35	.40	.45	.50
17	15	.0000	.0000	.0000	.0000	.0000	.0000	.0000	.0001	.0003	.0010
	16	.0000	.0000	.0000	.0000	.0000	.0000	.0000	.0000	.0000	.0001
	17	.0000	.0000	.0000	.0000	.0000	.0000	.0000	.0000	.0000	.0000
18	0	.3972	.1501	.0536	.0180	.0056	.0016	.0004	.0001	.0000	.0000
	1	.3763	.3002	.1704	.0811	.0338	.0126	.0042	.0012	.0003	.0001
	2	.1683	.2835	.2556	.1723	.0958	.0458	.0190	.0069	.0022	.0006
	3	.0473	.1680	.2406	.2297	.1704	.1046	.0547	.0246	.0095	.0031
	4	.0093	.0700	.1592	.2153	.2130	.1681	.1104	.0614	.0291	.0117
	5	.0014	.0218	.0787	.1507	.1988	.2017	.1664	.1146	.0666	.0327
	6	.0002	.0052	.0301	.0816	.1436	.1873	.1941	.1655	.1181	.0708
	7	.0000	.0010	.0091	.0350	.0820	.1376	.1792	.1892	.1657	.1214
	8	.0000	.0002	.0022	.0120	.0376	.0811	.1327	.1734	.1864	.1669
	9	.0000	.0000	.0004	.0033	.0139	.0386	.0794	.1284	.1694	.1855
	10	.0000	.0000	.0001	.0008	.0042	.0149	.0385	.0771	.1248	.1669
	11	.0000	.0000	.0000	.0001	.0010	.0046	.0151	.0374	.0742	.1214
	12	.0000	.0000	.0000	.0000	.0002	.0012	.0047	.0145	.0354	.0708
	13	.0000	.0000	.0000	.0000	.0000	.0002	.0012	.0045	.0134	.0327
	14	.0000	.0000	.0000	.0000	.0000	.0000	.0002	.0011	.0039	.0117
	15	.0000	.0000	.0000	.0000	.0000	.0000	.0000	.0002	.0009	.0031
	16	.0000	.0000	.0000	.0000	.0000	.0000	.0000	.0000	.0001	.0006
	17	.0000	.0000	.0000	.0000	.0000	.0000	.0000	.0000	.0000	.0001
	18	.0000	.0000	.0000	.0000	.0000	.0000	.0000	.0000	.0000	.0000
19	0	.3774	.1351	.0456	.0144	.0042	.0011	.0003	.0001	.0000	.0000
	1	.3774	.2852	.1529	.0685	.0268	.0093	.0029	.0008	.0002	.0000
	2	.1787	.2852	.2428	.1540	.0803	.0358	.0138	.0046	.0013	.0003
	3	.0533	.1796	.2428	.2182	.1517	.0869	.0422	.0175	.0062	.0018
	4	.0112	.0798	.1714	.2182	.2023	.1491	.0909	.0467	.0203	.0074
	5	.0018	.0266	.0907	.1636	.2023	.1916	.1468	.0933	.0497	.0222
	6	.0002	.0069	.0374	.0955	.1574	.1916	.1844	.1451	.0949	.0518
	7	.0000	.0014	.0122	.0443	.0974	.1525	.1844	.1797	.1443	.0961
	8	.0000	.0002	.0032	.0166	.0487	.0981	.1489	.1797	.1771	.1442
	9	.0000	.0000	.0007	.0051	.0198	.0514	.0980	.1464	.1771	.1762
	10	.0000	.0000	.0001	.0013	.0066	.0220	.0528	.0976	.1449	.1762
	11	.0000	.0000	.0000	.0003	.0018	.0077	.0233	.0532	.0970	.1442
	12	.0000	.0000	.0000	.0000	.0004	.0022	.0083	.0237	.0529	.0961
	13	.0000	.0000	.0000	.0000	.0001	.0005	.0024	.0085	.0233	.0518
	14	.0000	.0000	.0000	.0000	.0000	.0001	.0006	.0024	.0082	.0222
	15	.0000	.0000	.0000	.0000	.0000	.0000	.0001	.0005	.0022	.0074
	16	.0000	.0000	.0000	.0000	.0000	.0000	.0000	.0001	.0005	.0018
	17	.0000	.0000	.0000	.0000	.0000	.0000	.0000	.0000	.0001	.0003
	18	.0000	.0000	.0000	.0000	.0000	.0000	.0000	.0000	.0000	.0000
	19	.0000	.0000	.0000	.0000	.0000	.0000	.0000	.0000	.0000	.0000
20	0	.3585	.1216	.0388	.0115	.0032	.0008	.0002	.0000	.0000	.0000
	1	.3774	.2702	.1368	.0576	.0211	.0068	.0020	.0005	.0001	.0000
	2	.1887	.2852	.2293	.1369	.0669	.0278	.0100	.0031	.0008	.0002
	3	.0596	.1901	.2428	.2054	.1339	.0716	.0323	.0123	.0040	.0011
	4	.0133	.0898	.1821	.2182	.1897	.1304	.0738	.0350	.0139	.0046
	5	.0022	.0319	.1028	.1746	.2023	.1789	.1272	.0746	.0365	.0148
	6	.0003	.0089	.0454	.1091	.1686	.1916	.1712	.1244	.0746	.0370
	7	.0000	.0020	.0160	.0545	.1124	.1643	.1844	.1659	.1221	.0739
	8	.0000	.0004	.0046	.0222	.0609	.1144	.1614	.1797	.1623	.1201
	9	.0000	.0001	.0011	.0074	.0271	.0654	.1158	.1597	.1771	.1602
	10	.0000	.0000	.0002	.0020	.0099	.0308	.0686	.1171	.1593	.1762
	11	.0000	.0000	.0000	.0005	.0030	.0120	.0336	.0710	.1185	.1602
	12	.0000	.0000	.0000	.0001	.0008	.0039	.0136	.0355	.0727	.1201
	13	.0000	.0000	.0000	.0000	.0002	.0010	.0045	.0146	.0366	.0739
	14	.0000	.0000	.0000	.0000	.0000	.0002	.0012	.0049	.0150	.0370
	15	.0000	.0000	.0000	.0000	.0000	.0000	.0003	.0013	.0049	.0148
	16	.0000	.0000	.0000	.0000	.0000	.0000	.0000	.0003	.0013	.0046
	17	.0000	.0000	.0000	.0000	.0000	.0000	.0000	.0000	.0002	.0011
	18	.0000	.0000	.0000	.0000	.0000	.0000	.0000	.0000	.0000	.0002
	19	.0000	.0000	.0000	.0000	.0000	.0000	.0000	.0000	.0000	.0000
	20	.0000	.0000	.0000	.0000	.0000	.0000	.0000	.0000	.0000	.0000

Table II

SUMMED BINOMIAL DISTRIBUTION FUNCTION

$$\sum_{x=x'}^{x=n} C_x{}^n p^x (1-p)^{n-x}$$

(See Chapter VIII)

Entries in the table are values of $\sum_{x=x'}^{x=n} C_x{}^n p^x (1-p)^{n-x}$ for the indicated values of n, x' and p. When $p > 0.5$, the value of $\sum_{x=x'}^{x=n} C_x{}^n p^x (1-p)^{n-x}$ for a given n, x' and p is equal to one minus the tabular entry for the given n, with $n - x' + 1$ in place of the given value of x', and $1 - p$ in place of the given value of p.

n	x'	.05	.10	.15	.20	.25	.30	.35	.40	.45	.50
2	1	.0975	.1900	.2775	.3600	.4375	.5100	.5775	.6400	.6975	.7500
	2	.0025	.0100	.0225	.0400	.0625	.0900	.1225	.1600	.2025	.2500
3	1	.1426	.2710	.3859	.4880	.5781	.6570	.7254	.7840	.8336	.8750
	2	.0072	.0280	.0608	.1040	.1562	.2160	.2818	.3520	.4252	.5000
	3	.0001	.0010	.0034	.0080	.0156	.0270	.0429	.0640	.0911	.1250
4	1	.1855	.3439	.4780	.5904	.6836	.7599	.8215	.8704	.9085	.9375
	2	.0140	.0523	.1095	.1808	.2617	.3483	.4370	.5248	.6090	.6875
	3	.0005	.0037	.0120	.0272	.0508	.0837	.1265	.1792	.2415	.3125
	4	.0000	.0001	.0005	.0016	.0039	.0081	.0150	.0256	.0410	.0625
5	1	.2262	.4095	.5563	.6723	.7627	.8319	.8840	.9222	.9497	.9688
	2	.0226	.0815	.1648	.2627	.3672	.4718	.5716	.6630	.7438	.8125
	3	.0012	.0086	.0266	.0579	.1035	.1631	.2352	.3174	.4069	.5000
	4	.0000	.0005	.0022	.0067	.0156	.0308	.0540	.0870	.1312	.1875
	5	.0000	.0000	.0001	.0003	.0010	.0024	.0053	.0102	.0185	.0312
6	1	.2649	.4686	.6229	.7379	.8220	.8824	.9246	.9533	.9723	.9844
	2	.0328	.1143	.2235	.3446	.4661	.5798	.6809	.7667	.8364	.8906
	3	.0022	.0158	.0473	.0989	.1694	.2557	.3529	.4557	.5585	.6562
	4	.0001	.0013	.0059	.0170	.0376	.0705	.1174	.1792	.2553	.3438
	5	.0000	.0001	.0004	.0016	.0046	.0109	.0223	.0410	.0692	.1094
	6	.0000	.0000	.0000	.0001	.0002	.0007	.0018	.0041	.0083	.0156
7	1	.3017	.5217	.6794	.7903	.8665	.9176	.9510	.9720	.9848	.9922
	2	.0444	.1497	.2834	.4233	.5551	.6706	.7662	.8414	.8976	.9375
	3	.0038	.0257	.0738	.1480	.2436	.3529	.4677	.5801	.6836	.7734
	4	.0002	.0027	.0121	.0333	.0706	.1260	.1998	.2898	.3917	.5000
	5	.0000	.0002	.0012	.0047	.0129	.0288	.0556	.0963	.1529	.2266
	6	.0000	.0000	.0001	.0004	.0013	.0038	.0090	.0188	.0357	.0625
	7	.0000	.0000	.0000	.0000	.0001	.0002	.0006	.0016	.0037	.0078
8	1	.3366	.5695	.7275	.8322	.8999	.9424	.9681	.9832	.9916	.9961
	2	.0572	.1869	.3428	.4967	.6329	.7447	.8309	.8936	.9368	.9648
	3	.0058	.0381	.1052	.2031	.3215	.4482	.5722	.6846	.7799	.8555
	4	.0004	.0050	.0214	.0563	.1138	.1941	.2936	.4059	.5230	.6367
	5	.0000	.0004	.0029	.0104	.0273	.0580	.1061	.1737	.2604	.3633
	6	.0000	.0000	.0002	.0012	.0042	.0113	.0253	.0498	.0885	.1445
	7	.0000	.0000	.0000	.0001	.0004	.0013	.0036	.0085	.0181	.0352
	8	.0000	.0000	.0000	.0000	.0000	.0001	.0002	.0007	.0017	.0039

Linear interpolation with respect to p will generally not be accurate to more than two decimal places, and sometimes less.

For extensive tables of $\sum_{x=x'}^{x=n} C_x{}^n p^x (1-p)^{n-x}$ see *Tables of the Binomial Probability Distribution*, National Bureau of Standards, Applied Mathematics Series 6, Washington, D. C., 1950.

SUMMED BINOMIAL DISTRIBUTION FUNCTION

$$\sum_{x=x'}^{x=n} C_x^{\,n} p^x (1-p)^{n-x}$$

n	x'	.05	.10	.15	.20	.25	.30	.35	.40	.45	.50
9	1	.3698	.6126	.7684	.8658	.9249	.9596	.9793	.9899	.9954	.9980
	2	.0712	.2252	.4005	.5638	.6997	.8040	.8789	.9295	.9615	.9805
	3	.0084	.0530	.1409	.2618	.3993	.5372	.6627	.7682	.8505	.9102
	4	.0006	.0083	.0339	.0856	.1657	.2703	.3911	.5174	.6386	.7461
	5	.0000	.0009	.0056	.0196	.0489	.0988	.1717	.2666	.3786	.5000
	6	.0000	.0001	.0006	.0031	.0100	.0253	.0536	.0994	.1658	.2539
	7	.0000	.0000	.0000	.0003	.0013	.0043	.0112	.0250	.0498	.0898
	8	.0000	.0000	.0000	.0000	.0001	.0004	.0014	.0038	.0091	.0195
	9	.0000	.0000	.0000	.0000	.0000	.0000	.0001	.0003	.0008	.0020
10	1	.4013	.6513	.8031	.8926	.9437	.9718	.9865	.9940	.9975	.9990
	2	.0861	.2639	.4557	.6242	.7360	.8507	.9140	.9536	.9767	.9893
	3	.0115	.0702	.1798	.3222	.4744	.6172	.7384	.8327	.9004	.9453
	4	.0010	.0128	.0500	.1209	.2241	.3504	.4862	.6177	.7340	.8281
	5	.0001	.0016	.0099	.0328	.0781	.1503	.2485	.3669	.4956	.6230
	6	.0000	.0001	.0014	.0064	.0197	.0473	.0949	.1662	.2616	.3770
	7	.0000	.0000	.0001	.0009	.0035	.0106	.0260	.0548	.1020	.1719
	8	.0000	.0000	.0000	.0001	.0004	.0016	.0048	.0123	.0274	.0547
	9	.0000	.0000	.0000	.0000	.0000	.0001	.0005	.0017	.0045	.0107
	10	.0000	.0000	.0000	.0000	.0000	.0000	.0000	.0001	.0003	.0010
11	1	.4312	.6862	.8327	.9141	.9578	.9802	.9912	.9964	.9986	.9995
	2	.1019	.3026	.5078	.6779	.8029	.8870	.9394	.9698	.9861	.9941
	3	.0152	.0896	.2212	.3826	.5448	.6873	.7999	.8811	.9348	.9673
	4	.0016	.0185	.0694	.1611	.2867	.4304	.5744	.7037	.8089	.8867
	5	.0001	.0028	.0159	.0504	.1146	.2103	.3317	.4672	.6029	.7256
	6	.0000	.0003	.0027	.0117	.0343	.0782	.1487	.2465	.3669	.5000
	7	.0000	.0000	.0003	.0020	.0076	.0216	.0501	.0994	.1738	.2744
	8	.0000	.0000	.0000	.0002	.0012	.0043	.0122	.0293	.0610	.1133
	9	.0000	.0000	.0000	.0000	.0001	.0006	.0020	.0059	.0148	.0327
	10	.0000	.0000	.0000	.0000	.0000	.0000	.0002	.0007	.0022	.0059
	11	.0000	.0000	.0000	.0000	.0000	.0000	.0000	.0000	.0002	.0005
12	1	.4596	.7176	.8578	.9313	.9683	.9862	.9943	.9978	.9992	.9998
	2	.1184	.3410	.5565	.7251	.8416	.9150	.9576	.9804	.9917	.9968
	3	.0196	.1109	.2642	.4417	.6093	.7472	.8487	.9166	.9579	.9807
	4	.0022	.0256	.0922	.2054	.3512	.5075	.6533	.7747	.8655	.9270
	5	.0002	.0043	.0239	.0726	.1576	.2763	.4167	.5618	.6956	.8062
	6	.0000	.0005	.0046	.0194	.0544	.1178	.2127	.3348	.4731	.6128
	7	.0000	.0001	.0007	.0039	.0143	.0386	.0846	.1582	.2607	.3872
	8	.0000	.0000	.0001	.0006	.0028	.0095	.0255	.0573	.1117	.1938
	9	.0000	.0000	.0000	.0001	.0004	.0017	.0056	.0153	.0356	.0730
	10	.0000	.0000	.0000	.0000	.0000	.0002	.0008	.0028	.0079	.0193
	11	.0000	.0000	.0000	.0000	.0000	.0000	.0001	.0003	.0011	.0032
	12	.0000	.0000	.0000	.0000	.0000	.0000	.0000	.0000	.0001	.0002
13	1	.4867	.7458	.8791	.9450	.9762	.9903	.9963	.9987	.9996	.9999
	2	.1354	.3787	.6017	.7664	.8733	.9363	.9704	.9874	.9951	.9983
	3	.0245	.1339	.2704	.4983	.6674	.7975	.8868	.9421	.9731	.9888
	4	.0031	.0342	.0967	.2527	.4157	.5794	.7217	.8314	.9071	.9539
	5	.0003	.0065	.0260	.0991	.2060	.3457	.4995	.6470	.7721	.8666
	6	.0000	.0009	.0053	.0300	.0802	.1654	.2841	.4256	.5732	.7095
	7	.0000	.0001	.0013	.0070	.0243	.0624	.1295	.2288	.3563	.5000
	8	.0000	.0000	.0002	.0012	.0056	.0182	.0462	.0977	.1788	.2905
	9	.0000	.0000	.0000	.0002	.0010	.0040	.0126	.0321	.0698	.1334
	10	.0000	.0000	.0000	.0000	.0001	.0007	.0025	.0078	.0203	.0461
	11	.0000	.0000	.0000	.0000	.0000	.0001	.0003	.0013	.0041	.0112
	12	.0000	.0000	.0000	.0000	.0000	.0000	.0000	.0001	.0005	.0017
	13	.0000	.0000	.0000	.0000	.0000	.0000	.0000	.0000	.0000	.0001

Table II continued $$\sum_{x=x'}^{x=n} C_x{}^n p^x (1-p)^{n-x}$$

n	x'	.05	.10	.15	.20	.25	.30	.35	.40	.45	.50
14	1	.5123	.7712	.8972	.9560	.9822	.9932	.9976	.9992	.9998	.9999
	2	.1530	.4154	.6433	.8021	.8990	.9525	.9795	.9919	.9971	.9991
	3	.0301	.1584	.3521	.5519	.7189	.8392	.9161	.9602	.9830	.9935
	4	.0042	.0441	.1465	.3018	.4787	.6448	.7795	.8757	.9368	.9713
	5	.0004	.0092	.0467	.1298	.2585	.4158	.5773	.7207	.8328	.9102
	6	.0000	.0015	.0115	.0439	.1117	.2195	.3595	.5141	.6627	.7880
	7	.0000	.0002	.0022	.0116	.0383	.0933	.1836	.3075	.4539	.6047
	8	.0000	.0000	.0003	.0024	.0103	.0315	.0753	.1501	.2586	.3953
	9	.0000	.0000	.0000	.0004	.0022	.0083	.0243	.0583	.1189	.2120
	10	.0000	.0000	.0000	.0000	.0003	.0017	.0060	.0175	.0426	.0898
	11	.0000	.0000	.0000	.0000	.0000	.0002	.0011	.0039	.0114	.0287
	12	.0000	.0000	.0000	.0000	.0000	.0000	.0001	.0006	.0022	.0065
	13	.0000	.0000	.0000	.0000	.0000	.0000	.0000	.0001	.0003	.0009
	14	.0000	.0000	.0000	.0000	.0000	.0000	.0000	.0000	.0000	.0001
15	1	.5367	.7941	.9126	.9648	.9866	.9953	.9984	.9995	.9999	1.0000
	2	.1710	.4510	.6814	.8329	.9198	.9647	.9858	.9948	.9983	.9995
	3	.0362	.1841	.3958	.6020	.7639	.8732	.9383	.9729	.9893	.9963
	4	.0055	.0556	.1773	.3518	.5387	.7031	.8273	.9095	.9576	.9824
	5	.0006	.0127	.0617	.1642	.3135	.4845	.6481	.7827	.8796	.9408
	6	.0001	.0022	.0168	.0611	.1484	.2784	.4357	.5968	.7392	.8491
	7	.0000	.0003	.0036	.0181	.0566	.1311	.2452	.3902	.5478	.6964
	8	.0000	.0000	.0006	.0042	.0173	.0500	.1132	.2131	.3465	.5000
	9	.0000	.0000	.0001	.0008	.0042	.0152	.0422	.0950	.1818	.3036
	10	.0000	.0000	.0000	.0001	.0008	.0037	.0124	.0338	.0769	.1509
	11	.0000	.0000	.0000	.0000	.0001	.0007	.0028	.0093	.0255	.0592
	12	.0000	.0000	.0000	.0000	.0000	.0001	.0005	.0019	.0063	.0176
	13	.0000	.0000	.0000	.0000	.0000	.0000	.0001	.0003	.0011	.0037
	14	.0000	.0000	.0000	.0000	.0000	.0000	.0000	.0000	.0001	.0005
	15	.0000	.0000	.0000	.0000	.0000	.0000	.0000	.0000	.0000	.0000
16	1	.5599	.8147	.9257	.9719	.9900	.9967	.9990	.9997	.9999	1.0000
	2	.1892	.4853	.7161	.8593	.9365	.9739	.9902	.9967	.9990	.9997
	3	.0429	.2108	.4386	.6482	.8029	.9006	.9549	.9817	.9934	.9979
	4	.0070	.0684	.2101	.4019	.5950	.7541	.8661	.9349	.9719	.9894
	5	.0009	.0170	.0791	.2018	.3698	.5501	.7108	.8334	.9147	.9616
	6	.0001	.0033	.0235	.0817	.1897	.3402	.5100	.6712	.8024	.8949
	7	.0000	.0005	.0056	.0267	.0796	.1753	.3119	.4728	.6340	.7228
	8	.0000	.0001	.0011	.0070	.0271	.0744	.1594	.2839	.4371	.5982
	9	.0000	.0000	.0002	.0015	.0075	.0257	.0671	.1423	.2559	.4018
	10	.0000	.0000	.0000	.0002	.0016	.0071	.0229	.0583	.1241	.2272
	11	.0000	.0000	.0000	.0000	.0003	.0016	.0062	.0191	.0486	.1051
	12	.0000	.0000	.0000	.0000	.0000	.0003	.0013	.0049	.0149	.0384
	13	.0000	.0000	.0000	.0000	.0000	.0000	.0002	.0009	.0035	.0106
	14	.0000	.0000	.0000	.0000	.0000	.0000	.0000	.0001	.0006	.0021
	15	.0000	.0000	.0000	.0000	.0000	.0000	.0000	.0000	.0001	.0003
	16	.0000	.0000	.0000	.0000	.0000	.0000	.0000	.0000	.0000	.0000
17	1	.5819	.8332	.9369	.9775	.9925	.9977	.9993	.9998	1.0000	1.0000
	2	.2078	.5182	.7475	.8818	.9499	.9807	.9933	.9979	.9994	.9999
	3	.0503	.2382	.4802	.6904	.8363	.9226	.9673	.9877	.9959	.9988
	4	.0088	.0826	.2444	.4511	.6470	.7981	.8972	.9536	.9816	.9936
	5	.0012	.0221	.0987	.2418	.4261	.6113	.7652	.8740	.9404	.9755
	6	.0001	.0047	.0319	.1057	.2347	.4032	.5803	.7361	.8529	.9283
	7	.0000	.0008	.0083	.0377	.1071	.2248	.3812	.5522	.7098	.8338
	8	.0000	.0001	.0017	.0109	.0402	.1046	.2128	.3595	.5257	.6855
	9	.0000	.0000	.0003	.0026	.0124	.0403	.0994	.1989	.3374	.5000
	10	.0000	.0000	.0000	.0005	.0031	.0127	.0383	.0919	.1834	.3145
	11	.0000	.0000	.0000	.0001	.0006	.0032	.0120	.0348	.0826	.1662
	12	.0000	.0000	.0000	.0000	.0001	.0007	.0030	.0106	.0301	.0717
	13	.0000	.0000	.0000	.0000	.0000	.0001	.0006	.0025	.0086	.0245
	14	.0000	.0000	.0000	.0000	.0000	.0000	.0001	.0005	.0019	.0064
	15	.0000	.0000	.0000	.0000	.0000	.0000	.0000	.0001	.0003	.0012
	16	.0000	.0000	.0000	.0000	.0000	.0000	.0000	.0000	.0000	.0001
	17	0000	0000	0000	0000	0000	0000	0000	0000	0000	0000

SUMMED BINOMIAL DISTRIBUTION FUNCTION

$$\sum_{x=x'}^{x=n} C_x{}^n p^x (1-p)^{n-x}$$

n	x′	.05	.10	.15	.20	.25	*p* .30	.35	.40	.45	.50
18	1	.6028	.8499	.9464	.9820	.9944	.9984	.9996	.9999	1.0000	1.0000
	2	.2265	.5497	.7759	.9009	.9605	.9858	.9954	.9987	.9997	.9999
	3	.0581	.2662	.5203	.7287	.8647	.9400	.9764	.9918	.9975	.9993
	4	.0109	.0982	.2798	.4990	.6943	.8354	.9217	.9672	.9880	.9962
	5	.0015	.0282	.1206	.2836	.4813	.6673	.8114	.9058	.9589	.9846
	6	.0002	.0064	.0419	.1329	.2825	.4656	.6450	.7912	.8923	.9519
	7	.0000	.0012	.0118	.0513	.1390	.2783	.4509	.6257	.7742	.8811
	8	.0000	.0002	.0027	.0163	.0569	.1407	.2717	.4366	.6085	.7597
	9	.0000	.0000	.0005	.0043	.0193	.0596	.1391	.2632	.4222	.5927
	10	.0000	.0000	.0001	.0009	.0054	.0210	.0597	.1347	.2527	.4073
	11	.0000	.0000	.0000	.0002	.0012	.0061	.0212	.0576	.1280	.2403
	12	.0000	.0000	.0000	.0000	.0002	.0014	.0062	.0203	.0537	.1189
	13	.0000	.0000	.0000	.0000	.0000	.0003	.0014	.0058	.0183	.0481
	14	.0000	.0000	.0000	.0000	.0000	.0000	.0003	.0013	.0049	.0154
	15	.0000	.0000	.0000	.0000	.0000	.0000	.0000	.0002	.0010	.0038
	16	.0000	.0000	.0000	.0000	.0000	.0000	.0000	.0000	.0001	.0007
	17	.0000	.0000	.0000	.0000	.0000	.0000	.0000	.0000	.0000	.0001
	18	.0000	.0000	.0000	.0000	.0000	.0000	.0000	.0000	.0000	.0000
19	1	.6226	.8649	.9544	.9856	.9958	.9989	.9997	.9999	1.0000	1.0000
	2	.2453	.5797	.8015	.9171	.9690	.9896	.9969	.9992	.9998	1.0000
	3	.0665	.2946	.5587	.7631	.8887	.9538	.9830	.9945	.9985	.9996
	4	.0132	.1150	.1444	.5449	.7369	.8668	.9409	.9770	.9923	.9978
	5	.0020	.0352	.1444	.3267	.5346	.7178	.8500	.9304	.9720	.9904
	6	.0002	.0086	.0537	.1631	.3322	.5261	.7032	.8371	.9223	.9682
	7	.0000	.0017	.0163	.0676	.1749	.3345	.5188	.6919	.8273	.9165
	8	.0000	.0003	.0041	.0233	.0775	.1820	.3344	.5122	.6831	.8204
	9	.0000	.0000	.0008	.0067	.0287	.0839	.1855	.3325	.5060	.6762
	10	.0000	.0000	.0001	.0016	.0089	.0326	.0875	.1861	.3290	.5000
	11	.0000	.0000	.0000	.0003	.0023	.0105	.0347	.0885	.1841	.3238
	12	.0000	.0000	.0000	.0000	.0005	.0028	.0114	.0352	.0871	.1796
	13	.0000	.0000	.0000	.0000	.0001	.0006	.0031	.0116	.0342	.0835
	14	.0000	.0000	.0000	.0000	.0000	.0001	.0007	.0031	.0109	.0318
	15	.0000	.0000	.0000	.0000	.0000	.0000	.0001	.0006	.0028	.0096
	16	.0000	.0000	.0000	.0000	.0000	.0000	.0000	.0001	.0005	.0022
	17	.0000	.0000	.0000	.0000	.0000	.0000	.0000	.0000	.0001	.0004
	18	.0000	.0000	.0000	.0000	.0000	.0000	.0000	.0000	.0000	.0000
	19	.0000	.0000	.0000	.0000	.0000	.0000	.0000	.0000	.0000	.0000
20	1	.6415	.8784	.9612	.9885	.9968	.9992	.9998	1.0000	1.0000	1.0000
	2	.2642	.6083	.8244	.9308	.9757	.9924	.9979	.9995	.9999	1.0000
	3	.0755	.3231	.5951	.7939	.9087	.9645	.9879	.9964	.9991	.9998
	4	.0159	.1330	.3523	.5886	.7748	.8929	.9556	.9840	.9951	.9987
	5	.0026	.0432	.1702	.3704	.5852	.7625	.8818	.9490	.9811	.9941
	6	.0003	.0113	.0673	.1958	.3828	.5836	.7546	.8744	.9447	.9793
	7	.0000	.0024	.0219	.0867	.2142	.3920	.5834	.7500	.8701	.9423
	8	.0000	.0004	.0059	.0321	.1018	.2277	.3990	.5841	.7480	.8684
	9	.0000	.0001	.0013	.0100	.0409	.1133	.2376	.4044	.5857	.7483
	10	.0000	.0000	.0002	.0026	.0139	.0480	.1218	.2447	.4086	.5881
	11	.0000	.0000	.0000	.0006	.0039	.0171	.0532	.1275	.2493	.4119
	12	.0000	.0000	.0000	.0001	.0009	.0051	.0196	.0565	.1308	.2517
	13	.0000	.0000	.0000	.0000	.0002	.0013	.0060	.0210	.0580	.1316
	14	.0000	.0000	.0000	.0000	.0000	.0003	.0015	.0065	.0214	.0577
	15	.0000	.0000	.0000	.0000	.0000	.0000	.0003	.0016	.0064	.0207
	16	.0000	.0000	.0000	.0000	.0000	.0000	.0000	.0003	.0015	.0059
	17	.0000	.0000	.0000	.0000	.0000	.0000	.0000	.0000	.0003	.0013
	18	.0000	.0000	.0000	.0000	.0000	.0000	.0000	.0000	.0000	.0002
	19	.0000	.0000	.0000	.0000	.0000	.0000	.0000	.0000	.0000	.0000
	20	.0000	.0000	.0000	.0000	.0000	.0000	.0000	.0000	.0000	.0000

Table III

355

INCOMPLETE BETA FUNCTION RATIO, $I_p(x, n - x + 1)$

$$I_p(x, n - x + 1) = \frac{B_p(x, n - x + 1)}{B(x, n - x + 1)} = \sum_{i=x}^{n} C_i^{\,n} p^i (1 - p)^{n-i}.$$

[Eq. 8.5.3.]

Entries in the table are values of p corresponding to the indicated values of x and $n - x + 1$ for which $I_p(x, n - x + 1)$ has the indicated fixed value.*

$$I_p(x, n - x + 1) = 0.100$$

x	1	2	3	4	5	6	10	15	20	30	60
					$n - x + 1$						
1	.1000	.0513	.0345	.0260	.0209	.0174	.0105	.0070	.0053	.0035	.0018
2	.3162	.1958	.1426	.1122	.0926	.0788	.0495	.0337	.0256	.0173	.0088
3	.4642	.3205	.2466	.2009	.1696	.1468	.0957	.0667	.0512	.0349	.0179
4	.5623	.4161	.3332	.2786	.2397	.2104	.1416	.1006	.0781	.0539	.0280
5	.6310	.4897	.4038	.3446	.3010	.2673	.1851	.1339	.1050	.0733	.0385
6	.6813	.5474	.4618	.4006	.3542	.3177	.2256	.1659	.1312	.0926	.0492
7	.7197	.5938	.5099	.4483	.4005	.3623	.2629	.1962	.1566	.1116	.0600
8	.7499	.6316	.5504	.4892	.4410	.4018	.2973	.2248	.1809	.1302	.0708
9	.7743	.6632	.5848	.5247	.4766	.4369	.3288	.2518	.2042	.1483	.0815
10	.7943	.6898	.6145	.5557	.5080	.4683	.3579	.2772	.2264	.1658	.0921
11	.8111	.7125	.6402	.5830	.5360	.4965	.3848	.3011	.2476	.1828	.1026
12	.8254	.7322	.6628	.6072	.5611	.5219	.4095	.3236	.2678	.1993	.1129
13	.8377	.7493	.6827	.6288	.5836	.5450	.4325	.3448	.2870	.2152	.1231
14	.8483	.7644	.7004	.6481	.6040	.5660	.4538	.3648	.3053	.2305	.1331
15	.8577	.7778	.7163	.6656	.6225	.5851	.4736	.3837	.3228	.2454	.1430
20	.8912	.8271	.7758	.7322	.6941	.6603	.5548	.4639	.3991	.3124	.1896
30	.9261	.8802	.8421	.8086	.7785	.7510	.6603	.5754	.5107	.4175	.2706
60	.9624	.9377	.9164	.8970	.8790	.8620	.8019	.7395	.6869	.6024	.4416
∞	1.0000	1.0000	1.0000	1.0000	1.0000	1.0000	1.0000	1.0000	1.0000	1.0000	1.0000

$$I_p(x, n - x + 1) = 0.050$$

x	1	2	3	4	5	6	10	15	20	30	60
					$n - x + 1$						
1	.0500	.0253	.0170	.0127	.0102	.0085	.0051	.0034	.0026	.0017	.0009
2	.2236	.1354	.0976	.0764	.0628	.0534	.0333	.0227	.0172	.0116	.0059
3	.3684	.2486	.1893	.1532	.1288	.1111	.0719	.0499	.0382	.0260	.0133
4	.4729	.3426	.2713	.2253	.1929	.1688	.1127	.0797	.0617	.0425	.0220
5	.5493	.4182	.3413	.2892	.2514	.2224	.1527	.1099	.0859	.0598	.0313
6	.6070	.4793	.4003	.3449	.3035	.2712	.1909	.1396	.1101	.0774	.0410
7	.6518	.5293	.4504	.3934	.3498	.3152	.2267	.1682	.1338	.0950	.0508
8	.6877	.5709	.4931	.4356	.3909	.3548	.2601	.1956	.1568	.1124	.0608
9	.7169	.6058	.5299	.4727	.4274	.3904	.2912	.2216	.1791	.1295	.0708
10	.7411	.6356	.5619	.5054	.4600	.4226	.3201	.2464	.2005	.1462	.0808
11	.7616	.6613	.5899	.5343	.4892	.4516	.3469	.2698	.2211	.1625	.0907
12	.7791	.6837	.6146	.5602	.5156	.4781	.3719	.2921	.2408	.1784	.1005
13	.7942	.7033	.6366	.5834	.5394	.5022	.3952	.3131	.2597	.1938	.1102
14	.8074	.7206	.6562	.6044	.5611	.5242	.4168	.3331	.2778	.2088	.1198
15	.8190	.7360	.6738	.6233	.5809	.5444	.4371	.3520	.2951	.2233	.1293
20	.8609	.7933	.7405	.6964	.6582	.6246	.5210	.4332	.3714	.2894	.1745
30	.9050	.8559	.8161	.7815	.7507	.7228	.6318	.5481	.4848	.3946	.2542
60	.9513	.9246	.9019	.8815	.8627	.8450	.7834	.7202	.6674	.5833	.4252
∞	1.0000	1.0000	1.0000	1.0000	1.0000	1.0000	1.0000	1.0000	1.0000	1.0000	1.0000

*More extensive tables of this type are given in C. M. Thompson, "Percentage Points of the Incomplete Beta Function," *Biometrika*, Vol. 32, pp. 168-181.

INCOMPLETE BETA FUNCTION RATIO, $I_p(x, n - x + 1)$

$I_p(x, n - x + 1) = 0.025$

x	$n - x + 1$										
	1	2	3	4	5	6	10	15	20	30	60
1	.0250	.0126	.0084	.0063	.0051	.0042	.0025	.0017	.0013	.0008	.0004
2	.1581	.0943	.0676	.0527	.0433	.0367	.0228	.0155	.0117	.0079	.0040
3	.2924	.1941	.1466	.1181	.0990	.0852	.0549	.0380	.0291	.0198	.0101
4	.3976	.2836	.2228	.1840	.1570	.1370	.0909	.0641	.0495	.0340	.0176
5	.4782	.3588	.2904	.2449	.2120	.1871	.1276	.0915	.0713	.0495	.0259
6	.5407	.4213	.3491	.2993	.2624	.2338	.1634	.1189	.0936	.0656	.0346
7	.5904	.4735	.3999	.3476	.3079	.2767	.1975	.1459	.1157	.0819	.0437
8	.6306	.5175	.4439	.3903	.3489	.3158	.2298	.1720	.1375	.0983	.0530
9	.6637	.5550	.4822	.4281	.3857	.3514	.2602	.1971	.1588	.1144	.0623
10	.6915	.5872	.5159	.4619	.4190	.3838	.2886	.2211	.1794	.1304	.0717
11	.7151	.6152	.5455	.4920	.4490	.4134	.3153	.2440	.1993	.1460	.0811
12	.7354	.6397	.5719	.5191	.4762	.4404	.3402	.2659	.2185	.1613	.0905
13	.7530	.6613	.5954	.5435	.5010	.4652	.3636	.2867	.2370	.1762	.0998
14	.7684	.6805	.6165	.5657	.5236	.4880	.3854	.3065	.2548	.1908	.1090
15	.7820	.6977	.6356	.5858	.5444	.5090	.4059	.3253	.2718	.2049	.1181
20	.8316	.7618	.7084	.6641	.6262	.5930	.4917	.4070	.3478	.2700	.1620
30	.8843	.8330	.7919	.7567	.7255	.6974	.6067	.5242	.4624	.3750	.2403
60	.9404	.9120	.8883	.8671	.8476	.8295	.7668	.7030	.6502	.5666	.4111
∞	1.0000	1.0000	1.0000	1.0000	1.0000	1.0000	1.0000	1.0000	1.0000	1.0000	1.0000

$I_p(x, n - x + 1) = 0.010$

x	$n - x + 1$										
	1	2	3	4	5	6	10	15	20	30	60
1	.0100	.0050	.0033	.0025	.0020	.0017	.0010	.0007	.0005	.0003	.0002
2	.1000	.0589	.0420	.0327	.0268	.0227	.0141	.0095	.0072	.0049	.0025
3	.2154	.1409	.1056	.0847	.0708	.0608	.0390	.0269	.0206	.0140	.0071
4	.3162	.2221	.1731	.1423	.1210	.1053	.0695	.0488	.0376	.0258	.0133
5	.3981	.2943	.2363	.1982	.1710	.1504	.1019	.0728	.0566	.0392	.0204
6	.4642	.3566	.2932	.2500	.2183	.1940	.1346	.0975	.0765	.0535	.0282
7	.5180	.4101	.3437	.2971	.2622	.2349	.1665	.1224	.0968	.0683	.0363
8	.5623	.4560	.3883	.3396	.3024	.2729	.1971	.1468	.1170	.0833	.0448
9	.5995	.4956	.4277	.3778	.3391	.3080	.2263	.1705	.1370	.0984	.0534
10	.6310	.5302	.4627	.4122	.3726	.3403	.2540	.1935	.1565	.1133	.0621
11	.6579	.5605	.4938	.4433	.4031	.3700	.2801	.2156	.1755	.1281	.0709
12	.6813	.5872	.5217	.4715	.4310	.3975	.3047	.2369	.1940	.1427	.0797
13	.7017	.6109	.5468	.4971	.4566	.4228	.3280	.2572	.2119	.1570	.0885
14	.7197	.6321	.5695	.5204	.4801	.4462	.3498	.2767	.2292	.1710	.0972
15	.7356	.6512	.5901	.5417	.5018	.4679	.3705	.2953	.2460	.1846	.1059
20	.7943	.7232	.6695	.6256	.5882	.5557	.4578	.3770	.3211	.2482	.1481
30	.8577	.8043	.7623	.7265	.6951	.6670	.5772	.4965	.4366	.3526	.2246
60	.9261	.8961	.8712	.8492	.8292	.8106	.7468	.6826	.6299	.5471	.3948
∞	1.0000	1.0000	1.0000	1.0000	1.0000	1.0000	1.0000	1.0000	1.0000	1.0000	1.0000

Table III continued

357

INCOMPLETE BETA FUNCTION RATIO, $I_p(x, n - x + 1)$

$$I_p(x, n - x + 1) = 0.005$$

x	1	2	3	4	$n - x + 1$ 5	6	10	15	20	30	60
1	.0050	.0025	.0017	.0013	.0010	.0008	.0005	.0003	.0003	.0002	.0001
2	.0707	.0414	.0294	.0229	.0187	.0158	.0098	.0067	.0050	.0034	.0017
3	.1710	.1109	.0828	.0663	.0553	.0475	.0303	.0209	.0160	.0108	.0055
4	.2659	.1851	.1436	.1177	.0999	.0868	.0571	.0400	.0308	.0211	.0109
5	.3466	.2540	.2030	.1697	.1461	.1283	.0866	.0617	.0479	.0332	.0172
6	.4135	.3151	.2578	.2191	.1909	.1693	.1170	.0846	.0663	.0463	.0243
7	.4691	.3685	.3074	.2649	.2332	.2085	.1471	.1078	.0852	.0600	.0318
8	.5157	.4150	.3518	.3067	.2725	.2454	.1764	.1310	.1042	.0741	.0397
9	.5550	.4557	.3915	.3448	.3087	.2799	.2046	.1537	.1232	.0883	.0478
10	.5887	.4914	.4270	.3794	.3421	.3118	.2316	.1759	.1420	.1026	.0561
11	.6178	.5230	.4590	.4108	.3727	.3415	.2572	.1974	.1604	.1168	.0644
12	.6430	.5510	.4877	.4395	.4009	.3690	.2815	.2181	.1783	.1308	.0728
13	.6653	.5760	.5137	.4656	.4268	.3945	.3046	.2381	.1957	.1446	.0812
14	.6849	.5984	.5372	.4896	.4508	.4182	.3264	.2572	.2127	.1582	.0897
15	.7024	.6186	.5587	.5116	.4729	.4402	.3470	.2757	.2291	.1715	.0980
20	.7673	.6957	.6423	.5988	.5620	.5302	.4349	.3570	.3034	.2339	.1391
30	.8381	.7837	.7412	.7053	.6738	.6458	.5569	.4776	.4191	.3376	.2142
60	.9155	.8844	.8589	.8364	.8160	.7972	.7328	.6684	.6159	.5338	.3838
∞	1.0000	1.0000	1.0000	1.0000	1.0000	1.0000	1.0000	1.0000	1.0000	1.0000	1.0000

Table IV

BINOMIAL COEFFICIENTS

n	$\binom{n}{0}$	$\binom{n}{1}$	$\binom{n}{2}$	$\binom{n}{3}$	$\binom{n}{4}$	$\binom{n}{5}$	$\binom{n}{6}$	$\binom{n}{7}$	$\binom{n}{8}$	$\binom{n}{9}$	$\binom{n}{10}$
0	1										
1	1	1									
2	1	2	1								
3	1	3	3	1							
4	1	4	6	4	1						
5	1	5	10	10	5	1					
6	1	6	15	20	15	6	1				
7	1	7	21	35	35	21	7	1			
8	1	8	28	56	70	56	28	8	1		
9	1	9	36	84	126	126	84	36	9	1	
10	1	10	45	120	210	252	210	120	45	10	1
11	1	11	55	165	330	462	462	330	165	55	11
12	1	12	66	220	495	792	924	792	495	220	66
13	1	13	78	286	715	1287	1716	1716	1287	715	286
14	1	14	91	364	1001	2002	3003	3432	3003	2002	1001
15	1	15	105	455	1365	3003	5005	6435	6435	5005	3003
16	1	16	120	560	1820	4368	8008	11440	12870	11440	8008
17	1	17	136	680	2380	6188	12376	19448	24310	24310	19448
18	1	18	153	816	3060	8568	18564	31824	43758	48620	43758
19	1	19	171	969	3876	11628	27132	50388	75582	92378	92378
20	1	20	190	1140	4845	15504	38760	77520	125970	167960	184756

$$C_r{}^n = \binom{n}{r} = \frac{n!}{r!(n - r)!} = \binom{n}{n - r} \quad , \quad \binom{n}{0} = 1.$$

$$(q + p)^n = q^n + C_1{}^n pq^{n-1} + \cdots + C_r{}^n p^r q^{n-r} + \cdots + p^n.$$

Table V

VALUES OF $\sqrt{npq}$

$$(q = 1 - p)$$

n	.05	.10	.15	.20	.25	.30	.35	.40	.45	.50
1	.2179	.3000	.3571	.4000	.4330	.4583	.4770	.4899	.4975	.5000
2	.3082	.4243	.5050	.5657	.6124	.6481	.6745	.6928	.7036	.7071
3	.3775	.5196	.6185	.6928	.7500	.7937	.8261	.8485	.8617	.8660
4	.4359	.6000	.7141	.8000	.8660	.9165	.9539	.9798	.9950	1.0000
5	.4873	.6708	.7984	.8944	.9682	1.0247	1.0665	1.0954	1.1124	1.1180
6	.5339	.7348	.8746	.9798	1.0607	1.1225	1.1683	1.2000	1.2186	1.2247
7	.5766	.7937	.9447	1.0583	1.1456	1.2124	1.2619	1.2961	1.3162	1.3229
8	.6164	.8485	1.0100	1.1314	1.2247	1.2961	1.3491	1.3856	1.4071	1.4142
9	.6538	.9000	1.0712	1.2000	1.2990	1.3748	1.4309	1.4697	1.4925	1.5000
10	.6892	.9487	1.1292	1.2649	1.3693	1.4491	1.5083	1.5492	1.5732	1.5811
11	.7228	.9950	1.1843	1.3266	1.4361	1.5199	1.5819	1.6248	1.6500	1.6583
12	.7550	1.0392	1.2369	1.3856	1.5000	1.5875	1.6523	1.6971	1.7234	1.7321
13	.7858	1.0817	1.2874	1.4422	1.5612	1.6523	1.7197	1.7664	1.7937	1.8028
14	.8155	1.1225	1.3360	1.4967	1.6202	1.7146	1.7847	1.8330	1.8615	1.8708
15	.8441	1.1619	1.3829	1.5492	1.6771	1.7748	1.8473	1.8974	1.9268	1.9365
16	.8718	1.2000	1.4283	1.6000	1.7321	1.8330	1.9079	1.9596	1.9900	2.0000
17	.8986	1.2369	1.4722	1.6492	1.7854	1.8894	1.9666	2.0199	2.0512	2.0616
18	.9247	1.2728	1.5149	1.6971	1.8371	1.9442	2.0236	2.0785	2.1107	2.1213
19	.9500	1.3077	1.5564	1.7436	1.8875	1.9975	2.0791	2.1354	2.1685	2.1794
20	.9747	1.3416	1.5968	1.7888	1.9364	2.0493	2.1330	2.1908	2.2248	2.2360

Table VI

VALUES OF $\sqrt{pq}$

$$(q = 1 - p)$$

p	$\sqrt{pq}$	p	$\sqrt{pq}$	p	$\sqrt{pq}$	p	$\sqrt{pq}$	p	$\sqrt{pq}$
.005	.0705	.105	.3066	.205	.4037	.305	.4604	.405	.4909
.010	.0995	.110	.3129	.210	.4073	.310	.4625	.410	.4918
.015	.1216	.115	.3190	.215	.4108	.315	.4645	.415	.4927
.020	.1400	.120	.3250	.220	.4142	.320	.4665	.420	.4936
.025	.1561	.125	.3307	.225	.4176	.325	.4684	.425	.4943
.030	.1706	.130	.3363	.230	.4208	.330	.4702	.430	.4951
.035	.1838	.135	.3417	.235	.4240	.335	.4720	.435	.4958
.040	.1960	.140	.3470	.240	.4271	.340	.4737	.440	.4964
.045	.2073	.145	.3521	.245	.4301	.345	.4754	.445	.4970
.050	.2179	.150	.3571	.250	.4330	.350	.4770	.450	.4975
.055	.2280	.155	.3619	.255	.4359	.355	.4785	.455	.4980
.060	.2375	.160	.3666	.260	.4386	.360	.4800	.460	.4984
.065	.2465	.165	.3712	.265	.4413	.365	.4814	.465	.4988
.070	.2551	.170	.3756	.270	.4440	.370	.4828	.470	.4991
.075	.2634	.175	.3800	.275	.4465	.375	.4841	.475	.4994
.080	.2713	.180	.3842	.280	.4490	.380	.4854	.480	.4996
.085	.2789	.185	.3883	.285	.4514	.385	.4866	.485	.4998
.090	.2862	.190	.3923	.290	.4538	.390	.4877	.490	.4999
.095	.2932	.195	.3962	.295	.4560	.395	.4889	.495	.5000
.100	.3000	.200	.4000	.300	.4583	.400	.4899	.500	.5000

Table VII

POISSON DISTRIBUTION FUNCTION $e^{-m}m^x/x!$

(See Chapter IX)

Entries in the table are values of $e^{-m}m^x/x!$ for the indicated values of x and m.

x	0.1	0.2	0.3	0.4	0.5	0.6	0.7	0.8	0.9	1.0
0	.9048	.8187	.7408	.6703	.6065	.5488	.4966	.4493	.4066	.3679
1	.0905	.1637	.2222	.2681	.3033	.3293	.3476	.3595	.3659	.3679
2	.0045	.0164	.0333	.0536	.0758	.0988	.1217	.1438	.1647	.1839
3	.0002	.0011	.0033	.0072	.0126	.0198	.0284	.0383	.0494	.0613
4	.0000	.0001	.0002	.0007	.0016	.0030	.0050	.0077	.0111	.0153
5	.0000	.0000	.0000	.0001	.0002	.0004	.0007	.0012	.0020	.0031
6	.0000	.0000	.0000	.0000	.0000	.0000	.0001	.0002	.0003	.0005
7	.0000	.0000	.0000	.0000	.0000	.0000	.0000	.0000	.0000	.0001

x	1.1	1.2	1.3	1.4	1.5	1.6	1.7	1.8	1.9	2.0
0	.3329	.3012	.2725	.2466	.2231	.2019	.1827	.1653	.1496	.1353
1	.3662	.3614	.3543	.3452	.3347	.3230	.3106	.2975	.2842	.2707
2	.2014	.2169	.2303	.2417	.2510	.2584	.2640	.2678	.2700	.2707
3	.0738	.0867	.0998	.1128	.1255	.1378	.1496	.1607	.1710	.1804
4	.0203	.0260	.0324	.0395	.0471	.0551	.0636	.0723	.0812	.0902
5	.0045	.0062	.0084	.0111	.0141	.0176	.0216	.0260	.0309	.0361
6	.0008	.0012	.0018	.0026	.0035	.0047	.0061	.0078	.0098	.0120
7	.0001	.0002	.0003	.0005	.0008	.0011	.0015	.0020	.0027	.0034
8	.0000	.0000	.0001	.0001	.0001	.0002	.0003	.0005	.0006	.0009
9	.0000	.0000	.0000	.0000	.0000	.0000	.0001	.0001	.0001	.0002

x	2.1	2.2	2.3	2.4	2.5	2.6	2.7	2.8	2.9	3.0
0	.1225	.1108	.1003	.0907	.0821	.0743	.0672	.0608	.0550	.0498
1	.2572	.2438	.2306	.2177	.2052	.1931	.1815	.1703	.1596	.1494
2	.2700	.2681	.2652	.2613	.2565	.2510	.2450	.2384	.2314	.2240
3	.1890	.1966	.2033	.2090	.2138	.2176	.2205	.2225	.2237	.2240
4	.0992	.1082	.1169	.1254	.1336	.1414	.1488	.1557	.1622	.1680
5	.0417	.0476	.0538	.0602	.0668	.0735	.0804	.0872	.0940	.1008
6	.0146	.0174	.0206	.0241	.0278	.0319	.0362	.0407	.0455	.0504
7	.0044	.0055	.0068	.0083	.0099	.0118	.0139	.0163	.0188	.0216
8	.0011	.0015	.0019	.0025	.0031	.0038	.0047	.0057	.0068	.0081
9	.0003	.0004	.0005	.0007	.0009	.0011	.0014	.0018	.0022	.0027
10	.0001	.0001	.0001	.0002	.0002	.0003	.0004	.0005	.0006	.0008
11	.0000	.0000	.0000	.0000	.0000	.0001	.0001	.0001	.0002	.0002
12	.0000	.0000	.0000	.0000	.0000	.0000	.0000	.0000	.0000	.0001

x	3.1	3.2	3.3	3.4	3.5	3.6	3.7	3.8	3.9	4.0
0	.0450	.0408	.0369	.0334	.0302	.0273	.0247	.0224	.0202	.0183
1	.1397	.1304	.1217	.1135	.1057	.0984	.0915	.0850	.0789	.0733
2	.2165	.2087	.2008	.1929	.1850	.1771	.1692	.1615	.1539	.1465
3	.2237	.2226	.2209	.2186	.2158	.2125	.2087	.2046	.2001	.1954
4	.1734	.1781	.1823	.1858	.1888	.1912	.1931	.1944	.1951	.1954
5	.1075	.1140	.1203	.1264	.1322	.1377	.1429	.1477	.1522	.1563
6	.0555	.0608	.0662	.0716	.0771	.0826	.0881	.0936	.0989	.1042
7	.0246	.0278	.0312	.0348	.0385	.0425	.0466	.0508	.0551	.0595
8	.0095	.0111	.0129	.0148	.0169	.0191	.0215	.0241	.0269	.0298
9	.0033	.0040	.0047	.0056	.0066	.0076	.0089	.0102	.0116	.0132
10	.0010	.0013	.0016	.0019	.0023	.0028	.0033	.0039	.0045	.0053
11	.0003	.0004	.0005	.0006	.0007	.0009	.0011	.0013	.0016	.0019
12	.0001	.0001	.0001	.0002	.0002	.0003	.0003	.0004	.0005	.0006
13	.0000	.0000	.0000	.0000	.0001	.0001	.0001	.0001	.0002	.0002
14	.0000	.0000	.0000	.0000	.0000	.0000	.0000	.0000	.0000	.0001

For extensive tables of $e^{-m}m^x/x!$ see Molina, E. C.: *Poisson's Exponential Binomial Limit*, D. Van Nostrand, New York, 1942.

POISSON DISTRIBUTION FUNCTION $e^{-m}m^x/x!$

m

x	4.1	4.2	4.3	4.4	4.5	4.6	4.7	4.8	4.9	5.0
0	.0166	.0150	.0136	.0123	.0111	.0101	.0091	.0082	.0074	.0067
1	.0679	.0630	.0583	.0540	.0500	.0462	.0427	.0395	.0365	.0337
2	.1393	.1323	.1254	.1188	.1125	.1063	.1005	.0948	.0894	.0842
3	.1904	.1852	.1798	.1743	.1687	.1631	.1574	.1517	.1460	.1404
4	.1951	.1944	.1933	.1917	.1898	.1875	.1849	.1820	.1789	.1755
5	.1600	.1633	.1662	.1687	.1708	.1725	.1738	.1747	.1753	.1755
6	.1093	.1143	.1191	.1237	.1281	.1323	.1362	.1398	.1432	.1462
7	.0640	.0686	.0732	.0778	.0824	.0869	.0914	.0959	.1002	.1044
8	.0328	.0360	.0393	.0428	.0463	.0500	.0537	.0575	.0614	.0653
9	.0150	.0168	.0188	.0209	.0232	.0255	.0280	.0307	.0334	.0363
10	.0061	.0071	.0081	.0092	.0104	.0118	.0132	.0147	.0164	.0181
11	.0023	.0027	.0032	.0037	.0043	.0049	.0056	.0064	.0073	.0082
12	.0008	.0009	.0011	.0014	.0016	.0019	.0022	.0026	.0030	.0034
13	.0002	.0003	.0004	.0005	.0006	.0007	.0008	.0009	.0011	.0013
14	.0001	.0001	.0001	.0001	.0002	.0002	.0003	.0003	.0004	.0005
15	.0000	.0000	.0000	.0000	.0001	.0001	.0001	.0001	.0001	.0002

m

x	5.1	5.2	5.3	5.4	5.5	5.6	5.7	5.8	5.9	6.0
0	.0061	.0055	.0050	.0045	.0041	.0037	.0033	.0030	.0027	.0025
1	.0311	.0287	.0265	.0244	.0225	.0207	.0191	.0176	.0162	.0149
2	.0793	.0746	.0701	.0659	.0618	.0580	.0544	.0509	.0477	.0446
3	.1348	.1293	.1239	.1185	.1133	.1082	.1033	.0985	.0938	.0892
4	.1719	.1681	.1641	.1600	.1558	.1515	.1472	.1428	.1383	.1339
5	.1753	.1748	.1740	.1728	.1714	.1697	.1678	.1656	.1632	.1606
6	.1490	.1515	.1537	.1555	.1571	.1584	.1594	.1601	.1605	.1606
7	.1086	.1125	.1163	.1200	.1234	.1267	.1298	.1326	.1353	.1377
8	.0692	.0731	.0771	.0810	.0849	.0887	.0925	.0962	.0998	.1033
9	.0392	.0423	.0454	.0486	.0519	.0552	.0586	.0620	.0654	.0688
10	.0200	.0220	.0241	.0262	.0285	.0309	.0334	.0359	.0386	.0413
11	.0093	.0104	.0116	.0129	.0143	.0157	.0173	.0190	.0207	.0225
12	.0039	.0045	.0051	.0058	.0065	.0073	.0082	.0092	.0102	.0113
13	.0015	.0018	.0021	.0024	.0028	.0032	.0036	.0041	.0046	.0052
14	.0006	.0007	.0008	.0009	.0011	.0013	.0015	.0017	.0019	.0022
15	.0002	.0002	.0003	.0003	.0004	.0005	.0006	.0007	.0008	.0009
16	.0001	.0001	.0001	.0001	.0001	.0002	.0002	.0002	.0003	.0003
17	.0000	.0000	.0000	.0000	.0000	.0001	.0001	.0001	.0001	.0001

m

x	6.1	6.2	6.3	6.4	6.5	6.6	6.7	6 8	6.9	7.0
0	.0022	.0020	.0018	.0017	.0015	.0014	.0012	.0011	.0010	.0009
1	.0137	.0126	.0116	.0106	.0098	.0090	.0082	.0076	.0070	.0064
2	.0417	.0390	.0364	.0340	.0318	.0296	.0276	.0258	.0240	.0223
3	.0848	.0806	.0765	.0726	.0688	.0652	.0617	.0584	.0552	.0521
4	.1294	.1249	.1205	.1162	.1118	.1076	.1034	.0992	.0952	.0912
5	.1579	.1549	.1519	.1487	.1454	.1420	.1385	.1349	.1314	.1277
6	.1605	.1601	.1595	.1586	.1575	.1562	.1546	.1529	.1511	.1490
7	.1399	.1418	.1435	.1450	.1462	.1472	.1480	.1486	.1489	.1490
8	.1066	.1099	.1130	.1160	.1188	.1215	.1240	.1263	.1284	.1304
9	.0723	.0757	.0791	.0825	.0858	.0891	.0923	.0954	.0985	.1014
10	.0441	.0469	.0498	.0528	.0558	.0588	.0618	.0649	.0679	.0710
11	.0245	.0265	.0285	.0307	.0330	.0353	.0377	.0401	.0426	.0452
12	.0124	.0137	.0150	.0164	.0179	.0194	.0210	.0227	.0245	.0264
13	.0058	.0065	.0073	.0081	.0089	.0098	.0108	.0119	.0130	.0142
14	.0025	.0029	.0033	.0037	.0041	.0046	.0052	.0058	.0064	.0071
15	.0010	.0012	.0014	.0016	.0018	.0020	.0023	.0026	.0029	.0033
16	.0004	.0005	.0005	.0006	.0007	.0008	.0010	.0011	.0013	.0014
17	.0001	.0002	.0002	.0002	.0003	.0003	.0004	.0004	.0005	.0006
18	.0000	.0001	.0001	.0001	.0001	.0001	.0001	.0002	.0002	.0002
19	.0000	.0000	.0000	.0000	.0000	.0000	.0000	.0001	.0001	.0001

Table VII continued 361

POISSON DISTRIBUTION FUNCTION $e^{-m}m^x/x!$

					m					
x	7.1	7.2	7.3	7.4	7.5	7.6	7.7	7.8	7.9	8.0
0	.0008	.0007	.0007	.0006	.0006	.0005	.0005	.0004	.0004	.0003
1	.0059	.0054	.0049	.0045	.0041	.0038	.0035	.0032	.0029	.0027
2	.0208	.0194	.0180	.0167	.0156	.0145	.0134	.0125	.0116	.0107
3	.0492	.0464	.0438	.0413	.0389	.0366	.0345	.0324	.0305	.0286
4	.0874	.0836	.0799	.0764	.0729	.0696	.0663	.0632	.0602	.0573
5	.1241	.1204	.1167	.1130	.1094	.1057	.1021	.0986	.0951	.0916
6	.1468	.1445	.1420	.1394	.1367	.1339	.1311	.1282	.1252	.1221
7	.1489	.1486	.1481	.1474	.1465	.1454	.1442	.1428	.1413	.1396
8	.1321	.1337	.1351	.1363	.1373	.1382	.1388	.1392	.1395	.1396
9	.1042	.1070	.1096	.1121	.1144	.1167	.1187	.1207	.1224	.1241
10	.0740	.0770	.0800	.0829	.0858	.0887	.0914	.0941	.0967	.0993
11	.0478	.0504	.0531	.0558	.0585	.0613	.0640	.0667	.0695	.0722
12	.0283	.0303	.0323	.0344	.0366	.0388	.0411	.0434	.0457	.0481
13	.0154	.0168	.0181	.0196	.0211	.0227	.0243	.0260	.0278	.0296
14	.0078	.0086	.0095	.0104	.0113	.0123	.0134	.0145	.0157	.0169
15	.0037	.0041	.0046	.0051	.0057	.0062	.0069	.0075	.0083	.0090
16	.0016	.0019	.0021	.0024	.0026	.0030	.0033	.0037	.0041	.0045
17	.0007	.0008	.0009	.0010	.0012	.0013	.0015	.0017	.0019	.0021
18	.0003	.0003	.0004	.0004	.0005	.0006	.0006	.0007	.0008	.0009
19	.0001	.0001	.0001	.0002	.0002	.0002	.0003	.0003	.0003	.0004
20	.0000	.0000	.0001	.0001	.0001	.0001	.0001	.0001	.0001	.0002
21	.0000	.0000	.0000	.0000	.0000	.0000	.0000	.0000	.0001	.0001

					m					
x	8.1	8.2	8.3	8.4	8.5	8.6	8.7	8.8	8.9	9.0
0	.0003	.0003	.0002	.0002	.0002	.0002	.0002	.0002	.0001	.0001
1	.0025	.0023	.0021	.0019	.0017	.0016	.0014	.0013	.0012	.0011
2	.0100	.0092	.0086	.0079	.0074	.0068	.0063	.0058	.0054	.0050
3	.0269	.0252	.0237	.0222	.0208	.0195	.0183	.0171	.0160	.0150
4	.0544	.0517	.0491	.0466	.0443	.0420	.0398	.0377	.0357	.0337
5	.0882	.0849	.0816	.0784	.0752	.0722	.0692	.0663	.0635	.0607
6	.1191	.1160	.1128	.1097	.1066	.1034	.1003	.0972	.0941	.0911
7	.1378	.1358	.1338	.1317	.1294	.1271	.1247	.1222	.1197	.1171
8	.1395	.1392	.1388	.1382	.1375	.1366	.1356	.1344	.1332	.1318
9	.1256	.1269	.1280	.1290	.1299	.1306	.1311	.1315	.1317	.1318
10	.1017	.1040	.1063	.1084	.1104	.1123	.1140	.1157	.1172	.1186
11	.0749	.0776	.0802	.0828	.0853	.0878	.0902	.0925	.0948	.0970
12	.0505	.0530	.0555	.0579	.0604	.0629	.0654	.0679	.0703	.0728
13	.0315	.0334	.0354	.0374	.0395	.0416	.0438	.0459	.0481	.0504
14	.0182	.0196	.0210	.0225	.0240	.0256	.0272	.0289	.0306	.0324
15	.0098	.0107	.0116	.0126	.0136	.0147	.0158	.0169	.0182	.0194
16	.0050	.0055	.0060	.0066	.0072	.0079	.0086	.0093	.0101	.0109
17	.0024	.0026	.0029	.0033	.0036	.0040	.0044	.0048	.0053	.0058
18	.0011	.0012	.0014	.0015	.0017	.0019	.0021	.0024	.0026	.0029
19	.0005	.0005	.0006	.0007	.0008	.0009	.0010	.0011	.0012	.0014
20	.0002	.0002	.0002	.0003	.0003	.0004	.0004	.0005	.0005	.0006
21	.0001	.0001	.0001	.0001	.0001	.0002	.0002	.0002	.0002	.0003
22	.0000	.0000	.0000	.0000	.0001	.0001	.0001	.0001	.0001	.0001

					m					
x	9.1	9.2	9.3	9.4	9.5	9.6	9.7	9.8	9.9	10
0	.0001	.0001	.0001	.0001	.0001	.0001	.0001	.0001	.0001	.0000
1	.0010	.0009	.0009	.0008	.0007	.0007	.0006	.0005	.0005	.0005
2	.0046	.0043	.0040	.0037	.0034	.0031	.0029	.0027	.0025	.0023
3	.0140	.0131	.0123	.0115	.0107	.0100	.0093	.0087	.0081	.0076
4	.0319	.0302	.0285	.0269	.0254	.0240	.0226	.0213	.0201	.0189
5	.0581	.0555	.0530	.0506	.0483	.0460	.0439	.0418	.0398	.0378
6	.0881	.0851	.0822	.0793	.0764	.0736	.0709	.0682	.0656	.0631
7	.1145	.1118	.1091	.1064	.1037	.1010	.0982	.0955	.0928	.0901
8	.1302	.1286	.1269	.1251	.1232	.1212	.1191	.1170	.1148	.1126
9	.1317	.1315	.1311	.1306	.1300	.1293	.1284	.1274	.1263	.1251

POISSON DISTRIBUTION FUNCTION $e^{-m}m^x/x!$

x	9.1	9.2	9.3	9.4	9.5	9.6	9.7	9.8	9.9	10
10	.1198	.1210	.1219	.1228	.1235	.1241	.1245	.1249	.1250	.1251
11	.0991	.1012	.1031	.1049	.1067	.1083	.1098	.1112	.1125	.1137
12	.0752	.0776	.0799	.0822	.0844	.0866	.0888	.0908	.0928	.0948
13	.0526	.0549	.0572	.0594	.0617	.0640	.0662	.0685	.0707	.0729
14	.0342	.0361	.0380	.0399	.0419	.0439	.0459	.0479	.0500	.0521
15	.0208	.0221	.0235	.0250	.0265	.0281	.0297	.0313	.0330	.0347
16	.0118	.0127	.0137	.0147	.0157	.0168	.0180	.0192	.0204	.0217
17	.0063	.0069	.0075	.0081	.0088	.0095	.0103	.0111	.0119	.0128
18	.0032	.0035	.0039	.0042	.0046	.0051	.0055	.0060	.0065	.0071
19	.0015	.0017	.0019	.0021	.0023	.0026	.0028	.0031	.0034	.0037
20	.0007	.0008	.0009	.0010	.0011	.0012	.0014	.0015	.0017	.0019
21	.0003	.0003	.0004	.0004	.0005	.0006	.0006	.0007	.0008	.0009
22	.0001	.0001	.0002	.0002	.0002	.0002	.0003	.0003	.0004	.0004
23	.0000	.0001	.0001	.0001	.0001	.0001	.0001	.0001	.0002	.0002
24	.0000	.0000	.0000	.0000	.0000	.0000	.0000	.0001	.0001	.0001

x	11	12	13	14	15	16	17	18	19	20
0	.0000	.0000	.0000	.0000	.0000	.0000	.0000	.0000	.0000	
1	.0002	.0001	.0000	.0000	.0000	.0000	.0000	.0000	.0000	
2	.0010	.0004	.0002	.0001	.0000	.0000	.0000	.0000	.0000	
3	.0037	.0018	.0008	.0004	.0002	.0001	.0000	.0000	.0000	.0000
4	.0102	.0053	.0027	.0013	.0006	.0003	.0001	.0001	.0000	.0000
5	.0224	.0127	.0070	.0037	.0019	.0010	.0005	.0002	.0001	.0001
6	.0411	.0255	.0152	.0087	.0048	.0026	.0014	.0007	.0004	.0002
7	.0646	.0437	.0281	.0174	.0104	.0060	.0034	.0018	.0010	.0005
8	.0888	.0655	.0457	.0304	.0194	.0120	.0072	.0042	.0024	.0013
9	.1085	.0874	.0661	.0473	.0324	.0213	.0135	.0083	.0050	.0029
10	.1194	.1048	.0859	.0663	.0486	.0341	.0230	.0150	.0095	.0058
11	.1194	.1144	.1015	.0844	.0663	.0496	.0355	.0245	.0164	.0106
12	.1094	.1144	.1099	.0984	.0829	.0661	.0504	.0368	.0259	.0176
13	.0926	.1056	.1099	.1060	.0956	.0814	.0658	.0509	.0378	.0271
14	.0728	.0905	.1021	.1060	.1024	.0930	.0800	.0655	.0514	.0387
15	.0534	.0724	.0885	.0989	.1024	.0992	.0906	.0786	.0650	.0516
16	.0367	.0543	.0719	.0866	.0960	.0992	.0963	.0884	.0772	.0646
17	.0237	.0383	.0550	.0713	.0847	.0934	.0963	.0936	.0863	.0760
18	.0145	.0256	.0397	.0554	.0706	.0830	.0909	.0936	.0911	.0844
19	.0084	.0161	.0272	.0409	.0557	.0699	.0814	.0887	.0911	.0888
20	.0046	.0097	.0177	.0286	.0418	.0559	.0692	.0798	.0866	.0888
21	.0024	.0055	.0109	.0191	.0299	.0426	.0560	.0684	.0783	.0846
22	.0012	.0030	.0065	.0121	.0204	.0310	.0433	.0560	.0676	.0769
23	.0006	.0016	.0037	.0074	.0133	.0216	.0320	.0438	.0559	.0669
24	.0003	.0008	.0020	.0043	.0083	.0144	.0226	.0328	.0442	.0557
25	.0001	.0004	.0010	.0024	.0050	.0092	.0154	.0237	.0336	.0446
26	.0000	.0002	.0005	.0013	.0029	.0057	.0101	.0164	.0246	.0343
27	.0000	.0001	.0002	.0007	.0016	.0034	.0063	.0109	.0173	.0254
28	.0000	.0000	.0001	.0003	.0009	.0019	.0038	.0070	.0117	.0181
29	.0000	.0000	.0001	.0002	.0004	.0011	.0023	.0044	.0077	.0125
30	.0000	.0000	.0000	.0001	.0002	.0006	.0013	.0026	.0049	.0083
31	.0000	.0000	.0000	.0000	.0001	.0003	.0007	.0015	.0030	.0054
32	.0000	.0000	.0000	.0000	.0001	.0001	.0004	.0009	.0018	.0034
33	.0000	.0000	.0000	.0000	.0000	.0001	.0002	.0005	.0010	.0020
34	.0000	.0000	.0000	.0000	.0000	.0000	.0001	.0002	.0006	.0012
35	.0000	.0000	.0000	.0000	.0000	.0000	.0000	.0001	.0003	.0007
36	.0000	.0000	.0000	.0000	.0000	.0000	.0000	.0001	.0002	.0004
37	.0000	.0000	.0000	.0000	.0000	.0000	.0000	.0000	.0001	.0002
38	.0000	.0000	.0000	.0000	.0000	.0000	.0000	.0000	.0000	.0001
39	.0000	.0000	.0000	.0000	.0000	.0000	.0000	.0000	.0000	.0001

Table VIII 363

SUMMED POISSON DISTRIBUTION FUNCTION

$$\sum_{x=x'}^{x=\infty} e^{-m}m^x/x!$$

(See Chapter IX)

Entries in the table are values of $\sum_{x=x'}^{x=\infty} e^{-m}m^x/x!$ for the indicated values of x' and m.

x'					m					
	0.1	0.2	0.3	0.4	0.5	0.6	0.7	0.8	0.9	1.0
0	1.0000	1.0000	1.0000	1.0000	1.0000	1.0000	1.0000	1.0000	1.0000	1.0000
1	.0952	.1813	.2592	.3297	.3935	.4512	.5034	.5507	.5934	.6321
2	.0047	.0175	.0369	.0616	.0902	.1219	.1558	.1912	.2275	.2642
3	.0002	.0011	.0036	.0079	.0144	.0231	.0341	.0474	.0629	.0803
4	.0000	.0001	.0003	.0008	.0018	.0034	.0058	.0091	.0135	.0190
5	.0000	.0000	.0000	.0001	.0002	.0004	.0008	.0014	.0023	.0037
6	.0000	.0000	.0000	.0000	.0000	.0000	.0001	.0002	.0003	.0006
7	.0000	.0000	.0000	.0000	.0000	.0000	.0000	.0000	.0000	.0001

x'					m					
	1.1	1.2	1.3	1.4	1.5	1.6	1.7	1.8	1.9	2.0
0	1.0000	1.0000	1.0000	1.0000	1.0000	1.0000	1.0000	1.0000	1.0000	1.0000
1	.6671	.6988	.7275	.7534	.7769	.7981	.8173	.8347	.8504	.8647
2	.3010	.3374	.3732	.4082	.4422	.4751	.5068	.5372	.5663	.5940
3	.0996	.1205	.1429	.1665	.1912	.2166	.2428	.2694	.2963	.3233
4	.0257	.0338	.0431	.0537	.0656	.0788	.0932	.1087	.1253	.1429
5	.0054	.0077	.0107	.0143	.0186	.0237	.0296	.0364	.0441	.0527
6	.0010	.0015	.0022	.0032	.0045	.0060	.0080	.0104	.0132	.0166
7	.0001	.0003	.0004	.0006	.0009	.0013	.0019	.0026	.0034	.0045
8	.0000	.0000	.0001	.0001	.0002	.0003	.0004	.0006	.0008	.0011
9	.0000	.0000	.0000	.0000	.0000	.0000	.0001	.0001	.0002	.0002

x'					m					
	2.1	2.2	2.3	2.4	2.5	2.6	2.7	2.8	2.9	3.0
0	1.0000	1.0000	1.0000	1.0000	1.0000	1.0000	1.0000	1.0000	1.0000	1.0000
1	.8775	.8892	.8997	.9093	.9179	.9257	.9328	.9392	.9450	.9502
2	.6204	.6454	.6691	.6916	.7127	.7326	.7513	.7689	.7854	.8009
3	.3504	.3773	.4040	.4303	.4562	.4816	.5064	.5305	.5540	.5768
4	.1614	.1806	.2007	.2213	.2424	.2640	.2859	.3081	.3304	.3528
5	.0621	.0725	.0838	.0959	.1088	.1226	.1371	.1523	.1682	.1847
6	.0204	.0249	.0300	.0357	.0420	.0490	.0567	.0651	.0742	.0839
7	.0059	.0075	.0094	.0116	.0142	.0172	.0206	.0244	.0287	.0335
8	.0015	.0020	.0026	.0033	.0042	.0053	.0066	.0081	.0099	.0119
9	.0003	.0005	.0006	.0009	.0011	.0015	.0019	.0024	.0031	.0038
10	.0001	.0001	.0001	.0002	.0003	.0004	.0005	.0007	.0009	.0011
11	.0000	.0000	.0000	.0000	.0001	.0001	.0001	.0002	.0002	.0003
12	.0000	.0000	.0000	.0000	.0000	.0000	.0000	.0000	.0001	.0001

x'					m					
	3.1	3.2	3.3	3.4	3.5	3.6	3.7	3.8	3.9	4.0
0	1.0000	1.0000	1.0000	1.0000	1.0000	1.0000	1.0000	1.0000	1.0000	1.0000
1	.9550	.9592	.9631	.9666	.9698	.9727	.9753	.9776	.9798	.9817
2	.8153	.8288	.8414	.8532	.8641	.8743	.8838	.8926	.9008	.9084
3	.5988	.6201	.6406	.6603	.6792	.6973	.7146	.7311	.7469	.7619
4	.3752	.3975	.4197	.4416	.4634	.4848	.5058	.5265	.5468	.5665
5	.2018	.2194	.2374	.2558	.2746	.2936	.3128	.3322	.3516	.3712
6	.0943	.1054	.1171	.1295	.1424	.1559	.1699	.1844	.1994	.2149
7	.0388	.0446	.0510	.0579	.0653	.0733	.0818	.0909	.1005	.1107
8	.0142	.0168	.0198	.0231	.0267	.0308	.0352	.0401	.0454	.0511
9	.0047	.0057	.0069	.0083	.0099	.0117	.0137	.0160	.0185	.0214
10	.0014	.0018	.0022	.0027	.0033	.0040	.0048	.0058	.0069	.0081
11	.0004	.0005	.0006	.0008	.0010	.0013	.0016	.0019	.0023	.0028
12	.0001	.0001	.0002	.0002	.0003	.0004	.0005	.0006	.0007	.0009
13	.0000	.0000	.0000	.0001	.0001	.0001	.0001	.0002	.0002	.0003
14	.0000	.0000	.0000	.0000	.0000	.0000	.0000	.0000	.0001	.0001

For extensive tables of $\sum_{x=x'}^{x=\infty} e^{-m}m^x/x!$ see Molina, E. C.: *Poisson's Exponential Binomial Limit*, D. Van Nostrand, New York, 1942.

$$\sum_{x=x'}^{x=\infty} e^{-m}m^x/x!$$

Table VIII continued

x'	4.1	4.2	4.3	4.4	4.5	4.6	4.7	4.8	4.9	5.0
0	1.0000	1.0000	1.0000	1.0000	1.0000	1.0000	1.0000	1.0000	1.0000	1.0000
1	.9834	.9850	.9864	.9877	.9889	.9899	.9909	.9918	.9926	.9933
2	.9155	.9220	.9281	.9337	.9389	.9437	.9482	.9523	.9561	.9596
3	.7762	.7898	.8026	.8149	.8264	.8374	.8477	.8575	.8667	.8753
4	.5858	.6046	.6228	.6406	.6577	.6743	.6903	.7058	.7207	.7350
5	.3907	.4102	.4296	.4488	.4679	.4868	.5054	.5237	.5418	.5595
6	.2307	.2469	.2633	.2801	.2971	.3142	.3316	.3490	.3665	.3840
7	.1214	.1325	.1442	.1564	.1689	.1820	.1954	.2092	.2233	.2378
8	.0573	.0639	.0710	.0786	.0866	.0951	.1040	.1133	.1231	.1334
9	.0245	.0279	.0317	.0358	.0403	.0451	.0503	.0558	.0618	.0681
10	.0095	.0111	.0129	.0149	.0171	.0195	.0222	.0251	.0283	.0318
11	.0034	.0041	.0048	.0057	.0067	.0078	.0090	.0104	.0120	.0137
12	.0011	.0014	.0017	.0020	.0024	.0029	.0034	.0040	.0047	.0055
13	.0003	.0004	.0005	.0007	.0008	.0010	.0012	.0014	.0017	.0020
14	.0001	.0001	.0002	.0002	.0003	.0003	.0004	.0005	.0006	.0007
15	.0000	.0000	.0000	.0001	.0001	.0001	.0001	.0001	.0002	.0002
16	.0000	.0000	.0000	.0000	.0000	.0000	.0000	.0000	.0001	.0001

x'	5.1	5.2	5.3	5.4	5.5	5.6	5.7	5.8	5.9	6.0
0	1.0000	1.0000	1.0000	1.0000	1.0000	1.0000	1.0000	1.0000	1.0000	1.0000
1	.9939	.9945	.9950	.9955	.9959	.9963	.9967	.9970	.9973	.9975
2	.9628	.9658	.9686	.9711	.9734	.9756	.9776	.9794	.9811	.9826
3	.8835	.8912	.8984	.9052	.9116	.9176	.9232	.9285	.9334	.9380
4	.7487	.7619	.7746	.7867	.7983	.8094	.8200	.8300	.8396	.8488
5	.5769	.5939	.6105	.6267	.6425	.6579	.6728	.6873	.7013	.7149
6	.4016	.4191	.4365	.4539	.4711	.4881	.5050	.5217	.5381	.5543
7	.2526	.2676	.2829	.2983	.3140	.3297	.3456	.3616	.3776	.3937
8	.1440	.1551	.1665	.1783	.1905	.2030	.2159	.2290	.2424	.2560
9	.0748	.0819	.0894	.0974	.1056	.1143	.1234	.1328	.1426	.1528
10	.0356	.0397	.0441	.0488	.0538	.0591	.0648	.0708	.0772	.0839
11	.0156	.0177	.0200	.0225	.0253	.0282	.0314	.0349	.0386	.0426
12	.0063	.0073	.0084	.0096	.0110	.0125	.0141	.0160	.0179	.0201
13	.0024	.0028	.0033	.0038	.0045	.0051	.0059	.0068	.0078	.0088
14	.0008	.0010	.0012	.0014	.0017	.0020	.0023	.0027	.0031	.0036
15	.0003	.0003	.0004	.0005	.0006	.0007	.0009	.0010	.0012	.0014
16	.0001	.0001	.0001	.0002	.0002	.0002	.0003	.0004	.0004	.0005
17	.0000	.0000	.0000	.0001	.0001	.0001	.0001	.0001	.0001	.0002
18	.0000	.0000	.0000	.0000	.0000	.0000	.0000	.0000	.0000	.0001

x'	6.1	6.2	6.3	6.4	6.5	6.6	6.7	6.8	6.9	7.0
0	1.0000	1.0000	1.0000	1.0000	1.0000	1.0000	1.0000	1.0000	1.0000	1.0000
1	.9978	.9980	.9982	.9983	.9985	.9986	.9988	.9989	.9990	.9991
2	.9841	.9854	.9866	.9877	.9887	.9897	.9905	.9913	.9920	.9927
3	.9423	.9464	.9502	.9537	.9570	.9600	.9629	.9656	.9680	.9704
4	.8575	.8658	.8736	.8811	.8882	.8948	.9012	.9072	.9129	.9182
5	.7281	.7408	.7531	.7649	.7763	.7873	.7978	.8080	.8177	.8270
6	.5702	.5859	.6012	.6163	.6310	.6453	.6594	.6730	.6863	.6993
7	.4098	.4258	.4418	.4577	.4735	.4892	.5047	.5201	.5353	.5503
8	.2699	.2840	.2983	.3127	.3272	.3419	.3567	.3715	.3864	.4013
9	.1633	.1741	.1852	.1967	.2084	.2204	.2327	.2452	.2580	.2709
10	.0910	.0984	.1061	.1142	.1226	.1314	.1404	.1498	.1505	.1695
11	.0469	.0514	.0563	.0614	.0668	.0726	.0786	.0849	.0916	.0985
12	.0224	.0250	.0277	.0307	.0339	.0373	.0409	.0448	.0490	.0534
13	.0100	.0113	.0127	.0143	.0160	.0179	.0199	.0221	.0245	.0270
14	.0042	.0048	.0055	.0063	.0071	.0080	.0091	.0102	.0115	.0128
15	.0016	.0019	.0022	.0026	.0030	.0034	.0039	.0044	.0050	.0057
16	.0006	.0007	.0008	.0010	.0012	.0014	.0016	.0018	.0021	.0024
17	.0002	.0003	.0003	.0004	.0004	.0005	.0006	.0007	.0008	.0010
18	.0001	.0001	.0001	.0001	.0002	.0002	.0002	.0003	.0003	.0004
19	.0000	.0000	.0000	.0000	.0001	.0001	.0001	.0001	.0001	.0001

Table VIII continued
$$\sum_{x=x'}^{x=\infty} e^{-m} m^x / x!$$

x'	7.1	7.2	7.3	7.4	7.5	7.6	7.7	7.8	7.9	8.0
					m					
0	1.0000	1.0000	1.0000	1.0000	1.0000	1.0000	1.0000	1.0000	1.0000	1.0000
1	.9992	.9993	.9993	.9994	.9994	.9995	.9995	.9996	.9996	.9997
2	.9933	.9939	.9944	.9949	.9953	.9957	.9961	.9964	.9967	.9970
3	.9725	.9745	.9764	.9781	.9797	.9812	.9826	.9839	.9851	.9862
4	.9233	.9281	.9326	.9368	.9409	.9446	.9482	.9515	.9547	.9576
5	.8359	.8445	.8527	.8605	.8679	.8751	.8819	.8883	.8945	.9004
6	.7119	.7241	.7360	.7474	.7586	.7693	.7797	.7897	.7994	.8088
7	.5651	.5796	.5940	.6080	.6218	.6354	.6486	.6616	.6743	.6866
8	.4162	.4311	.4459	.4607	.4754	.4900	.5044	.5188	.5330	.5470
9	.2840	.2973	.3108	.3243	.3380	.3518	.3657	.3796	.3935	.4075
10	.1798	.1904	.2012	.2123	.2236	.2351	.2469	.2589	.2710	.2834
11	.1058	.1133	.1212	.1293	.1378	.1465	.1555	.1648	.1743	.1841
12	.0580	.0629	.0681	.0735	.0792	.0852	.0915	.0980	.1048	.1119
13	.0297	.0327	.0358	.0391	.0427	.0464	.0504	.0546	.0591	.0638
14	.0143	.0159	.0176	.0195	.0216	.0238	.0261	.0286	.0313	.0342
15	.0065	.0073	.0082	.0092	.0103	.0114	.0127	.0141	.0156	.0173
16	.0028	.0031	.0036	.0041	.0046	.0052	.0059	.0066	.0074	.0082
17	.0011	.0013	.0015	.0017	.0020	.0022	.0026	.0029	.0033	.0037
18	.0004	.0005	.0006	.0007	.0008	.0009	.0011	.0012	.0014	.0016
19	.0002	.0002	.0002	.0003	.0003	.0004	.0004	.0005	.0006	.0006
20	.0001	.0001	.0001	.0001	.0001	.0001	.0002	.0002	.0002	.0003
21	.0000	.0000	.0000	.0000	.0000	.0000	.0001	.0001	.0001	.0001

x'	8.1	8.2	8.3	8.4	8.5	8.6	8.7	8.8	8.9	9.0
					m					
0	1.0000	1.0000	1.0000	1.0000	1.0000	1.0000	1.0000	1.0000	1.0000	1.0000
1	.9997	.9997	.9998	.9998	.9998	.9998	.9998	.9998	.9999	.9999
2	.9972	.9975	.9977	.9979	.9981	.9982	.9984	.9985	.9987	.9988
3	.9873	.9882	.9891	.9900	.9907	.9914	.9921	.9927	.9932	.9938
4	.9604	.9630	.9654	.9677	.9699	.9719	.9738	.9756	.9772	.9788
5	.9060	.9113	.9163	.9211	.9256	.9299	.9340	.9379	.9416	.9450
6	.8178	.8264	.8347	.8427	.8504	.8578	.8648	.8716	.8781	.8843
7	.6987	.7104	.7219	.7330	.7438	.7543	.7645	.7744	.7840	.7932
8	.5609	.5746	.5881	.6013	.6144	.6272	.6398	.6522	.6643	.6761
9	.4214	.4353	.4493	.4631	.4769	.4906	.5042	.5177	.5311	.5443
10	.2959	.3085	.3212	.3341	.3470	.3600	.3731	.3863	.3994	.4126
11	.1942	.2045	.2150	.2257	.2366	.2478	.2591	.2706	.2822	.2940
12	.1193	.1269	.1348	.1429	.1513	.1600	.1689	.1780	.1874	.1970
13	.0687	.0739	.0793	.0850	.0909	.0971	.1035	.1102	.1171	.1242
14	.0372	.0405	.0439	.0476	.0514	.0555	.0597	.0642	.0689	.0739
15	.0190	.0209	.0229	.0251	.0274	.0299	.0325	.0353	.0383	.0415
16	.0092	.0102	.0113	.0125	.0138	.0152	.0168	.0184	.0202	.0220
17	.0042	.0047	.0053	.0059	.0066	.0074	.0082	.0091	.0101	.0111
18	.0018	.0021	.0023	.0027	.0030	.0034	.0038	.0043	.0048	.0053
19	.0008	.0009	.0010	.0011	.0013	.0015	.0017	.0019	.0022	.0024
20	.0003	.0003	.0004	.0005	.0005	.0006	.0007	.0008	.0009	.0011
21	.0001	.0001	.0002	.0002	.0002	.0002	.0003	.0003	.0004	.0004
22	.0000	.0000	.0001	.0001	.0001	.0001	.0001	.0001	.0002	.0002
23	.0000	.0000	.0000	.0000	.0000	.0000	.0000	.0000	.0001	.0001

x'	9.1	9.2	9.3	9.4	9.5	9.6	9.7	9.8	9.9	10
					m					
0	1.0000	1.0000	1.0000	1.0000	1.0000	1.0000	1.0000	1.0000	1.0000	1.0000
1	.9999	.9999	.9999	.9999	.9999	.9999	.9999	.9999	1.0000	1.0000
2	.9989	.9990	.9991	.9991	.9992	.9993	.9993	.9994	.9995	.9995
3	.9942	.9947	.9951	.9955	.9958	.9962	.9965	.9967	.9970	.9972
4	.9802	.9816	.9828	.9840	.9851	.9862	.9871	.9880	.9889	.9897
5	.9483	.9514	.9544	.9571	.9597	.9622	.9645	.9667	.9688	.9707
6	.8902	.8959	.9014	.9065	.9115	.9162	.9207	.9250	.9290	.9329
7	.8022	.8108	.8192	.8273	.8351	.8426	.8498	.8567	.8634	.8699
8	.6877	.6990	.7101	.7208	.7313	.7416	.7515	.7612	.7706	.7798
9	.5574	.5704	.5832	.5958	.6082	.6204	.6324	.6442	.6558	.6672

SUMMED POISSON DISTRIBUTION FUNCTION

$$\sum_{x=x'}^{x=\infty} e^{-m}m^x/x!$$

					m					
x'	9.1	9.2	9.3	9.4	9.5	9.6	9.7	9.8	9.9	10
10	.4258	.4389	.4521	.4651	.4782	.4911	.5040	.5168	.5295	.5421
11	.3059	.3180	.3301	.3424	.3547	.3671	.3795	.3920	.4045	.4170
12	.2068	.2168	.2270	.2374	.2480	.2588	.2697	.2807	.2919	.3032
13	.1316	.1393	.1471	.1552	.1636	.1721	.1809	.1899	.1991	.2084
14	.0790	.0844	.0900	.0958	.1019	.1081	.1147	.1214	.1284	.1355
15	.0448	.0483	.0520	.0559	.0600	.0643	.0688	.0735	.0784	.0835
16	.0240	.0262	.0285	.0309	.0335	.0362	.0391	.0421	.0454	.0487
17	.0122	.0135	.0148	.0162	.0177	.0194	.0211	.0230	.0249	.0270
18	.0059	.0066	.0073	.0081	.0089	.0098	.0108	.0119	.0130	.0143
19	.0027	.0031	.0034	.0038	.0043	.0048	.0053	.0059	.0065	.0072
20	.0012	.0014	.0015	.0017	.0020	.0022	.0025	.0028	.0031	.0035
21	.0005	.0006	.0007	.0008	.0009	.0010	.0011	.0013	.0014	.0016
22	.0002	.0002	.0003	.0003	.0004	.0004	.0005	.0005	.0006	.0007
23	.0001	.0001	.0001	.0001	.0001	.0002	.0002	.0002	.0003	.0003
24	.0000	.0000	.0000	.0000	.0001	.0001	.0001	.0001	.0001	.0001

					m					
x'	11	12	13	14	15	16	17	18	19	20
0	1.0000	1.0000	1.0000	1.0000	1.0000	1.0000	1.0000	1.0000	1.0000	1.0000
1	1.0000	1.0000	1.0000	1.0000	1.0000	1.0000	1.0000	1.0000	1.0000	1.0000
2	.9998	.9999	1.0000	1.0000	1.0000	1.0000	1.0000	1.0000	1.0000	1.0000
3	.9988	.9995	.9998	.9999	1.0000	1.0000	1.0000	1.0000	1.0000	1.0000
4	.9951	.9977	.9990	.9995	.9998	.9999	1.0000	1.0000	1.0000	1.0000
5	.9849	.9924	.9963	.9982	.9991	.9996	.9998	.9999	1.0000	1.0000
6	.9625	.9797	.9893	.9945	.9972	.9986	.9993	.9997	.9998	.9999
7	.9214	.9542	.9741	.9858	.9924	.9960	.9979	.9990	.9995	.9997
8	.8568	.9105	.9460	.9684	.9820	.9900	.9946	.9971	.9985	.9992
9	.7680	.8450	.9002	.9379	.9626	.9780	.9874	.9929	.9961	.9979
10	.6595	.7576	.8342	.8906	.9301	.9567	.9739	.9846	.9911	.9950
11	.5401	.6528	.7483	.8243	.8815	.9226	.9509	.9696	.9817	.9892
12	.4207	.5384	.6468	.7400	.8152	.8730	.9153	.9451	.9653	.9786
13	.3113	.4240	.5369	.6415	.7324	.8069	.8650	.9083	.9394	.9610
14	.2187	.3185	.4270	.5356	.6368	.7255	.7991	.8574	.9016	.9339
15	.1460	.2280	.3249	.4296	.5343	.6325	.7192	.7919	.8503	.8951
16	.0926	.1556	.2364	.3306	.4319	.5333	.6285	.7133	.7852	.8435
17	.0559	.1013	.1645	.2441	.3359	.4340	.5323	.6250	.7080	.7789
18	.0322	.0630	.1095	.1728	.2511	.3407	.4360	.5314	.6216	.7030
19	.0177	.0374	.0698	.1174	.1805	.2577	.3450	.4378	.5305	.6186
20	.0093	.0213	.0427	.0765	.1248	.1878	.2637	.3491	.4394	.5297
21	.0047	.0116	.0250	.0479	.0830	.1318	.1945	.2693	.3528	.4409
22	.0023	.0061	.0141	.0288	.0531	.0892	.1385	.2009	.2745	.3563
23	.0010	.0030	.0076	.0167	.0327	.0582	.0953	.1449	.2069	.2794
24	.0005	.0015	.0040	.0093	.0195	.0367	.0633	.1011	.1510	.2125
25	.0002	.0007	.0020	.0050	.0112	.0223	.0406	.0683	.1067	.1568
26	.0001	.0003	.0010	.0026	.0062	.0131	.0252	.0446	.0731	.1122
27	.0000	.0001	.0005	.0013	.0033	.0075	.0152	.0282	.0486	.0779
28	.0000	.0001	.0002	.0006	.0017	.0041	.0088	.0173	.0313	.0525
29	.0000	.0000	.0001	.0003	.0009	.0022	.0050	.0103	.0195	.0343
30	.0000	.0000	.0000	.0001	.0004	.0011	.0027	.0059	.0118	.0218
31	.0000	.0000	.0000	.0001	.0002	.0006	.0014	.0033	.0070	.0135
32	.0000	.0000	.0000	.0000	.0001	.0003	.0007	.0018	.0040	.0081
33	.0000	.0000	.0000	.0000	.0000	.0001	.0004	.0010	.0022	.0047
34	.0000	.0000	.0000	.0000	.0000	.0001	.0002	.0005	.0012	.0027
35	.0000	.0000	.0000	.0000	.0000	.0000	.0001	.0002	.0006	.0015
36	.0000	.0000	.0000	.0000	.0000	.0000	.0000	.0001	.0003	.0008
37	.0000	.0000	.0000	.0000	.0000	.0000	.0000	.0001	.0002	.0004
38	.0000	.0000	.0000	.0000	.0000	.0000	.0000	.0000	.0001	.0002
39	.0000	.0000	.0000	.0000	.0000	.0000	.0000	.0000	.0000	.0001
40	.0000	.0000	.0000	.0000	.0000	.0000	.0000	.0000	.0000	.0001

Table IX 367

NORMAL DISTRIBUTION

$$\psi(t) = (1/\sqrt{2\pi})e^{-t^2/2}. \qquad \text{(See Chapter X.)}$$

$$\psi^{(n)}(t) = d^n\psi(t)/dt^n = n^{\text{th}} \text{ derivative of } \psi(t) \text{ with respect to } t.$$

$$\alpha = \int_{-t}^{t} \psi(\tau)\, d\tau. \quad \Psi(t) = \int_{-\infty}^{t} \psi(\tau)\, d\tau = 0.5 + (\alpha/2), \quad (t \geqq 0).$$

t	$\alpha/2$	$\psi(t)$	$\psi^{(1)}(t)$	$\psi^{(2)}(t)$	$\psi^{(3)}(t)$	$\psi^{(4)}(t)$	$\psi^{(5)}(t)$	$\psi^{(6)}(t)$
.00	.0000	.3989	− .0000	− .3989	.0000	1.197	− .0000	−5.984
.01	.0040	.3989	− .0040	− .3989	.0120	1.197	− .0598	−5.982
.02	.0080	.3989	− .0080	− .3987	.0239	1.196	− .1196	−5.976
.03	.0120	.3988	− .0120	− .3984	.0359	1.194	− .1793	−5.965
.04	.0160	.3986	− .0159	− .3980	.0478	1.192	− .2389	−5.951
.05	.0199	.3984	− .0199	− .3975	.0597	1.189	− .2983	−5.932
.06	.0239	.3982	− .0239	− .3968	.0716	1.186	− .3575	−5.909
.07	.0279	.3980	− .0279	− .3960	.0834	1.182	− .4165	−5.882
.08	.0319	.3977	− .0318	− .3951	.0952	1.178	− .4752	−5.851
.09	.0359	.3973	− .0358	− .3941	.1070	1.173	− .5335	−5.816
.10	.0398	.3970	− .0397	− .3930	.1187	1.167	− .5915	−5.776
.11	.0438	.3965	− .0436	− .3917	.1303	1.161	− .6490	−5.733
.12	.0478	.3961	− .0475	− .3904	.1419	1.154	− .7061	−5.686
.13	.0517	.3956	− .0514	− .3889	.1534	1.147	− .7627	−5.635
.14	.0557	.3951	− .0553	− .3873	.1648	1.139	− .8188	−5.580
.15	.0596	.3945	− .0592	− .3856	.1762	1.130	− .8743	−5.521
.16	.0636	.3939	− .0630	− .3838	.1874	1.121	− .9292	−5.458
.17	.0675	.3932	− .0668	− .3819	.1986	1.112	− .9835	−5.392
.18	.0714	.3925	− .0707	− .3798	.2097	1.102	−1.037	−5.322
.19	.0753	.3918	− .0744	− .3777	.2206	1.091	−1.090	−5.248
.20	.0793	.3910	− .0782	− .3754	.2315	1.080	−1.142	−5.171
.21	.0832	.3902	− .0820	− .3730	.2422	1.068	−1.193	−5.091
.22	.0871	.3894	− .0857	− .3706	.2529	1.056	−1.244	−5.007
.23	.0910	.3885	− .0894	− .3680	.2634	1.043	−1.293	−4.919
.24	.0948	.3876	− .0930	− .3653	.2737	1.030	−1.342	−4.829
.25	.0987	.3867	− .0967	− .3625	.2840	1.017	−1.390	−4.735
.26	.1026	.3857	− .1003	− .3596	.2941	1.002	−1.437	−4.638
.27	.1064	.3847	− .1039	− .3566	.3040	.9878	−1.483	−4.539
.28	.1103	.3836	− .1074	− .3535	.3138	.9727	−1.528	−4.436
.29	.1141	.3825	− .1109	− .3504	.3235	.9572	−1.571	−4.330
.30	.1179	.3814	− .1144	− .3471	.3330	.9413	−1.614	−4.222
.31	.1217	.3802	− .1179	− .3437	.3423	.9250	−1.656	−4.111
.32	.1255	.3790	− .1213	− .3402	.3515	.9082	−1.696	−3.998
.33	.1293	.3778	− .1247	− .3367	.3605	.8910	−1.736	−3.882
.34	.1331	.3765	− .1280	− .3330	.3693	.8735	−1.774	−3.764
.35	.1368	.3752	− .1313	− .3293	.3779	.8556	−1.811	−3.644
.36	.1406	.3739	− .1346	− .3255	.3864	.8373	−1.847	−3.521
.37	.1443	.3725	− .1378	− .3216	.3947	.8186	−1.882	−3.397
.38	.1480	.3712	− .1410	− .3176	.4028	.7996	−1.915	−3.271
.39	.1517	.3697	− .1442	− .3135	.4107	.7803	−1.947	−3.142
.40	.1554	.3683	− .1473	− .3094	.4184	.7607	−1.978	−3.012
.41	.1591	.3668	− .1504	− .3059	.4259	.7408	−2.007	−2.881
.42	.1628	.3653	− .1534	− .3008	.4332	.7206	−2.035	−2.748
.43	.1664	.3637	− .1564	− .2965	.4403	.7001	−2.062	−2.614
.44	.1700	.3621	− .1593	− .2920	.4472	.6793	−2.088	−2.478
.45	.1736	.3605	− .1622	− .2875	.4539	.6583	−2.112	−2.341
.46	.1772	.3589	− .1651	− .2830	.4603	.6371	−2.134	−2.204
.47	.1808	.3572	− .1679	− .2783	.4666	.6156	−2.156	−2.065
.48	.1844	.3555	− .1707	− .2736	.4727	.5940	−2.176	−1.926
.49	.1879	.3538	− .1734	− .2689	.4785	.5721	−2.194	−1.785

Many extensive tables of $\psi(t)$ and related functions are available. For example, values of $\psi(t)$ and α to 15 decimal places are given in *Tables of Probability Functions*, Vol. II, Federal Works Agency, sponsored by National Bureau of Standards, New York, 1942; values of the second through the eighth derivatives of $\psi(t)$ to 5 decimal places are given in J. W. Glover, *Tables of Applied Mathematics in Finance, Insurance, Statistics*, George Wahr, Ann Arbor, Michigan, 1930.

Table IX continued

NORMAL DISTRIBUTION

t	$\alpha/2$	$\psi(t)$	$\psi^{(1)}(t)$	$\psi^{(2)}(t)$	$\psi^{(3)}(t)$	$\psi^{(4)}(t)$	$\psi^{(5)}(t)$	$\psi^{(6)}(t)$
.50	.1915	.3521	−.1760	−.2641	.4841	.5501	−2.211	−1.645
.51	.1950	.3503	−.1786	−.2592	.4895	.5279	−2.227	−1.504
.52	.1985	.3485	−.1812	−.2543	.4947	.5056	−2.241	−1.362
.53	.2019	.3467	−.1837	−.2493	.4996	.4831	−2.254	−1.221
.54	.2054	.3448	−.1862	−.2443	.5043	.4605	−2.266	−1.079
.55	.2088	.3429	−.1886	−.2392	.5088	.4378	−2.276	−.9371
.56	.2123	.3410	−.1910	−.2341	.5131	.4150	−2.285	−.7954
.57	.2157	.3391	−.1933	−.2289	.5171	.3921	−2.292	−.6540
.58	.2190	.3372	−.1956	−.2238	.5209	.3691	−2.298	−.5130
.59	.2224	.3352	−.1978	−.2185	.5245	.3461	−2.302	−.3724
.60	.2257	.3332	−.1999	−.2133	.5278	.3231	−2.305	−.2324
.61	.2291	.3312	−.2020	−.2080	.5309	.3000	−2.307	−.0930
.62	.2324	.3292	−.2041	−.2027	.5338	.2770	−2.307	.0455
.63	.2357	.3271	−.2061	−.1973	.5365	.2539	−2.306	.1832
.64	.2389	.3251	−.2080	−.1919	.5389	.2309	−2.303	.3199
.65	.2422	.3230	−.2099	−.1865	.5411	.2078	−2.299	.4555
.66	.2454	.3209	−.2118	−.1811	.5431	.1849	−2.294	.5899
.67	.2486	.3187	−.2136	−.1757	.5448	.1620	−2.288	.7230
.68	.2517	.3166	−.2153	−.1702	.5463	.1391	−2.280	.8547
.69	.2549	.3144	−.2170	−.1647	.5476	.1164	−2.271	.9849
.70	.2580	.3123	−.2186	−.1593	.5486	.0937	−2.260	1.114
.71	.2611	.3101	−.2201	−.1538	.5495	.0712	−2.248	1.241
.72	.2642	.3079	−.2217	−.1483	.5501	.0487	−2.235	1.366
.73	.2673	.3056	−.2231	−.1428	.5504	.0265	−2.221	1.489
.74	.2704	.3034	−.2245	−.1373	.5506	.0043	−2.206	1.610
.75	.2734	.3011	−.2259	−.1318	.5505	−.0176	−2.189	1.730
.76	.2764	.2989	−.2271	−.1262	.5502	−.0394	−2.171	1.847
.77	.2794	.2966	−.2284	−.1207	.5497	−.0611	−2.152	1.962
.78	.2823	.2943	−.2296	−.1153	.5490	−.0825	−2.132	2.075
.79	.2852	.2920	−.2307	−.1098	.5481	−.1037	−2.110	2.186
.80	.2881	.2897	−.2318	−.1043	.5469	−.1247	−2.088	2.294
.81	.2910	.2874	−.2328	−.0988	.5456	−.1455	−2.065	2.400
.82	.2939	.2850	−.2337	−.0934	.5440	−.1660	−2.040	2.503
.83	.2967	.2827	−.2346	−.0880	.5423	−.1862	−2.014	2.603
.84	.2995	.2803	−.2355	−.0825	.5403	−.2063	−1.988	2.701
.85	.3023	.2780	−.2363	−.0771	.5381	−.2260	−1.960	2.796
.86	.3051	.2756	−.2370	−.0718	.5358	−.2455	−1.932	2.889
.87	.3078	.2732	−.2377	−.0664	.5332	−.2646	−1.903	2.979
.88	.3106	.2709	−.2384	−.0611	.5305	−.2835	−1.872	3.065
.89	.3133	.2685	−.2389	−.0558	.5276	−.3021	−1.841	3.149
.90	.3159	.2661	−.2395	−.0506	.5245	−.3203	−1.810	3.230
.91	.3186	.2637	−.2400	−.0453	.5212	−.3383	−1.777	3.308
.92	.3212	.2613	−.2404	−.0401	.5177	−.3559	−1.743	3.383
.93	.3238	.2589	−.2408	−.0350	.5140	−.3731	−1.709	3.455
.94	.3264	.2565	−.2411	−.0299	.5102	−.3901	−1.674	3.524
.95	.3289	.2541	−.2414	−.0248	.5062	−.4066	−1.639	3.590
.96	.3315	.2516	−.2416	−.0197	.5021	−.4228	−1.602	3.653
.97	.3340	.2492	−.2418	−.0147	.4978	−.4387	−1.566	3.712
.98	.3365	.2468	−.2419	−.0098	.4933	−.4541	−1.528	3.768
.99	.3389	.2444	−.2419	−.0049	.4887	−.4692	−1.490	3.822

$\psi^{(n)}(t) = (-1)^n H_n(t)\, \psi(t)$, where $H_n(t)$ are the Hermite polynomials:

$$H_0(t) = 1, \quad H_1(t) = t, \quad H_2(t) = t^2 - 1, \quad H_3(t) = t^3 - 3t,$$

$$H_4(t) = t^4 - 6t^2 + 3, \quad H_5(t) = t^5 - 10t^3 + 15t,$$

$$H_6(t) = t^6 - 15t^4 + 45t^2 - 15.$$

Table IX continued   369

NORMAL DISTRIBUTION

t	$\alpha/2$	$\psi(t)$	$\psi^{(1)}(t)$	$\psi^{(2)}(t)$	$\psi^{(3)}(t)$	$\psi^{(4)}(t)$	$\psi^{(5)}(t)$	$\psi^{(6)}(t)$
1.00	.3413	.2420	−.2420	.0000	.4839	−.4839	−1.452	3.872
1.01	.3438	.2396	−.2419	.0048	.4790	−.4983	−1.413	3.918
1.02	.3461	.2371	−.2419	.0096	.4740	−.5122	−1.373	3.962
1.03	.3485	.2347	−.2418	.0143	.4688	−.5257	−1.334	4.002
1.04	.3508	.2323	−.2416	.0190	.4635	−.5389	−1.293	4.040
1.05	.3531	.2299	−.2414	.0236	.4580	−.5516	−1.253	4.074
1.06	.3554	.2275	−.2411	.0281	.4524	−.5639	−1.212	4.104
1.07	.3577	.2251	−.2408	.0326	.4467	−.5758	−1.171	4.132
1.08	.3599	.2227	−.2405	.0371	.4409	−.5873	−1.129	4.156
1.09	.3621	.2203	−.2401	.0414	.4350	−.5984	−1.088	4.178
1.10	.3643	.2179	−.2396	.0458	.4290	−.6091	−1.046	4.196
1.11	.3665	.2155	−.2392	.0500	.4228	−.6193	−1.004	4.211
1.12	.3686	.2131	−.2386	.0542	.4166	−.6292	−.9616	4.223
1.13	.3708	.2107	−.2381	.0583	.4102	−.6386	−.9193	4.232
1.14	.3729	.2083	−.2375	.0624	.4038	−.6476	−.8770	4.238
1.15	.3749	.2059	−.2368	.0664	.3973	−.6561	−.8346	4.240
1.16	.3770	.2036	−.2361	.0704	.3907	−.6643	−.7922	4.240
1.17	.3790	.2012	−.2354	.0742	.3840	−.6720	−.7498	4.237
1.18	.3810	.1989	−.2347	.0780	.3772	−.6792	−.7074	4.231
1.19	.3830	.1965	−.2339	.0818	.3704	−.6861	−.6652	4.222
1.20	.3849	.1942	−.2330	.0854	.3635	−.6926	−.6230	4.210
1.21	.3869	.1919	−.2322	.0890	.3566	−.6986	−.5810	4.196
1.22	.3888	.1895	−.2312	.0926	.3496	−.7042	−.5391	4.179
1.23	.3907	.1872	−.2303	.0960	.3425	−.7094	−.4974	4.159
1.24	.3925	.1849	−.2293	.0994	.3354	−.7141	−.4559	4.136
1.25	.3944	.1826	−.2283	.1027	.3282	−.7185	−.4147	4.111
1.26	.3962	.1804	−.2273	.1060	.3210	−.7224	−.3737	4.083
1.27	.3980	.1781	−.2262	.1092	.3138	−.7259	−.3331	4.053
1.28	.3997	.1758	−.2251	.1123	.3065	−.7291	−.2927	4.020
1.29	.4015	.1736	−.2239	.1153	.2992	−.7318	−.2527	3.985
1.30	.4032	.1714	−.2228	.1182	.2918	−.7341	−.2130	3.948
1.31	.4049	.1691	−.2216	.1211	.2845	−.7361	−.1737	3.908
1.32	.4066	.1669	−.2204	.1239	.2771	−.7376	−.1349	3.866
1.33	.4082	.1647	−.2191	.1267	.2697	−.7388	−.0964	3.822
1.34	.4099	.1626	−.2178	.1293	.2624	−.7395	−.0584	3.776
1.35	.4115	.1604	−.2165	.1319	.2550	−.7399	−.0209	3.728
1.36	.4131	.1582	−.2152	.1344	.2476	−.7400	.0161	3.678
1.37	.4147	.1561	−.2138	.1369	.2402	−.7396	.0527	3.626
1.38	.4162	.1539	−.2124	.1392	.2328	−.7389	.0887	3.572
1.39	.4177	.1518	−.2110	.1415	.2254	−.7378	.1241	3.517
1.40	.4192	.1497	−.2096	.1437	.2180	−.7364	.1590	3.460
1.41	.4207	.1476	−.2082	.1459	.2107	−.7347	.1933	3.401
1.42	.4222	.1456	−.2067	.1480	.2033	−.7326	.2270	3.340
1.43	.4236	.1435	−.2052	.1500	.1960	−.7301	.2601	3.279
1.44	.4251	.1415	−.2037	.1519	.1887	−.7274	.2926	3.216
1.45	.4265	.1394	−.2022	.1537	.1815	−.7243	.3244	3.151
1.46	.4279	.1374	−.2006	.1555	.1742	−.7209	.3556	3.085
1.47	.4292	.1354	−.1991	.1572	.1670	−.7172	.3861	3.018
1.48	.4306	.1334	−.1975	.1588	.1599	−.7132	.4159	2.950
1.49	.4319	.1315	−.1959	.1604	.1528	−.7089	.4451	2.881

NORMAL DISTRIBUTION

t	$\alpha/2$	$\psi(t)$	$\psi^{(1)}(t)$	$\psi^{(2)}(t)$	$\psi^{(3)}(t)$	$\psi^{(4)}(t)$	$\psi^{(5)}(t)$	$\psi^{(6)}(t)$
1.50	.4332	.1295	−.1943	.1619	.1457	−.7043	.4736	2.811
1.51	.4345	.1276	−.1927	.1633	.1387	−.6994	.5013	2.740
1.52	.4357	.1257	−.1910	.1647	.1317	−.6942	.5283	2.668
1.53	.4370	.1238	−.1894	.1660	.1248	−.6888	.5547	2.595
1.54	.4382	.1219	−.1877	.1672	.1180	−.6831	.5803	2.522
1.55	.4394	.1200	−.1860	.1683	.1111	−.6772	.6051	2.448
1.56	.4406	.1182	−.1843	.1694	.1044	−.6710	.6292	2.374
1.57	.4418	.1163	−.1826	.1704	.0977	−.6646	.6526	2.299
1.58	.4429	.1145	−.1809	.1714	.0911	−.6580	.6752	2.223
1.59	.4441	.1127	−.1792	.1722	.0846	−.6511	.6970	2.147
1.60	.4452	.1109	−.1775	.1730	.0781	−.6441	.7181	2.071
1.61	.4463	.1092	−.1757	.1738	.0717	−.6368	.7385	1.995
1.62	.4474	.1074	−.1740	.1745	.0654	−.6293	.7580	1.918
1.63	.4484	.1057	−.1722	.1751	.0591	−.6216	.7768	1.842
1.64	.4495	.1040	−.1705	.1757	.0529	−.6138	.7949	1.765
1.65	.4505	.1023	−.1687	.1762	.0468	−.6057	.8121	1.689
1.66	.4515	.1006	−.1670	.1766	.0408	−.5975	.8286	1.612
1.67	.4525	.0989	−.1652	.1770	.0349	−.5891	.8444	1.536
1.68	.4535	.0973	−.1634	.1773	.0290	−.5806	.8594	1.459
1.69	.4545	.0957	−.1617	.1776	.0233	−.5720	.8736	1.383
1.70	.4554	.0940	−.1599	.1778	.0176	−.5632	.8870	1.308
1.71	.4564	.0925	−.1581	.1779	.0120	−.5542	.8997	1.233
1.72	.4573	.0909	−.1563	.1780	.0065	−.5452	.9117	1.158
1.73	.4582	.0893	−.1545	.1780	.0011	−.5360	.9229	1.083
1.74	.4591	.0878	−.1528	.1780	−.0042	−.5267	.9333	1.010
1.75	.4599	.0863	−.1510	.1780	−.0094	−.5173	.9431	.9363
1.76	.4608	.0848	−.1492	.1778	−.0146	−.5079	.9521	.8636
1.77	.4616	.0833	−.1474	.1777	−.0196	−.4983	.9603	.7916
1.78	.4625	.0818	−.1457	.1774	−.0245	−.4887	.9679	.7204
1.79	.4633	.0804	−.1439	.1772	−.0294	−.4789	.9748	.6499
1.80	.4641	.0790	−.1421	.1769	−.0341	−.4692	.9809	.5801
1.81	.4649	.0775	−.1403	.1765	−.0388	−.4593	.9864	.5113
1.82	.4656	.0761	−.1386	.1761	−.0433	−.4494	.9911	.4433
1.83	.4664	.0748	−.1368	.1756	−.0477	−.4395	.9952	.3762
1.84	.4671	.0734	−.1351	.1751	−.0521	−.4295	.9987	.3101
1.85	.4678	.0721	−.1333	.1746	−.0563	−.4195	1.001	.2450
1.86	.4686	.0707	−.1316	.1740	−.0605	−.4095	1.004	.1809
1.87	.4693	.0694	−.1298	.1734	−.0645	−.3995	1.005	.1178
1.88	.4699	.0681	−.1281	.1727	−.0685	−.3894	1.006	.0559
1.89	.4706	.0669	−.1264	.1720	−.0723	−.3793	1.006	−.0050
1.90	.4713	.0656	−.1247	.1713	−.0761	−.3693	1.006	−.0647
1.91	.4719	.0644	−.1230	.1705	−.0797	−.3592	1.005	−.1232
1.92	.4726	.0632	−.1213	.1697	−.0832	−.3492	1.003	−.1805
1.93	.4732	.0620	−.1196	.1688	−.0867	−.3392	1.001	−.2367
1.94	.4738	.0608	−.1179	.1679	−.0900	−.3292	.9986	−.2916
1.95	.4744	.0596	−.1162	.1670	−.0933	−.3192	.9955	−.3452
1.96	.4750	.0584	−.1145	.1661	−.0964	−.3093	.9917	−.3975
1.97	.4756	.0573	−.1129	.1651	−.0994	−.2994	.9875	−.4486
1.98	.4761	.0562	−.1112	.1641	−.1024	−.2895	.9828	−.4984
1.99	.4767	.0551	−.1096	.1630	−.1052	−.2797	.9775	−.5468

Table IX continued 371

NORMAL DISTRIBUTION

t	$\alpha/2$	$\psi(t)$	$\psi^{(1)}(t)$	$\psi^{(2)}(t)$	$\psi^{(3)}(t)$	$\psi^{(4)}(t)$	$\psi^{(5)}(t)$	$\psi^{(6)}(t)$
2.00	.4772	.0540	−.1080	.1620	−.1080	−.2700	.9718	−.5939
2.01	.4778	.0529	−.1064	.1609	−.1106	−.2603	.9657	−.6397
2.02	.4783	.0519	−.1048	.1598	−.1132	−.2506	.9591	−.6841
2.03	.4788	.0508	−.1032	.1586	−.1157	−.2411	.9520	−.7271
2.04	.4793	.0498	−.1016	.1575	−.1180	−.2316	.9445	−.7688
2.05	.4798	.0488	−.1000	.1563	−.1203	−.2222	.9366	−.8091
2.06	.4803	.0478	−.0985	.1550	−.1225	−.2129	.9283	−.8480
2.07	.4808	.0468	−.0969	.1538	−.1245	−.2036	.9197	−.8855
2.08	.4812	.0459	−.0954	.1526	−.1265	−.1945	.9106	−.9217
2.09	.4817	.0449	−.0939	.1513	−.1284	−.1854	.9012	−.9565
2.10	.4821	.0440	−.0924	.1500	−.1302	−.1765	.8915	−.9899
2.11	.4826	.0431	−.0909	.1487	−.1320	−.1676	.8814	−1.022
2.12	.4830	.0422	−.0894	.1474	−.1336	−.1588	.8711	−1.053
2.13	.4834	.0413	−.0879	.1460	−.1351	−.1502	.8604	−1.082
2.14	.4838	.0404	−.0865	.1446	−.1366	−.1416	.8494	−1.110
2.15	.4842	.0396	−.0850	.1433	−.1380	−.1332	.8382	−1.136
2.16	.4846	.0387	−.0836	.1419	−.1393	−.1249	.8267	−1.161
2.17	.4850	.0379	−.0822	.1405	−.1405	−.1167	.8150	−1.185
2.18	.4854	.0371	−.0808	.1391	−.1416	−.1086	.8030	−1.208
2.19	.4857	.0363	−.0794	.1377	−.1426	−.1006	.7908	−1.229
2.20	.4861	.0355	−.0780	.1362	−.1436	−.0927	.7784	−1.249
2.21	.4864	.0347	−.0767	.1348	−.1445	−.0850	.7659	−1.267
2.22	.4868	.0339	−.0753	.1333	−.1453	−.0774	.7531	−1.285
2.23	.4871	.0332	−.0740	.1319	−.1460	−.0700	.7402	−1.301
2.24	.4875	.0325	−.0727	.1304	−.1467	−.0626	.7271	−1.316
2.25	.4878	.0317	−.0714	.1289	−.1473	−.0554	.7139	−1.329
2.26	.4881	.0310	−.0701	.1275	−.1478	−.0484	.7005	−1.341
2.27	.4884	.0303	−.0689	.1260	−.1483	−.0414	.6870	−1.353
2.28	.4887	.0297	−.0676	.1245	−.1486	−.0346	.6735	−1.362
2.29	.4890	.0290	−.0664	.1230	−.1490	−.0279	.6598	−1.371
2.30	.4893	.0283	−.0652	.1215	−.1492	−.0214	.6460	−1.379
2.31	.4896	.0277	−.0639	.1200	−.1494	−.0150	.6322	−1.385
2.32	.4898	.0270	−.0628	.1185	−.1495	−.0088	.6183	−1.391
2.33	.4901	.0264	−.0616	.1170	−.1496	−.0027	.6044	−1.395
2.34	.4904	.0258	−.0604	.1155	−.1496	.0033	.5904	−1.398
2.35	.4906	.0252	−.0593	.1141	−.1495	.0092	.5765	−1.400
2.36	.4909	.0246	−.0581	.1126	−.1494	.0149	.5624	−1.402
2.37	.4911	.0241	−.0570	.1111	−.1492	.0204	.5484	−1.402
2.38	.4913	.0235	−.0559	.1096	−.1490	.0258	.5344	−1.401
2.39	.4916	.0229	−.0548	.1081	−.1487	.0311	.5204	−1.399
2.40	.4918	.0224	−.0537	.1066	−.1483	.0362	.5064	−1.397
2.41	.4920	.0219	−.0527	.1051	−.1480	.0412	.4925	−1.393
2.42	.4922	.0213	−.0516	.1036	−.1475	.0461	.4786	−1.389
2.43	.4925	.0208	−.0506	.1022	−.1470	.0508	.4647	−1.383
2.44	.4927	.0203	−.0496	.1007	−.1465	.0554	.4509	−1.377
2.45	.4929	.0198	−.0486	.0992	−.1459	.0598	.4372	−1.370
2.46	.4931	.0194	−.0476	.0978	−.1453	.0641	.4235	−1.362
2.47	.4932	.0189	−.0466	.0963	−.1446	.0683	.4099	−1.354
2.48	.4934	.0184	−.0457	.0949	−.1439	.0723	.3964	−1.345
2.49	.4936	.0180	−.0447	.0935	−.1432	.0762	.3830	−1.335

NORMAL DISTRIBUTION

t	$\alpha/2$	$\psi(t)$	$\psi^{(1)}(t)$	$\psi^{(2)}(t)$	$\psi^{(3)}(t)$	$\psi^{(4)}(t)$	$\psi^{(5)}(t)$	$\psi^{(6)}(t)$
2.50	.4938	.0175	− .0438	.0920	− .1424	.0800	.3697	−1.324
2.51	.4940	.0171	− .0429	.0906	− .1416	.0836	.3566	−1.313
2.52	.4941	.0167	− .0420	.0892	− .1408	.0871	.3435	−1.301
2.53	.4943	.0163	− .0411	.0878	− .1399	.0905	.3305	−1.289
2.54	.4945	.0158	− .0403	.0864	− .1389	.0937	.3177	−1.276
2.55	.4946	.0154	− .0394	.0850	− .1380	.0968	.3050	−1.262
2.56	.4948	.0151	− .0386	.0836	− .1370	.0998	.2925	−1.248
2.57	.4949	.0147	− .0377	.0823	− .1360	.1027	.2801	−1.233
2.58	.4951	.0143	− .0369	.0809	− .1350	.1054	.2678	−1.218
2.59	.4952	.0139	− .0361	.0796	− .1339	.1080	.2557	−1.202
2.60	.4953	.0136	− .0353	.0782	− .1328	.1105	.2438	−1.186
2.61	.4955	.0132	− .0345	.0769	− .1317	.1129	.2320	−1.170
2.62	.4956	.0129	− .0338	.0756	− .1305	.1152	.2204	−1.153
2.63	.4957	.0126	− .0330	.0743	− .1294	.1173	.2089	−1.136
2.64	.4959	.0122	− .0323	.0730	− .1282	.1194	.1976	−1.119
2.65	.4960	.0119	− .0316	.0717	− .1270	.1213	.1866	−1.101
2.66	.4961	.0116	− .0309	.0705	− .1258	.1231	.1756	−1.083
2.67	.4962	.0113	− .0302	.0692	− .1245	.1248	.1649	−1.064
2.68	.4963	.0110	− .0295	.0680	− .1233	.1264	.1544	−1.046
2.69	.4964	.0107	− .0288	.0668	− .1220	.1279	.1440	−1.027
2.70	.4965	.0104	− .0281	.0656	− .1207	.1293	.1338	−1.008
2.71	.4966	.0101	− .0275	.0644	− .1194	.1306	.1238	− .9883
2.72	.4967	.0099	− .0268	.0632	− .1181	.1317	.1141	− .9689
2.73	.4968	.0096	− .0262	.0620	− .1168	.1328	.1045	− .9493
2.74	.4969	.0093	− .0256	.0608	− .1154	.1338	.0951	− .9296
2.75	.4970	.0091	− .0250	.0597	− .1141	.1347	.0859	− .9098
2.76	.4971	.0088	− .0244	.0585	− .1127	.1356	.0769	− .8899
2.77	.4972	.0086	− .0238	.0574	− .1114	.1363	.0681	− .8699
2.78	.4973	.0084	− .0233	.0563	− .1100	.1369	.0595	− .8499
2.79	.4974	.0081	− .0227	.0552	− .1087	.1375	.0511	− .8298
2.80	.4974	.0079	− .0222	.0541	− .1073	.1379	.0429	− .8097
2.81	.4975	.0077	− .0216	.0531	− .1059	.1383	.0349	− .7896
2.82	.4976	.0075	− .0211	.0520	− .1045	.1386	.0271	− .7695
2.83	.4977	.0073	− .0206	.0510	− .1031	.1389	.0195	− .7495
2.84	.4977	.0071	− .0201	.0500	− .1017	.1390	.0121	− .7294
2.85	.4978	.0069	− .0196	.0490	− .1003	.1391	.0049	− .7095
2.86	.4979	.0067	− .0191	.0480	− .0990	.1391	− .0021	− .6896
2.87	.4979	.0065	− .0186	.0470	− .0976	.1391	− .0089	− .6698
2.88	.4980	.0063	− .0182	.0460	− .0962	.1389	− .0155	− .6501
2.89	.4981	.0061	− .0177	.0451	− .0948	.1388	− .0219	− .6305
2.90	.4981	.0060	− .0173	.0441	− .0934	.1385	− .0281	− .6110
2.91	.4982	.0058	− .0168	.0432	− .0920	.1382	− .0341	− .5917
2.92	.4982	.0056	− .0164	.0423	− .0906	.1378	− .0399	− .5725
2.93	.4983	.0055	− .0160	.0414	− .0893	.1374	− .0456	− .5535
2.94	.4984	.0053	− .0156	.0405	− .0879	.1369	− .0510	− .5346
2.95	.4984	.0051	− .0152	.0396	− .0865	.1364	− .0563	− .5159
2.96	.4985	.0050	− .0148	.0388	− .0852	.1358	− .0613	− .4974
2.97	.4985	.0048	− .0144	.0379	− .0838	.1352	− .0662	− .4791
2.98	.4986	.0047	− .0140	.0371	− .0825	.1345	− .0709	− .4610
2.99	.4986	.0046	− .0137	.0363	− .0811	.1337	− .0754	− .4431

Table IX continued 373

NORMAL DISTRIBUTION

t	$\alpha/2$	$\psi(t)$	$\psi^{(1)}(t)$	$\psi^{(2)}(t)$	$\psi^{(3)}(t)$	$\psi^{(4)}(t)$	$\psi^{(5)}(t)$	$\psi^{(6)}(t)$
3.00	.4987	.0044	−.0133	.0355	−.0798	.1330	−.0798	−.4255
3.01	.4987	.0043	−.0129	.0347	−.0785	.1321	−.0839	−.4080
3.02	.4987	.0042	−.0126	.0339	−.0771	.1313	−.0879	−.3908
3.03	.4988	.0040	−.0123	.0331	−.0758	.1304	−.0918	−.3739
3.04	.4988	.0039	−.0119	.0324	−.0745	.1294	−.0954	−.3572
3.05	.4989	.0038	−.0116	.0316	−.0732	.1285	−.0989	−.3407
3.06	.4989	.0037	−.0113	.0309	−.0720	.1275	−.1022	−.3245
3.07	.4989	.0036	−.0110	.0302	−.0707	.1264	−.1054	−.3086
3.08	.4990	.0035	−.0107	.0295	−.0694	.1254	−.1084	−.2929
3.09	.4990	.0034	−.0104	.0288	−.0682	.1243	−.1113	−.2775
3.10	.4990	.0033	−.0101	.0281	−.0669	.1231	−.1140	−.2624
3.11	.4991	.0032	−.0098	.0275	−.0657	.1220	−.1165	−.2476
3.12	.4991	.0031	−.0096	.0268	−.0645	.1208	−.1189	−.2330
3.13	.4991	.0030	−.0093	.0262	−.0633	.1196	−.1212	−.2188
3.14	.4992	.0029	−.0091	.0256	−.0621	.1184	−.1233	−.2048
3.15	.4992	.0028	−.0088	.0249	−.0609	.1171	−.1253	−.1911
3.16	.4992	.0027	−.0086	.0243	−.0598	.1159	−.1271	−.1777
3.17	.4992	.0026	−.0083	.0237	−.0586	.1146	−.1288	−.1646
3.18	.4993	.0025	−.0081	.0232	−.0575	.1133	−.1304	−.1518
3.19	.4993	.0025	−.0079	.0226	−.0564	.1120	−.1319	−.1393
3.20	.4993	.0024	−.0076	.0220	−.0552	.1107	−.1332	−.1271
3.21	.4993	.0023	−.0074	.0215	−.0541	.1093	−.1344	−.1152
3.22	.4994	.0022	−.0072	.0210	−.0531	.1080	−.1355	−.1036
3.23	.4994	.0022	−.0070	.0204	−.0520	.1066	−.1365	−.0923
3.24	.4994	.0021	−.0068	.0199	−.0509	.1053	−.1373	−.0813
3.25	.4994	.0020	−.0066	.0194	−.0499	.1039	−.1381	−.0705
3.26	.4994	.0020	−.0064	.0189	−.0488	.1025	−.1388	−.0601
3.27	.4995	.0019	−.0062	.0184	−.0478	.1011	−.1393	−.0500
3.28	.4995	.0018	−.0060	.0180	−.0468	.0997	−.1398	−.0401
3.29	.4995	.0018	−.0059	.0175	−.0458	.0983	−.1401	−.0306
3.30	.4995	.0017	−.0057	.0170	−.0449	.0969	−.1404	−.0213
3.31	.4995	.0017	−.0055	.0166	−.0439	.0955	−.1405	−.0123
3.32	.4995	.0016	−.0054	.0162	−.0429	.0941	−.1406	−.0036
3.33	.4996	.0016	−.0052	.0157	−.0420	.0927	−.1406	.0048
3.34	.4996	.0015	−.0050	.0153	−.0411	.0913	−.1405	.0129
3.35	.4996	.0015	−.0049	.0149	−.0402	.0899	−.1403	.0208
3.36	.4996	.0014	−.0047	.0145	−.0393	.0885	−.1401	.0284
3.37	.4996	.0014	−.0046	.0141	−.0384	.0871	−.1398	.0357
3.38	.4996	.0013	−.0045	.0138	−.0376	.0857	−.1394	.0428
3.39	.4997	.0013	−.0043	.0134	−.0367	.0843	−.1389	.0495
3.40	.4997	.0012	−.0042	.0130	−.0359	.0829	−.1384	.0561
3.41	.4997	.0012	−.0041	.0127	−.0350	.0815	−.1378	.0623
3.42	.4997	.0012	−.0039	.0123	−.0342	.0801	−.1372	.0684
3.43	.4997	.0011	−.0038	.0120	−.0334	.0788	−.1364	.0741
3.44	.4997	.0011	−.0037	.0116	−.0327	.0774	−.1357	.0796
3.45	.4997	.0010	−.0036	.0113	−.0319	.0761	−.1349	.0849
3.46	.4997	.0010	−.0035	.0110	−.0311	.0747	−.1340	.0900
3.47	.4997	.0010	−.0034	.0107	−.0304	.0734	−.1331	.0948
3.48	.4997	.0009	−.0033	.0104	−.0297	.0721	−.1321	.0994
3.49	.4998	.0009	−.0032	.0101	−.0290	.0707	−.1311	.1037

NORMAL DISTRIBUTION

t	$\alpha/2$	$\psi(t)$	$\psi^{(1)}(t)$	$\psi^{(2)}(t)$	$\psi^{(3)}(t)$	$\psi^{(4)}(t)$	$\psi^{(5)}(t)$	$\psi^{(6)}(t)$
3.50	.4998	.0009	− .0031	.0098	− .0283	.0694	− .1300	.1078
3.51	.4998	.0008	− .0030	.0095	− .0276	.0681	− .1289	.1118
3.52	.4998	.0008	− .0029	.0093	− .0269	.0669	− .1278	.1155
3.53	.4998	.0008	− .0028	.0090	− .0262	.0656	− .1266	.1190
3.54	.4998	.0008	− .0027	.0087	− .0256	.0643	− .1254	.1223
3.55	.4998	.0007	− .0026	.0085	− .0249	.0631	− .1242	.1254
3.56	.4998	.0007	− .0025	.0082	− .0243	.0618	− .1229	.1283
3.57	.4998	.0007	− .0024	.0080	− .0237	.0606	− .1216	.1310
3.58	.4998	.0007	− .0024	.0078	− .0231	.0594	− .1203	.1335
3.59	.4998	.0006	− .0023	.0075	− .0225	.0582	− .1189	.1359
3.60	.4998	.0006	− .0022	.0073	− .0219	.0570	− .1176	.1380
3.61	.4998	.0006	− .0021	.0071	− .0214	.0559	− .1162	.1400
3.62	.4999	.0006	− .0021	.0069	− .0208	.0547	− .1148	.1419
3.63	.4999	.0005	− .0020	.0067	− .0203	.0536	− .1133	.1435
3.64	.4999	.0005	− .0019	.0065	− .0198	.0524	− .1119	.1450
3.65	.4999	.0005	− .0019	.0063	− .0192	.0513	− .1104	.1464
3.66	.4999	.0005	− .0018	.0061	− .0187	.0502	− .1090	.1476
3.67	.4999	.0005	− .0017	.0059	− .0182	.0492	− .1075	.1487
3.68	.4999	.0005	− .0017	.0057	− .0177	.0481	− .1060	.1496
3.69	.4999	.0004	− .0016	.0056	− .0173	.0470	− .1045	.1504
3.70	.4999	.0004	− .0016	.0054	− .0168	.0460	− .1030	.1510
3.71	.4999	.0004	− .0015	.0052	− .0164	.0450	− .1015	.1516
3.72	.4999	.0004	− .0015	.0051	− .0159	.0440	− .0999	.1520
3.73	.4999	.0004	− .0014	.0049	− .0155	.0430	− .0984	.1522
3.74	.4999	.0004	− .0014	.0048	− .0150	.0420	− .0969	.1524
3.75	.4999	.0004	− .0013	.0046	− .0146	.0410	− .0954	.1525
3.76	.4999	.0003	− .0013	.0045	− .0142	.0401	− .0939	.1524
3.77	.4999	.0003	− .0012	.0043	− .0138	.0392	− .0923	.1523
3.78	.4999	.0003	− .0012	.0042	− .0134	.0382	− .0908	.1520
3.79	.4999	.0003	− .0011	.0041	− .0131	.0373	− .0893	.1517
3.80	.4999	.0003	− .0011	.0039	− .0127	.0365	− .0878	.1512
3.81	.4999	.0003	− .0011	.0038	− .0123	.0356	− .0863	.1507
3.82	.4999	.0003	− .0010	.0037	− .0120	.0347	− .0848	.1501
3.83	.4999	.0003	− .0010	.0036	− .0116	.0339	− .0833	.1494
3.84	.4999	.0003	− .0010	.0034	− .0113	.0331	− .0818	.1487
3.85	.4999	.0002	− .0009	.0033	− .0110	.0323	− .0803	.1478
3.86	.4999	.0002	− .0009	.0032	− .0107	.0315	− .0788	.1469
3.87	.4999	.0002	− .0009	.0031	− .0104	.0307	− .0774	.1459
3.88	.4999	.0002	− .0008	.0030	− .0100	.0299	− .0759	.1449
3.89	.4999	.0002	− .0008	.0029	− .0098	.0292	− .0745	.1438
3.90	.5000	.0002	− .0008	.0028	− .0095	.0284	− .0730	.1426
3.91	.5000	.0002	− .0007	.0027	− .0092	.0277	− .0716	.1414
3.92	.5000	.0002	− .0007	.0026	− .0089	.0270	− .0702	.1402
3.93	.5000	.0002	− .0007	.0026	− .0086	.0263	− .0688	.1389
3.94	.5000	.0002	− .0007	.0025	− .0084	.0256	− .0674	.1375
3.95	.5000	.0002	− .0006	.0024	− .0081	.0250	− .0660	.1361
3.96	.5000	.0002	− .0006	.0023	− .0079	.0243	− .0647	.1347
3.97	.5000	.0002	− .0006	.0022	− .0076	.0237	− .0634	.1332
3.98	.5000	.0001	− .0006	.0022	− .0074	.0230	− .0620	.1317
3.99	.5000	.0001	− .0006	.0021	− .0072	.0224	− .0607	.1302

Table IX continued   375

NORMAL DISTRIBUTION

t	$\alpha/2$	$\psi(t)$	$\psi^{(1)}(t)$	$\psi^{(2)}(t)$	$\psi^{(3)}(t)$	$\psi^{(4)}(t)$	$\psi^{(5)}(t)$	$\psi^{(6)}(t)$
4.00	.5000	.0001	−.0005	.0020	−.0070	.0218	−.0594	.1286
4.05	.5000	.0001	−.0004	.0017	−.0059	.0190	−.0532	.1204
4.10	.5000	.0001	−.0004	.0014	−.0051	.0165	−.0474	.1118
4.15	.5000	.0001	−.0003	.0012	−.0043	.0143	−.0420	.1031
4.20	.5000	.0001	−.0002	.0010	−.0036	.0123	−.0371	.0943
4.25	.5000	.0000	−.0002	.0008	−.0031	.0105	−.0326	.0858
4.30	.5000	.0000	−.0002	.0007	−.0026	.0090	−.0285	.0775
4.35	.5000	.0000	−.0001	.0006	−.0022	.0077	−.0248	.0696
4.40	.5000	.0000	−.0001	.0005	−.0018	.0065	−.0215	.0621
4.45	.5000	.0000	−.0001	.0004	−.0015	.0055	−.0186	.0552
4.50	.5000	.0000	−.0001	.0003	−.0012	.0047	−.0160	.0487
4.55	.5000	.0000	−.0001	.0003	−.0010	.0039	−.0137	.0428
4.60	.5000	.0000	−.0000	.0002	−.0009	.0033	−.0117	.0375
4.65	.5000	.0000	−.0000	.0002	−.0007	.0027	−.0100	.0326
4.70	.5000	.0000	−.0000	.0001	−.0006	.0023	−.0084	.0283
4.75	.5000	.0000	−.0000	.0001	−.0005	.0019	−.0071	.0244
4.80	.5000	.0000	−.0000	.0001	−.0004	.0016	−.0060	.0210
4.85	.5000	.0000	−.0000	.0001	−.0003	.0013	−.0050	.0179
4.90	.5000	.0000	−.0000	.0001	−.0003	.0011	−.0042	.0153
4.95	.5000	.0000	−.0000	.0000	−.0002	.0009	−.0035	.0130
5.00	.5000	.0000	−.0000	.0000	−.0002	.0007	−.0029	.0109

IMPORTANT CONSTANTS

	N	$\log_{10} N$
$\pi =$	3.14159265	0.4971499
$2\pi =$	6.28318531	0.7981799
$\pi/2 =$	1.57079633	0.1961199
$1/\pi =$	0.31830989	$9.5028501 - 10$
$1/2\pi =$	0.15915494	$9.2018201 - 10$
$\pi^2 =$	9.86960440	0.9942997
$1/\pi^2 =$	0.10132118	$9.0057003 - 10$
$\sqrt{\pi} =$	1.77245385	0.2485749
$\sqrt{2\pi} =$	2.50662827	0.3990899
$\sqrt{\pi/2} =$	1.25331414	0.0980599
$1/\sqrt{\pi} =$	0.56418958	$9.7514251 - 10$
$\sqrt{2/\pi} =$	0.79788456	$9.9019401 - 10$
$1/\sqrt{2\pi} =$	0.39894228	$9.6009101 - 10$
$\sqrt{2} =$	1.41421356	0.1505150

$e = 2.71828183$
$M = \log_{10} e = 0.43429448$
$1/M = \log_e 10 = 2.30258509$
$\log_e \pi = 1.14472989$

Table X

F-DISTRIBUTION

(See §13.65-13.69)

Entries in the table are values of F_ϵ corresponding to the indicated values of m_1 and m_2 for which

$$\int_0^{F_\epsilon} h_{m_1, m_2}(F)\, dF = 1 - \epsilon, \qquad F_\epsilon > 1,$$

ϵ having the indicated fixed value.

$$\epsilon = 0.01$$

m_2	1	2	3	4	5	6	7	8	9	10	11	12
1	4052	4999	5404	5625	5764	5859	5928	5981	6022	6056	6082	6106
2	98.50	99.01	99.17	99.25	99.30	99.32	99.34	99.36	99.38	99.40	99.41	99.42
3	34.12	30.82	29.46	28.71	28.24	27.91	27.67	27.49	27.34	27.23	27.13	27.05
4	21.20	18.00	16.69	15.98	15.52	15.21	14.98	14.80	14.66	14.54	14.45	14.37
5	16.26	13.27	12.06	11.39	10.97	10.67	10.45	10.27	10.15	10.05	9.96	9.89
6	13.74	10.92	9.78	9.15	8.75	8.46	8.26	8.10	7.98	7.87	7.79	7.72
7	12.25	9.55	8.45	7.85	7.46	7.19	7.00	6.84	6.71	6.62	6.54	6.47
8	11.26	8.65	7.59	7.01	6.63	6.37	6.19	6.03	5.91	5.82	5.74	5.67
9	10.56	8.02	6.99	6.42	6.06	5.80	5.62	5.47	5.35	5.26	5.18	5.11
10	10.04	7.56	6.55	5.99	5.64	5.39	5.21	5.06	4.95	4.85	4.78	4.71
11	9.65	7.20	6.22	5.67	5.32	5.07	4.88	4.74	4.63	4.54	4.46	4.40
12	9.33	6.93	5.95	5.41	5.06	4.82	4.65	4.50	4.39	4.30	4.22	4.16
13	9.07	6.70	5.74	5.20	4.86	4.62	4.44	4.30	4.19	4.10	4.02	3.96
14	8.86	6.51	5.56	5.04	4.70	4.46	4.28	4.14	4.03	3.94	3.86	3.80
15	8.68	6.36	5.42	4.89	4.56	4.32	4.14	4.00	3.89	3.80	3.73	3.67
16	8.53	6.23	5.29	4.77	4.44	4.20	4.03	3.89	3.78	3.69	3.61	3.55
17	8.40	6.11	5.18	4.67	4.34	4.10	3.93	3.79	3.68	3.59	3.52	3.46
18	8.28	6.01	5.09	4.58	4.25	4.02	3.85	3.71	3.60	3.51	3.44	3.37
19	8.18	5.93	5.01	4.50	4.17	3.94	3.77	3.63	3.52	3.43	3.36	3.30
20	8.10	5.85	4.94	4.43	4.10	3.87	3.71	3.56	3.45	3.37	3.30	3.23
21	8.02	5.78	4.88	4.37	4.04	3.81	3.65	3.51	3.40	3.31	3.24	3.17
22	7.94	5.72	4.82	4.31	3.99	3.76	3.59	3.45	3.35	3.26	3.18	3.12
23	7.88	5.66	4.76	4.26	3.94	3.71	3.54	3.41	3.30	3.21	3.14	3.07
24	7.82	5.61	4.72	4.22	3.90	3.67	3.50	3.36	3.25	3.17	3.09	3.03
25	7.77	5.57	4.68	4.18	3.86	3.63	3.46	3.32	3.21	3.13	3.05	2.99
26	7.72	5.53	4.64	4.14	3.82	3.59	3.42	3.29	3.17	3.09	3.02	2.96
27	7.68	5.49	4.60	4.11	3.78	3.56	3.39	3.26	3.14	3.06	2.98	2.92
28	7.64	5.45	4.57	4.07	3.75	3.53	3.36	3.23	3.11	3.03	2.95	2.90
29	7.60	5.42	4.54	4.04	3.73	3.50	3.33	3.20	3.08	3.00	2.92	2.87
30	7.56	5.39	4.51	4.02	3.70	3.47	3.30	3.17	3.06	2.98	2.90	2.84
32	7.50	5.34	4.46	3.97	3.66	3.42	3.25	3.12	3.01	2.94	2.86	2.80
34	7.44	5.29	4.42	3.93	3.61	3.38	3.21	3.08	2.97	2.89	2.82	2.76
36	7.39	5.25	4.38	3.89	3.58	3.35	3.18	3.04	2.94	2.86	2.78	2.72
38	7.35	5.21	4.34	3.86	3.54	3.32	3.15	3.02	2.91	2.82	2.75	2.69
40	7.31	5.18	4.31	3.83	3.51	3.29	3.12	2.99	2.88	2.80	2.73	2.66
42	7.27	5.15	4.29	3.80	3.49	3.26	3.10	2.96	2.86	2.77	2.70	2.64
44	7.24	5.12	4.26	3.78	3.46	3.24	3.07	2.94	2.84	2.75	2.68	2.62
46	7.21	5.10	4.24	3.76	3.44	3.22	3.05	2.92	2.82	2.73	2.66	2.60
48	7.19	5.08	4.22	3.74	3.42	3.20	3.04	2.90	2.80	2.71	2.64	2.58
50	7.17	5.06	4.20	3.72	3.41	3.18	3.02	2.88	2.78	2.70	2.62	2.56
55	7.12	5.01	4.16	3.68	3.37	3.15	2.98	2.85	2.75	2.66	2.59	2.53
60	7.08	4.98	4.13	3.65	3.34	3.12	2.95	2.82	2.72	2.63	2.56	2.50
65	7.04	4.95	4.10	3.62	3.31	3.09	2.93	2.79	2.70	2.61	2.54	2.47
70	7.01	4.92	4.08	3.60	3.29	3.07	2.91	2.77	2.67	2.59	2.51	2.45
80	6.96	4.88	4.04	3.56	3.25	3.04	2.87	2.74	2.64	2.55	2.48	2.41
100	6.90	4.82	3.98	3.51	3.20	2.99	2.82	2.69	2.59	2.51	2.43	2.36
125	6.84	4.78	3.94	3.47	3.17	2.95	2.79	2.65	2.56	2.47	2.40	2.33
150	6.81	4.75	3.91	3.44	3.14	2.92	2.76	2.62	2.53	2.44	2.37	2.30
200	6.76	4.71	3.88	3.41	3.11	2.90	2.73	2.60	2.50	2.41	2.34	2.28
400	6.70	4.66	3.83	3.36	3.06	2.85	2.69	2.55	2.46	2.37	2.29	2.23
1000	6.66	4.62	3.80	3.34	3.04	2.82	2.66	2.53	2.43	2.34	2.26	2.20
∞	6.64	4.60	3.78	3.32	3.02	2.80	2.64	2.51	2.41	2.32	2.24	2.18

This table is taken from Snedecor, G. W., *Statistical Methods*, Iowa State College Press, Ames, Iowa, (4th ed.) 1946, by kind permission of the author and publishers.

Table X continued   377

F-DISTRIBUTION

$$\epsilon = 0.01$$

m_2	14	16	20	24	30	40	50	75	100	200	500	∞
						m_1						
1	6142	6169	6208	6234	6258	6286	6302	6323	6334	6352	6361	6366
2	99.43	99.44	99.45	99.46	99.47	99.48	99.48	99.49	99.49	99.49	99.50	99.50
3	26.92	26.83	26.69	26.60	26.50	26.41	26.35	26.27	26.23	26.18	26.14	26.12
4	14.24	14.15	14.02	13.93	13.83	13.74	13.69	13.61	13.57	13.52	13.48	13.46
5	9.77	9.68	9.55	9.47	9.38	9.29	9.24	9.17	9.13	9.07	9.04	9.02
6	7.60	7.52	7.39	7.31	7.23	7.14	7.09	7.02	6.99	6.94	6.90	6.88
7	6.35	6.27	6.15	6.07	5.98	5.90	5.85	5.78	5.75	5.70	5.67	5.65
8	5.56	5.48	5.36	5.28	5.20	5.11	5.06	5.00	4.96	4.91	4.88	4.86
9	5.00	4.92	4.80	4.73	4.64	4.56	4.51	4.45	4.41	4.36	4.33	4.31
10	4.60	4.52	4.41	4.33	4.25	4.17	4.12	4.05	4.01	3.96	3.93	3.91
11	4.29	4.21	4.10	4.02	3.94	3.86	3.80	3.74	3.70	3.66	3.62	3.60
12	4.05	3.98	3.86	3.78	3.70	3.61	3.56	3.49	3.46	3.41	3.38	3.36
13	3.85	3.78	3.67	3.59	3.51	3.42	3.37	3.30	3.27	3.21	3.18	3.16
14	3.70	3.62	3.51	3.43	3.34	3.26	3.21	3.14	3.11	3.06	3.02	3.00
15	3.56	3.48	3.36	3.29	3.20	3.12	3.07	3.00	2.97	2.92	2.89	2.87
16	3.45	3.37	3.25	3.18	3.10	3.01	2.96	2.89	2.86	2.80	2.77	2.75
17	3.35	3.27	3.16	3.08	3.00	2.92	2.86	2.79	2.76	2.70	2.67	2.65
18	3.27	3.19	3.07	3.00	2.91	2.83	2.78	2.71	2.68	2.62	2.59	2.57
19	3.19	3.12	3.00	2.92	2.84	2.76	2.70	2.63	2.60	2.54	2.51	2.49
20	3.13	3.05	2.94	2.86	2.77	2.69	2.63	2.56	2.53	2.47	2.44	2.42
21	3.07	2.99	2.88	2.80	2.72	2.63	2.58	2.51	2.47	2.42	2.38	2.36
22	3.02	2.94	2.83	2.75	2.67	2.58	2.53	2.46	2.42	2.37	2.33	2.30
23	2.97	2.89	2.78	2.70	2.62	2.53	2.48	2.41	2.37	2.32	2.28	2.26
24	2.93	2.85	2.74	2.66	2.58	2.49	2.44	2.36	2.33	2.27	2.23	2.21
25	2.89	2.81	2.70	2.62	2.54	2.45	2.40	2.32	2.29	2.23	2.19	2.17
26	2.86	2.77	2.66	2.58	2.50	2.41	2.36	2.28	2.25	2.19	2.15	2.13
27	2.83	2.74	2.63	2.55	2.47	2.38	2.33	2.25	2.21	2.16	2.12	2.10
28	2.80	2.71	2.60	2.52	2.44	2.35	2.30	2.22	2.18	2.13	2.09	2.06
29	2.77	2.68	2.57	2.49	2.41	2.32	2.27	2.19	2.15	2.10	2.06	2.03
30	2.74	2.66	2.55	2.47	2.38	2.29	2.24	2.16	2.13	2.07	2.03	2.01
32	2.70	2.62	2.51	2.42	2.34	2.25	2.20	2.12	2.08	2.02	1.98	1.96
34	2.66	2.58	2.47	2.38	2.30	2.21	2.15	2.08	2.04	1.98	1.94	1.91
36	2.62	2.54	2.43	2.35	2.26	2.17	2.12	2.04	2.00	1.94	1.90	1.87
38	2.59	2.51	2.40	2.32	2.22	2.14	2.08	2.00	1.97	1.90	1.86	1.84
40	2.56	2.49	2.37	2.29	2.20	2.11	2.05	1.97	1.94	1.88	1.84	1.81
42	2.54	2.46	2.35	2.26	2.17	2.08	2.02	1.94	1.91	1.85	1.80	1.78
44	2.52	2.44	2.32	2.24	2.15	2.06	2.00	1.92	1.88	1.82	1.78	1.75
46	2.50	2.42	2.30	2.22	2.13	2.04	1.98	1.90	1.86	1.80	1.76	1.72
48	2.48	2.40	2.28	2.20	2.11	2.02	1.96	1.88	1.84	1.78	1.73	1.70
50	2.46	2.39	2.26	2.18	2.10	2.00	1.94	1.86	1.82	1.76	1.71	1.68
55	2.43	2.35	2.23	2.15	2.06	1.96	1.90	1.82	1.78	1.71	1.66	1.64
60	2.40	2.32	2.20	2.12	2.03	1.93	1.87	1.79	1.74	1.68	1.63	1.60
65	2.37	2.30	2.18	2.09	2.00	1.90	1.84	1.76	1.71	1.64	1.60	1.56
70	2.35	2.28	2.15	2.07	1.98	1.88	1.82	1.74	1.69	1.62	1.56	1.53
80	2.32	2.24	2.11	2.03	1.94	1.84	1.78	1.70	1.65	1.57	1.52	1.49
100	2.26	2.19	2.06	1.98	1.89	1.79	1.73	1.64	1.59	1.51	1.46	1.43
125	2.23	2.15	2.03	1.94	1.85	1.75	1.68	1.59	1.54	1.46	1.40	1.37
150	2.20	2.12	2.00	1.91	1.83	1.72	1.66	1.56	1.51	1.43	1.37	1.33
200	2.17	2.09	1.97	1.88	1.79	1.69	1.62	1.53	1.48	1.39	1.33	1.28
400	2.12	2.04	1.92	1.84	1.74	1.64	1.57	1.47	1.42	1.32	1.24	1.19
1000	2.09	2.01	1.89	1.81	1.71	1.61	1.54	1.44	1.38	1.28	1.19	1.11
∞	2.07	1.99	1.87	1.79	1.69	1.59	1.52	1.41	1.36	1.25	1.15	1.00

Table X continued

F-DISTRIBUTION

$$\epsilon = 0.05$$

m_2	1	2	3	4	5	6	7	8	9	10	11	12
						m_1						
1	161	200	216	225	230	234	237	239	241	242	243	244
2	18.51	19.00	19.16	19.25	19.30	19.33	19.36	19.37	19.38	19.39	19.40	19.41
3	10.13	9.55	9.28	9.12	9.01	8.94	8.88	8.84	8.81	8.78	8.76	8.74
4	7.71	6.94	6.59	6.39	6.26	6.16	6.09	6.04	6.00	5.96	5.93	5.91
5	6.61	5.79	5.41	5.19	5.05	4.95	4.88	4.82	4.78	4.74	4.70	4.68
6	5.99	5.14	4.76	4.53	4.39	4.28	4.21	4.15	4.10	4.06	4.03	4.00
7	5.59	4.74	4.35	4.12	3.97	3.87	3.79	3.72	3.68	3.63	3.60	3.57
8	5.32	4.46	4.07	3.84	3.69	3.58	3.50	3.44	3.39	3.34	3.31	3.28
9	5.12	4.26	3.86	3.63	3.48	3.37	3.29	3.23	3.18	3.13	3.10	3.07
10	4.96	4.10	3.71	3.48	3.33	3.22	3.14	3.07	3.02	2.97	2.94	2.91
11	4.84	3.98	3.59	3.36	3.20	3.09	3.01	2.95	2.90	2.86	2.82	2.79
12	4.75	3.88	3.49	3.26	3.11	3.00	2.92	2.85	2.80	2.76	2.72	2.69
13	4.67	3.80	3.41	3.18	3.02	2.92	2.84	2.77	2.72	2.67	2.63	2.60
14	4.60	3.74	3.34	3.11	2.96	2.85	2.77	2.70	2.65	2.60	2.56	2.53
15	4.54	3.68	3.29	3.06	2.90	2.79	2.70	2.64	2.59	2.55	2.51	2.48
16	4.49	3.63	3.24	3.01	2.85	2.74	2.66	2.59	2.54	2.49	2.45	2.42
17	4.45	3.59	3.20	2.96	2.81	2.70	2.62	2.55	2.50	2.45	2.41	2.38
18	4.41	3.56	3.16	2.93	2.77	2.66	2.58	2.51	2.46	2.41	2.37	2.34
19	4.38	3.52	3.13	2.90	2.74	2.63	2.55	2.48	2.43	2.38	2.34	2.31
20	4.35	3.49	3.10	2.87	2.71	2.60	2.52	2.45	2.40	2.35	2.31	2.28
21	4.32	3.47	3.07	2.84	2.68	2.57	2.49	2.42	2.37	2.32	2.28	2.25
22	4.30	3.44	3.05	2.82	2.66	2.55	2.47	2.40	2.35	2.30	2.26	2.23
23	4.28	3.42	3.03	2.80	2.64	2.53	2.45	2.38	2.32	2.28	2.24	2.20
24	4.26	3.40	3.01	2.78	2.62	2.51	2.43	2.36	2.30	2.26	2.22	2.18
25	4.24	3.38	2.99	2.76	2.60	2.49	2.41	2.34	2.28	2.24	2.20	2.16
26	4.22	3.37	2.98	2.74	2.59	2.47	2.39	2.32	2.27	2.22	2.18	2.15
27	4.21	3.35	2.96	2.73	2.57	2.46	2.37	2.30	2.25	2.20	2.16	2.13
28	4.20	3.34	2.95	2.71	2.56	2.44	2.36	2.29	2.24	2.19	2.15	2.12
29	4.18	3.33	2.93	2.70	2.54	2.43	2.35	2.28	2.22	2.18	2.14	2.10
30	4.17	3.32	2.92	2.69	2.53	2.42	2.34	2.27	2.21	2.16	2.12	2.09
32	4.15	3.30	2.90	2.67	2.51	2.40	2.32	2.25	2.19	2.14	2.10	2.07
34	4.13	3.28	2.88	2.65	2.49	2.38	2.30	2.23	2.17	2.12	2.08	2.05
36	4.11	3.26	2.86	2.63	2.48	2.36	2.28	2.21	2.15	2.10	2.06	2.03
38	4.10	3.25	2.85	2.62	2.46	2.35	2.26	2.19	2.14	2.09	2.05	2.02
40	4.08	3.23	2.84	2.61	2.45	2.34	2.25	2.18	2.12	2.07	2.04	2.00
42	4.07	3.22	2.83	2.59	2.44	2.32	2.24	2.17	2.11	2.06	2.02	1.99
44	4.06	3.21	2.82	2.58	2.43	2.31	2.23	2.16	2.10	2.05	2.01	1.98
46	4.05	3.20	2.81	2.57	2.42	2.30	2.22	2.14	2.09	2.04	2.00	1.97
48	4.04	3.19	2.80	2.56	2.41	2.30	2.21	2.14	2.08	2.03	1.99	1.96
50	4.03	3.18	2.79	2.56	2.40	2.29	2.20	2.13	2.07	2.02	1.98	1.95
55	4.02	3.17	2.78	2.54	2.38	2.27	2.18	2.11	2.05	2.00	1.97	1.93
60	4.00	3.15	2.76	2.52	2.37	2.25	2.17	2.10	2.04	1.99	1.95	1.92
65	3.99	3.14	2.75	2.51	2.36	2.24	2.15	2.08	2.02	1.98	1.94	1.90
70	3.98	3.13	2.74	2.50	2.35	2.23	2.14	2.07	2.01	1.97	1.93	1.89
80	3.96	3.11	2.72	2.48	2.33	2.21	2.12	2.05	1.99	1.95	1.91	1.88
100	3.94	3.09	2.70	2.46	2.30	2.19	2.10	2.03	1.97	1.92	1.88	1.85
125	3.92	3.07	2.68	2.44	2.29	2.17	2.08	2.01	1.95	1.90	1.86	1.83
150	3.91	3.06	2.67	2.43	2.27	2.16	2.07	2.00	1.94	1.89	1.85	1.82
200	3.89	3.04	2.65	2.41	2.26	2.14	2.05	1.98	1.92	1.87	1.83	1.80
400	3.86	3.02	2.62	2.39	2.23	2.12	2.03	1.96	1.90	1.85	1.81	1.78
1000	3.85	3.00	2.61	2.38	2.22	2.10	2.02	1.95	1.89	1.84	1.80	1.76
∞	3.84	3.00	2.60	2.37	2.21	2.10	2.01	1.94	1.88	1.83	1.79	1.75

Table X continued   379

F-DISTRIBUTION

$$\epsilon = 0.05$$

						m_1						
m_2	14	16	20	24	30	40	50	75	100	200	500	∞
1	245	246	248	249	250	251	252	253	253	254	254	254
2	19.42	19.43	19.44	19.45	19.46	19.47	19.47	19.48	19.49	19.49	19.50	19.50
3	8.71	8.69	8.66	8.64	8.62	8.60	8.58	8.57	8.56	8.54	8.54	8.53
4	5.87	5.84	5.80	5.77	5.74	5.71	5.70	5.68	5.66	5.65	5.64	5.63
5	4.64	4.60	4.56	4.53	4.50	4.46	4.44	4.42	4.40	4.38	4.37	4.36
6	3.96	3.92	3.87	3.84	3.81	3.77	3.75	3.72	3.71	3.69	3.68	3.67
7	3.52	3.49	3.44	3.41	3.38	3.34	3.32	3.29	3.28	3.25	3.24	3.23
8	3.23	3.20	3.15	3.12	3.08	3.05	3.03	3.00	2.98	2.96	2.94	2.93
9	3.02	2.98	2.93	2.90	2.86	2.82	2.80	2.77	2.76	2.73	2.72	2.71
10	2.86	2.82	2.77	2.74	2.70	2.67	2.64	2.61	2.59	2.56	2.55	2.54
11	2.74	2.70	2.65	2.61	2.57	2.53	2.50	2.47	2.45	2.42	2.41	2.40
12	2.64	2.60	2.54	2.50	2.46	2.42	2.40	2.36	2.35	2.32	2.31	2.30
13	2.55	2.51	2.46	2.42	2.38	2.34	2.32	2.28	2.26	2.24	2.22	2.21
14	2.48	2.44	2.39	2.35	2.31	2.27	2.24	2.21	2.19	2.16	2.14	2.13
15	2.43	2.39	2.33	2.29	2.25	2.21	2.18	2.15	2.12	2.10	2.08	2.07
16	2.37	2.33	2.28	2.24	2.20	2.16	2.13	2.09	2.07	2.04	2.02	2.01
17	2.33	2.29	2.23	2.19	2.15	2.11	2.08	2.04	2.02	1.99	1.97	1.96
18	2.29	2.25	2.19	2.15	2.11	2.07	2.04	2.00	1.98	1.95	1.93	1.92
19	2.26	2.21	2.15	2.11	2.07	2.02	2.00	1.96	1.94	1.91	1.90	1.88
20	2.23	2.18	2.12	2.08	2.04	1.99	1.96	1.92	1.90	1.87	1.85	1.84
21	2.20	2.15	2.09	2.05	2.00	1.96	1.93	1.89	1.87	1.84	1.82	1.81
22	2.18	2.13	2.07	2.03	1.98	1.93	1.91	1.87	1.84	1.81	1.80	1.78
23	2.14	2.10	2.04	2.00	1.96	1.91	1.88	1.84	1.82	1.79	1.77	1.76
24	2.13	2.09	2.02	1.98	1.94	1.89	1.86	1.82	1.80	1.76	1.74	1.73
25	2.11	2.06	2.00	1.96	1.92	1.87	1.84	1.80	1.77	1.74	1.72	1.71
26	2.10	2.05	1.99	1.95	1.90	1.85	1.82	1.78	1.76	1.72	1.70	1.69
27	2.08	2.03	1.97	1.93	1.88	1.84	1.80	1.76	1.74	1.71	1.68	1.67
28	2.06	2.02	1.96	1.92	1.87	1.81	1.78	1.75	1.72	1.69	1.67	1.65
29	2.05	2.00	1.94	1.90	1.85	1.80	1.77	1.73	1.71	1.68	1.65	1.64
30	2.04	1.99	1.93	1.89	1.84	1.79	1.76	1.72	1.69	1.66	1.64	1.62
32	2.02	1.97	1.91	1.86	1.82	1.76	1.74	1.69	1.67	1.64	1.61	1.59
34	2.00	1.95	1.89	1.84	1.80	1.74	1.71	1.67	1.64	1.61	1.59	1.57
36	1.98	1.93	1.87	1.82	1.78	1.72	1.69	1.65	1.62	1.59	1.56	1.55
38	1.96	1.92	1.85	1.80	1.76	1.71	1.67	1.63	1.60	1.57	1.54	1.53
40	1.95	1.90	1.84	1.79	1.74	1.69	1.66	1.61	1.59	1.55	1.53	1.51
42	1.94	1.89	1.82	1.78	1.73	1.68	1.64	1.60	1.57	1.54	1.51	1.49
44	1.92	1.88	1.81	1.76	1.72	1.66	1.63	1.58	1.56	1.52	1.50	1.48
46	1.91	1.87	1.80	1.75	1.71	1.65	1.62	1.57	1.54	1.51	1.48	1.46
48	1.90	1.86	1.79	1.74	1.70	1.64	1.61	1.56	1.53	1.50	1.47	1.45
50	1.90	1.85	1.78	1.74	1.69	1.63	1.60	1.55	1.52	1.48	1.46	1.44
55	1.88	1.83	1.76	1.72	1.67	1.61	1.58	1.52	1.50	1.46	1.43	1.41
60	1.86	1.81	1.75	1.70	1.65	1.59	1.56	1.50	1.48	1.44	1.41	1.39
65	1.85	1.80	1.73	1.68	1.63	1.57	1.54	1.49	1.46	1.42	1.39	1.37
70	1.84	1.79	1.72	1.67	1.62	1.56	1.53	1.47	1.45	1.40	1.37	1.35
80	1.82	1.77	1.70	1.65	1.60	1.54	1.51	1.45	1.42	1.38	1.35	1.32
100	1.79	1.75	1.68	1.63	1.57	1.51	1.48	1.42	1.39	1.34	1.30	1.28
125	1.77	1.72	1.65	1.60	1.55	1.49	1.45	1.39	1.36	1.31	1.27	1.25
150	1.76	1.71	1.64	1.59	1.54	1.47	1.44	1.37	1.34	1.29	1.25	1.22
200	1.74	1.69	1.62	1.57	1.52	1.45	1.42	1.35	1.32	1.26	1.22	1.19
400	1.72	1.67	1.60	1.54	1.49	1.42	1.38	1.32	1.28	1.22	1.16	1.13
1000	1.70	1.65	1.58	1.53	1.47	1.41	1.36	1.30	1.26	1.19	1.13	1.08
∞	1.69	1.64	1.57	1.52	1.46	1.40	1.35	1.28	1.24	1.17	1.11	1.00

Table XI

z-DISTRIBUTION

(See §13.71)

Entries in the table are values of z_ϵ corresponding to the indicated values of m_1 and m_2 for which

$$\int_{-\infty}^{z_\epsilon} G_{m_1,m_2}(z)\, dz = 1 - \epsilon, \qquad z_\epsilon \geqq 0,$$

ϵ having the indicated fixed value.

$$\epsilon = 0.001$$

m_2	m_1 1	2	3	4	5	6	8	12	24	∞
1	6.4562	6.5612	6.5966	6.6201	6.6323	6.6405	6.6508	6.6611	6.6715	6.6819
2	3.4531	3.4534	3.4535	3.4535	3.4535	3.4535	3.4536	3.4536	3.4536	3.4536
3	2.5604	2.5003	2.4748	2.4603	2.4511	2.4446	2.4361	2.4272	2.4179	2.4081
4	2.1529	2.0574	2.0143	1.9892	1.9728	1.9612	1.9459	1.9294	1.9118	1.8927
5	1.9255	1.8002	1.7513	1.7184	1.6964	1.6808	1.6596	1.6370	1.6123	1.5845
6	1.7849	1.6479	1.5828	1.5433	1.5177	1.4986	1.4730	1.4449	1.4134	1.3783
7	1.6874	1.5384	1.4662	1.4221	1.3927	1.3711	1.3417	1.3090	1.2721	1.2296
8	1.6177	1.4587	1.3809	1.3332	1.3008	1.2770	1.2443	1.2077	1.1662	1.1169
9	1.5646	1.3982	1.3160	1.2653	1.2304	1.2047	1.1694	1.1293	1.0830	1.0279
10	1.5232	1.3509	1.2650	1.2116	1.1748	1.1475	1.1098	1.0668	1.0165	.9557
11	1.4900	1.3128	1.2238	1.1683	1.1297	1.1012	1.0614	1.0157	.9619	.8957
12	1.4627	1.2814	1.1900	1.1326	1.0926	1.0628	1.0213	.9733	.9162	.8450
13	1.4400	1.2553	1.1616	1.1026	1.0614	1.0306	.9875	.9374	.8774	.8014
14	1.4208	1.2332	1.1376	1.0772	1.0348	1.0031	.9586	.9066	.8439	.7635
15	1.4043	1.2141	1.1169	1.0553	1.0119	.9795	.9336	.8800	.8147	.7301
16	1.3900	1.1976	1.0989	1.0362	.9920	.9588	.9119	.8567	.7891	.7005
17	1.3775	1.1832	1.0832	1.0195	.9745	.9407	.8927	.8361	.7664	.6740
18	1.3665	1.1704	1.0693	1.0047	.9590	.9246	.8757	.8178	.7462	.6502
19	1.3567	1.1591	1.0569	.9915	.9442	.9103	.8605	.8014	.7277	.6285
20	1.3480	1.1489	1.0458	.9798	.9329	.8974	.8469	.7867	.7115	.6086
21	1.3401	1.1398	1.0358	.9691	.9217	.8858	.8346	.7735	.6964	.5904
22	1.3329	1.1315	1.0268	.9595	.9116	.8753	.8234	.7612	.6828	.5738
23	1.3264	1.1240	1.0186	.9507	.9024	.8657	.8132	.7501	.6704	.5583
24	1.3205	1.1171	1.0111	.9427	.8939	.8569	.8038	.7400	.6589	.5440
25	1.3151	1.1108	1.0041	.9354	.8862	.8489	.7953	.7306	.6483	.5307
26	1.3101	1.1050	.9978	.9286	.8791	.8415	.7873	.7220	.6385	.5183
27	1.3055	1.0997	.9920	.9223	.8725	.8346	.7800	.7140	.6294	.5066
28	1.3013	1.0947	.9866	.9165	.8664	.8282	.7732	.7066	.6209	.4957
29	1.2973	1.0903	.9815	.9112	.8607	.8223	.7679	.6997	.6129	.4853
30	1.2936	1.0859	.9768	.9061	.8554	.8168	.7610	.6932	.6056	.4756
60	1.2413	1.0248	.9100	.8345	.7798	.7377	.6760	.5992	.4955	.3198
∞	1.1910	.9663	.8453	.7648	.7059	.6599	.5917	.5044	.3786	0

Table XI is reprinted from Table VI of Fisher: *Statistical Methods for Research Workers*, published by Oliver and Boyd Ltd., Edinburgh, by permission of the author and publishers.

Table XI continued   381

z-DISTRIBUTION

$$\epsilon = 0.01$$

m_2	1	2	3	4	5	6	8	12	24	∞
1	4.1535	4.2585	4.2974	4.3175	4.3297	4.3379	4.3482	4.3585	4.3689	4.3794
2	2.2950	2.2976	2.2984	2.2988	2.2991	2.2992	2.2994	2.2997	2.2999	2.3001
3	1.7649	1.7140	1.6915	1.6786	1.6703	1.6645	1.6569	1.6489	1.6404	1.6314
4	1.5270	1.4452	1.4075	1.3856	1.3711	1.3609	1.3473	1.3327	1.3170	1.3000
5	1.3943	1.2929	1.2449	1.2164	1.1974	1.1838	1.1656	1.1457	1.1239	1.0997
6	1.3103	1.1955	1.1401	1.1068	1.0843	1.0680	1.0460	1.0218	.9948	.9643
7	1.2526	1.1281	1.0672	1.0300	1.0048	.9864	.9614	.9335	.9020	.8658
8	1.2106	1.0787	1.0135	.9734	.9459	.9259	.8983	.8673	.8319	.7904
9	1.1786	1.0411	.9724	.9299	.9006	.8791	.8494	.8157	.7769	.7305
10	1.1535	1.0114	.9399	.8954	.8646	.8419	.8104	.7744	.7324	.6816
11	1.1333	.9874	.9136	.8674	.8354	.8116	.7785	.7405	.6958	.6408
12	1.1166	.9677	.8919	.8443	.8111	.7864	.7520	.7122	.6649	.6061
13	1.1027	.9511	.8737	.8248	.7907	.7652	.7295	.6882	.6386	.5761
14	1.0909	.9370	.8581	.8082	.7732	.7471	.7103	.6675	.6159	.5500
15	1.0807	.9249	.8448	.7939	.7582	.7314	.6937	.6496	.5961	.5269
16	1.0719	.9144	.8331	.7814	.7450	.7177	.6791	.6339	.5786	.5064
17	1.0641	.9051	.8229	.7705	.7335	.7057	.6663	.6199	.5630	.4879
18	1.0572	.8970	.8138	.7607	.7232	.6950	.6549	.6075	.5491	.4712
19	1.0511	.8897	.8057	.7521	.7140	.6854	.6447	.5964	.5366	.4560
20	1.0457	.8831	.7985	.7443	.7058	.6768	.6355	.5864	.5253	.4421
21	1.0408	.8772	.7920	.7372	.6984	.6690	.6272	.5773	.5150	.4294
22	1.0363	.8719	.7860	.7309	.6916	.6620	.6196	.5691	.5056	.4176
23	1.0322	.8670	.7806	.7251	.6855	.6555	.6127	.5615	.4969	.4068
24	1.0285	.8626	.7757	.7197	.6799	.6496	.6064	.5545	.4890	.3967
25	1.0251	.8585	.7712	.7148	.6747	.6442	.6006	.5481	.4816	.3872
26	1.0220	.8548	.7670	.7103	.6699	.6392	.5952	.5422	.4748	.3784
27	1.0191	.8513	.7631	.7062	.6655	.6346	.5902	.5367	.4685	.3701
28	1.0164	.8481	.7595	.7023	.6614	.6303	.5856	.5316	.4626	.3624
29	1.0139	.8451	.7562	.6987	.6576	.6263	.5813	.5269	.4570	.3550
30	1.0116	.8423	.7531	.6954	.6540	.6226	.5773	.5224	.4519	.3481
60	.9784	.8025	.7086	.6472	.6028	.5687	.5189	.4574	.3746	.2352
∞	.9462	.7636	.6651	.5999	.5522	.5152	.4604	.3908	.2913	0

Table XI continued

z-DISTRIBUTION

$\epsilon = 0.05$

m_2	1	2	3	4	5	6	8	12	24	∞
1	2.5421	2.6479	2.6870	2.7071	2.7194	2.7276	2.7380	2.7484	2.7588	2.7693
2	1.4592	1.4722	1.4765	1.4787	1.4800	1.4808	1.4819	1.4830	1.4840	1.4851
3	1.1577	1.1284	1.1137	1.1051	1.0994	1.0953	1.0899	1.0842	1.0781	1.0716
4	1.0212	.9690	.9429	.9272	.9168	.9093	.8993	.8885	.8767	.8639
5	.9441	.8777	.8441	.8236	.8097	.7997	.7862	.7714	.7550	.7368
6	.8948	.8188	.7798	.7558	.7394	.7274	.7112	.6931	.6729	.6499
7	.8606	.7777	.7347	.7080	.6896	.6761	.6576	.6369	.6134	.5862
8	.8355	.7475	.7014	.6725	.6525	.6378	.6175	.5945	.5682	.5371
9	.8163	.7242	.6757	.6450	.6238	.6080	.5862	.5613	.5324	.4979
10	.8012	.7058	.6553	.6232	.6009	.5843	.5611	.5346	.5035	.4657
11	.7889	.6909	.6387	.6055	.5822	.5648	.5406	.5126	.4795	.4387
12	.7788	.6786	.6250	.5907	.5666	.5487	.5234	.4941	.4592	.4156
13	.7703	.6682	.6134	.5783	.5535	.5350	.5089	.4785	.4419	.3957
14	.7630	.6594	.6036	.5677	.5423	.5233	.4964	.4649	.4269	.3782
15	.7568	.6518	.5950	.5585	.5326	.5131	.4855	.4532	.4138	.3628
16	.7514	.6451	.5876	.5505	.5241	.5042	.4760	.4428	.4022	.3490
17	.7466	.6393	.5811	.5434	.5166	.4964	.4676	.4337	.3919	.3366
18	.7424	.6341	.5753	.5371	.5099	.4894	.4602	.4255	.3827	.3253
19	.7386	.6295	.5701	.5315	.5040	.4832	.4535	.4182	.3743	.3151
20	.7352	.6254	.5654	.5265	.4986	.4776	.4474	.4116	.3668	.3057
21	.7322	.6216	.5612	.5219	.4938	.4725	.4420	.4055	.3599	.2971
22	.7294	.6182	.5574	.5178	.4894	.4679	.4370	.4001	.3536	.2892
23	.7269	.6151	.5540	.5140	.4854	.4636	.4325	.3950	.3478	.2818
24	.7246	.6123	.5508	.5106	.4817	.4598	.4283	.3904	.3425	.2749
25	.7225	.6097	.5478	.5074	.4783	.4562	.4244	.3862	.3376	.2685
26	.7205	.6073	.5451	.5045	.4752	.4529	.4209	.3823	.3330	.2625
27	.7187	.6051	.5427	.5017	.4723	.4499	.4176	.3786	.3287	.2569
28	.7171	.6030	.5403	.4992	.4696	.4471	.4146	.3752	.3248	.2516
29	.7155	.6011	.5382	.4969	.4671	.4444	.4117	.3720	.3211	.2466
30	.7141	.5994	.5362	.4947	.4648	.4420	.4090	.3691	.3176	.2419
60	.6933	.5738	.5073	.4632	.4311	.4064	.3702	.3255	.2654	.1644
∞	.6729	.5486	.4787	.4319	.3974	.3706	.3309	.2804	.2085	0

Table XII

t-DISTRIBUTION;

INTERVAL $-t_\epsilon$ TO t_ϵ CONTAINING ALL BUT "FRACTION" ϵ OF DISTRIBUTION

(See §§13.59-13.63.)

Entries in the table are values of t_ϵ, for the indicated ϵ and m, such that

$$\int_{-t_\epsilon}^{t_\epsilon} g_m(t)\, dt = 2 \int_{0}^{t_\epsilon} g_m(t)\, dt = 1 - \epsilon.$$

m	.90	.80	.70	.60	.50	.40	ϵ .30	.20	.10	.05	.02	.01	.001
1	0.158	0.325	0.510	0.727	1.000	1.376	1.963	3.078	6.314	12.71	31.82	63.66	636.6
2	0.142	0.289	0.445	0.617	0.816	1.061	1.386	1.886	2.920	4.303	6.965	9.925	31.60
3	0.137	0.277	0.424	0.584	0.765	0.978	1.250	1.638	2.353	3.182	4.541	5.841	12.94
4	0.134	0.271	0.414	0.569	0.741	0.941	1.190	1.533	2.132	2.776	3.747	4.604	8.610
5	0.132	0.267	0.408	0.559	0.727	0.920	1.156	1.476	2.015	2.571	3.365	4.032	6.859
6	0.131	0.265	0.404	0.553	0.718	0.906	1.134	1.440	1.943	2.447	3.143	3.707	5.959
7	0.130	0.263	0.402	0.549	0.711	0.896	1.119	1.415	1.895	2.365	2.998	3.499	5.405
8	0.130	0.262	0.399	0.546	0.706	0.889	1.108	1.397	1.860	2.306	2.896	3.355	5.041
9	0.129	0.261	0.398	0.543	0.703	0.883	1.100	1.383	1.833	2.262	2.821	3.250	4.781
10	0.129	0.260	0.397	0.542	0.700	0.879	1.093	1.372	1.812	2.228	2.764	3.169	4.587
11	0.129	0.260	0.396	0.540	0.697	0.876	1.088	1.363	1.796	2.201	2.718	3.106	4.437
12	0.128	0.259	0.395	0.539	0.695	0.873	1.083	1.356	1.782	2.179	2.681	3.055	4.318
13	0.128	0.259	0.394	0.538	0.694	0.870	1.079	1.350	1.771	2.160	2.650	3.012	4.221
14	0.128	0.258	0.393	0.537	0.692	0.868	1.076	1.345	1.761	2.145	2.624	2.977	4.140
15	0.128	0.258	0.393	0.536	0.691	0.866	1.074	1.341	1.753	2.131	2.602	2.947	4.073
16	0.128	0.258	0.392	0.535	0.690	0.865	1.071	1.337	1.746	2.120	2.583	2.921	4.015
17	0.128	0.257	0.392	0.534	0.689	0.863	1.069	1.333	1.740	2.110	2.567	2.898	3.965
18	0.127	0.257	0.392	0.534	0.688	0.862	1.067	1.330	1.734	2.101	2.552	2.878	3.922
19	0.127	0.257	0.391	0.533	0.688	0.861	1.066	1.328	1.729	2.093	2.539	2.861	3.883
20	0.127	0.257	0.391	0.533	0.687	0.860	1.064	1.325	1.725	2.086	2.528	2.845	3.850
21	0.127	0.257	0.391	0.532	0.686	0.859	1.063	1.323	1.721	2.080	2.518	2.831	3.819
22	0.127	0.256	0.390	0.532	0.686	0.858	1.061	1.321	1.717	2.074	2.508	2.819	3.792
23	0.127	0.256	0.390	0.532	0.685	0.858	1.060	1.319	1.714	2.069	2.500	2.807	3.767
24	0.127	0.256	0.390	0.531	0.685	0.857	1.059	1.318	1.711	2.064	2.492	2.797	3.745
25	0.127	0.256	0.390	0.531	0.684	0.856	1.058	1.316	1.708	2.060	2.485	2.787	3.725
26	0.127	0.256	0.390	0.531	0.684	0.856	1.058	1.315	1.706	2.056	2.479	2.779	3.707
27	0.127	0.256	0.389	0.531	0.684	0.855	1.057	1.314	1.703	2.052	2.473	2.771	3.690
28	0.127	0.256	0.389	0.530	0.683	0.855	1.056	1.313	1.701	2.048	2.467	2.763	3.674
29	0.127	0.256	0.389	0.530	0.683	0.854	1.055	1.311	1.699	2.045	2.462	2.756	3.659
30	0.127	0.256	0.389	0.530	0.683	0.854	1.055	1.310	1.697	2.042	2.457	2.750	3.646
40	0.126	0.255	0.388	0.529	0.681	0.851	1.050	1.303	1.684	2.021	2.423	2.704	3.551
60	0.126	0.254	0.387	0.527	0.679	0.848	1.046	1.296	1.671	2.000	2.390	2.660	3.460
120	0.126	0.254	0.386	0.526	0.677	0.845	1.041	1.289	1.658	1.980	2.358	2.617	3.373
∞	0.126	0.253	0.385	0.524	0.674	0.842	1.036	1.282	1.645	1.960	2.326	2.576	3.291

Table XII is reprinted from Table IV of Fisher: *Statistical Methods for Research Workers*, published by Oliver and Boyd Ltd., Edinburgh, by permission of the author and publishers.

Table XIII

t-DISTRIBUTION;
CUMULATIVE DISTRIBUTION FUNCTION

(See §13.63.)

Entries in the table are values of $\int_{-\infty}^{t_0} g_m(t)\, dt$ for indicated values of t_0 and m.

t_0	1	2	3	4	m 6	8	10	15	20
0	.500	.500	.500	.500	.500	.500	.500	.500	.500
0.2	.563	.570	.573	.574	.576	.577	.577	.578	.578
0.4	.621	.636	.642	.645	.648	.650	.651	.653	.653
0.6	.672	.695	.705	.710	.715	.717	.719	.721	.722
0.8	.715	.746	.759	.766	.773	.777	.779	.782	.783
1.0	.750	.789	.804	.813	.822	.827	.830	.833	.835
1.2	.779	.824	.842	.852	.862	.868	.871	.876	.878
1.4	.803	.852	.872	.883	.894	.900	.904	.909	.912
1.6	.822	.875	.896	.908	.920	.926	.930	.935	.937
1.8	.839	.893	.915	.927	.939	.945	.949	.954	.956
2.0	.852	.908	.930	.942	.954	.960	.963	.968	.970
2.2	.864	.921	.942	.954	.965	.970	.974	.978	.980
2.4	.874	.931	.952	.963	.973	.978	.981	.985	.987
2.6	.883	.939	.960	.970	.980	.984	.987	.990	.991
2.8	.891	.946	.966	.976	.984	.988	.991	.993	.994
3.0	.898	.952	.971	.980	.988	.992	.993	.996	.996
3.2	.904	.957	.975	.984	.991	.994	.995	.997	.998
3.4	.909	.962	.979	.986	.993	.995	.997	.998	.999
3.6	.914	.965	.982	.989	.994	.996	.998	.999	.999
3.8	.918	.969	.984	.990	.996	.997	.998	.999	.999
4.0	.922	.971	.986	.992	.996	.998	.999	.999	1.000
4.2	.926	.974	.988	.993	.997	.998	.999	1.000	
4.4	.929	.976	.989	.994	.998	.999	.999		
4.6	.932	.978	.990	.995	.998	.999	1.000		
4.8	.935	.980	.991	.996	.998	.999			
5.0	.937	.981	.992	.996	.999	1.000			

A more extensive table of this type is given by "Student" in *Metron*, Vol: 5, Part 3, 1925, pp: 114-117.

NOMOGRAM FOR CALCULATING $1 - (1 - p)^n$

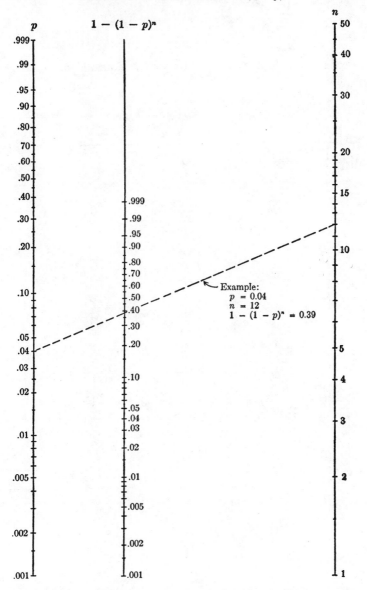

Chart for Obtaining the Probability $1 - (1 - p)^n$ of at Least One Occurrence of an Event in n Independent Trials When the Probability of Occurrence in Each Trial is p.
(Use straight-edge across scales).

Table XIV

χ^2-DISTRIBUTION

(See §13.26-13.30)

Entries in this table give values of $\chi_0{}^2$, for the indicated values of m and ϵ, such that

$$\int_{w=\chi_0{}^2}^{w=\infty} f_m(w) \, dw = \epsilon.$$

Degrees of Freedom, m	.99	.98	.95	ϵ .90	.80	.70	.50
				$\chi_0{}^2$			
1	0.000	0.001	0.004	0.016	0.064	0.148	0.455
2	0.020	0.040	0.103	0.211	0.446	0.713	1.386
3	0.115	0.185	0.352	0.584	1.005	1.424	2.366
4	0.297	0.429	0.711	1.064	1.649	2.195	3.357
5	0.554	0.752	1.145	1.610	2.343	3.000	4.351
6	0.872	1.134	1.635	2.204	3.070	3.828	5.348
7	1.239	1.564	2.167	2.833	3.822	4.671	6.346
8	1.646	2.032	2.733	3.490	4.594	5.527	7.344
9	2.088	2.532	3.325	4.168	5.380	6.393	8.343
10	2.558	3.059	3.940	4.865	6.179	7.267	9.342
11	3.053	3.609	4.575	5.578	6.989	8.148	10.341
12	3.571	4.178	5.226	6.304	7.807	9.034	11.340
13	4.107	4.765	5.892	7.042	8.634	9.926	12.340
14	4.660	5.368	6.571	7.790	9.467	10.821	13.339
15	5.229	5.985	7.261	8.547	10.307	11.721	14.339
16	5.812	6.614	7.962	9.312	11.152	12.624	15.338
17	6.408	7.255	8.672	10.085	12.002	13.531	16.338
18	7.015	7.906	9.390	10.865	12.857	14.440	17.338
19	7.633	8.567	10.117	11.651	13.716	15.352	18.338
20	8.260	9.237	10.851	12.443	14.578	16.266	19.337
21	8.897	9.915	11.591	13.240	15.445	17.182	20.337
22	9.542	10.600	12.338	14.041	16.314	18.101	21.337
23	10.196	11.293	13.091	14.848	17.187	19.021	22.337
24	10.856	11.992	13.848	15.659	18.062	19.943	23.337
25	11.524	12.697	14.611	16.473	18.940	20.867	24.337
26	12.198	13.409	15.379	17.292	19.820	21.792	25.336
27	12.879	14.125	16.151	18.114	20.703	22.719	26.336
28	13.565	14.847	16.928	18.939	21.588	23.647	27.336
29	14.256	15.574	17.708	19.768	22.475	24.577	28.336
30	14.953	16.306	18.493	20.599	23.364	25.508	29.336

Table XIV is reprinted from Table III of Fisher: *Statistical Methods for Research Workers*, published by Oliver and Boyd Ltd., Edinburgh, by permission of the author and publishers.

Table XIV continued   387

χ^2-DISTRIBUTION

Degrees of Freedom, m	.30	.20	.10	ϵ .05	.02	.01	.001
				χ_0^2			
1	1.074	1.642	2.706	3.841	5.412	6.635	10.827
2	2.408	3.219	4.605	5.991	7.824	9.210	13.815
3	3.665	4.642	6.251	7.815	9.837	11.341	16.268
4	4.878	5.989	7.779	9.488	11.668	13.277	18.465
5	6.064	7.289	9.236	11.070	13.388	15.086	20.517
6	7.231	8.558	10.645	12.592	15.033	16.812	22.457
7	8.383	9.803	12.017	14.067	16.622	18.475	24.322
8	9.524	11.030	13.362	15.507	18.168	20.090	26.125
9	10.656	12.242	14.684	16.919	19.679	21.666	27.877
10	11.781	13.442	15.987	18.307	21.161	23.209	29.588
11	12.899	14.631	17.275	19.675	22.618	24.725	31.264
12	14.011	15.812	18.549	21.026	24.054	26.217	32.909
13	15.119	16.985	19.812	22.362	25.472	27.688	34.528
14	16.222	18.151	21.064	23.685	26.873	29.141	36.123
15	17.322	19.311	22.307	24.996	28.259	30.578	37.697
16	18.418	20.465	23.542	26.296	29.633	32.000	39.252
17	19.511	21.615	24.769	27.587	30.995	33.409	40.790
18	20.601	22.760	25.989	28.869	32.346	34.805	42.312
19	21.689	23.900	27.204	30.144	33.687	36.191	43.820
20	22.775	25.038	28.412	31.410	35.020	37.566	45.315
21	23.858	26.171	29.615	32.671	36.343	38.932	46.797
22	24.939	27.301	30.813	33.924	37.659	40.289	48.268
23	26.018	28.429	32.007	35.172	38.968	41.638	49.728
24	27.096	29.553	33.196	36.415	40.270	42.980	51.179
25	28.172	30.675	34.382	37.652	41.566	44.314	52.620
26	29.246	31.795	35.563	38.885	42.856	45.642	54.052
27	30.319	32.912	36.741	40.113	44.140	46.963	55.476
28	31.391	34.027	37.916	41.337	45.419	48.278	56.893
29	32.461	35.139	39.087	42.557	46.693	49.588	58.302
30	33.530	36.250	40.256	43.773	47.962	50.892	59.703

Table XV

VALUES OF e^{-x}

x	e^{-x}	x	e^{-x}	x	e^{-x}	x	e^{-x}
0.00	1.00000	0.50	.60653	1.00	.36788	1.50	.22313
0.01	0.99005	0.51	.60050	1.01	.36422	1.51	.22091
0.02	.98020	0.52	.59452	1.02	.36059	1.52	.21871
0.03	.97045	0.53	.58860	1.03	.35701	1.53	.21654
0.04	.96079	0.54	.58275	1.04	.35345	1.54	.21438
0.05	.95123	0.55	.57695	1.05	.34994	1.55	.21225
0.06	.94176	0.56	.57121	1.06	.34646	1.56	.21014
0.07	.93239	0.57	.56553	1.07	.34301	1.57	.20805
0.08	.92312	0.58	.55990	1.08	.33960	1.58	.20598
0.09	.91393	0.59	.55433	1.09	.33622	1.59	.20393
0.10	.90484	0.60	.54881	1.10	.33287	1.60	.20190
0.11	.89583	0.61	.54335	1.11	.32956	1.61	.19989
0.12	.88692	0.62	.53794	1.12	.32628	1.62	.19790
0.13	.87810	0.63	.53259	1.13	.32303	1.63	.19593
0.14	.86936	0.64	.52729	1.14	.31982	1.64	.19398
0.15	.86071	0.65	.52205	1.15	.31664	1.65	.19205
0.16	.85214	0.66	.51685	1.16	.31349	1.66	.19014
0.17	.84366	0.67	.51171	1.17	.31037	1.67	.18825
0.18	.83527	0.68	.50662	1.18	.30728	1.68	.18637
0.19	.82696	0.69	.50158	1.19	.30422	1.69	.18452
0.20	.81873	0.70	.49659	1.20	.30119	1.70	.18268
0.21	.81058	0.71	.49164	1.21	.29820	1.71	.18087
0.22	.80252	0.72	.48675	1.22	.29523	1.72	.17907
0.23	.79453	0.73	.48191	1.23	.29229	1.73	.17728
0.24	.78663	0.74	.47711	1.24	.28938	1.74	.17552
0.25	.77880	0.75	.47237	1.25	.28650	1.75	.17377
0.26	.77105	0.76	.46767	1.26	.28365	1.76	.17204
0.27	.76338	0.77	.46301	1.27	.28083	1.77	.17033
0.28	.75578	0.78	.45841	1.28	.27804	1.78	.16864
0.29	.74826	0.79	.45384	1.29	.27527	1.79	.16696
0.30	.74082	0.80	.44933	1.30	.27253	1.80	.16530
0.31	.73345	0.81	.44486	1.31	.26982	1.81	.16365
0.32	.72615	0.82	.44043	1.32	.26714	1.82	.16203
0.33	.71892	0.83	.43605	1.33	.26448	1.83	.16041
0.34	.71177	0.84	.43171	1.34	.26185	1.84	.15882
0.35	.70469	0.85	.42741	1.35	.25924	1.85	.15724
0.36	.69768	0.86	.42316	1.36	.25666	1.86	.15567
0.37	.69073	0.87	.41895	1.37	.25411	1.87	.15412
0.38	.68386	0.88	.41478	1.38	.25158	1.88	.15259
0.39	.67706	0.89	.41066	1.39	.24908	1.89	.15107
0.40	.67032	0.90	.40657	1.40	.24660	1.90	.14957
0.41	.66365	0.91	.40252	1.41	.24414	1.91	.14808
0.42	.65705	0.92	.39852	1.42	.24171	1.92	.14661
0.43	.65051	0.93	.39455	1.43	.23931	1.93	.14515
0.44	.64404	0.94	.39063	1.44	.23693	1.94	.14370
0.45	.63763	0.95	.38674	1.45	.23457	1.95	.14227
0.46	.63128	0.96	.38289	1.46	.23224	1.96	.14086
0.47	.62500	0.97	.37908	1.47	.22993	1.97	.13946
0.48	.61878	0.98	.37531	1.48	.22764	1.98	.13807
0.49	.61263	0.99	.37158	1.49	.22537	1.99	.13670

For extensive tables of e^{-x} see G. F. Becker and C. E. Van Orstrand: *Hyperbolic Functions*, Smithsonian Mathematical Tables, Washington, D. C., 1909.

Table XV continued 389

VALUES OF e^{-x}

x	e^{-x}	x	e^{-x}	x	e^{-x}	x	e^{-x}
2.00	.13534	2.40	.09072	2.80	.06081	4.00	.01832
2.01	.13399	2.41	.08982	2.81	.06020	4.10	.01657
2.02	.13266	2.42	.08892	2.82	.05961	4.20	.01500
2.03	.13134	2.43	.08804	2.83	.05901	4.30	.01357
2.04	.13003	2.44	.08716	2.84	.05843	4.40	.01228
2.05	.12873	2.45	.08629	2.85	.05784	4.50	.01111
2.06	.12745	2.46	.08544	2.86	.05727	4.60	.01005
2.07	.12619	2.47	.08458	2.87	.05670	4.70	.00910
2.08	.12493	2.48	.08374	2.88	.05613	4.80	.00823
2.09	.12369	2.49	.08291	2.89	.05558	4.90	.00745
2.10	.12246	2.50	.08208	2.90	.05502	5.00	.00674
2.11	.12124	2.51	.08127	2.91	.05448	5.10	.00610
2.12	.12003	2.52	.08046	2.92	.05393	5.20	.00552
2.13	.11884	2.53	.07966	2.93	.05340	5.30	.00499
2.14	.11765	2.54	.07887	2.94	.05287	5.40	.00452
2.15	.11648	2.55	.07808	2.95	.05234	5.50	.00409
2.16	.11533	2.56	.07730	2.96	.05182	5.60	.00370
2.17	.11418	2.57	.07654	2.97	.05130	5.70	.00335
2.18	.11304	2.58	.07577	2.98	.05079	5.80	.00303
2.19	.11192	2.59	.07502	2.99	.05029	5.90	.00274
2.20	.11080	2.60	.07427	3.00	.04979	6.00	.00248
2.21	.10970	2.61	.07353	3.05	.04736	6.25	.00193
2.22	.10861	2.62	.07280	3.10	.04505	6.50	.00150
2.23	.10753	2.63	.07208	3.15	.04285	6.75	.00117
2.24	.10646	2.64	.07136	3.20	.04076	7.00	.00091
2.25	.10540	2.65	.07065	3.25	.03877	7.50	.00055
2.26	.10435	2.66	.06995	3.30	.03688	8.00	.00034
2.27	.10331	2.67	.06925	3.35	.03508	8.50	.00020
2.28	.10228	2.68	.06856	3.40	.03337	9.00	.00012
2.29	.10127	2.69	.06788	3.45	.03175	9.50	.00007
2.30	.10026	2.70	.06721	3.50	.03020	10.00	.00005
2.31	.09926	2.71	.06654	3.55	.02872		
2.32	.09827	2.72	.06587	3.60	.02732		
2.33	.09730	2.73	.06522	3.65	.02599		
2.34	.09633	2.74	.06457	3.70	.02472		
2.35	.09537	2.75	.06393	3.75	.02352		
2.36	.09442	2.76	.06329	3.80	.02237		
2.37	.09348	2.77	.06266	3.85	.02128		
2.38	.09255	2.78	.06204	3.90	.02024		
2.39	.09163	2.79	.06142	3.95	.01925		

Table XVI

FACTORIALS AND LOGARITHMS OF FACTORIALS

n	$n!$	$\log_{10} n!$	n	$n!$	$\log_{10} n!$
			50	3.0414×10^{64}	64.48307
1	1.0000	0.00000	51	1.5511×10^{66}	66.19064
2	2.0000	0.30103	52	8.0658×10^{67}	67.90665
3	6.0000	0.77815	53	4.2749×10^{69}	69.63092
4	2.4000×10	1.38021	54	2.3084×10^{71}	71.36332
5	1.2000×10^{2}	2.07918	55	1.2696×10^{73}	73.10368
6	7.2000×10^{2}	2.85733	56	7.1100×10^{74}	74.85187
7	5.0400×10^{3}	3.70243	57	4.0527×10^{76}	76.60774
8	4.0320×10^{4}	4.60552	58	2.3506×10^{78}	78.37117
9	3.6288×10^{5}	5.55976	59	1.3868×10^{80}	80.14202
10	3.6288×10^{6}	6.55976	60	8.3210×10^{81}	81.92017
11	3.9917×10^{7}	7.60116	61	5.0758×10^{83}	83.70550
12	4.7900×10^{8}	8.68034	62	3.1470×10^{85}	85.49790
13	6.2270×10^{9}	9.79428	63	1.9826×10^{87}	87.29724
14	8.7178×10^{10}	10.94041	64	1.2689×10^{89}	89.10342
15	1.3077×10^{12}	12.11650	65	8.2477×10^{90}	90.91633
16	2.0923×10^{13}	13.32062	66	5.4434×10^{92}	92.73587
17	3.5569×10^{14}	14.55107	67	3.6471×10^{94}	94.56195
18	6.4024×10^{15}	15.80634	68	2.4800×10^{96}	96.39446
19	1.2165×10^{17}	17.08509	69	1.7112×10^{98}	98.23331
20	2.4329×10^{18}	18.38612	70	1.1979×10^{100}	100.07841
21	5.1091×10^{19}	19.70834	71	8.5048×10^{101}	101.92966
22	1.1240×10^{21}	21.05077	72	6.1234×10^{103}	103.78700
23	2.5852×10^{22}	22.41249	73	4.4701×10^{105}	105.65032
24	6.2045×10^{23}	23.79271	74	3.3079×10^{107}	107.51955
25	1.5511×10^{25}	25.19065	75	2.4809×10^{109}	109.39461
26	4.0329×10^{26}	26.60562	76	1.8855×10^{111}	111.27543
27	1.0889×10^{28}	28.03698	77	1.4518×10^{113}	113.16192
28	3.0489×10^{29}	29.48414	78	1.1324×10^{115}	115.05401
29	8.8418×10^{30}	30.94654	79	8.9462×10^{116}	116.95164
30	2.6525×10^{32}	32.42366	80	7.1569×10^{118}	118.85473
31	8.2228×10^{33}	33.91502	81	5.7971×10^{120}	120.76321
32	2.6313×10^{35}	35.42017	82	4.7536×10^{122}	122.67703
33	8.6833×10^{36}	36.93869	83	3.9455×10^{124}	124.59610
34	2.9523×10^{38}	38.47016	84	3.3142×10^{126}	126.52038
35	1.0333×10^{40}	40.01423	85	2.8171×10^{128}	128.44980
36	3.7199×10^{41}	41.57054	86	2.4227×10^{130}	130.38430
37	1.3764×10^{43}	43.13874	87	2.1078×10^{132}	132.32382
38	5.2302×10^{44}	44.71852	88	1.8548×10^{134}	134.26830
39	2.0398×10^{46}	46.30959	89	1.6508×10^{136}	136.21769
40	8.1592×10^{47}	47.91165	90	1.4857×10^{138}	138.17194
41	3.3453×10^{49}	49.52443	91	1.3520×10^{140}	140.13098
42	1.4050×10^{51}	51.14768	92	1.2438×10^{142}	142.09476
43	6.0415×10^{52}	52.78115	93	1.1568×10^{144}	144.06325
44	2.6583×10^{54}	54.42460	94	1.0874×10^{146}	146.03638
45	1.1962×10^{56}	56.07781	95	1.0330×10^{148}	148.01410
46	5.5026×10^{57}	57.74057	96	9.9168×10^{149}	149.99637
47	2.5862×10^{59}	59.41267	97	9.6193×10^{151}	151.98314
48	1.2414×10^{61}	61.09391	98	9.4269×10^{153}	153.97437
49	6.0828×10^{62}	62.78410	99	9.3326×10^{155}	155.97000
			100	9.3326×10^{157}	157.97000

Table XVII

GAMMA FUNCTION $\Gamma(n)$

n	0	1	2	3	4	5	6	7	8	9
1.0	1.0000	.9943	.9888	.9835	.9784	.9735	.9687	.9642	.9597	.9555
1.1	.9514	.9474	.9436	.9399	.9364	.9330	.9298	.9267	.9237	.9209
1.2	.9182	.9156	.9131	.9108	.9085	.9064	.9044	.9025	.9007	.8990
1.3	.8975	.8960	.8946	.8934	.8922	.8912	.8902	.8893	.8885	.8879
1.4	.8873	.8868	.8864	.8860	.8858	.8857	.8856	.8856	.8857	.8859
1.5	.8862	.8866	.8870	.8876	.8882	.8889	.8896	.8905	.8914	.8924
1.6	.8935	.8947	.8959	.8972	.8986	.9001	.9017	.9033	.9050	.9068
1.7	.9086	.9106	.9126	.9147	.9168	.9191	.9214	.9238	.9262	.9288
1.8	.9314	.9341	.9368	.9397	.9426	.9456	.9487	.9518	.9551	.9584
1.9	.9618	.9652	.9688	.9724	.9761	.9799	.9837	.9877	.9917	.9958
2.0	1.0000	1.0043	1.0086	1.0131	1.0176	1.0222	1.0269	1.0316	1.0365	1.0415
2.1	1.0465	1.0516	1.0568	1.0621	1.0675	1.0730	1.0786	1.0842	1.0900	1.0959
2.2	1.1018	1.1078	1.1140	1.1202	1.1266	1.1330	1.1395	1.1462	1.1529	1.1598
2.3	1.1667	1.1738	1.1809	1.1882	1.1956	1.2031	1.2107	1.2184	1.2262	1.2341
2.4	1.2422	1.2503	1.2586	1.2670	1.2756	1.2842	1.2930	1.3019	1.3109	1.3201
2.5	1.3293	1.3388	1.3483	1.3580	1.3678	1.3777	1.3878	1.3981	1.4084	1.4190
2.6	1.4296	1.4404	1.4514	1.4625	1.4738	1.4852	1.4968	1.5085	1.5204	1.5325
2.7	1.5447	1.5571	1.5696	1.5824	1.5953	1.6084	1.6216	1.6351	1.6487	1.6625
2.8	1.6765	1.6907	1.7051	1.7196	1.7344	1.7494	1.7646	1.7799	1.7955	1.8113
2.9	1.8274	1.8436	1.8600	1.8767	1.8936	1.9108	1.9281	1.9457	1.9636	1.9816
3.0	2.000	2.019	2.037	2.057	2.076	2.095	2.115	2.136	2.156	2.177
3.1	2.198	2.219	2.240	2.262	2.284	2.307	2.330	2.353	2.376	2.400
3.2	2.424	2.448	2.473	2.498	2.524	2.549	2.575	2.602	2.629	2.656
3.3	2.683	2.711	2.740	2.768	2.798	2.827	2.857	2.888	2.918	2.950
3.4	2.981	3.013	3.046	3.079	3.112	3.146	3.181	3.216	3.251	3.287
3.5	3.323	3.360	3.398	3.436	3.474	3.513	3.553	3.593	3.634	3.675
3.6	3.717	3.760	3.803	3.846	3.891	3.936	3.981	4.028	4.075	4.122
3.7	4.171	4.220	4.269	4.320	4.371	4.423	4.476	4.529	4.583	4.638
3.8	4.694	4.751	4.808	4.867	4.926	4.986	5.047	5.108	5.171	5.235
3.9	5.299	5.365	5.431	5.499	5.567	5.637	5.707	5.779	5.851	5.925
4.0	6.000	6.076	6.153	6.231	6.311	6.391	6.473	6.556	6.640	6.726
4.1	6.813	6.901	6.990	7.081	7.173	7.267	7.362	7.458	7.556	7.656
4.2	7.757	7.859	7.963	8.069	8.176	8.285	8.396	8.508	8.622	8.738
4.3	8.855	8.975	9.096	9.219	9.344	9.471	9.600	9.731	9.864	9.999
4.4	10.136	10.275	10.417	10.561	10.707	10.855	11.005	11.158	11.314	11.471
4.5	11.632	11.795	11.960	12.128	12.299	12.472	12.648	12.827	13.009	13.194
4.6	13.381	13.572	13.766	13.962	14.162	14.366	14.572	14.782	14.995	15.211
4.7	15.431	15.655	15.882	16.113	16.348	16.586	16.829	17.075	17.325	17.579
4.8	17.84	18.10	18.37	18.64	18.91	19.20	19.48	19.77	20.06	20.36
4.9	20.67	20.98	21.29	21.61	21.94	22.27	22.60	22.94	23.29	23.64

For example, $\Gamma(2.83) = 1.7196$.

$$\Gamma(n) = \int_0^\infty x^{n-1} e^{-x}\, dx. \quad \Gamma(n) = (n-1)\cdot\Gamma(n-1), \quad n > 1.$$

$\Gamma(n) = (n-1)!$ when n is a positive integer.

Table XVIII

COMMON LOGARITHMS OF GAMMA FUNCTION $\Gamma(n)$

$$\Gamma(n) = \int_0^\infty x^{n-1}e^{-x}\,dx = \int_0^1 [\log_e (1/x)]^{n-1}\,dx.$$

n	$\log_{10}\Gamma(n)+10$	n	$\log_{10}\Gamma(n)+10$	n	$\log_{10}\Gamma(n)+10$	n	$\log_{10}\Gamma(n)+10$	n	$\log_{10}\Gamma(n)+10$
1.01	9.9975	1.21	9.9617	1.41	9.9478	1.61	9.9517	1.81	9.9704
1.02	9.9951	1.22	9.9605	1.42	9.9476	1.62	9.9523	1.82	9.9717
1.03	9.9928	1.23	9.9594	1.43	9.9475	1.63	9.9529	1.83	9.9730
1.04	9.9905	1.24	9.9583	1.44	9.9473	1.64	9.9536	1.84	9.9743
1.05	9.9883	1.25	9.9573	1.45	9.9473	1.65	9.9543	1.85	9.9757
1.06	9.9862	1.26	9.9564	1.46	9.9472	1.66	9.9550	1.86	9.9771
1.07	9.9841	1.27	9.9554	1.47	9.9473	1.67	9.9558	1.87	9.9786
1.08	9.9821	1.28	9.9546	1.48	9.9473	1.68	9.9566	1.88	9.9800
1.09	9.9802	1.29	9.9538	1.49	9.9474	1.69	9.9575	1.89	9.9815
1.10	9.9783	1.30	9.9530	1.50	9.9475	1.70	9.9584	1.90	9.9831
1.11	9.9765	1.31	9.9523	1.51	9.9477	1.71	9.9593	1.91	9.9846
1.12	9.9748	1.32	9.9516	1.52	9.9479	1.72	9.9603	1.92	9.9862
1.13	9.9731	1.33	9.9510	1.53	9.9482	1.73	9.9613	1.93	9.9878
1.14	9.9715	1.34	9.9505	1.54	9.9485	1.74	9.9623	1.94	9.9895
1.15	9.9699	1.35	9.9500	1.55	9.9488	1.75	9.9633	1.95	9.9912
1.16	9.9684	1.36	9.9495	1.56	9.9492	1.76	9.9644	1.96	9.9929
1.17	9.9669	1.37	9.9491	1.57	9.9496	1.77	9.9656	1.97	9.9946
1.18	9.9655	1.38	9.9487	1.58	9.9501	1.78	9.9667	1.98	9.9964
1.19	9.9642	1.39	9.9483	1.59	9.9506	1.79	9.9679	1.99	9.9982
1.20	9.9629	1.40	9.9481	1.60	9.9511	1.80	9.9691	2.00	10.0000

$$\Gamma(n+1) = n\cdot\Gamma(n), \qquad n > 0. \qquad \Gamma(2) = \Gamma(1) = 1.$$

Table XIX

FACTORIALS AND THEIR RECIPROCALS

n	$n!$	n	$n!$
1	1	11	39916800
2	2	12	479001600
3	6	13	6227020800
4	24	14	87178291200
5	120	15	1307674368000
6	720	16	20922789888000
7	5040	17	355687428096000
8	40320	18	6402373705728000
9	362880	19	121645100408832000
10	3628800	20	2432902008176640000

n	$1/n!$	n	$1/n!$
1	1.	11	$.25052 \times 10^{-7}$
2	0.5	12	$.20877 \times 10^{-8}$
3	.16667	13	$.16059 \times 10^{-9}$
4	$.41667 \times 10^{-1}$	14	$.11471 \times 10^{-10}$
5	$.83333 \times 10^{-2}$	15	$.76472 \times 10^{-12}$
6	$.13889 \times 10^{-2}$	16	$.47795 \times 10^{-13}$
7	$.19841 \times 10^{-3}$	17	$.28115 \times 10^{-14}$
8	$.24802 \times 10^{-4}$	18	$.15619 \times 10^{-15}$
9	$.27557 \times 10^{-5}$	19	$.82206 \times 10^{-17}$
10	$.27557 \times 10^{-6}$	20	$.41103 \times 10^{-18}$

Table XX

TWO-SIDED TOLERANCE FACTORS FOR NORMAL POPULATIONS

The entries in this table are factors K such that the probability is γ that at least a proportion P of the population distribution will be included in the interval $(\bar{x} - Ks, \bar{x} + Ks)$, where $\bar{x}$ and s are the mean and unbiased standard deviation of a sample of size n. γ is the confidence coefficient. Here $s^2 = \sum_{i=1}^{n} (x_i - \bar{x})^n/(n-1)$.

	$\gamma = 0.75$				$\gamma = 0.90$			
		P				P		
n	0.75	0.90	0.95	0.99	0.75	0.90	0.95	0.99
5	1.82	2.60	3.09	4.03	2.45	3.49	4.15	5.42
6	1.70	2.43	2.89	3.78	2.20	3.13	3.72	4.87
7	1.62	2.32	2.76	3.61	2.03	2.90	3.45	4.52
8	1.57	2.24	2.66	3.49	1.92	2.74	3.26	4.28
9	1.52	2.18	2.59	3.40	1.84	2.63	3.12	4.10
10	1.49	2.13	2.54	3.33	1.78	2.54	3.02	3.96
15	1.40	1.99	2.38	3.12	1.59	2.28	2.71	3.56
20	1.35	1.92	2.29	3.01	1.51	2.15	2.56	3.37
30	1.30	1.86	2.21	2.90	1.42	2.02	2.41	3.17
40	1.27	1.82	2.17	2.85	1.37	1.96	2.33	3.07
50	1.26	1.79	2.14	2.81	1.34	1.92	2.28	3.00
150	1.20	1.72	2.05	2.70	1.25	1.78	2.13	2.80
∞	1.15	1.64	1.96	2.58	1.15	1.64	1.96	2.58

	$\gamma = 0.95$				$\gamma = 0.99$			
		P				P		
n	0.75	0.90	0.95	0.99	0.75	0.90	0.95	0.99
5	3.00	4.28	5.08	6.63	4.64	6.61	7.86	10.26
6	2.60	3.71	4.41	5.78	3.74	5.34	6.34	8.30
7	2.36	3.37	4.01	5.25	3.23	4.61	5.49	7.19
8	2.20	3.14	3.73	4.89	2.90	4.15	4.94	6.47
9	2.08	2.97	3.53	4.63	2.68	3.82	4.55	5.97
10	1.99	2.84	3.38	4.43	2.51	3.58	4.26	5.59
15	1.74	2.48	2.95	3.88	2.06	2.94	3.51	4.60
20	1.62	2.31	2.75	3.62	1.86	2.66	3.17	4.16
30	1.50	2.14	2.55	3.35	1.67	2.38	2.84	3.73
40	1.44	2.05	2.44	3.21	1.57	2.25	2.68	3.52
50	1.40	2.00	2.38	3.13	1.51	2.16	2.58	3.38
150	1.28	1.82	2.18	2.86	1.33	1.90	2.27	2.98
∞	1.15	1.64	1.96	2.58	1.15	1.64	1.96	2.58

Table XXI

ONE-SIDED TOLERANCE FACTORS FOR NORMAL POPULATIONS

The entries in this table are factors K such that the probability is γ that at least a proportion P of the population distribution will be included in the interval $(\bar{x} - Ks, +\infty)$, or the interval $(-\infty, \bar{x} + Ks)$, where $\bar{x}$ and s are the mean and unbiased standard deviation of a sample of size n. γ is the confidence coefficient. Here $s^2 = \sum_{i=1}^{n} (x_i - \bar{x})^n/(n - 1)$.

	$\gamma = 0.75$				$\gamma = 0.90$			
	P				P			
n	0.75	0.90	0.95	0.99	0.75	0.90	0.95	0.99
5	1.15	1.96	2.46	3.42	1.70	2.74	3.40	4.67
6	1.09	1.86	2.34	3.24	1.54	2.49	3.09	4.24
7	1.04	1.79	2.25	3.13	1.44	2.33	2.89	3.97
8	1.01	1.74	2.19	3.04	1.36	2.22	2.76	3.78
9	0.98	1.70	2.14	2.98	1.30	2.13	2.65	3.64
10	0.96	1.67	2.10	2.93	1.26	2.06	2.57	3.53
15	0.90	1.58	1.99	2.78	1.12	1.87	2.33	3.21
20	0.86	1.53	1.93	2.70	1.05	1.76	2.21	3.05
30	0.82	1.48	1.87	2.61	0.97	1.66	2.08	2.88
40	0.80	1.44	1.83	2.57	0.92	1.60	2.01	2.79
50	0.79	1.43	1.81	2.54	0.89	1.56	1.96	2.74

	$\gamma = 0.95$				$\gamma = 0.99$			
	P				P			
n	0.75	0.90	0.95	0.99	0.75	0.90	0.95	0.99
5	2.15	3.41	4.20	5.74				
6	1.90	3.01	3.71	5.06	2.85	4.41	5.41	7.33
7	1.73	2.76	3.40	4.64	2.49	3.86	4.73	6.41
8	1.62	2.58	3.19	4.35	2.25	3.50	4.29	5.81
9	1.53	2.45	3.03	4.14	2.08	3.24	3.97	5.39
10	1.46	2.36	2.91	3.98	1.95	3.05	3.74	5.08
15	1.27	2.07	2.57	3.52	1.60	2.52	3.10	4.22
20	1.17	1.93	2.40	3.30	1.42	2.28	2.81	3.83
30	1.06	1.78	2.22	3.06	1.25	2.03	2.52	3.45
40	1.00	1.70	2.13	2.94	1.15	1.90	2.36	3.25
50	0.96	1.65	2.06	2.86	1.10	1.82	2.27	3.12

Table XXII

RANDOM NUMBERS* (UNIFORM)

10	09	73	25	33	76	52	01	35	86	34	67	35	48	76	80	95	90	91	17	39	29	27	49	45
37	54	20	48	05	64	89	47	42	96	24	80	52	40	37	20	63	61	04	02	00	82	29	16	65
08	42	26	89	53	19	64	50	93	03	23	20	90	25	60	15	95	33	47	64	35	08	03	36	06
99	01	90	25	29	09	37	67	07	15	38	31	13	11	65	88	67	67	43	97	04	43	62	76	59
12	80	79	99	70	80	15	73	61	47	64	03	23	66	53	98	95	11	68	77	12	17	17	68	33
66	06	57	47	17	34	07	27	68	50	36	69	73	61	70	65	81	33	98	85	11	19	92	91	70
31	06	01	08	05	45	57	18	24	06	35	30	34	26	14	86	79	90	74	39	23	40	30	97	32
85	26	97	76	02	02	05	16	56	92	68	66	57	48	18	73	05	38	52	47	18	62	38	85	79
63	57	33	21	35	05	32	54	70	48	90	55	35	75	48	28	46	82	87	09	83	49	12	56	24
73	79	64	57	53	03	52	96	47	78	35	80	83	42	82	60	93	52	03	44	35	27	38	84	35
98	52	01	77	67	14	90	56	86	07	22	10	94	05	58	60	97	09	34	33	50	50	07	39	98
11	80	50	54	31	39	80	82	77	32	50	72	56	82	48	29	40	52	42	01	52	77	56	78	51
83	45	29	96	34	06	28	89	80	83	13	74	67	00	78	18	47	54	06	10	68	71	17	78	17
88	68	54	02	00	86	50	75	84	01	36	76	66	79	51	90	36	47	64	93	29	60	91	10	62
99	59	46	73	48	87	51	76	49	69	91	82	60	89	28	93	78	56	13	68	23	47	83	41	13
65	48	11	76	74	17	46	85	09	50	58	04	77	69	74	73	03	95	71	86	40	21	81	65	44
80	12	43	56	35	17	72	70	80	15	45	31	82	23	74	21	11	57	82	53	14	38	55	37	63
74	35	09	98	17	77	40	27	72	14	43	23	60	02	10	45	52	16	42	37	96	28	60	26	55
69	91	62	68	03	66	25	22	91	48	36	93	68	72	03	76	62	11	39	90	94	40	05	64	18
09	89	32	05	05	14	22	56	85	14	46	42	75	67	88	96	29	77	88	22	54	38	21	45	98
91	49	91	45	23	68	47	92	76	86	46	16	28	35	54	94	75	08	99	23	37	08	92	00	48
80	33	69	45	98	26	94	03	68	58	70	29	73	41	35	53	14	03	33	40	42	05	08	23	41
44	10	48	19	49	85	15	74	79	54	32	97	92	65	75	57	60	04	08	81	22	22	20	64	13
12	55	07	37	42	11	10	00	20	40	12	86	07	46	97	96	64	48	94	39	28	70	72	58	15
63	60	64	93	29	16	50	53	44	84	40	21	95	25	63	43	65	17	70	82	07	20	73	17	90
61	19	69	04	46	26	45	74	77	74	51	92	43	37	29	65	39	45	95	93	42	58	26	05	27
15	47	44	52	66	95	27	07	99	53	59	36	78	38	48	82	39	61	01	18	33	21	15	94	66
94	55	72	85	73	67	89	75	43	87	54	62	24	44	31	91	19	04	25	92	92	92	74	59	73
42	48	11	62	13	97	34	40	87	21	16	86	84	87	67	03	07	11	20	59	25	70	14	66	70
23	52	37	83	17	73	20	88	98	37	68	93	59	14	16	26	25	22	96	63	05	52	28	25	62
04	49	35	24	94	75	24	63	38	24	45	86	25	10	25	61	96	27	93	35	65	33	71	24	72
00	54	99	76	54	64	05	18	81	59	96	11	96	38	96	54	69	28	23	91	23	28	72	95	29
35	96	31	53	07	26	89	80	93	54	33	35	13	54	62	77	97	45	00	24	90	10	33	93	33
59	80	80	83	91	45	42	72	68	42	83	60	94	97	00	13	02	12	48	92	78	56	52	01	06
46	05	88	52	36	01	39	09	22	86	77	28	14	40	77	93	91	08	36	47	70	61	74	29	41
32	17	90	05	97	87	37	92	52	41	05	56	70	70	07	86	74	31	71	57	85	39	41	18	38
69	23	46	14	06	20	11	74	52	04	15	95	66	00	00	18	74	39	24	23	97	11	89	63	38
19	56	54	14	30	01	75	87	53	79	40	41	92	15	85	66	67	43	68	06	84	96	28	52	07
45	15	51	49	38	19	47	60	72	46	43	66	79	45	43	59	04	79	00	33	20	82	66	95	41
94	86	43	19	94	36	16	81	08	51	34	88	88	15	53	01	54	03	54	56	05	01	45	11	76
98	08	62	48	26	45	24	02	84	04	44	99	90	88	96	39	09	47	34	07	35	44	13	18	80
33	18	51	62	32	41	94	15	09	49	89	43	54	85	81	88	69	54	19	94	37	54	87	30	43
80	95	10	04	06	96	38	27	07	74	20	15	12	33	87	25	01	62	52	98	94	62	46	11	71
79	75	24	91	40	71	96	12	82	96	69	86	10	25	91	74	85	22	05	39	00	38	75	95	79
18	63	33	25	37	98	14	50	65	71	31	01	02	46	74	05	45	56	14	27	77	93	89	19	36
74	02	94	39	02	77	55	73	22	70	97	79	01	71	19	52	52	75	80	21	80	81	45	17	48
54	17	84	56	11	80	99	33	71	43	05	33	51	29	69	56	12	71	92	55	36	04	09	03	24
11	66	44	98	83	52	07	98	48	27	59	38	17	15	39	09	97	33	34	40	88	46	12	33	56
48	32	47	79	28	31	24	96	47	10	02	29	53	68	70	32	30	75	75	46	15	02	00	99	94
69	07	49	41	38	87	63	79	19	76	35	58	40	44	01	10	51	82	16	15	01	84	87	69	38

This table is reproduced with permission from *A Million Random Digits*, published by the RAND Corporation.

Table XXII continued

RANDOM NUMBERS

09	18	82	00	97	32	82	53	95	27	04	22	08	63	04	83	38	98	73	74	64	27	85	80	44
90	04	58	54	97	51	98	15	06	54	94	93	88	19	97	91	87	07	61	50	68	47	66	46	59
73	18	95	02	07	47	67	72	62	69	62	29	06	44	64	27	12	46	70	18	41	36	18	27	60
75	76	87	64	90	20	97	18	17	49	90	42	91	22	72	95	37	50	58	71	93	82	34	31	78
54	01	64	40	56	66	28	13	10	03	00	68	22	73	98	20	71	45	32	95	07	70	61	78	13
08	35	86	99	10	78	54	24	27	85	13	66	15	88	73	04	61	89	75	53	31	22	30	84	20
28	30	60	32	64	81	33	31	05	91	40	51	00	78	93	32	60	46	04	75	94	11	90	18	40
53	84	08	62	33	81	59	41	36	28	51	21	59	02	90	28	46	66	87	95	77	76	22	07	91
91	75	75	37	41	61	61	36	22	69	50	26	39	02	12	55	78	17	65	14	83	48	34	70	55
89	41	59	26	94	00	39	75	83	91	12	60	71	76	46	48	94	97	23	06	94	54	13	74	08
77	51	30	38	20	86	83	42	99	01	68	41	48	27	74	51	90	81	39	80	72	89	35	55	07
19	50	23	71	74	69	97	92	02	88	55	21	02	97	73	74	28	77	52	51	65	34	46	74	15
21	81	85	93	13	93	27	88	17	57	05	68	67	31	56	07	08	28	50	46	31	85	33	84	52
51	47	46	64	99	68	10	72	36	21	94	04	99	13	45	42	83	60	91	91	08	00	74	54	49
99	55	96	83	31	62	53	52	41	70	69	77	71	28	30	74	81	97	81	42	43	86	07	28	34
33	71	34	80	07	93	58	47	28	69	51	92	66	47	21	58	30	32	98	22	93	17	49	39	72
85	27	48	68	93	11	30	32	92	70	28	83	43	41	37	73	51	59	04	00	71	14	84	36	43
84	13	38	96	40	44	03	55	21	66	73	85	27	00	91	61	22	26	05	61	62	32	71	84	23
56	73	21	62	34	17	39	59	61	31	10	12	39	16	22	85	49	65	75	60	81	60	41	88	80
65	13	85	68	06	87	64	88	52	61	34	31	36	58	61	45	87	52	10	69	85	64	44	72	77
38	00	10	21	76	81	71	91	17	11	71	60	29	29	37	74	21	96	40	49	65	58	44	96	98
37	40	29	63	97	01	30	47	75	86	56	27	11	00	86	47	32	46	26	05	40	03	03	74	38
97	12	54	03	48	87	08	33	14	17	21	81	53	92	50	75	23	76	20	47	15	50	12	95	78
21	82	64	11	34	47	14	33	40	72	64	63	88	59	02	49	13	90	64	41	03	85	65	45	52
73	13	54	27	42	95	71	90	90	35	85	79	47	42	96	08	78	98	81	56	64	69	11	92	02
07	63	87	79	29	03	06	11	80	72	96	20	74	41	56	23	82	19	95	38	04	71	36	69	94
60	52	88	34	41	07	95	41	98	14	59	17	52	06	95	05	53	35	21	39	61	21	20	64	55
83	59	63	56	55	06	95	89	29	83	05	12	80	97	19	77	43	35	37	83	92	30	15	04	98
10	85	06	27	46	99	59	91	05	07	13	49	90	63	19	53	07	57	18	39	06	41	01	93	62
39	82	09	89	52	43	62	26	31	47	64	42	18	08	14	43	80	00	93	51	31	02	47	31	67
59	58	00	64	78	75	56	97	88	00	88	83	55	44	86	23	76	80	61	56	04	11	10	84	08
38	50	80	73	41	23	79	34	87	63	90	82	29	70	22	17	71	90	42	07	95	95	44	99	53
30	69	27	06	68	94	68	81	61	27	56	19	68	00	91	82	06	76	34	00	05	46	26	92	00
65	44	39	56	59	18	28	82	74	37	49	63	22	40	41	08	33	76	56	76	96	29	99	08	36
27	26	75	02	64	13	19	27	22	94	07	47	74	46	06	17	98	54	89	11	97	34	13	03	58
91	30	70	69	91	19	07	22	42	10	36	69	95	37	28	28	82	53	57	93	28	97	66	62	52
68	43	49	46	88	84	47	31	36	22	62	12	69	84	08	12	84	38	25	90	09	81	59	31	46
48	90	81	58	77	54	74	52	45	91	35	70	00	47	54	83	82	45	26	92	54	13	05	51	60
06	91	34	51	97	42	67	27	86	01	11	88	30	95	28	63	01	19	89	01	14	97	44	03	44
10	45	51	60	19	14	21	03	37	12	91	34	23	78	21	88	32	58	08	51	43	66	77	08	83
12	88	39	73	43	65	02	76	11	84	04	28	50	13	92	17	97	41	50	77	90	71	22	67	69
21	77	83	09	76	38	80	73	69	61	31	64	94	20	96	63	28	10	20	23	08	81	64	74	49
19	52	35	95	15	65	12	25	96	59	86	28	36	82	58	69	57	21	37	98	16	43	59	15	29
67	24	55	26	70	35	58	31	65	63	79	24	68	66	86	76	46	33	42	22	26	65	59	08	02
60	58	44	73	77	07	50	03	79	92	45	13	42	65	29	26	76	08	36	37	41	32	64	43	44
53	85	34	13	77	36	06	69	48	50	58	83	87	38	59	49	36	47	33	31	96	24	04	36	42
24	63	73	87	36	74	38	48	93	42	52	62	30	79	92	12	36	91	86	01	03	74	28	38	73
83	08	01	24	51	38	99	22	28	15	07	75	95	17	77	97	37	72	75	85	51	97	23	78	67
16	44	42	43	34	36	15	19	90	73	27	49	37	09	39	85	13	03	25	52	54	84	65	47	59
60	79	01	81	57	57	17	86	57	62	11	16	17	85	76	45	81	95	29	79	65	13	00	48	60

Table XXII continued   397

RANDOM NUMBERS

```
03 99 11 04 61    93 71 61 68 94    66 08 32 46 53    84 60 95 82 32    88 61 81 91 61
38 55 59 55 54    32 88 65 97 80    08 35 56 08 60    29 73 54 77 62    71 29 92 38 53
17 54 67 37 04    92 05 24 62 15    55 12 12 92 81    59 07 60 79 36    27 95 45 89 09
32 64 35 28 61    95 81 90 68 31    00 91 19 89 36    76 35 59 37 79    80 86 30 05 14
69 57 26 87 77    39 51 03 59 05    14 06 04 06 19    29 54 96 96 16    33 56 46 07 80

24 12 26 65 91    27 69 90 64 94    14 84 54 66 72    61 95 87 71 00    90 89 97 57 54
61 19 63 02 31    92 96 26 17 73    41 83 95 53 82    17 26 77 09 43    78 03 87 02 67
30 53 22 17 04    10 27 41 22 02    39 68 52 33 09    10 06 16 88 29    55 98 66 64 85
03 78 89 75 99    75 86 72 07 17    74 41 65 31 66    35 20 83 33 74    87 53 90 88 23
48 22 86 33 79    85 78 34 76 19    53 15 26 74 33    35 66 35 29 72    16 81 86 03 11

60 36 59 46 53    35 07 53 39 49    42 61 42 92 97    01 91 82 83 16    98 95 37 32 31
83 79 94 24 02    56 62 33 44 42    34 99 44 13 74    70 07 11 47 36    09 95 81 80 65
32 96 00 74 05    36 40 98 32 32    99 38 54 16 00    11 13 30 75 86    15 91 70 62 53
19 32 25 38 45    57 62 05 26 06    66 49 76 86 46    78 13 86 65 59    19 64 09 94 13
11 22 09 47 47    07 39 93 74 08    48 50 92 39 29    27 48 24 54 76    85 24 43 51 59

31 75 15 72 60    68 98 00 53 39    15 47 04 83 55    88 65 12 25 96    03 15 21 91 21
88 49 29 93 82    14 45 40 45 04    20 09 49 89 77    74 84 39 34 13    22 10 97 85 08
30 93 44 77 44    07 48 18 38 28    73 78 80 65 33    28 59 72 04 05    94 20 52 03 80
22 88 84 88 93    27 49 99 87 48    60 53 04 51 28    74 02 28 46 17    82 03 71 02 68
78 21 21 69 93    35 90 29 13 86    44 37 21 54 86    65 74 11 40 14    87 48 13 72 20

41 84 98 45 47    46 85 05 23 26    34 67 75 83 00    74 91 06 43 45    19 32 58 15 49
46 35 23 30 49    69 24 89 34 60    45 30 50 75 21    61 31 83 18 55    14 41 37 09 51
11 08 79 62 94    14 01 33 17 92    59 74 76 72 77    76 50 33 45 13    39 66 37 75 44
52 70 10 83 37    56 30 38 73 15    16 52 06 96 76    11 65 49 98 93    02 18 16 81 61
57 27 53 68 98    81 30 44 85 85    68 65 22 73 76    92 85 25 58 66    88 44 80 35 84

20 85 77 31 56    70 28 42 43 26    79 37 59 52 20    01 15 96 32 67    10 62 24 83 91
15 63 38 49 24    90 41 59 36 14    33 52 12 66 65    55 82 34 76 41    86 22 53 17 04
92 69 44 82 97    39 90 40 21 15    59 58 94 90 67    66 82 14 15 75    49 76 70 40 37
77 61 31 90 19    88 15 20 00 80    20 55 49 14 09    96 27 74 82 57    50 81 69 76 16
38 68 83 24 86    45 13 46 35 45    59 40 47 20 59    43 94 75 16 80    43 85 25 96 93

25 16 30 18 89    70 01 41 50 21    41 29 06 73 12    71 85 71 59 57    68 97 11 14 03
65 25 10 76 29    37 23 93 32 95    05 87 00 11 19    92 78 42 63 40    18 47 76 56 22
36 81 54 36 25    18 63 73 75 09    82 44 49 90 05    04 92 17 37 01    14 70 79 39 97
64 39 71 16 92    05 32 78 21 62    20 24 78 17 59    45 19 72 53 32    83 74 52 25 67
04 51 52 56 24    95 09 66 79 46    48 46 08 55 58    15 19 11 87 82    16 93 03 33 61

83 76 16 08 73    43 25 38 41 45    60 83 32 59 83    01 29 14 13 49    20 36 80 71 26
14 38 70 63 45    80 85 40 92 79    43 52 90 63 18    38 38 47 47 61    41 19 63 74 80
51 32 19 22 46    80 08 87 70 74    88 72 25 67 36    66 16 44 94 31    66 91 93 16 78
72 47 20 00 08    80 89 01 80 02    94 81 33 19 00    54 15 58 34 36    35 35 25 41 31
05 46 65 53 06    93 12 81 84 64    74 45 79 05 61    72 84 81 18 34    79 98 26 84 16

39 52 87 24 84    82 47 42 55 93    48 54 53 52 47    18 61 91 36 74    18 61 11 92 41
81 61 61 87 11    53 34 24 42 76    75 12 21 17 24    74 62 77 37 07    58 31 91 59 97
07 58 61 61 20    82 64 12 28 20    92 90 41 31 41    32 39 21 97 63    61 19 96 79 40
90 76 70 42 35    13 57 41 72 00    69 90 26 37 42    78 46 42 25 01    18 62 79 08 72
40 18 82 81 93    29 59 38 86 27    94 97 21 15 98    62 09 53 67 87    00 44 15 89 97

34 41 48 21 57    86 88 75 50 87    19 15 20 00 23    12 30 28 07 83    32 62 46 86 91
63 43 97 53 63    44 98 91 68 22    36 02 40 08 67    76 37 84 16 05    65 96 17 34 88
67 04 90 90 70    93 39 94 55 47    94 45 87 42 84    05 04 14 98 07    20 28 83 40 60
79 49 50 41 46    52 16 29 02 86    54 15 83 42 43    46 97 83 54 82    59 36 29 59 38
91 70 43 05 52    04 73 72 10 31    75 05 19 30 29    47 66 56 43 82    99 78 29 34 78
```

RANDOM NUMBERS

94 01 54 68 74	32 44 44 82 77	59 82 09 61 63	64 65 42 58 43	41 14 54 28 20
74 10 88 82 22	88 57 07 40 15	25 70 49 10 35	01 75 51 47 50	48 96 83 86 03
62 88 08 78 73	95 16 05 92 21	22 30 49 03 14	72 87 71 73 34	39 28 30 41 49
11 74 81 21 02	80 58 04 18 67	17 71 05 96 21	06 55 40 78 50	73 95 07 95 52
17 94 40 56 00	60 47 80 33 43	25 85 25 89 05	57 21 63 96 18	49 85 69 93 26
66 06 74 27 92	95 04 35 26 80	46 78 05 64 87	09 97 15 94 81	37 00 62 21 86
54 24 49 10 30	45 54 77 08 18	59 84 99 61 69	61 45 92 16 47	87 41 71 71 98
30 94 55 75 89	31 73 25 72 60	47 67 00 76 54	46 37 62 53 66	94 74 64 95 80
69 17 03 74 03	86 99 59 03 07	94 30 47 18 03	26 82 50 55 11	12 45 99 13 14
08 34 58 89 75	35 84 18 57 71	08 10 55 99 87	87 11 22 14 76	14 71 37 11 81
27 76 74 35 84	85 30 18 89 77	29 49 06 97 14	73 03 54 12 07	74 69 90 93 10
13 02 51 43 38	54 06 61 52 43	47 72 46 67 33	47 43 14 39 05	31 04 85 66 99
80 21 73 62 92	98 52 52 43 35	24 43 22 48 96	43 27 75 88 74	11 46 61 60 82
10 87 56 20 04	90 39 16 11 05	57 41 10 63 68	53 85 63 07 43	08 67 08 47 41
54 12 75 73 26	26 62 91 90 87	24 47 28 87 79	30 54 02 78 86	61 73 27 54 54
60 31 14 28 24	37 30 14 26 78	45 99 04 32 42	17 37 45 20 03	70 70 77 02 14
49 73 97 14 84	92 00 39 80 86	76 66 87 32 09	59 20 21 19 73	02 90 23 32 50
78 62 65 15 94	16 45 39 46 14	39 01 49 70 66	83 01 20 98 32	25 57 17 76 28
66 69 21 39 86	99 83 70 05 82	81 23 24 49 87	09 50 49 64 12	90 19 37 95 68
44 07 12 80 91	07 36 29 77 03	76 44 74 25 37	98 52 49 78 31	65 70 40 95 14
41 46 88 51 49	49 55 41 79 94	14 92 43 96 50	95 29 40 05 56	70 48 10 69 05
94 55 93 75 59	49 67 85 31 19	70 31 20 56 82	66 98 63 40 99	74 47 42 07 40
41 61 57 03 60	64 11 45 86 60	90 85 06 46 18	80 62 05 17 90	11 43 63 80 72
50 27 39 31 13	41 79 48 68 61	24 78 18 96 83	55 41 18 56 67	77 53 59 98 92
41 39 68 05 04	90 67 00 82 89	40 90 20 50 69	95 08 30 67 83	28 10 25 78 16
25 80 72 42 60	71 52 97 89 20	72 68 20 73 85	90 72 65 71 66	98 88 40 85 83
06 17 09 79 65	88 30 29 80 41	21 44 34 18 08	68 98 48 36 20	89 74 79 88 82
60 80 85 44 44	74 41 28 11 05	01 17 62 88 38	36 42 11 64 89	18 05 95 10 61
80 94 04 48 93	10 40 83 62 22	80 58 27 19 44	92 63 84 03 33	67 05 41 60 67
19 51 69 01 20	46 75 97 16 43	13 17 75 52 92	21 03 68 28 08	77 50 19 74 27
49 38 65 44 80	23 60 42 35 54	21 78 54 11 01	91 17 81 01 74	29 42 09 04 38
06 31 28 89 40	15 99 56 93 21	47 45 86 48 09	98 18 98 18 51	29 65 18 42 15
60 94 20 03 07	11 89 79 26 74	40 40 56 80 32	96 71 75 42 44	10 70 14 13 93
92 32 99 89 32	78 28 44 63 47	71 20 99 20 61	39 44 89 31 36	25 72 20 85 64
77 93 66 35 74	31 38 45 19 24	85 56 12 96 71	58 13 71 78 20	22 75 13 65 18
38 10 17 77 56	11 65 71 38 97	95 88 95 70 67	47 64 81 38 85	70 66 99 34 06
39 64 16 94 57	91 33 92 25 02	92 61 38 97 19	11 94 75 62 03	19 32 42 05 04
84 05 44 04 55	99 39 66 36 80	67 66 76 06 31	69 18 19 68 45	38 52 51 16 00
47 46 80 35 77	57 64 96 32 66	24 70 07 15 94	14 00 42 31 53	69 24 90 57 47
43 32 13 13 70	28 97 72 38 96	76 47 96 85 62	62 34 20 75 89	08 89 90 59 85
64 28 16 18 26	18 55 56 49 37	13 17 33 33 65	78 85 11 64 99	87 06 41 30 75
66 84 77 04 95	32 35 00 29 85	86 71 63 87 46	26 31 37 74 63	55 38 77 26 81
72 46 13 32 30	21 52 95 34 24	92 58 10 22 62	78 43 86 62 76	18 39 67 35 38
21 03 29 10 50	13 05 81 62 18	12 47 05 65 00	15 29 27 61 39	59 52 65 21 13
95 36 26 70 11	06 65 11 61 36	01 01 60 08 57	55 01 85 63 74	35 82 47 17 08
40 71 29 73 80	10 40 45 54 52	34 03 06 07 26	75 21 11 02 71	36 63 36 84 24
58 27 56 17 64	97 58 65 47 16	50 25 94 63 45	87 19 54 60 92	26 78 76 09 39
89 51 41 17 88	68 22 42 34 17	73 95 97 61 45	30 34 24 02 77	11 04 97 20 49
15 47 25 06 69	48 13 93 67 32	46 87 43 70 88	73 46 50 98 19	58 86 93 52 20
12 12 08 61 24	51 24 74 43 02	60 88 35 21 09	21 43 73 67 86	49 22 67 78 37

Table XXII continued   399

RANDOM NUMBERS

19	61	27	84	30	11	66	19	47	70	77	60	36	56	69	86	86	81	26	65	30	01	27	59	89
39	14	17	74	00	28	00	06	42	38	73	25	87	17	94	31	34	02	62	56	66	45	33	70	16
64	75	68	04	57	08	74	71	28	36	03	46	95	06	78	03	27	44	34	23	66	67	78	25	56
92	90	15	18	78	56	44	12	29	98	29	71	83	84	47	06	45	32	53	11	07	56	55	37	71
03	55	19	00	70	09	48	39	40	50	45	93	81	81	35	36	90	84	33	21	11	07	35	18	03
98	88	46	62	09	06	83	05	36	56	14	66	35	63	46	71	43	00	49	09	19	81	80	57	07
27	36	98	68	82	53	47	30	75	41	53	63	37	08	63	03	74	81	28	22	19	36	04	90	88
59	06	67	59	74	63	33	52	04	83	43	51	43	74	81	58	27	82	69	67	49	32	54	39	51
91	64	79	37	83	64	16	94	90	22	98	58	80	94	95	49	82	95	90	68	38	83	10	48	38
83	60	59	24	19	39	54	20	77	72	71	56	87	56	73	35	18	58	97	59	44	90	17	42	91
24	89	58	85	30	70	77	43	54	39	46	75	87	04	72	70	20	79	26	75	91	62	36	12	75
35	72	02	65	56	95	59	62	00	94	73	75	08	57	88	34	26	40	17	03	46	83	36	52	48
14	14	15	34	10	38	64	90	63	43	57	25	66	13	42	72	70	97	53	18	90	37	93	75	62
27	41	67	56	70	92	17	67	25	35	93	11	95	60	77	06	88	61	82	44	92	34	43	13	74
82	07	10	74	29	81	00	74	77	49	40	74	45	69	74	23	33	68	88	21	53	84	11	05	36
21	44	58	27	93	24	83	19	32	41	14	19	97	62	68	70	88	36	80	02	03	82	91	74	43
72	51	37	64	00	52	22	59	23	48	62	30	89	84	81	29	74	43	31	65	33	14	16	10	20
71	47	94	50	27	76	16	05	74	11	13	78	01	36	32	52	30	87	77	62	88	87	43	36	97
83	21	05	14	66	09	08	85	03	95	26	74	30	53	06	21	70	67	00	01	99	43	98	07	67
68	74	99	51	48	94	89	77	86	36	96	75	00	90	24	94	53	89	11	43	96	69	36	18	86
05	18	47	57	63	47	07	58	81	58	05	31	35	34	39	14	90	80	88	30	60	09	62	15	51
13	65	16	25	46	96	89	22	52	40	47	51	15	84	83	87	34	27	88	18	07	85	53	92	69
00	56	62	12	20	00	29	22	40	69	25	07	22	95	19	52	54	85	40	91	21	28	22	12	96
50	95	81	76	95	58	07	26	89	90	60	32	99	59	55	71	58	66	34	17	35	94	76	78	07
57	62	16	45	47	46	85	03	79	81	38	52	70	90	37	64	75	60	33	24	04	98	68	36	66
09	28	22	58	44	79	13	97	84	35	35	42	84	35	61	69	79	96	33	14	12	99	19	35	16
23	39	49	42	06	93	43	23	78	36	94	91	92	68	46	02	55	57	44	10	94	91	54	81	99
05	28	03	74	70	93	62	20	43	45	15	09	21	95	10	18	09	41	66	13	78	23	45	00	01
95	49	19	79	76	38	30	63	21	92	82	63	95	46	24	72	43	49	26	06	23	19	17	46	93
78	52	10	01	04	18	24	87	55	83	90	32	65	07	85	54	03	46	62	51	35	77	41	46	92
96	34	54	45	79	85	93	24	40	53	75	70	42	08	40	86	58	38	39	44	52	45	67	37	66
77	96	33	11	51	32	36	49	16	91	47	35	74	03	38	23	43	52	40	65	08	45	89	53	66
07	52	01	12	94	23	23	80	17	48	41	69	06	73	28	54	81	43	77	77	10	05	74	23	32
38	42	30	23	09	70	70	38	57	36	46	14	81	42	58	29	23	61	21	52	05	08	86	58	25
02	46	36	55	33	21	19	96	05	55	33	92	80	18	17	07	39	68	92	15	30	72	22	21	02
15	88	09	22	61	17	29	28	81	90	61	78	14	88	98	92	52	52	12	83	88	58	16	00	98
71	92	60	08	19	59	14	40	02	24	30	57	09	01	94	18	32	90	69	99	26	85	71	92	38
64	42	52	81	08	16	55	41	60	16	00	04	28	32	29	10	33	33	61	68	65	61	79	48	34
79	78	22	39	24	49	44	03	04	32	81	07	73	15	43	95	21	66	48	65	13	65	85	10	81
35	33	77	45	38	44	55	36	46	72	90	96	04	18	49	93	86	54	46	08	92	17	63	48	51
05	24	92	93	29	19	71	59	40	82	14	73	88	66	67	43	70	86	63	54	93	69	22	55	27
56	46	39	93	80	38	79	38	57	74	19	05	61	39	39	46	06	22	76	47	66	14	66	32	10
96	29	63	31	21	54	19	63	41	08	75	81	48	59	86	71	17	11	51	02	28	99	26	31	65
98	38	03	62	69	60	01	40	72	01	62	44	84	65	85	42	17	58	83	50	46	18	24	91	26
52	56	76	43	50	16	31	55	39	69	80	39	58	11	14	54	35	86	45	78	47	26	91	57	47
78	49	89	08	30	25	95	59	92	36	43	28	69	10	64	99	96	99	51	44	64	42	47	73	77
49	55	32	42	41	08	15	08	95	35	08	70	39	10	41	77	32	38	10	79	45	12	79	36	86
32	15	10	70	75	83	15	51	02	52	73	10	08	86	18	23	89	18	74	18	45	41	72	02	68
11	31	45	03	63	26	86	02	77	99	49	41	68	35	34	19	18	70	80	59	76	67	70	21	10
12	36	47	12	10	87	05	25	02	41	90	78	59	78	89	81	39	95	81	30	64	43	90	56	14

Table XXIII

RANDOM NORMAL NUMBERS, $\mu = 0$, $\sigma = 1$*

01	02	03	04	05	06	07	08	09	10
0.464	0.137	2.455	−0.323	−0.068	0.296	−0.288	1.298	0.241	−0.957
0.060	−2.526	−0.531	−0.194	0.543	−1.558	0.187	−1.190	0.022	0.525
1.486	−0.354	−0.634	0.697	0.926	1.375	0.785	−0.963	−0.853	−1.865
1.022	−0.472	1.279	3.521	0.571	−1.851	0.194	1.192	−0.501	−0.273
1.394	−0.555	0.046	0.321	2.945	1.974	−0.258	0.412	0.439	−0.035
0.906	−0.513	−0.525	0 595	0.881	−0.934	1.579	0.161	−1.885	0.371
1.179	−1.055	0.007	0.769	0.971	0.712	1.090	−0.631	−0.255	−0.702
−1.501	−0.488	−0.162	−0.136	1.033	0.203	0.448	0.748	−0.423	−0.432
−0.690	0.756	−1.618	−0.345	−0.511	−2.051	−0.457	−0.218	0.857	−0.465
1.372.	0.225	0.378	0.761	0.181	−0.736	0.960	−1.530	−0.260	0.120
−0.482	1.678	−0.057	−1.229	−0.486	0.856	−0.491	−1.983	−2.830	−0.238
−1.376	−0.150	1.356	−0.561	−0.256	−0.212	0.219	0.779	0.953	−0.869
−1.010	0.598	−0.918	1.598	0.065	0.415	−0.169	0.313	−0.973	−1.016
−0.005	−0.899	0.012	−0.725	1.147	−0.121	1.096	0.481	−1.691	0.417
1.393	−1.163	−0.911	1.231	−0.199	−0.246	1.239	−2.574	−0.558	0.056
−1.787	−0.261	1.237	1.046	−0.508	−1.630	−0.146	−0.392	−0.627	0.561
−0.105	−0.357	−1.384	0.360	−0.992	−0.116	−1.698	−2.832	−1.108	−2.357
−1.339	1.827	−0.959	0.424	0.969	−1.141	−1.041	0.362	−1.726	1.956
1.041	0.535	0.731	1.377	0.983	−1.330	1.620	−1.040	0.524	−0.281
0.279	−2.056	0.717	−0.873	−1.096	−1.396	1.047	0.089	−0.573	0.932
−1.805	−2.008	−1.633	0.542	0.250	−0.166	0.032	0.079	0.471	−1.029
−1.186	1.180	1.114	0.882	1.265	−0.202	0.151	−0.376	−0.310	0.479
0.658	−1.141	1.151	−1.210	−0.927	0.425	0.290	−0.902	0.610	2.709
−0.439	0.358	−1.939	0.891	−0.227	0.602	0.873	−0.437	−0.220	−0.057
−1.399	−0.230	0.385	−0.649	−0.577	0.237	−0.289	0.513	0.738	−0.300
0.199	0.208	−1.083	−0.219	−0.291	1.221	1.119	0.004	−2.015	−0.594
0.159	0.272	−0.313	0.084	−2.828	−0.439	−0.792	−1.275	−0.623	−1.047
2.273	0.606	0.606	−0.747	0.247	1.291	0.063	−1.793	−0.699	−1.347
0.041	−0.307	0.121	0.790	−0.584	0.541	0.484	−0.986	0.481	0.996
−1.132	−2.098	0.921	0.145	0.446	−1.661	1.045	−1.363	−0.586	−1.023
0.768	0.079	−1.473	0.034	−2.127	0.665	0.084	−0.880	−0.579	0.551
0.375	−1.658	−0.851	0.234	−0.656	0.340	−0.086	−0.158	−0.120	0.418
−0.513	−0.344	0.210	−0.736	1.041	0.008	0.427	−0.831	0.191	0.074
0.292	−0.521	1.266	−1.206	−0.899	0.110	−0.528	−0.813	0.071	0.524
1.026	2.990	−0.574	−0.491	−1.114	1.297	−1.433	−1.345	−3.001	0.479
−1.334	1.278	−0.568	−0.109	−0.515	−0.566	2.923	0.500	0.359	0.326
−0.287	−0.144	−0.254	0.574	−0.451	−1.181	−1.190	−0.318	−0.094	1.114
0.161	−0.886	−0.921	−0.509	1.410	−0.518	0.192	−0.432	1.501	1.068
−1.346	0.193	−1.202	0.394	−1.045	0.843	0.942	1.045	0.031	0.772
1.250	−0.199	−0.288	1.810	1.378	0.584	1.216	0.733	0.402	0.226
0.630	−0.537	0.782	0.060	0.499	−0.431	1.705	1.164	0.884	−0.298
0.375	−1.941	0.247	−0.491	0.665	−0.135	−0.145	−0.498	0.457	1.064
−1.420	0.489	−1.711	−1.186	0.754	−0.732	−0.066	1.006	−0.798	0.162
−0.151	−0.243	−0.430	−0.762	0.298	1.049	1.810	2.885	−0.768	−0.129
−0.309	0.531	0.416	−1.541	1.456	2.040	−0.124	0.196	0.023	−1.204
0.424	−0.444	0.593	0.993	−0.106	0.116	0.484	−1.272	1.066	1.097
0.593	0.658	−1.127	−1.407	−1.579	−1.616	1.458	1.262	0.736	−0.916
0.862	−0.885	−0.142	−0.504	0.532	1.381	0.022	−0.281	−0.342	1.222
0.235	−0.628	−0.023	−0.463	−0.899	−0.394	−0.538	1.707	−0.188	−1.153
−0.853	0.402	0.777	0.833	0.410	−0.349	−1.094	0.580	1.395	1.298

This table is reproduced with permission from *A Million Random Digits*, published by the RAND Corporation.

Table XXIII *continued*   401

RANDOM NORMAL NUMBERS, $\mu = 0$, $\sigma = 1$

11	12	13	14	15	16	17	18	19	20
−1.329	−0.238	−0.838	−0.988	−0.445	0.964	−0.266	−0.322	−1.726	2.252
1.284	−0.229	1.058	0.090	0.050	0.523	0.016	0.277	1.639	0.554
0.619	0.628	0.005	0.973	−0.058	0.150	−0.635	−0.917	0.313	−1.203
0.699	−0.269	0.722	−0.994	−0.807	−1.203	1.163	1.244	1.306	−1.210
0.101	0.202	−0.150	0.731	0.420	0.116	−0.496	−0.037	−2.466	0.794
−1.381	0.301	0.522	0.233	0.791	−1.017	−0.182	0.926	−1.096	1.001
−0.574	1.366	−1.843	0.746	0.890	0.824	−1.249	−0.806	−0.240	0.217
0.096	0.210	1.091	0.990	0.900	−0.837	−1.097	−1.238	0.030	−0.311
1.389	−0.236	0.094	3.282	0.295	−0.416	0.313	0.720	0.007	0.354
1.249	0.706	1.453	0.366	−2.654	−1.400	0.212	0.307	−1.145	0.639
0.756	−0.397	−1.772	−0.257	1.120	1.188	−0.527	0.709	0.479	0.317
−0.860	0.412	−0.327	0.178	0.524	−0.672	−0.831	0.758	0.131	0.771
−0.778	−0.979	0.236	−1.033	1.497	−0.661	0.906	1.169	−1.582	1.303
0.037	0.062	0.426	1.220	0.471	0.784	−0.719	0.465	1.559	−1.326
2.619	−0.440	0.477	1.063	0.320	1.406	−0.701	−0.128	0.518	−0.676
−0.420	−0.287	−0.050	−0.481	1.521	−1.367	0.609	0.292	0.048	0.592
1.048	0.220	1.121	−1.789	−1.211	−0.871	−0.740	0.513	−0.558	−0.395
1.000	−0.638	1.261	0.510	−0.150	0.034	0.054	−0.055	0.639	−0.825
0.170	−1.131	−0.985	0.102	−0.939	−1.457	1.766	1.087	−1.275	2.362
0.389	−0.435	0.171	0.891	1.158	1.041	1.048	−0.324	−0.404	1.060
−0.305	0.838	−2.019	−0.540	0.905	1.195	−1.190	0.106	0.571	0.298
−0.321	−0.039	1.799	−1.032	−2.225	−0.148	0.758	−0.862	0.158	−0.726
1.900	1.572	−0.244	−1.721	1.130	0.495	−0.484	0.014	−0.778	−1.483
−0.778	−0.288	−0.224	−1.324	−0.072	0.890	−0.410	0.752	0.376	−0.224
0.617	−1.718	−0.183	−0.100	1.719	0.696	−1.339	−0.614	1.071	−0.386
−1.430	−0.953	0.770	−0.007	−1.872	1.075	−0.913	−1.168	1.775	0.238
0.267	−0.048	0.972	0.734	−1.408	−1.955	−0.848	2.002	0.232	−1.273
0.978	−0.520	−0.368	1.690	−1.479	0.985	1.475	−0.098	−1.633	2.399
−1.235	−1.168	0.325	1.421	2.652	−0.486	−1.253	0.270	−1.103	0.118
−0.258	0.638	2.309	0.741	−0.161	−0.679	0.336	1.973	0.370	−2.277
0.243	0.629	−1.516	−0.157	0.693	1.710	0.800	−0.265	1.218	0.655
−0.292	−1.455	−1.451	1.492	−0.713	0.821	−0.031	−0.780	1.330	0.977
−0.505	0.389	0.544	−0.042	1.615	−1.440	−0.989	−0.580	0.156	0.052
0.397	−0.287	1.712	0.289	−0.904	0.259	−0.600	−1.635	−0.009	−0.799
−0.605	−0.470	0.007	0.721	−1.117	0.635	0.592	−1.362	−1.441	0.672
1.360	0.182	−1.476	−0.599	−0.875	0.292	−0.700	0.058	−0.340	−0.639
0.480	−0.699	1.615	−0.225	1.014	−1.370	−1.097	0.294	0.309	−1.389
−0.027	−0.487	−1.000	−0.015	0.119	−1.990	−0.687	−1.964	−0.366	1.759
−1.482	−0.815	−0.121	1.884	−0.185	0.601	0.793	0.430	−1.181	0.426
−1.256	−0.567	−0.994	1.011	−1.071	−0.623	−0.420	−0.309	1.362	0.863
−1.132	2.039	1.934	−0.222	0.386	1.100	0.284	1.597	−1.718	−0.560
−0.780	−0.239	−0.497	−0.434	−0.284	−0.241	−0.333	1.348	−0.478	−0.169
−0.859	−0.215	0.241	1.471	0.389	−0.952	0.245	0.781	1.093	−0.240
0.447	1.479	0.067	0.426	−0.370	−0.675	−0.972	0.225	0.815	0.389
0.269	0.735	−0.066	−0.271	−1.439	1.036	−0.306	−1.439	−0.122	−0.336
0.097	−1.883	−0.218	0.202	−0.357	0.019	1.631	1.400	0.223	−0.793
−0.686	1.596	−0.286	0.722	0.655	−0.275	1.245	−1.504	0.066	−1.280
0.957	0.057	−1.153	0.701	−0.280	1.747	−0.745	1.338	−1.421	0.386
0.976	−1.789	−0.696	−1.799	−0.354	0.071	2.355	0.135	−0.598	1.883
0.274	0.226	−0.909	−0.572	0.181	1.115	0.496	0.453	−1.218	−0.115

402 Table XXIII continued

RANDOM NORMAL NUMBERS, $\mu = 0$, $\sigma = 1$

21	22	23	24	25	26	27	28	29	30
−1.752	−0.329	−1.256	0.318	1.531	0.349	−0.958	−0.059	0.415	−1.084
−0.291	0.085	1.701	−1.087	−0.443	−0.292	0.248	−0.539	−1.382	0.318
−0.933	0.130	0.634	0.899	1.409	−0.883	−0.095	0.229	0.129	0.367
−0.450	−0.244	0.072	1.028	1.730	−0.056	−1.488	−0.078	−2.361	−0.992
0.512	−0.882	0.490	−1.304	−0.266	0.757	−0.361	0.194	−1.078	0.529
−0.702	0.472	0.429	−0.664	−0.592	1.443	−1.515	−1.209	−1.043	0.278
0.284	0.039	−0.518	1.351	1.473	0.889	0.300	0.339	−0.206	1.392
−0.509	1.420	−0.782	−0.429	−1.266	0.627	−1.165	0.819	−0.261	0.409
−1.776	−1.033	1.977	0.014	0.702	−0.435	−0.816	1.131	0.656	0.061
−0.044	1.807	0.342	−2.510	1.071	−1.220	−0.060	−0.764	0.079	−0.964
0.263	−0.578	1.612	−0.148	−0.383	−1.007	−0.414	0.638	−0.186	0.507
0.986	0.439	−0.192	−0.132	0.167	0.883	−0.400	−1.440	−0.385	−1.414
−0.441	−0.852	−1.446	−0.605	−0.348	1.018	0.963	−0.004	2.504	−0.847
−0.866	0.489	0.097	0.379	0.192	−0.842	0.065	1.420	0.426	−1.191
−1.215	0.675	1.621	0.394	−1.447	2.199	−0.321	−0.540	−0.037	0.185
−0.475	−1.210	0.183	0.526	0.495	1.297	−1.613	1.241	−1.016	−0.090
1.200	0.131	2.502	0.344	−1.060	−0.909	−1.695	−0.666	−0.838	−0.866
−0.498	−1.202	−0.057	−1.354	−1.441	−1.590	0.987	0.441	0.637	−1.116
−0.743	0.894	−0.028	1.119	−0.598	0.279	2.241	0.830	0.267	−0.156
0.779	−0.780	−0.954	0.705	−0.361	−0.734	1.365	1.297	−0.142	−1.387
−0.206	−0.195	1.017	−1.167	−0.079	−0.452	0.058	−1.068	−0.394	−0.406
−0.092	−0.927	−0.439	0.256	0.503	0.338	1.511	−0.465	−0.118	−0.454
−1.222	−1.582	1.786	−0.517	−1.080	−0.409	−0.474	−1.890	0.247	0.575
0.068	0.075	−1.383	−0.084	0.159	1.276	1.141	0.186	−0.973	−0.266
0.183	1.600	−0.335	1.553	0.889	0.896	−0.035	0.461	0.486	1.246
−0.811	−2.904	0.618	0.588	0.533	0.803	−0.696	0.690	0.820	0.557
−1.010	1.149	1.033	0.336	1.306	0.835	1.523	0.296	−0.426	0.004
1.453	1.210	−0.043	0.220	−0.256	−1.161	−2.030	−0.046	0.243	1.082
0.759	−0.838	−0.877	−0.177	1.183	−0.218	−3.154	−0.963	−0.822	−1.114
0.287	0.278	−0.454	0.897	−0.122	0.013	0.346	0.921	0.238	−0.586
−0.669	0.035	−2.077	1.077	0.525	−0.154	−1.036	0.015	−0.220	0.882
0.392	0.106	−1.430	−0.204	−0.326	0.825	−0.432	−0.094	−1.566	0.679
−0.337	0.199	−0.160	0.625	−0.891	−1.464	−0.318	1.297	0.932	−0.032
0.369	−1.990	−1.190	0.666	−1.614	0.082	0.922	−0.139	−0.833	0.091
−1.694	0.710	−0.655	−0.546	1.654	0.134	0.466	0.033	−0.039	0.838
0.985	0.340	0.276	0.911	−0.170	−0.551	1.000	−0.838	0.275	−0.304
−1.063	−0.594	−1.526	−0.787	0.873	−0.405	−1.324	0.162	−0.163	−2.716
0.033	−1.527	1.422	0.308	0.845	−0.151	0.741	0.064	1.212	0.823
0.597	0.362	−3.760	1.159	0.874	−0.794	−0.915	1.215	1.627	−1.248
−1.601	−0.570	0.133	−0.660	1.485	0.682	−0.898	0.686	0.658	0.346
−0.266	−1.309	0.597	0.989	0.934	1.079	−0.656	−0.999	−0.036	−0.537
0.901	1.531	−0.889	−1.019	0.084	1.531	−0.144	−1.920	0.678	−0.402
−1.433	−1.008	−0.990	0.090	0.207		−0.745	0.638	1.469	1.214
1.327	0.763	−1.724	−0.709	−1.100	−1.346	−0.946	−0.157	0.522	−1.264
−0.248	0.788	−0.577	0.122	−0.536	0.293	1.207	−2.243	1.642	1.353
−0.401	−0.679	0.921	0.476	1.121	−0.864	0.128	−0.551	−0.872	1.511
0.344	−0.324	0.686	−1.487	−0.126	0.803	−0.961	0.183	−0.358	−0.184
0.441	−0.372	−1.336	0.062	1.506	−0.315	−0.112	−0.452	1.594	−0.264
0.824	0.040	−1.734	0.251	0.054	−0.379	1.298	−0.126	0.104	−0.529
1.385	1.320	−0.509	−0.381	−1.671	−0.524	−0.805	1.348	0.676	0.799

RANDOM NORMAL NUMBERS, $\mu = 0$, $\sigma = 1$

	32	33	34	35	36	37	38	39	40
.556	0.119	−0.078	0.164	−0.455	0.077	−0.043	−0.299	0.249	−0.182
0.647	1.029	1.186	0.887	1.204	−0.657	0.644	−0.410	−0.652	−0.165
0.329	0.407	1.169	−2.072	1.661	0.891	0.233	−1.628	−0.762	−0.717
−1.188	1.171	−1.170	−0.291	0.863	−0.045	−0.205	0.574	−0.926	1.407
−0.917	−0.616	−1.589	1.184	0.266	0.559	−1.833	−0.572	−0.648	−1.090
0.414	0.469	−0.182	0.397	1.649	1.198	0.067	−1.526	−0.081	−0.192
0.107	−0.187	1.343	0.472	−0.112	1.182	0.548	2.748	0.249	0.154
−0.497	1.907	0.191	0.136	−0.475	0.458	0.183	−1.640	−0.058	1.278
0.501	0.083	−0.321	1.133	1.126	−0.299	1.299	1.617	1.581	2.455
−1.382	−0.738	1.225	1.564	−0.363	−0.548	1.070	0.390	−1.398	0.524
−0.590	0.699	−0.162	−0.011	1.049	−0.689	1.225	0.339	−0.539	−0.445
−1.125	1.111	−1.065	0.534	0.102	0.425	−1.026	0.695	−0.057	0.795
0.849	0.169	−0.351	0.584	2.177	0.009	−0.696	−0.426	−0.692	−1.638
−1.233	−0.585	0.306	0.773	1.304	−1.304	0.282	−1.705	0.187	−0.880
0.104	−0.468	0.185	0.498	−0.624	−0.322	−0.875	1.478	−0.691	−0.281
0.261	−1.883	−0.181	1.675	−0.324	−1.029	−0.185	0.004	−0.101	−1.187
−0.007	1.280	0.568	−1.270	1.405	1.731	2.072	1.686	0.728	−0.417
0.794	−0.111	0.040	−0.536	−0.976	2.192	1.609	−0.190	−0.279	−1.611
0.431	−2.300	−1.081	−1.370	2.943	0.653	−2.523	0.756	0.886	−0.983
−0.149	1.294	−0.580	0.482	−1.449	−1.067	1.996	−0.274	0.721	0.490
−0.216	−1.647	1.043	0.481	−0.011	−0.587	−0.916	−1.016	−1.040	−1.117
1.604	−0.851	−0.317	−0.686	−0.008	1.939	0.078	−0.465	0.533	0.652
−0.212	0.005	0.535	0.837	0.362	1.103	0.219	0.488	1.332	−0.200
0.007	−0.076	1.484	0.455	−0.207	−0.554	1.120	0.913	−0.681	1.751
−0.217	0.937	0.860	0.323	1.321	−0.492	−1.386	−0.003	−0.230	0.539
−0.649	0.300	−0.698	0.900	0.569	0.842	0.804	1.025	0.603	−1.546
−1.541	0.193	2.047	−0.552	1.190	−0.087	2.062	−2.173	−0.791	−0.520
0.274	−0.530	0.112	0.385	0.656	0.436	0.882	0.312	−2.265	−0.218
0.876	−1.498	−0.128	−0.387	−1.259	−0.856	−0.353	0.714	0.863	1.169
−0.859	−1.083	1.288	−0.078	−0.081	0.210	0.572	1.194	−1.118	−1.543
−0.015	−0.567	0.113	2.127	−0.719	3.256	−0.721	−0.663	−0.779	−0.930
−1.529	−0.231	1.223	0.300	−0.995	−0.651	0.505	0.138	−0.064	1.341
0.278	−0.058	−2.740	−0.296	−1.180	0.574	1.452	0.846	−0.243	−1.208
1.428	0.322	2.302	−0.852	0.782	−1.322	−0.092	−0.546	0.560	−1.430
0.770	−1.874	0.347	0.994	−0.485	−1.179	0.048	−1.324	1.061	0.449
−0.303	−0.629	0.764	0.013	−1.192	−0.475	−1.085	−0.880	1.738	−1.225
−0.263	−2.105	0.509	−0.645	1.362	0.504	−0.755	1.274	1.448	0.604
0.997	−1.187	−0.242	0.121	2.510	−1.935	0.350	0.073	0.458	−0.446
−0.063	−0.475	−1.802	−0.476	0.193	−1.199	0.339	0.364	−0.684	1.353
−0.168	1.904	−0.485	−0.032	−0.554	0.056	−0.710	−0.778	0.722	−0.024
0.366	−0.491	0.301	−0.008	−0.894	−0.945	0.384	−1.748	−1.118	0.394
0.436	−0.464	0.539	0.942	−0.458	0.445	−1.883	1.228	1.113	−0.218
0.597	−1.471	−0.434	0.705	−0.788	0.575	0.086	0.504	1.445	−0.513
−0.805	−0.624	1.344	0.649	−1.124	0.680	−0.986	1.845	−1.152	−0.393
1.681	−1.910	0.440	0.067	−1.502	−0.755	−0.989	−0.054	−2.320	0.474
−0.007	−0.459	1.940	0.220	−1.259	−1.729	0.137	−0.520	−0.412	2.847
0.209	−0.633	0.299	0.174	1.975	−0.271	0.119	−0.199	0.007	2.315
1.254	1.672	−1.186	−1.310	0.474	0.878	−0.725	−0.191	0.642	−1.212
−1.016	−0.697	0.017	−0.263	−0.047	−1.294	−0.339	2.257	−0.078	−0.049
−1.169	−0.355	1.086	−0.199	0.031	0.396	−0.143	1.572	0.276	0.027

Table XXIII conti

RANDOM NORMAL NUMBERS, $\mu = 0$, $\sigma = 1$

41	42	43	44	45	46	47	48	49	50
−0.856	−0.063	0.787	−2.052	−1.192	−0.831	1.623	1.135	0.759	−0.189
−0.276	−1.110	0.752	−1.378	−0.583	0.360	0.365	1.587	0.621	1.344
0.379	−0.440	0.858	1.453	−1.356	0.503	−1.134	1.950	−1.816	−0.283
1.468	0.131	0.047	0.355	0.162	−1.491	−0.739	−1.182	−0.533	−0.497
−1.805	−0.772	1.286	−0.636	−1.312	−1.045	1.559	−0.871	−0.102	−0.123
2.285	0.554	0.418	−0.577	−1.489	−1.255	0.092	−0.597	−1.051	−0.980
−0.602	0.399	1.121	−1.026	0.087	1.018	−1.437	0.661	0.091	−0.637
0.229	−0.584	0.705	0.124	0.341	1.320	−0.824	−1.541	−0.163	2.329
1.382	−1.454	1.537	−1.299	0.363	−0.356	−0.025	0.294	2.194	−0.395
0.978	0.109	1.434	−1.094	−0.265	−0.857	−1.421	−1.773	0.570	−0.053
−0.678	−2.335	1.202	−1.697	0.547	−0.201	−0.373	−1.363	−0.081	0.958
−0.366	−1.084	−0.626	0.798	1.706	−1.160	−0.838	1.462	0.636	0.570
−1.074	−1.379	0.086	−0.331	−0.288	−0.309	−1.527	−0.408	0.183	0.856
−0.600	−0.096	0.696	0.446	1.417	−2.140	0.599	−0.157	1.485	1.387
0.918	1.163	−1.445	0.759	0.878	−1.781	−0.056	−2.141	−0.234	0.975
−0.791	−0.528	0.946	1.673	−0.680	−0.784	1.494	−0.086	−1.071	−1.196
0.598	−0.352	0.719	−0.341	0.056	−1.041	1.429	0.235	0.314	−1.693
0.567	−1.156	−0.125	−0.534	0.711	−0.511	0.187	−0.644	−1.090	−1.281
0.963	0.052	0.037	0.637	−1.335	0.055	0.010	−0.860	−0.621	0.713
0.489	−0.209	1.659	0.054	1.635	0.169	0.794	−1.550	1.845	−0.388
−1.627	−0.017	0.699	0.661	−0.073	0.188	1.183	−1.054	−1.615	−0.765
−1.096	1.215	0.320	0.738	1.865	−1.169	−0.667	−0.674	−0.062	1.378
−2.532	1.031	−0.799	1.665	−2.756	−0.151	−0.704	0.602	−0.672	1.264
0.024	−1.183	−0.927	−0.629	0.204	−0.825	0.496	2.543	0.262	−0.785
0.192	0.125	0.373	−0.931	−0.079	0.186	−0.306	0.621	−0.292	1.131
−1.324	−1.229	−0.648	−0.430	0.811	0.868	0.787	1.845	−0.374	−0.651
−0.726	−0.746	1.572	−1.420	1.509	−0.361	−0.310	−3.117	1.637	0.642
−1.618	1.082	−0.319	0.300	1.524	−0.418	−1.712	0.358	−1.032	0.537
1.695	0.843	2.049	0.388	−0.297	1.077	−0.462	0.655	0.940	−0.354
0.790	0.605	−3.077	1.009	−0.906	−1.004	0.693	−1.098	1.300	0.549
1.792	−0.895	−0.136	−1.765	1.077	0.418	−0.150	0.808	0.697	0.435
0.771	−0.741	−0.492	−0.770	−0.458	−0.021	1.385	−1.225	−0.066	−1.471
−1.438	0.423	−1.211	0.723	−0.731	0.883	−2.109	−2.455	−0.210	1.644
−0.294	1.266	−1.994	−0.730	0.545	0.397	1.069	−0.383	−0.097	−0.985
−1.966	0.909	0.400	0.685	−0.800	1.759	0.268	1.387	−0.414	1.615
−0.999	1.587	1.423	0.937	−0.943	0.090	1.185	−1.204	0.300	−1.354
0.581	0.481	−2.400	0.000	0.231	0.079	−2.842	−0.846	−0.508	−0.516
0.370	−1.452	−0.580	−1.462	−0.972	1.116	−0.994	0.374	−3.336	−0.058
0.834	−1.227	−0.709	−1.039	−0.014	−0.383	−0.512	−0.347	0.881	−0.638
−0.376	−0.813	0.660	−1.029	−0.137	0.371	0.376	0.968	1.338	−0.786
−1.621	0.815	−0.544	−0.376	−0.852	0.436	1.562	0.815	−1.048	0.188
0.163	−0.161	2.501	−0.265	−0.285	1.934	1.070	0.215	−0.876	0.073
1.786	−0.538	−0.437	0.324	0.105	−0.421	−0.410	−0.947	0.700	−1.006
2.140	1.218	−0.351	−0.068	0.254	0.448	−1.461	0.784	0.317	1.013
0.064	0.410	0.368	0.419	−0.982	1.371	0.100	−0.505	0.856	0.890
0.789	−0.131	1.330	0.506	−0.645	−1.414	2.426	1.389	−0.169	−0.194
−0.011	−0.372	−0.699	2.382	−1.395	−0.467	1.256	−0.585	−1.359	−1.804
−0.463	0.003	−1.470	1.493	0.960	0.364	−1.267	−0.007	0.616	0.624
−1.210	−0.669	0.009	1.284	−0.617	0.355	−0.589	−0.243	−0.015	−0.712
−1.157	0.481	0.560	1.287	1.129	−0.126	0.006	1.532	1.328	0.980

Table XXIII continued   405

RANDOM NORMAL NUMBERS, $\mu = 0$, $\sigma = 1$

51	52	53	54	55	56	57	58	59	60
0.240	1.774	0.210	−1.471	1.167	−1.114	0.182	−0.485	−0.318	1.156
0.627	−0.758	−0.930	1.641	0.162	−0.874	−0.235	0.203	−0.724	−0.155
−0.594	0.098	0.158	−0.722	1.385	−0.985	−1.707	0.175	0.449	0.654
1.082	−0.753	−1.944	−1.964	−2.131	−2.796	−1.286	0.807	−0.122	0.527
0.060	−0.014	1.577	−0.814	−0.633	0.275	−0.087	0.517	0.474	−1.432
−0.013	0.402	−0.086	−0.394	0.292	−2.862	−1.660	−1.658	1.610	−2.205
1.586	−0.833	1.444	−0.615	−1.157	−0.220	−0.517	−1.668	−2.036	−0.850
−0.405	−1.315	−1.355	−1.331	1.394	−0.381	−0.729	−0.447	−0.906	0.622
−0.329	1.701	0.427	0.627	−0.271	−0.971	−1.010	1.182	−0.143	0.844
0.992	0.708	−0.115	−1.630	0.596	0.499	−0.862	0.508	0.474	−0.974
0.296	−0.390	2.047	−0.363	0.724	0.788	−0.089	0.930	−0.497	0.058
−2.069	−1.422	−0.948	−1.742	−1.173	0.215	0.661	0.842	−0.984	−0.577
−0.211	−1.727	−0.277	1.592	−0.707	0.327	−0.527	0.912	0.571	−0.525
−0.467	1.848	−0.263	−0.862	0.706	−0.533	0.626	−0.200	−2.221	0.368
1.284	0.412	1.512	0.328	0.203	−1.231	−1.480	−0.400	−0.491	0.913
0.821	−1.503	−1.066	1.624	1.345	0.440	−1.416	0.301	−0.355	0.106
1.056	1.224	0.281	−0.098	1.868	−0.395	0.610	−1.173	−1.449	1.171
1.090	−0.790	0.882	1.687	−0.009	−2.053	−0.030	−0.421	1.253	−0.081
0.574	0.129	1.203	0.280	1.438	−2.052	−0.443	0.522	0.468	−1.211
−0.531	2.155	0.334	0.898	−1.114	0.243	1.026	0.391	−0.011	−0.024
0.896	0.181	−0.941	−0.511	0.648	−0.710	−0.181	−1.417	−0.585	0.087
0.042	0.579	−0.316	0.394	1.133	−0.305	−0.683	−1.318	−0.050	0.993
2.328	−0.243	0.534	0.241	0.275	0.060	0.727	−1.459	0.174	−1.072
0.486	−0.558	0.426	0.728	−0.360	−0.068	0.058	1.471	−0.051	0.337
−0.304	−0.309	0.646	0.309	−1.320	0.311	−1.407	−0.011	0.387	0.128
−2.319	−0.129	0.866	−0.424	0.236	0.419	−1.359	−1.088	−0.045	1.096
1.098	−0.875	0.659	−1.086	−0.424	−1.462	0.743	−0.787	1.472	1.677
−0.038	−0.118	−1.285	−0.545	−0.140	1.244	−1.104	0.146	0.058	1.245
−0.207	−0.746	1.681	0.137	0.104	−0.491	− 0.935	0.671	−0.448	−0.129
0.333	−1.386	1.840	1.089	0.837	−1.642	−0.273	−0.798	0.067	0.334
1.190	−0.547	−1.016	0.540	−0.993	0.443	−0.190	1.019	−1.021	−1.276
−1.416	−0.749	0.325	0.846	2.417	−0.479	−0.655	−1.326	−1.952	1.234
0.622	0.661	0.028	1.302	−0.032	−0.157	1.470	−0.766	0.697	−0.303
−1.134	0.499	0.538	0.564	−2.392	−1.398	0.010	1.874	1.386	0.000
0.725	−0.242	0.281	1.355	−0.036	0.204	−0.345	0.395	−0.753	1.645
−0.210	0.611	−0.219	0.450	0.308	0.993	−0.146	0.225	−1.496	0.246
0.219	0.302	0.000	−0.437	−2.127	0.883	−0.599	−1.516	0.826	1.242
−1.098	−0.252	−2.480	−0.973	0.712	−1.430	−0.167	−1.237	0.750	−0.763
0.144	0.489	−0.637	1.990	0.411	−0.563	0.027	1.278	2.105	−1.130
−1.738	−1.295	0.431	−0.503	2.327	−0.007	−1.293	−1.206	−0.066	1.370
−0.487	−0.097	−1.361	−0.340	0.204	0.938	−0.148	−1.099	−0.252	−0.384
−0.636	−0.626	1.967	1.677	−0.331	−0.440	−1.440	1.281	1.070	−1.167
−1.464	−1.493	0.945	0.180	−0.672	−0.035	−0.293	−0.905	0.196	−1.122
0.561	−0.375	−0.657	1.304	0.833	−1.159	1.501	1.265	0.438	−0.437
−0.525	−0.017	1.815	0.789	−1.908	−0.353	1.383	−1.208	−1.135	1.082
0.980	−0.111	−0.804	−1.078	−1.930	0.171	−1.318	2.377	−0.303	1.062
0.501	0.835	−0.518	−1.034	−1.493	0.712	0.421	−1.165	0.782	−1.484
1.081	−1.176	−0.542	0.321	0.688	0.670	−0.771	−0.090	−0.611	−0.813
−0.148	−1.203	−1.553	1.244	0.826	0.077	0.128	−0.772	1.683	0.318
0.096	−0.286	0.362	0.888	0.551	1.782	0.335	2.083	0.350	0.260

RANDOM NORMAL NUMBERS, $\mu = 0$, $\sigma = 1$

61	62	63	64	65	66	67	68	69	70
0.052	1.504	−1.350	−1.124	−0.521	0.515	0.839	0.778	0.438	−0.550
−0.315	−0.865	0.851	0.127	−0.379	1.640	−0.441	0.717	0.670	−0.301
0.938	−0.055	0.947	1.275	1.557	−1.484	−1.137	0.398	1.333	1.988
0.497	0.502	0.385	−0.467	2.468	−1.810	−1.438	0.283	1.740	0.420
2.308	−0.399	−1.798	0.018	0.780	1.030	0.806	−0.408	−0.547	−0.280
1.815	0.101	−0.561	0.236	0.166	0.227	−0.309	0.056	0.610	0.732
−0.421	0.432	0.586	1.059	0.278	−1.672	1.859	1.433	−0.919	−1.770
0.008	0.555	−1.310	−1.440	−0.142	−0.295	−0.630	−0.911	0.133	−0.308
1.191	−0.114	1.039	1.083	0.185	−0.492	0.419	−0.433	−1.019	−2.260
1.299	1.918	0.318	1.348	0.935	1.250	−0.175	−0.828	−0.336	0.726
0.012	−0.739	−1.181	−0.645	−0.736	1.801	−0.209	−0.389	0.867	−0.555
−0.586	−0.044	−0.983	0.332	0.371	−0.072	−1.212	1.047	−1.930	0.812
−0.122	1.515	0.338	−1.040	−0.008	0.467	−0.600	0.923	1.126	−0.752
0.879	0.516	−0.920	2.121	0.674	1.481	0.660	−0.986	1.644	−2.159
0.435	1.149	−0.065	1.391	0.707	0.548	−0.490	−1.139	0.249	−0.933
0.645	0.878	−0.904	0.896	−1.284	0.237	−0.378	−0.510	−1.123	−0.129
−0.514	−1.017	0.529	0.973	−1.202	0.005	−0.644	−0.167	−0.664	0.167
0.242	−0.427	−0.727	−1.150	−1.092	−0.736	0.925	−0.050	−0.200	−0.770
0.443	0.445	−1.287	−1.463	−0.650	0.412	−2.714	−0.903	−0.341	0.957
0.273	0.203	0.423	1.423	0.508	1.058	−0.828	0.143	−1.059	0.345
0.255	1.036	1.471	0.476	0.592	−0.658	0.677	0.155	1.068	−0.759
0.858	−0.370	0.522	−1.890	−0.389	0.609	1.210	0.489	−0.006	0.834
0.097	−1.709	1.790	−0.929	0.405	0.024	−0.036	0.580	−0.642	−1.121
0.520	0.889	−0.540	0.266	−0.354	0.524	−0.788	−0.497	−0.973	1.481
−0.311	−1.772	−0.496	1.275	−0.904	0.147	1.497	0.657	−0.469	−0.783
−0.604	0.857	−0.695	0.397	0.296	−0.285	0.191	0.158	1.672	1.190
−0.001	0.287	−0.868	−0.013	−1.576	−0.168	0.047	−0.159	0.086	−1.077
1.160	0.989	0.205	0.937	−0.099	−1.281	−0.276	0.845	0.752	0.663
1.579	−0.303	−1.174	−0.960	−0.470	−0.556	−0.689	1.535	−0.711	−0.743
−0.615	−0.154	0.008	1.353	−0.381	1.137	0.022	0.175	0.586	2.941
1.578	1.529	−0.294	−1.301	0.614	0.099	−0.700	−0.003	1.052	1.643
0.626	−0.447	−1.261	−2.029	0.182	−1.176	0.083	1.868	0.872	0.965
−0.493	−0.020	0.920	1.473	1.873	−0.289	0.410	0.394	0.881	0.054
−0.217	0.342	1.423	0.364	−0.119	0.509	−2.266	0.189	0.149	1.041
−0.792	0.347	−1.367	−0.632	−1.238	−0.136	−0.352	−0.157	−1.163	1.305
0.568	−0.226	0.391	−0.074	−0.312	0.400	1.583	0.481	−1.048	0.759
0.051	0.549	−2.192	1.257	−1.460	0.363	0.127	−1.020	−1.192	0.449
−0.891	0.490	0.279	0.372	−0.578	−0.836	2.285	−0.448	0.720	0.510
0.622	−0.126	−0.637	1.255	−0.354	0.032	−1.076	0.352	0.103	−0.496
0.623	0.819	−0.489	0.354	−0.943	−0.694	0.248	0.092	−0.673	−1.428
−1.208	−1.038	0.140	−0.762	−0.854	−0.249	2.431	0.067	−0.317	−0.874
−0.487	−2.117	0.195	2.154	1.041	−1.314	−0.785	−0.414	−0.695	2.310
0.522	0.314	−1.003	0.134	−1.748	−0.107	0.459	1.550	1.118	−1.004
0.838	0.613	0.227	0.308	−0.757	0.912	2.272	0.556	−0.041	0.008
−1.534	−0.407	1.202	1.251	−0.891	−1.588	−2.380	0.059	0.682	−0.878
−0.099	2.391	1.067	−2.060	−0.464	−0.103	3.486	1.121	0.632	−1.626
0.070	1.465	−0.080	−0.526	−1.090	−1.002	0.132	1.504	0.050	−0.393
0.115	−0.601	1.751	1.956	−0.196	0.400	−0.522	0.571	−0.101	−2.160
0.252	−0.329	−0.586	−0.118	−0.242	−0.521	0.818	−0.167	−0.469	0.430
0.017	0.185	0.377	1.883	−0.443	−0.039	−1.244	−0.820	−1.171	0.104

Table XXIII continued   407

RANDOM NORMAL NUMBERS, $\mu = 0$, $\sigma = 1$

71	72	73	74	75	76	77	78	79	80
2.988	0.423	−1.261	−1.893	0.187	−0.412	−0.228	0.002	−0.384	−1.032
0.760	0.995	−0.256	−0.505	0.750	−0.654	0.647	0.613	0.086	−0.118
−0.650	−0.927	−1.071	−0.796	1.130	−1.042	−0.181	−1.020	1.648	−1.327
−0.394	−0.452	0.893	1.410	1.133	0.319	0.537	−0.789	0.078	−0.062
−1.168	1.902	0.206	0.303	1.413	2.012	0.278	−0.566	−0.900	0.200
1.343	−0.377	−0.131	−0.585	0.053	0.137	−1.371	−0.175	−0.878	0.118
−0.733	−1.921	0.471	−1.394	−0.885	−0.523	0.553	0.344	−0.775	1.545
−0.172	−0.575	0.066	−0.310	1.795	−1.148	0.772	−1.063	0.818	0.302
1.457	0.862	1.677	−0.507	−1.691	−0.034	0.270	0.075	−0.554	1.420
−0.087	0.744	1.829	1.203	−0.436	−0.618	−0.200	−1.134	−1.352	−0.098
−0.092	1.043	−0.255	0.189	0.270	−1.034	−0.571	−0.336	−0.742	2.141
0.441	−0.379	−1.757	0.608	0.527	−0.338	−1.995	0.573	−0.034	−0.056
0.073	−0.250	0.531	−0.695	1.402	−0.462	−0.938	1.130	1.453	−0.106
0.637	0.276	−0.013	1.968	−0.205	0.486	0.727	1.416	0.963	1.349
−0.792	−1.778	1.284	−0.452	0.602	0.668	0.516	−0.210	0.040	−0.103
−1.223	1.561	−2.099	1.419	0.223	−0.482	1.098	0.513	0.418	−1.686
−0.407	1.587	0.335	−2.475	−0.284	1.567	−0.248	−0.759	1.792	−2.319
−0.462	−0.193	−0.012	−1.208	2.151	1.336	−1.968	−1.767	−0.374	0.783
1.457	0.883	1.001	−0.169	0.836	−1.236	1.632	−0.142	−0.222	0.340
−1.918	−1.246	−0.209	0.780	−0.330	−2.953	−0.447	−0.094	1.344	−0.196
−0.126	1.094	−1.206	−1.426	1.474	−1.080	0.000	0.764	1.476	−0.016
−0.306	−0.847	0.639	−0.262	−0.427	0.391	−1.298	−1.013	2.024	−0.539
0.477	1.595	−0.762	0.424	0.799	0.312	1.151	−1.095	1.199	−0.765
0.369	−0.709	1.283	−0.007	−1.440	−0.782	0.061	1.427	1.656	0.974
−0.579	0.606	−0.866	−0.715	−0.301	−0.180	0.188	0.668	−1.091	1.476
−0.418	−0.588	0.919	−0.083	1.084	0.944	0.253	−1.833	1.305	0.171
0.128	−0.834	0.009	0.742	0.539	−0.948	−1.055	−0.689	−0.338	1.091
−0.291	0.235	−0.971	−1.696	1.119	0.272	0.635	−0.792	−1.355	1.291
−1.024	1.212	−1.100	−0.348	1.741	0.035	1.268	0.192	0.729	−0.467
−0.378	1.026	0.093	0.468	−0.967	0.675	0.807	−2.109	−1.214	0.559
1.232	−0.815	0.608	1.429	−0.748	0.201	0.400	−1.230	−0.398	−0.674
1.793	−0.581	−1.076	0.512	−0.442	−1.488	−0.580	0.172	−0.891	0.311
0.766	0.310	−0.070	0.624	−0.389	1.035	−0.101	−0.926	0.816	−1.048
−0.606	−1.224	1.465	0.012	1.061	0.491	−1.023	1.948	0.866	−0.737
0.106	−2.715	0.363	0.343	−0.159	2.672	1.119	0.731	−1.012	−0.889
−0.060	0.444	1.596	−0.630	0.362	−0.306	1.163	−0.974	0.486	−0.373
2.081	1.161	−1.167	0.021	0.053	−0.094	0.381	−0.628	−2.581	−1.243
−1.727	−1.266	0.088	0.936	0.368	0.648	−0.799	1.115	−0.968	−2.588
0.091	1.364	1.677	0.644	1.505	0.440	−0.329	0.498	0.869	−0.965
−1.114	−0.239	−0.409	−0.334	−0.605	0.501	−1.921	−0.470	2.354	−0.660
0.189	−0.547	−1.758	−0.295	−0.279	−0.515	−1.053	0.553	−0.297	0.496
−0.065	−0.023	−0.267	−0.247	1.318	0.904	−0.712	−1.152	−0.543	0.176
−1.742	−0.599	0.430	−0.615	1.165	0.084	2.017	−1.207	2.614	1.490
0.732	0.188	2.343	0.526	−0.812	0.389	1.036	−0.023	0.229	−2.262
−1.490	0.014	0.167	1.422	0.015	0.069	0.133	0.897	−1.678	0.323
1.507	−0.571	−0.724	1.741	−0.152	−0.147	−0.158	−0.076	0.652	0.447
0.513	0.168	−0.076	−0.171	0.428	0.205	−0.865	0.107	1.023	0.077
−0.834	−1.121	1.441	0.492	0.559	1.724	−1.659	0.245	1.354	−0.041
0.258	1.880	−0.536	1.246	−0.188	−0.746	1.097	0.258	1.547	1.238
−0.818	0.273	0.159	−0.765	0.526	1.281	1.154	−0.687	−0.793	0.795

RANDOM NORMAL NUMBERS, $\mu = 0$, $\sigma = 1$

81	82	83	84	85	86	87	88	89	90
−0.713	−0.541	−0.571	−0.807	−1.560	1.000	0.140	−0.549	0.887	2.237
−0.117	0.530	−1.599	−1.602	0.412	−1.450	−1.217	1.074	−1.021	−0.424
1.187	−1.523	1.437	0.051	1.237	−0.798	1.616	−0.823	−1.207	1.258
−0.182	−0.186	0.517	1.438	0.831	−1.319	−0.539	−0.192	0.150	2.127
1.964	−0.629	−0.944	−0.028	0.948	1.005	0.242	−0.432	−0.329	0.113
0.230	1.523	1.658	0.753	0.724	0.183	−0.147	0.505	0.448	−0.053
0.839	−0.849	−0.145	−1.843	−1.276	0.481	−0.142	−0.534	0.403	0.370
−0.801	0.343	−1.822	0.447	−0.931	−0.824	−0.484	0.864	−1.069	0.860
−0.124	0.727	1.654	−0.182	−1.381	−1.146	−0.572	0.159	0.186	1.221
−0.088	0.032	−0.564	0.654	1.141	−0.056	−0.343	0.067	−0.267	−0.219
0.912	−1.114	−1.035	−1.070	−0.297	1.195	0.030	0.022	0.406	−0.414
1.397	−0.473	0.433	0.023	−1.204	1.254	0.551	−1.012	−0.789	0.906
−0.652	−0.029	0.064	0.511	1.117	−0.465	0.523	−0.083	0.386	0.259
1.236	−0.457	−1.354	−0.898	−0.270	−1.837	1.641	−0.657	−0.753	−1.686
−0.498	1.302	0.816	−0.936	1.404	0.555	2.450	−0.789	−0.120	0.505
−0.005	2.174	1.893	−1.361	−0.991	0.508	−0.823	0.918	0.524	0.488
0.115	−1.373	−0.900	−1.010	0.624	0.946	0.312	−1.384	0.224	2.343
0.167	0.254	1.219	1.153	−0.510	−0.007	−0.285	−0.631	−0.356	0.254
0.976	1.158	−0.469	1.099	0.509	−1.324	−0.102	−0.296	−0.907	0.449
0.653	−0.366	0.450	−2.653	−0.592	−0.510	0.983	0.023	−0.881	0.876
−0.150	−0.088	0.457	−0.448	0.605	0.668	−0.613	0.261	0.023	−0.050
0.060	0.276	0.229	−1.527	−0.316	−0.834	−1.652	−0.387	0.632	0.895
−0.678	0.547	0.243	−2.183	−0.368	1.158	−0.996	−0.705	−0.314	1.464
2.139	0.395	−0.376	−0.175	0.406	0.309	−1.02'	−0.460	−0.217	0.307
0.091	1.793	0.822	0.054	0.573	−0.729	−0.517	0.589	1.927	0.940
−0.003	0.344	1.242	−1.105	0.234	−1.222	−0.474	1.831	0.124	−0.840
−0.965	0.268	−1.543	0.690	0.917	2.017	−0.297	1.087	0.371	1.495
−0.076	−0.495	−0.103	0.646	2.427	−2.172	0.660	−1.541	−0.852	0.583
−0.365	−3.305	0.805	−0.418	−1.201	0.623	−0.223	0.109	0.205	−0.663
0.578	0.145	−1.438	1.122	−1.406	1.172	0.272	−2.245	1.207	1.227
−0.398	−0.304	0.529	−0.514	−0.681	−0.366	0.338	0.801	−0.301	−0.790
−0.951	−1.483	−0.613	−0.171	−0.459	1.231	−1.232	−0.497	−0.779	0.247
1.025	−0.039	−0.721	0.813	1.203	0.245	0.402	1.541	0.691	−1.420
−0.958	0.791	0.948	0.222	−0.704	−0.375	−0.246	−0.682	−0.871	0.056
1.097	−1.428	1.402	−1.425	−0.877	0.536	0.988	2.529	0.768	−1.321
0.377	2.240	0.854	−1.158	0.066	−1.222	0.821	−1.602	−0.760	−0.871
1.729	0.073	1.022	0.891	0.659	−1.040	0.251	−0.710	−1.734	−0.038
−1.329	−0.381	−0.515	1.484	−0.430	−0.466	−0.167	−0.788	−0.660	0.003
−0.132	0.391	2.205	−1.165	0.200	0.415	−0.765	0.239	−1.182	1.135
0.336	0.657	−0.805	0.150	−0.938	1.057	−1.090	1.604	−0.598	−0.760
0.124	−1.812	1.750	0.270	−0.114	0.517	−0.226	0.127	0.129	−0.751
−0.036	0.365	0.766	0.877	−0.804	−0.140	0.182	−0.483	−0.376	−0.564
−0.609	−0.019	−0.992	−1.193	−0.516	0.517	1.677	0.839	−1.134	0.675
−0.894	0.318	0.607	−0.865	0.526	−0.971	1.365	0.319	1.804	1.740
−0.357	−0.802	0.635	−0.491	−1.110	0.785	−0.042	−1.042	−0.572	0.243
−0.258	−0.383	−1.013	0.001	−1.673	0.561	−1.054	−0.106	−0.760	−1.009
2.245	−0.431	−0.496	0.796	0.193	1.202	−0.429	−0.217	0.333	−0.643
1.956	0.477	0.812	−0.117	0.606	−0.330	0.425	−0.232	0.802	0.656
1.358	0.139	0.199	−0.475	−0.120	0.184	−0.020	−1.326	0.517	−1.708
0.656	1.081	0.180	0.145	0.376	−1.363	−0.491	0.352	−1.477	1.280

Table XXIII continued   409

RANDOM NORMAL NUMBERS, $\mu = 0$, $\sigma = 1$

91	92	93	94	95	96	97	98	99	100
−0.181	0.583	−1.478	−0.181	0.281	−0.559	1.985	−1.122	−1.106	1.441
1.549	−1.183	−2.089	−1.997	−0.343	1.275	0.676	−0.212	1.252	0.163
0.978	−1.067	−2.640	0.134	0.328	−0.052	−0.030	−0.273	−0.570	1.026
−0.596	−0.420	−0.318	−0.057	−0.695	−1.148	0.333	−0.531	−2.037	−1.587
−0.440	0.032	0.163	1.029	0.079	1.148	0.762	−1.961	−0.674	−0.486
0.443	−1.100	0.728	−2.397	−0.543	0.872	−0.568	0.980	−0.174	0.728
−2.401	−1.375	−1.332	−2.177	−2.064	−0.245	−0.039	0.585	1.344	1.386
0.311	0.322	−0.158	0.359	0.103	0.371	0.735	0.011	2.091	0.490
−1.209	0.241	−1.488	−0.667	−1.772	−0.197	0.741	−1.303	−1.149	2.251
0.575	−1.227	−1.674	1.400	0.289	0.005	0.185	−1.072	0.431	−1.096
−0.190	0.272	1.216	0.227	1.358	0.215	−2.306	−1.301	−0.597	−1.401
−0.817	−0.769	−0.470	−0.633	0.187	−0.517	−0.888	−1.712	1.774	−0.162
0.265	−0.676	0.244	1.897	−0.629	−0.206	−1.419	1.049	0.266	−0.438
−0.221	0.678	2.149	1.486	−1.361	1.402	−0.028	0.493	0.744	0.195
−0.436	0.358	−0.602	0.107	0.085	0.573	0.529	1.577	0.239	1.898
−0.010	0.475	0.655	0.659	−0.029	−0.029	0.126	−1.335	−1.261	2.036
−0.244	1.654	1.335	−0.610	0.617	0.642	0.371	0.241	0.001	−1.799
−0.932	−1.275	−1.134	−1.246	−1.508	0.949	1.743	−0.271	−1.333	−1.875
−0.199	−1.285	−0.387	0.191	0.726	−0.151	0.064	−0.803	−0.062	0.780
−0.251	−0.431	−0.831	0.036	−0.464	−1.089	0.284	−0.451	1.693	1.004
1.074	−1.323	−1.659	−0.186	−0.612	1.612	−2.159	−1.210	0.596	−1.421
1.518	2.101	0.397	0.516	−1.169	−1.821	1.346	2.435	1.165	−0.428
0.935	−0.206	1.117	−0.241	−0.963	−0.099	0.412	−1.344	0.411	0.583
1.360	−0.380	0.031	1.066	0.893	0.431	−0.081	0.099	0.500	−2.441
0.115	−0.211	1.471	0.332	0.750	0.652	−0.812	1.383	−0.355	−0.638
0.082	−0.309	−0.355	−0.402	0.774	0.150	0.015	2.539	−0.756	−1.049
−1.492	0.259	0.323	0.697	−0.509	0.968	−0.053	1.033	−0.220	−2.322
−0.203	0.548	1.494	1.185	0.083	−1.196	−0.749	−1.105	1.324	0.689
1.857	−0.167	−1.531	1.551	0.848	0.120	0.415	−0.317	1.446	1.002
0.669	−1.017	−2.437	−0.558	−0.657	0.940	0.985	0.483	−0.361	0.095
0.128	1.463	−0.436	−0.239	−1.443	0.732	0.168	−0.144	−0.392	0.989
1.879	−2.456	0.029	0.429	0.618	−1.683	−2.262	0.034	−0.002	1.914
0.680	0.252	0.130	1.658	−1.023	0.407	−0.235	−0.224	−0.434	0.253
−0.631	0.225	−0.951	1.072	−0.285	−1.731	−0.427	−1.446	−0.873	0.619
−1.273	0.723	0.201	0.505	−0.370	−0.421	−0.015	−0.463	0.288	1.734
−0.643	−1.485	0.403	0.003	−0.243	0.000	0.964	−0.703	0.844	−0.686
−0.435	−2.162	−0.169	−1.311	−1.639	0.193	2.692	−1.994	0.326	0.562
−1.706	0.119	−1.566	0.637	−1.948	−1.068	0.935	0.738	0.650	0.491
−0.498	1.640	0.384	−0.945	−1.272	0.945	−1.013	−0.913	−0.469	2.250
−0.065	−0.005	0.618	−0.523	−0.055	1.071	0.758	−0.736	−0.959	0.598
0.190	−1.020	−1.104	0.936	−0.029	−1.004	−0.657	1.270	−0.060	−0.809
0.879	−0.642	1.155	−0.523	−0.757	−1.027	0.985	−1.222	1.078	0.163
0.559	1.094	1.587	−0.384	−1.701	0.418	0.327	0.669	0.019	0.782
−0.261	1.234	−0.505	−0.664	−0.446	−0.747	0.427	−0.369	0.089	−1.302
3.136	1.120	−0.591	2.515	−2.853	1.375	2.421	0.672	1.817	−0.067
−1.307	−0.586	−0.311	−0.026	1.633	−1.340	−1.209	0.110	−0.126	−0.288
1.455	1.099	−1.225	−0.817	0.667	−0.212	0.684	0.349	−1.161	−2.432
−0.443	−0.415	−0.660	0.098	0.435	−0.846	−0.375	−0.410	−1.747	−0.790
−0.326	0.798	0.349	0.524	0.690	−0.520	−0.522	0.602	−0.193	−0.535
- 1.027	−1.459	−0.840	−1.637	−0.462	0.607	−0.760	1.342	−1.916	0.424

Table XXIV

SQUARES, SQUARE-ROOTS, AND RECIPROCALS

n	n^2	$\sqrt{n}$	$\sqrt{10n}$	$1000/n$	n	n^2	$\sqrt{n}$	$\sqrt{10n}$	$1000/n$
					50	2 500	7.0711	22.361	20.000
1	1	1.0000	3.1623	1000.0	51	2 601	7.1414	22.583	19.608
2	4	1.4142	4.4721	500.00	52	2 704	7.2111	22.804	19.231
3	9	1.7321	5.4772	333.33	53	2 809	7.2801	23.022	18.868
4	16	2.0000	6.3246	250.00	54	2 916	7.3485	23.238	18.519
5	25	2.2361	7.0711	200.00	55	3 025	7.4162	23.452	18.182
6	36	2.4495	7.7460	166.67	56	3 136	7.4833	23.664	17.857
7	49	2.6458	8.3666	142.86	57	3 249	7.5498	23.875	17.544
8	64	2.8284	8.9443	125.00	58	3 364	7.6158	24.083	17.241
9	81	3.0000	9.4868	111.11	59	3 481	7.6811	24.290	16.949
10	100	3.1623	10.000	100.00	60	3 600	7.7460	24.495	16.667
11	121	3.3166	10.488	90.909	61	3 721	7.8103	24.698	16.393
12	144	3.4641	10.954	83.333	62	3 844	7.8740	24.900	16.129
13	169	3.6056	11.402	76.923	63	3 969	7.9373	25.100	15.873
14	196	3.7417	11.832	71.429	64	4 096	8.0000	25.298	15.625
15	225	3.8730	12.247	66.667	65	4 225	8.0623	25.495	15.385
16	256	4.0000	12.649	62.500	66	4 356	8.1240	25.690	15.152
17	289	4.1231	13.038	58.824	67	4 489	8.1854	25.884	14.925
18	324	4.2426	13.416	55.556	68	4 624	8.2462	26.077	14.706
19	361	4.3589	13.784	52.632	69	4 761	8.3066	26.268	14.493
20	400	4.4721	14.142	50.000	70	4 900	8.3666	26.458	14.286
21	441	4.5826	14.491	47.619	71	5 041	8.4262	26.646	14.085
22	484	4.6904	14.832	45.455	72	5 184	8.4853	26.833	13.889
23	529	4.7958	15.166	43.478	73	5 329	8.5440	27.019	13.699
24	576	4.8990	15.492	41.667	74	5 476	8.6023	27.203	13.514
25	625	5.0000	15.811	40.000	75	5 625	8.6603	27.386	13.333
26	676	5.0990	16.125	38.462	76	5 776	8.7178	27.568	13.158
27	729	5.1962	16.432	37.037	77	5 929	8.7750	27.749	12.987
28	784	5.2915	16.733	35.714	78	6 084	8.8318	27.928	12.821
29	841	5.3852	17.029	34.483	79	6 241	8.8882	28.107	12.658
30	900	5.4772	17.321	33.333	80	6 400	8.9443	28.284	12.500
31	961	5.5678	17.607	32.258	81	6 561	9.0000	28.461	12.346
32	1 024	5.6569	17.889	31.250	82	6 724	9.0554	28.636	12.195
33	1 089	5.7446	18.166	30.303	83	6 889	9.1104	28.810	12.048
34	1 156	5.8310	18.439	29.412	84	7 056	9.1652	28.983	11.905
35	1 225	5.9161	18.708	28.571	85	7 225	9.2195	29.155	11.765
36	1 296	6.0000	18.974	27.778	86	7 396	9.2736	29.326	11.628
37	1 369	6.0828	19.235	27.027	87	7 569	9.3274	29.496	11.494
38	1 444	6.1644	19.494	26.316	88	7 744	9.3808	29.665	11.364
39	1 521	6.2450	19.748	25.641	89	7 921	9.4340	29.833	11.236
40	1 600	6.3246	20.000	25.000	90	8 100	9.4868	30.000	11.111
41	1 681	6.4031	20.248	24.390	91	8 281	9.5394	30.166	10.989
42	1 764	6.4807	20.494	23.810	92	8 464	9.5917	30.332	10.870
43	1 849	6.5574	20.736	23.256	93	8 649	9.6437	30.496	10.753
44	1 936	6.6333	20.976	22.727	94	8 836	9.6954	30.659	10.638
45	2 025	6.7082	21.213	22.222	95	9 025	9.7468	30.822	10.526
46	2 116	6.7823	21.448	21.739	96	9 216	9.7980	30.984	10.417
47	2 209	6.8557	21.679	21.277	97	9 409	9.8489	31.145	10.309
48	2 304	6.9282	21.909	20.833	98	9 604	9.8995	31.305	10.204
49	2 401	7.0000	22.136	20.408	99	9 801	9.9499	31.464	10.101

Table XXIV continued   411

SQUARES, SQUARE-ROOTS, AND RECIPROCALS

n	n^2	$\sqrt{n}$	$\sqrt{10n}$	$1000/n$	n	n^2	$\sqrt{n}$	$\sqrt{10n}$	$1000/n$
100	10 000	10.000	31.623	10.000	150	22 500	12.247	38.730	6.6667
101	10 201	10.050	31.781	9.9010	151	22 801	12.288	38.859	6.6225
102	10 404	10.100	31.937	9.8039	152	23 104	12.329	38.987	6.5789
103	10 609	10.149	32.094	9.7087	153	23 409	12.369	39.115	6.5359
104	10 816	10.198	32.249	9.6154	154	23 716	12.410	39.243	6.4935
105	11 025	10.247	32.404	9.5238	155	24 025	12.450	39.370	6.4516
106	11 236	10.296	32.558	9.4340	156	24 336	12.490	39.497	6.4103
107	11 449	10.344	32.711	9.3458	157	24 649	12.530	39.623	6.3694
108	11 664	10.392	32.863	9.2593	158	24 964	12.570	39.749	6.3291
109	11 881	10.440	33.015	9.1743	159	25 281	12.610	39.875	6.2893
110	12 100	10.488	33.166	9.0909	160	25 600	12.649	40.000	6.2500
111	12 321	10.536	33.317	9.0090	161	25 921	12.689	40.125	6.2112
112	12 544	10.583	33.466	8.9286	162	26 244	12.728	40.249	6.1728
113	12 769	10.630	33.615	8.8496	163	26 569	12.767	40.373	6.1350
114	12 996	10.677	33.764	8.7719	164	26 896	12.806	40.497	6.0976
115	13 225	10.724	33.912	8.6957	165	27 225	12.845	40.620	6.0606
116	13 456	10.770	34.059	8.6207	166	27 556	12.884	40.743	6.0241
117	13 689	10.817	34.205	8.5470	167	27 889	12.923	40.866	5.9880
118	13 924	10.863	34.351	8.4746	168	28 224	12.961	40.988	5.9524
119	14 161	10.909	34.496	8.4034	169	28 561	13.000	41.110	5.9172
120	14 400	10.954	34.641	8.3333	170	28 900	13.038	41.231	5.8824
121	14 641	11.000	34.785	8.2645	171	29 241	13.077	41.352	5.8480
122	14 884	11.045	34.929	8.1967	172	29 584	13.115	41.473	5.8140
123	15 129	11.091	35.071	8.1301	173	29 929	13.153	41.593	5.7803
124	15 376	11.136	35.214	8.0645	174	30 276	13.191	41.713	5.7471
125	15 625	11.180	35.355	8.0000	175	30 625	13.229	41.833	5.7143
126	15 876	11.225	35.496	7.9365	176	30 976	13.267	41.952	5.6818
127	16 129	11.269	35.637	7.8740	177	31 329	13.304	42.071	5.6497
128	16 384	11.314	35.777	7.8125	178	31 684	13.342	42.190	5.6180
129	16 641	11.358	35.917	7.7519	179	32 041	13.379	42.308	5.5866
130	16 900	11.402	36.056	7.6923	180	32 400	13.416	42.426	5.5556
131	17 161	11.446	36.194	7.6336	181	32 761	13.454	42.544	5.5249
132	17 424	11.489	36.332	7.5758	182	33 124	13.491	42.661	5.4945
133	17 689	11.533	36.469	7.5188	183	33 489	13.528	42.779	5.4645
134	17 956	11.576	36.606	7.4627	184	33 856	13.565	42.895	5.4348
135	18 225	11.619	36.742	7.4074	185	34 225	13.601	43.012	5.4054
136	18 496	11.662	36.878	7.3529	186	34 596	13.638	43.128	5.3763
137	18 769	11.705	37.014	7.2993	187	34 969	13.675	43.244	5.3476
138	19 044	11.747	37.148	7.2464	188	35 344	13.711	43.359	5.3191
139	19 321	11.790	37.283	7.1942	189	35 721	13.748	43.474	5.2910
140	19 600	11.832	37.417	7.1429	190	36 100	13.784	43.589	5.2632
141	19 881	11.874	37.550	7.0922	191	36 481	13.820	43.704	5.2356
142	20 164	11.916	37.683	7.0423	192	36 864	13.856	43.818	5.2083
143	20 449	11.958	37.815	6.9930	193	37 249	13.892	43.932	5.1813
144	20 736	12.000	37.947	6.9444	194	37 636	13.928	44.045	5.1546
145	21 025	12.042	38.079	6.8966	195	38 025	13.964	44.159	5.1282
146	21 316	12.083	38.210	6.8493	196	38 416	14.000	44.272	5.1020
147	21 609	12.124	38.341	6.8027	197	38 809	14.036	44.385	5.0761
148	21 904	12.166	38.471	6.7568	198	39 204	14.071	44.497	5.0505
149	22 201	12.207	38.601	6.7114	199	39 601	14.107	44.609	5.0251

SQUARES, SQUARE-ROOTS, AND RECIPROCALS

n	n^2	$\sqrt{n}$	$\sqrt{10n}$	$1000/n$	n	n^2	$\sqrt{n}$	$\sqrt{10n}$	$1000/n$
200	40 000	14.142	44.721	5.0000	250	62 500	15.811	50.000	4.0000
201	40 401	14.177	44.833	4.9751	251	63 001	15.843	50.100	3.9841
202	40 804	14.213	44.944	4.9505	252	63 504	15.875	50.200	3.9683
203	41 209	14.248	45.056	4.9261	253	64 009	15.906	50.299	3.9526
204	41 616	14.283	45.166	4.9020	254	64 516	15.937	50.398	3.9370
205	42 025	14.318	45.277	4.8780	255	65 025	15.969	50.498	3.9216
206	42 436	14.353	45.387	4.8544	256	65 536	16.000	50.596	3.9063
207	42 849	14.387	45.497	4.8309	257	66 049	16.031	50.695	3.8911
208	43 264	14.422	45.607	4.8077	258	66 564	16.062	50.794	3.8760
209	43 681	14.457	45.717	4.7847	259	67 081	16.093	50.892	3.8610
210	44 100	14.491	45.826	4.7619	260	67 600	16.125	50.990	3.8462
211	44 521	14.526	45.935	4.7393	261	68 121	16.155	51.088	3.8314
212	44 944	14.560	46.043	4.7170	262	68 644	16.186	51.186	3.8168
213	45 369	14.595	46.152	4.6948	263	69 169	16.217	51.284	3.8023
214	45 796	14.629	46.260	4.6729	264	69 696	16.248	51.381	3.7879
215	46 225	14.663	46.368	4.6512	265	70 225	16.279	51.478	3.7736
216	46 656	14.697	46.476	4.6296	266	70 756	16.310	51.575	3.7594
217	47 089	14.731	46.583	4.6083	267	71 289	16.340	51.672	3.7453
218	47 524	14.765	46.690	4.5872	268	71 824	16.371	51.769	3.7313
219	47 961	14.799	46.797	4.5662	269	72 361	16.401	51.865	3.7175
220	48 400	14.832	46.904	4.5455	270	72 900	16.432	51.962	3.7037
221	48 841	14.866	47.011	4.5249	271	73 441	16.462	52.058	3.6900
222	49 284	14.900	47.117	4.5045	272	73 984	16.492	52.154	3.6765
223	49 729	14.933	47.223	4.4843	273	74 529	16.523	52.249	3.6630
224	50 176	14.967	47.329	4.4643	274	75 076	16.553	52.345	3.6496
225	50 625	15.000	47.434	4.4444	275	75 625	16.583	52.440	3.6364
226	51 076	15.033	47.539	4.4248	276	76 176	16.613	52.536	3.6232
227	51 529	15.067	47.645	4.4053	277	76 729	16.643	52.631	3.6101
228	51 984	15.100	47.749	4.3860	278	77 284	16.673	52.726	3.5971
229	52 441	15.133	47.854	4.3668	279	77 841	16.703	52.820	3.5842
230	52 900	15.166	47.958	4.3478	280	78 400	16.733	52.915	3.5714
231	53 361	15.199	48.062	4.3290	281	78 961	16.763	53.009	3.5587
232	53 824	15.232	48.166	4.3103	282	79 524	16.793	53.104	3.5461
233	54 289	15.264	48.270	4.2918	283	80 089	16.823	53.198	3.5336
234	54 756	15.297	48.374	4.2735	284	80 656	16.852	53.292	3.5211
235	55 225	15.330	48.477	4.2553	285	81 225	16.882	53.385	3.5088
236	55 696	15.362	48.580	4.2373	286	81 796	16.912	53.479	3.4965
237	56 169	15.395	48.683	4.2194	287	82 369	16.941	53.572	3.4843
238	56 644	15.427	48.785	4.2017	288	82 944	16.971	53.666	3.4722
239	57 121	15.460	48.888	4.1841	289	83 521	17.000	53.759	3.4602
240	57 600	15.492	48.990	4.1667	290	84 100	17.029	53.852	3.4483
241	58 081	15.524	49.092	4.1494	291	84 681	17.059	53.944	3.4364
242	58 564	15.556	49.194	4.1322	292	85 264	17.088	54.037	3.4247
243	59 049	15.588	49.295	4.1152	293	85 849	17.117	54.129	3.4130
244	59 536	15.621	49.396	4.0984	294	86 436	17.146	54.222	3.4014
245	60 025	15.652	49.497	4.0816	295	87 025	17.176	54.314	3.3898
246	60 516	15.684	49.598	4.0650	296	87 616	17.205	54.406	3.3784
247	61 009	15.716	49.699	4.0486	297	88 209	17.234	54.498	3.3670
248	61 504	15.748	49.800	4.0323	298	88 804	17.263	54.589	3.3557
249	62 001	15.780	49.900	4.0161	299	89 401	17.292	54.681	3.3445

Table XXIV continued   413

SQUARES, SQUARE-ROOTS, AND RECIPROCALS

n	n^2	$\sqrt{n}$	$\sqrt{10n}$	$1000/n$	n	n^2	$\sqrt{n}$	$\sqrt{10n}$	$1000/n$
300	90 000	17.321	54.772	3.3333	350	122 500	18.708	59.161	2.8571
301	90 601	17.349	54.863	3.3223	351	123 201	18.735	59.245	2.8490
302	91 204	17.378	54.955	3.3113	352	123 904	18.762	59.330	2.8409
303	91 809	17.407	55.045	3.3003	353	124 609	18.788	59.414	2.8329
304	92 416	17.436	55.136	3.2895	354	125 316	18.815	59.498	2.8249
305	93 025	17.464	55.227	3.2787	355	126 025	18.841	59.582	2.8169
306	93 636	17.493	55.317	3.2680	356	126 736	18.868	59.666	2.8090
307	94 249	17.521	55.408	3.2573	357	127 449	18.894	59.749	2.8011
308	94 864	17.550	55.498	3.2468	358	128 164	18.921	59.833	2.7933
309	95 481	17.578	55.588	3.2362	359	128 881	18.947	59.917	2.7855
310	96 100	17.607	55.678	3.2258	360	129 600	18.974	60.000	2.7778
311	96 721	17.635	55.767	3.2154	361	130 321	19.000	60.083	2.7701
312	97 344	17.664	55.857	3.2051	362	131 044	19.026	60.166	2.7624
313	97 969	17.692	55.946	3.1949	363	131 769	19.053	60.249	2.7548
314	98 596	17.720	56.036	3.1847	364	132 496	19.079	60.332	2.7473
315	99 225	17.748	56.125	3.1746	365	133 225	19.105	60.415	2.7397
316	99 856	17.776	56.214	3.1646	366	133 956	19.131	60.498	2.7322
317	100 489	17.804	56.303	3.1546	367	134 689	19.157	60.581	2.7248
318	101 124	17.833	56.391	3.1447	368	135 424	19.183	60.663	2.7174
319	101 761	17.861	56.480	3.1348	369	136 161	19.209	60.745	2.7100
320	102 400	17.889	56.569	3.1250	370	136 900	19.235	60.828	2.7027
321	103 041	17.916	56.657	3.1153	371	137 641	19.261	60.910	2.6954
322	103 684	17.944	56.745	3.1056	372	138 384	19.287	60.992	2.6882
323	104 329	17.972	56.833	3.0960	373	139 129	19.313	61.074	2.6810
324	104 976	18.000	56.921	3.0864	374	139 876	19.339	61.156	2.6738
325	105 625	18.028	57.009	3.0769	375	140 625	19.365	61.237	2.6667
326	106 276	18.055	57.096	3.0675	376	141 376	19.391	61.319	2.6596
327	106 929	18.083	57.184	3.0581	377	142 129	19.416	61.400	2.6525
328	107 584	18.111	57.271	3.0488	378	142 884	19.442	61.482	2.6455
329	108 241	18.138	57.359	3.0395	379	143 641	19.468	61.563	2.6385
330	108 900	18.166	57.446	3.0303	380	144 400	19.494	61.644	2.6316
331	109 561	18.193	57.533	3.0211	381	145 161	19.519	61.725	2.6247
332	110 224	18.221	57.619	3.0120	382	145 924	19.545	61.806	2.6178
333	110 889	18.248	57.706	3.0030	383	146 689	19.570	61.887	2.6110
334	111 556	18.276	57.793	2.9940	384	147 456	19.596	61.968	2.6042
335	112 225	18.303	57.879	2.9851	385	148 225	19.621	62.048	2.5974
336	112 896	18.330	57.966	2.9762	386	148 996	19.647	62.129	2.5907
337	113 569	18.358	58.052	2.9674	387	149 769	19.672	62.209	2.5840
338	114 244	18.385	58.138	2.9586	388	150 544	19.698	62.290	2.5773
339	114 921	18.412	58.224	2.9499	389	151 321	19.723	62.370	2.5707
340	115 600	18.439	58.310	2.9412	390	152 100	19.748	62.450	2.5641
341	116 281	18.466	58.395	2.9326	391	152 881	19.774	62.530	2.5575
342	116 964	18.493	58.481	2.9240	392	153 664	19.799	62.610	2.5510
343	117 649	18.520	58.566	2.9155	393	154 449	19.824	62.690	2.5445
344	118 336	18.547	58.652	2.9070	394	155 236	19.849	62.769	2.5381
345	119 025	18.574	58.737	2.8986	395	156 025	19.875	62.849	2.5316
346	119 716	18.601	58.822	2.8902	396	156 816	19.900	62.929	2.5253
347	120 409	18.628	58.907	2.8818	397	157 609	19.925	63.008	2.5189
348	121 104	18.655	58.992	2.8736	398	158 404	19.950	63.087	2.5126
349	121 801	18.682	59.076	2.8653	399	159 201	19.975	63.166	2.5063

SQUARES, SQUARE-ROOTS, AND RECIPROCALS

n	n^2	$\sqrt{n}$	$\sqrt{10n}$	$1000/n$	n	n^2	$\sqrt{n}$	$\sqrt{10n}$	$1000/n$
400	160 000	20.000	63.246	2.5000	450	202 500	21.213	67.082	2.2222
401	160 801	20.025	63.325	2.4938	451	203 401	21.237	67.157	2.2173
402	161 604	20.050	63.403	2.4876	452	204 304	21.260	67.231	2.2124
403	162 409	20.075	63.482	2.4814	453	205 209	21.284	67.305	2.2075
404	163 216	20.100	63.561	2.4752	454	206 116	21.307	67.380	2.2026
405	164 025	20.125	63.640	2.4691	455	207 025	21.331	67.454	2.1978
406	164.836	20.149	63.718	2.4631	456	207 936	21.354	67.528	2.1930
407	165 649	20.174	63.797	2.4570	457	208 849	21.378	67.602	2.1882
408	166 464	20.199	63.875	2.4510	458	209 764	21.401	67.676	2.1834
409	167 281	20.224	63.953	2.4450	459	210 681	21.424	67.750	2.1786
410	168 100	20.248	64.031	2.4390	460	211 600	21.448	67.823	2.1739
411	168 921	20.273	64.109	2.4331	461	212 521	21.471	67.897	2.1692
412	169 744	20.298	64.187	2.4272	462	213 444	21.494	67.971	2.1645
413	170 569	20.322	64.265	2.4213	463	214 369	21.517	68.044	2.1598
414	171 396	20.347	64.343	2.4155	464	215 296	21.541	68.118	2.1552
415	172 225	20.372	64.420	2.4096	465	216 225	21.564	68.191	2.1505
416	173 056	20.396	64.498	2.4038	466	217 156	21.587	68.264	2.1459
417	173 889	20.421	64.576	2.3981	467	218 089	21.610	68.337	2.1413
418	174 724	20.445	64.653	2.3923	468	219 024	21.633	68.411	2.1368
419	175 561	20.469	64.730	2.3866	469	219 961	21.656	68.484	2.1322
420	176 400	20.494	64.807	2.3810	470	220 900	21.679	68.557	2.1277
421	177 241	20.518	64.885	2.3753	471	221 841	21.703	68.629	2.1231
422	178 084	20.543	64.962	2.3697	472	222 784	21.726	68.702	2.1186
423	178 929	20.567	65.038	2.3641	473	223 729	21.749	68.775	2.1142
424	179 776	20.591	65.115	2.3585	474	224 676	21.772	68.848	2.1097
425	180 625	20.616	65.192	2.3529	475	225 625	21.794	68.920	2.1053
426	181 476	20.640	65.269	2.3474	476	226 576	21.817	68.993	2.1008
427	182 329	20.664	65.345	2.3419	477	227 529	21.840	69.065	2.0964
428	183 184	20.688	65.422	2.3364	478	228 484	21.863	69.138	2.0921
429	184 041	20.712	65.498	2.3310	479	229 441	21.886	69.210	2.0877
430	184 900	20.736	65.574	2.3256	480	230 400	21.909	69.282	2.0833
431	185 761	20.761	65.651	2.3202	481	231 361	21.932	69.354	2.0790
432	186 624	20.785	65.727	2.3148	482	232 324	21.955	69.426	2.0747
433	187 489	20.809	65.803	2.3095	483	233 289	21.977	69.498	2.0704
434	188 356	20.833	65.879	2.3041	484	234 256	22.000	69.570	2.0661
435	189 225	20.857	65.955	2.2989	485	235 225	22.023	69.642	2.0619
436	190 096	20.881	66.030	2.2936	486	236 196	22.045	69.714	2.0576
437	190 969	20.905	66.106	2.2883	487	237 169	22.068	69.785	2.0534
438	191 844	20.928	66.182	2.2831	488	238 144	22.091	69.857	2.0492
439	192 721	20.952	66.257	2.2779	489	239 121	22.113	69.929	2.0450
440	193 600	20.976	66.333	2.2727	490	240 100	22.136	70.000	2.0408
441	194 481	21.000	66.408	2.2676	491	241 081	22.159	70.071	2.0367
442	195 364	21.024	66.483	2.2624	492	242 064	22.181	70.143	2.0325
443	196 249	21.048	66.558	2.2573	493	243 049	22.204	70.214	2.0284
444	197 136	21.071	66.633	2.2523	494	244 036	22.226	70.285	2.0243
445	198 025	21.095	66.708	2.2472	495	245 025	22.249	70.356	2.0202
446	198 916	21.119	66.783	2.2422	496	246 016	22.271	70.427	2.0161
447	199 809	21.142	66.858	2.2371	497	247 009	22.294	70.498	2.0121
448	200 704	21.166	66.933	2.2321	498	248 004	22.316	70.569	2.0080
449	201 601	21.190	67.007	2.2272	499	249 001	22.338	70.640	2.0040

Table XXIV continued   415

SQUARES, SQUARE-ROOTS, AND RECIPROCALS

n	n^2	$\sqrt{n}$	$\sqrt{10n}$	$1000/n$	n	n^2	$\sqrt{n}$	$\sqrt{10n}$	$1000/n$
500	250 000	22.361	70.711	2.0000	550	302 500	23.452	74.162	1.8182
501	251 001	22.383	70.781	1.9960	551	303 601	23.473	74.229	1.8149
502	252 004	22.405	70.852	1.9920	552	304 704	23.495	74.297	1.8116
503	253 009	22.428	70.922	1.9881	553	305 809	23.516	74.364	1.8083
504	254 016	22.450	70.993	1.9841	554	306 916	23.537	74.431	1.8051
505	255 025	22.472	71.063	1.9802	555	308 025	23.558	74.498	1.8018
506	256 036	22.494	71.134	1.9763	556	309 136	23.580	74.565	1.7986
507	257 049	22.517	71.204	1.9724	557	310 249	23.601	74.632	1.7953
508	258 064	22.539	71.274	1.9685	558	311 364	23.622	74.699	1.7921
509	259 081	22.561	71.344	1.9646	559	312 481	23.643	74.766	1.7889
510	260 100	22.583	71.414	1.9608	560	313 600	23.664	74.833	1.7857
511	261 121	22.605	71.484	1.9569	561	314 721	23.685	74.900	1.7825
512	262 144	22.627	71.554	1.9531	562	315 844	23.707	74.967	1.7794
513	263 169	22.650	71.624	1.9493	563	316 969	23.728	75.033	1.7762
514	264 196	22.672	71.694	1.9455	564	318 096	23.749	75.100	1.7731
515	265 225	22.694	71.764	1.9417	565	319 225	23.770	75.166	1.7699
516	266 256	22.716	71.833	1.9380	566	320 356	23.791	75.233	1.7668
517	267 289	22.738	71.903	1.9342	567	321 489	23.812	75.299	1.7637
518	268 324	22.760	71.972	1.9305	568	322 624	23.833	75.366	1.7606
519	269 361	22.782	72.042	1.9268	569	323 761	23.854	75.432	1.7575
520	270 400	22.804	72.111	1.9231	570	324 900	23.875	75.498	1.7544
521	271 441	22.825	72.180	1.9194	571	326 041	23.896	75.565	1.7513
522	272 484	22.847	72.250	1.9157	572	327 184	23.917	75.631	1.7483
523	273 529	22.869	72.319	1.9120	573	328 329	23.937	75.697	1.7452
524	274 576	22.891	72.388	1.9084	574	329 476	23.958	75.763	1.7422
525	275 625	22.913	72.457	1.9048	575	330 625	23.979	75.829	1.7391
526	276 676	22.935	72.526	1.9011	576	331 776	24.000	75.895	1.7361
527	277 729	22.956	72.595	1.8975	577	332 929	24.021	75.961	1.7331
528	278 784	22.978	72.664	1.8939	578	334 084	24.042	76.026	1.7301
529	279 841	23.000	72.732	1.8904	579	335 241	24.062	76.092	1.7271
530	280 900	23.022	72.801	1.8868	580	336 400	24.083	76.158	1.7241
531	281 961	23.043	72.870	1.8832	581	337 561	24.104	76.223	1.7212
532	283 024	23.065	72.938	1.8797	582	338 724	24.125	76.289	1.7182
533	284 089	23.087	73.007	1.8762	583	339 889	24.145	76.354	1.7153
534	285 156	23.108	73.075	1.8727	584	341 056	24.166	76.420	1.7123
535	286 225	23.130	73.144	1.8692	585	342 225	24.187	76.485	1.7094
536	287 296	23.152	73.212	1.8657	586	343 396	24.207	76.551	1.7065
537	288 369	23.173	73.280	1.8622	587	344 569	24.228	76.616	1.7036
538	289 444	23.195	73.348	1.8587	588	345 744	24.249	76.681	1.7007
539	290 521	23.216	73.417	1.8553	589	346 921	24.269	76.746	1.6978
540	291 600	23.238	73.485	1.8519	590	348 100	24.290	76.811	1.6949
541	292 681	23.259	73.553	1.8484	591	349 281	24.310	76.877	1.6920
542	293 764	23.281	73.621	1.8450	592	350 464	24.331	76.942	1.6892
543	294 849	23.302	73.689	1.8416	593	351 649	24.352	77.006	1.6863
544	295 936	23.324	73.756	1.8382	594	352 836	24.372	77.071	1.6835
545	297 025	23.345	73.824	1.8349	595	354 025	24.393	77.136	1.6807
546	298 116	23.367	73.892	1.8315	596	355 216	24.413	77.201	1.6779
547	299 209	23.388	73.959	1.8282	597	356 409	24.434	77.266	1.6750
548	300 304	23.409	74.027	1.8248	598	357 604	24.454	77.330	1.6722
549	301 401	23.431	74.095	1.8215	599	358 801	24.474	77.395	1.6694

Table XXIV continued

SQUARES, SQUARE-ROOTS, AND RECIPROCALS

n	n^2	$\sqrt{n}$	$\sqrt{10n}$	$1000/n$	n	n^2	$\sqrt{n}$	$\sqrt{10n}$	$1000/n$
600	360 000	24.495	77.460	1.6667	650	422 500	25.495	80.623	1.5385
601	361 201	24.515	77.524	1.6639	651	423 801	25.515	80.685	1.5361
602	362 404	24.536	77.589	1.6611	652	425 104	25.534	80.747	1.5337
603	363 609	24.556	77.653	1.6584	653	426 409	25.554	80.808	1.5314
604	364 816	24.576	77.717	1.6556	654	427 716	25.573	80.870	1.5291
605	366 025	24.597	77.782	1.6529	655	429 025	25.593	80.932	1.5267
606	367 236	24.617	77.846	1.6502	656	430 336	25.613	80.994	1.5244
607	368 449	24.637	77.910	1.6474	657	431 649	25.632	81.056	1.5221
608	369 664	24.658	77.974	1.6447	658	432 964	25.652	81.117	1.5198
609	370 881	24.678	78.038	1.6420	659	434 281	25.671	81.179	1.5175
610	372 100	24.698	78.103	1.6393	660	435 600	25.690	81.240	1.5152
611	373 321	24.718	78.166	1.6367	661	436 921	25.710	81.302	1.5129
612	374 544	24.739	78.230	1.6340	662	438 244	25.729	81.363	1.5106
613	375 769	24.759	78.294	1.6313	663	439 569	25.749	81.425	1.5083
614	376 996	24.779	78.358	1.6287	664	440 896	25.768	81.486	1.5060
615	378 225	24.799	78.422	1.6260	665	442 225	25.788	81.548	1.5038
616	379 456	24.819	78.486	1.6234	666	443 556	25.807	81.609	1.5015
617	380 689	24.839	78.549	1.6207	667	444 889	25.826	81.670	1.4993
618	381 924	24.860	78.613	1.6181	668	446 224	25.846	81.731	1.4970
619	383 161	24.880	78.677	1.6155	669	447 561	25.865	81.792	1.4948
620	384 400	24.900	78.740	1.6129	670	448 900	25.884	81.854	1.4925
621	385 641	24.920	78.804	1.6103	671	450 241	25.904	81.915	1.4903
622	386 884	24.940	78.867	1.6077	672	451 584	25.923	81.976	1.4881
623	388 129	24.960	78.930	1.6051	673	452 929	25.942	82.037	1.4859
624	389 376	24.980	78.994	1.6026	674	454 276	25.962	82.098	1.4837
625	390 625	25.000	79.057	1.6000	675	455 625	25.981	82.158	1.4815
626	391 876	25.020	79.120	1.5974	676	456 976	26.000	82.219	1.4793
627	393 129	25.040	79.183	1.5949	677	458 329	26.019	82.280	1.4771
628	394 384	25.060	79.246	1.5924	678	459 684	26.038	82.341	1.4749
629	395 641	25.080	79.310	1.5898	679	461 041	26.058	82.401	1.4728
630	396 900	25.100	79.373	1.5873	680	462 400	26.077	82.462	1.4706
631	398 161	25.120	79.436	1.5848	681	463 761	26.096	82.523	1.4684
632	399 424	25.140	79.498	1.5823	682	465 124	26.115	82.583	1.4663
633	400 689	25.159	79.561	1.5798	683	466 489	26.134	82.644	1.4641
634	401 956	25.179	79.624	1.5773	684	467 856	26.153	82.704	1.4620
635	403 225	25.199	79.687	1.5748	685	469 225	26.173	82.765	1.4599
636	404 496	25.219	79.750	1.5723	686	470 596	26.192	82.825	1.4577
637	405 769	25.239	79.812	1.5699	687	471 969	26.211	82.885	1.4556
638	407 044	25.259	79.875	1.5674	688	473 344	26.230	82.946	1.4535
639	408 321	25.278	79.937	1.5649	689	474 721	26.249	83.006	1.4514
640	409 600	25.298	80.000	1.5625	690	476 100	26.268	83.066	1.4493
641	410 881	25.318	80.062	1.5601	691	477 481	26.287	83.126	1.4472
642	412 164	25.338	80.125	1.5576	692	478 864	26.306	83.187	1.4451
643	413 449	25.357	80.187	1.5552	693	480 249	26.325	83.247	1.4430
644	414 736	25.377	80.250	1.5528	694	481 636	26.344	83.307	1.4409
645	416 025	25.397	80.312	1.5504	695	483 025	26.363	83.367	1.4388
646	417 316	25.417	80.374	1.5480	696	484 416	26.382	83.427	1.4368
647	418 609	25.436	80.436	1.5456	697	485 809	26.401	83.487	1.4347
648	419 904	25.456	80.498	1.5432	698	487 204	26.420	83.546	1.4327
649	421 201	25.475	80.561	1.5408	699	488 601	26.439	83.606	1.4306

Table XXIV continued 417

SQUARES, SQUARE-ROOTS, AND RECIPROCALS

n	n^2	$\sqrt{n}$	$\sqrt{10n}$	$1000/n$	n	n^2	$\sqrt{n}$	$\sqrt{10n}$	$1000/n$
700	490 000	26.458	83.666	1.4286	750	562 500	27.386	86.603	1.3333
701	491 401	26.476	83.726	1.4265	751	564 001	27.404	86.660	1.3316
702	492 804	26.495	83.785	1.4245	752	565 504	27.423	86.718	1.3298
703	494 209	26.514	83.845	1.4225	753	567 009	27.441	86.776	1.3280
704	495 616	26.533	83.905	1.4205	754	568 516	27.459	86.833	1.3263
705	497 025	26.552	83.964	1.4184	755	570 025	27.477	86.891	1.3245
706	498 436	26.571	84.024	1.4164	756	571 536	27.495	86.948	1.3228
707	499 849	26.589	84.083	1.4144	757	573 049	27.514	87.006	1.3210
708	501 264	26.608	84.143	1.4124	758	574 564	27.532	87.063	1.3193
709	502 681	26.627	84.202	1.4104	759	576 081	27.550	87.121	1.3175
710	504 100	26.646	84.262	1.4085	760	577 600	27.568	87.178	1.3158
711	505 521	26.665	84.321	1.4065	761	579 121	27.586	87.235	1.3141
712	506 944	26.683	84.380	1.4045	762	580 644	27.604	87.293	1.3123
713	508 369	26.702	84.439	1.4025	763	582 169	27.622	87.350	1.3106
714	509 796	26.721	84.499	1.4006	764	583 696	27.641	87.407	1.3089
715	511 225	26.739	84.558	1.3986	765	585 225	27.659	87.464	1.3072
716	512 656	26.758	84.617	1.3966	766	586 756	27.677	87.521	1.3055
717	514 089	26.777	84.676	1.3947	767	588 289	27.695	87.579	1.3038
718	515 524	26.796	84.735	1.3928	768	589 824	27.713	87.636	1.3021
719	516 961	26.814	84.794	1.3908	769	591 361	27.731	87.693	1.3004
720	518 400	26.833	84.853	1.3889	770	592 900	27.749	87.750	1.2987
721	519 841	26.851	84.912	1.3870	771	594 441	27.767	87.807	1.2970
722	521 284	26.870	84.971	1.3850	772	595 984	27.785	87.864	1.2953
723	522 729	26.889	85.029	1.3831	773	597 529	27.803	87.920	1.2937
724	524 176	26.907	85.088	1.3812	774	599 076	27.821	87.977	1.2920
725	525 625	26.926	85.147	1.3793	775	600 625	27.839	88.034	1.2903
726	527 076	26.944	85.206	1.3774	776	602 176	27.857	88.091	1.2887
727	528 529	26.963	85.264	1.3755	777	603 729	27.875	88.148	1.2870
728	529 984	26.981	85.323	1.3736	778	605 284	27.893	88.204	1.2853
729	531 441	27.000	85.382	1.3717	779	606 841	27.911	88.261	1.2837
730	532 900	27.019	85.440	1.3699	780	608 400	27.928	88.318	1.2821
731	534 361	27.037	85.499	1.3680	781	609 961	27.946	88.374	1.2804
732	535 824	27.056	85.557	1.3661	782	611 524	27.964	88.431	1.2788
733	537 289	27.074	85.615	1.3643	783	613 089	27.982	88.487	1.2771
734	538 756	27.092	85.674	1.3624	784	614 656	28.000	88.544	1.2755
735	540 225	27.111	85.732	1.3605	785	616 225	28.018	88.600	1.2739
736	541 696	27.129	85.790	1.3587	786	617 796	28.036	88.657	1.2723
737	543 169	27.148	85.849	1.3569	787	619 369	28.054	88.713	1.2706
738	544 644	27.166	85.907	1.3550	788	620 944	28.071	88.769	1.2690
739	546 121	27.185	85.965	1.3532	789	622 521	28.089	88.826	1.2674
740	547 600	27.203	86.023	1.3514	790	624 100	28.107	88.882	1.2658
741	549 081	27.221	86.081	1.3495	791	625 681	28.125	88.938	1.2642
742	550 564	27.240	86.139	1.3477	792	627 264	28.142	88.994	1.2626
743	552 049	27.258	86.197	1.3459	793	628 849	28.160	89.051	1.2610
744	553 536	27.276	86.255	1.3441	794	630 436	28.178	89.107	1.2594
745	555 025	27.295	86.313	1.3423	795	632 025	28.196	89.163	1.2579
746	556 516	27.313	86.371	1.3405	796	633 616	28.213	89.219	1.2563
747	558 009	27.331	86.429	1.3387	797	635 209	28.231	89.275	1.2547
748	559 504	27.350	86.487	1.3369	798	636 804	28.249	89.331	1.2531
749	561 001	27.368	86.545	1.3351	799	638 401	28.267	89.387	1.2516

SQUARES, SQUARE-ROOTS, AND RECIPROCALS

n	n^2	$\sqrt{n}$	$\sqrt{10n}$	$1000/n$	n	n^2	$\sqrt{n}$	$\sqrt{10n}$	$1000/n$
800	640 000	28.284	89.443	1.2500	850	722 500	29.155	92.195	1.1765
801	641 601	28.302	89.499	1.2484	851	724 201	29.172	92.250	1.1751
802	643 204	28.320	89.554	1.2469	852	725 904	29.189	92.304	1.1737
803	644 809	28.337	89.610	1.2453	853	727 609	29.206	92.358	1.1723
804	646 416	28.355	89.666	1.2438	854	729 316	29.223	92.412	1.1710
805	648 025	28.373	89.722	1.2422	855	731 025	29.240	92.466	1.1696
806	649 636	28.390	89.778	1.2407	856	732 736	29.257	92.520	1.1682
807	651 249	28.408	89.833	1.2392	857	734 449	29.275	92.574	1.1669
808	652 864	28.425	89.889	1.2376	858	736 164	29.292	92.628	1.1655
809	654 481	28.443	89.944	1.2361	859	737 881	29.309	92.682	1.1641
810	656 100	28.461	90.000	1.2346	860	739 600	29.326	92.736	1.1628
811	657 721	28.478	90.056	1.2330	861	741 321	29.343	92.790	1.1614
812	659 344	28.496	90.111	1.2315	862	743 044	29.360	92.844	1.1601
813	660 969	28.513	90.167	1.2300	863	744 769	29.377	92.898	1.1587
814	662 596	28.531	90.222	1.2285	864	746 496	29.394	92.952	1.1574
815	664 225	28.548	90.277	1.2270	865	748 225	29.411	93.005	1.1561
816	665 856	28.566	90.333	1.2255	866	749 956	29.428	93.059	1.1547
817	667 489	28.583	90.388	1.2240	867	751 689	29.445	93.113	1.1534
818	669 124	28.601	90.443	1.2225	868	753 424	29.462	93.167	1.1521
819	670 761	28.618	90.499	1.2210	869	755 161	29.479	93.220	1.1507
820	672 400	28.636	90.554	1.2195	870	756 900	29.496	93.274	1.1494
821	674 041	28.653	90.609	1.2180	871	758 641	29.513	93.327	1.1481
822	675 684	28.671	90.664	1.2165	872	760 384	29.530	93.381	1.1468
823	677 329	28.688	90.719	1.2151	873	762 129	29.547	93.434	1.1455
824	678 976	28.705	90.774	1.2136	874	763 876	29.563	93.488	1.1442
825	680 625	28.723	90.830	1.2121	875	765 625	29.580	93.541	1.1429
826	682 276	28.740	90.885	1.2107	876	767 376	29.597	93.595	1.1416
827	683 929	28.758	90.940	1.2092	877	769 129	29.614	93.648	1.1403
828	685 584	28.775	90.995	1.2077	878	770 884	29.631	93.702	1.1390
829	687 241	28.792	91.049	1.2063	879	772 641	29.648	93.755	1.1377
830	688 900	28.810	91.104	1.2048	880	774 400	29.665	93.808	1.1364
831	690 561	28.827	91.159	1.2034	881	776 161	29.682	93.862	1.1351
832	692 224	28.844	91.214	1.2019	882	777 924	29.698	93.915	1.1338
833	693 889	28.862	91.269	1.2005	883	779 689	29.715	93.968	1.1325
834	695 556	28.879	91.324	1.1990	884	781 456	29.732	94.021	1.1312
835	697 225	28.896	91.378	1.1976	885	783 225	29.749	94.074	1.1299
836	698 896	28.914	91.433	1.1962	886	784 996	29.766	94.128	1.1287
837	700 569	28.931	91.488	1.1947	887	786 769	29.783	94.181	1.1274
838	702 244	28.948	91.542	1.1933	888	788 544	29.799	94.234	1.1261
839	703 921	28.966	91.597	1.1919	889	790 321	29.816	94.287	1.1249
840	705 600	28.983	91.652	1.1905	890	792 100	29.833	94.340	1.1236
841	707 281	29.000	91.706	1.1891	891	793 881	29.850	94.393	1.1223
842	708 964	29.017	91.761	1.1876	892	795 664	29.866	94.446	1.1211
843	710 649	29.034	91.815	1.1862	893	797 449	29.883	94.499	1.1198
844	712 336	29.052	91.869	1.1848	894	799 236	29.900	94.552	1.1186
845	714 025	29.069	91.924	1.1834	895	801 025	29.917	94.604	1.1173
846	715 716	29.086	91.978	1.1820	896	802 816	29.933	94.657	1.1161
847	717 409	29.103	92.033	1.1806	897	804 609	29.950	94.710	1.1148
848	719 104	29.120	92.087	1.1792	898	806 404	29.967	94.763	1.1136
849	720 801	29.138	92.141	1.1779	899	808 201	29.983	94.816	1.1123

Table XXIV continued 419

SQUARES, SQUARE-ROOTS, AND RECIPROCALS

n	n^2	$\sqrt{n}$	$\sqrt{10n}$	$1000/n$	n	n^2	$\sqrt{n}$	$\sqrt{10n}$	$1000/n$
900	810 000	30.000	94.868	1.1111	950	902 500	30.822	97.468	1.0526
901	811 801	30.017	94.921	1.1099	951	904 401	30.838	97.519	1.0515
902	813 604	30.033	94.974	1.1086	952	906 304	30.855	97.570	1.0504
903	815 409	30.050	95.026	1.1074	953	908 209	30.871	97.622	1.0493
904	817 216	30.067	95.079	1.1062	954	910 116	30.887	97.673	1.0482
905	819 025	30.083	95.131	1.1050	955	912 025	30.903	97.724	1.0471
906	820 836	30.100	95.184	1.1038	956	913 936	30.919	97.775	1.0460
907	822 649	30.116	95.237	1.1025	957	915 849	30.935	97.826	1.0449
908	824 464	30.133	95.289	1.1013	958	917 764	30.952	97.877	1.0438
909	826 281	30.150	95.341	1.1001	959	919 681	30.968	97.929	1.0428
910	828 100	30.166	95.394	1.0989	960	921 600	30.984	97.980	1.0417
911	829 921	30.183	95.446	1.0977	961	923 521	31.000	98.031	1.0406
912	831 744	30.199	95.499	1.0965	962	925 444	31.016	98.082	1.0395
913	833 569	30.216	95.551	1.0953	963	927 369	31.032	98.133	1.0384
914	835 396	30.232	95.603	1.0941	964	929 296	31.048	98.184	1.0373
915	837 225	30.249	95.656	1.0929	965	931 225	31.064	98.234	1.0363
916	839 056	30.265	95.708	1.0917	966	933 156	31.081	98.285	1.0352
917	840 889	30.282	95.760	1.0905	967	935 089	31.097	98.336	1.0341
918	842 724	30.299	95.812	1.0893	968	937 024	31.113	98.387	1.0331
919	844 561	30.315	95.864	1.0881	969	938 961	31.129	98.438	1.0320
920	846 400	30.332	95.917	1.0870	970	940 900	31.145	98.489	1.0309
921	848 241	30.348	95.969	1.0858	971	942 841	31.161	98.539	1.0299
922	850 084	30.364	96.021	1.0846	972	944 784	31.177	98.590	1.0288
923	851 929	30.381	96.073	1.0834	973	946 729	31.193	98.641	1.0277
924	853 776	30.397	96.125	1.0823	974	948 676	31.209	98.691	1.0267
925	855 625	30.414	96.177	1.0811	975	950 625	31.225	98.742	1.0256
926	857 476	30.430	96.229	1.0799	976	952 576	31.241	98.793	1.0246
927	859 329	30.447	96.281	1.0787	977	954 529	31.257	98.843	1.0235
928	861 184	30.463	96.333	1.0776	978	956 484	31.273	98.894	1.0225
929	863 041	30.480	96.385	1.0764	979	958 441	31.289	98.944	1.0215
930	864 900	30.496	96.437	1.0753	980	960 400	31.305	98.995	1.0204
931	866 761	30.512	96.488	1.0741	981	962 361	31.321	99.045	1.0194
932	868 624	30.529	96.540	1.0730	982	964 324	31.337	99.096	1.0183
933	870 489	30.545	96.592	1.0718	983	966 289	31.353	99.146	1.0173
934	872 356	30.561	96.644	1.0707	984	968 256	31.369	99.197	1.0163
935	874 225	30.578	96.695	1.0695	985	970 225	31.385	99.247	1.0152
936	876 096	30.594	96.747	1.0684	986	972 196	31.401	99.298	1.0142
937	877 969	30.610	96.799	1.0672	987	974 169	31.417	99.348	1.0132
938	879 844	30.627	96.850	1.0661	988	976 144	31.432	99.398	1.0121
939	881 721	30.643	96.902	1.0650	989	978 121	31.448	99.448	1.0111
940	883 600	30.659	96.954	1.0638	990	980 100	31.464	99.499	1.0101
941	885 481	30.676	97.005	1.0627	991	982 081	31.480	99.549	1.0091
942	887 364	30.692	97.057	1.0616	992	984 064	31.496	99.599	1.0081
943	889 249	30.708	97.108	1.0604	993	986 049	31.512	99.649	1.0070
944	891 136	30.725	97.160	1.0593	994	988 036	31.528	99.700	1.0060
945	893 025	30.741	97.211	1.0582	995	990 025	31.544	99.750	1.0050
946	894 916	30.757	97.263	1.0571	996	992 016	31.559	99.800	1.0040
947	896 809	30.773	97.314	1.0560	997	994 009	31.575	99.850	1.0030
948	898 704	30.790	97.365	1.0549	998	996 004	31.591	99.900	1.0020
949	900 601	30.806	97.417	1.0537	999	998 001	31.607	99.950	1.0010

420

Table XXV

NATURAL TRIGONOMETRIC FUNCTIONS

For degrees indicated in the left hand column use the column headings at the top. For degrees indicated in the right hand column use the column headings at the bottom.

Deg.	Rad	Sin	Cos	Tan	Ctn	Sec	Csc		
0	0.0000	0.0000	1.0000	0.0000	......	1.0000	......	1.5708	90
1	0.0175	0.0175	0.9998	0.0175	57.290	1.0002	57.299	1.5533	89
2	0.0349	0.0349	0.9994	0.0349	28.636	1.0006	28.654	1.5359	88
3	0.0524	0.0523	0.9986	0.0524	19.081	1.0014	19.107	1.5184	87
4	0.0698	0.0698	0.9976	0.0699	14.301	1.0024	14.336	1.5010	86
5	0.0873	0.0872	0.9962	0.0875	11.430	1.0038	11.474	1.4835	85
6	0.1047	0.1045	0.9945	0.1051	9.5144	1.0055	9.5668	1.4661	84
7	0.1222	0.1219	0.9925	0.1228	8.1443	1.0075	8.2055	1.4486	83
8	0.1396	0.1392	0.9903	0.1405	7.1154	1.0098	7.1853	1.4312	82
9	0.1571	0.1564	0.9877	0.1584	6.3138	1.0125	6.3925	1.4137	81
10	0.1745	0.1736	0.9848	0.1763	5.6713	1.0154	5.7588	1.3963	80
11	0.1920	0.1908	0.9816	0.1944	5.1446	1.0187	5.2408	1.3788	79
12	0.2094	0.2079	0.9781	0.2126	4.7046	1.0223	4.8097	1.3614	78
13	0.2269	0.2250	0.9744	0.2309	4.3315	1.0263	4.4454	1.3439	77
14	0.2443	0.2419	0.9703	0.2493	4.0108	1.0306	4.1336	1.3265	76
15	0.2618	0.2588	0.9659	0.2679	3.7321	1.0353	3.8637	1.3090	75
16	0.2793	0.2756	0.9613	0.2867	3.4874	1.0403	3.6280	1.2915	74
17	0.2967	0.2924	0.9563	0.3057	3.2709	1.0457	3.4203	1.2741	73
18	0.3142	0.3090	0.9511	0.3249	3.0777	1.0515	3.2361	1.2566	72
19	0.3316	0.3256	0.9455	0.3443	2.9042	1.0576	3.0716	1.2392	71
20	0.3491	0.3420	0.9397	0.3640	2.7475	1.0642	2.9238	1.2217	70
21	0.3665	0.3584	0.9336	0.3839	2.6051	1.0711	2.7904	1.2043	69
22	0.3840	0.3746	0.9272	0.4040	2.4751	1.0785	2.6695	1.1868	68
23	0.4014	0.3907	0.9205	0.4245	2.3559	1.0864	2.5593	1.1694	67
24	0.4189	0.4067	0.9135	0.4452	2.2460	1.0946	2.4586	1.1519	66
25	0.4363	0.4226	0.9063	0.4663	2.1445	1.1034	2.3662	1.1345	65
26	0.4538	0.4384	0.8988	0.4877	2.0503	1.1126	2.2812	1.1170	64
27	0.4712	0.4540	0.8910	0.5095	1.9626	1.1223	2.2027	1.0996	63
28	0.4887	0.4695	0.8829	0.5317	1.8807	1.1326	2.1301	1.0821	62
29	0.5061	0.4848	0.8746	0.5543	1.8040	1.1434	2.0627	1.0647	61
30	0.5236	0.5000	0.8660	0.5774	1.7321	1.1547	2.0000	1.0472	60
31	0.5411	0.5150	0.8572	0.6009	1.6643	1.1666	1.9416	1.0297	59
32	0.5585	0.5299	0.8480	0.6249	1.6003	1.1792	1.8871	1.0123	58
33	0.5760	0.5446	0.8387	0.6494	1.5399	1.1924	1.8361	0.9948	57
34	0.5934	0.5592	0.8290	0.6745	1.4826	1.2062	1.7883	0.9774	56
35	0.6109	0.5736	0.8192	0.7002	1.4281	1.2208	1.7434	0.9599	55
36	0.6283	0.5878	0.8090	0.7265	1.3764	1.2361	1.7013	0.9425	54
37	0.6458	0.6018	0.7986	0.7536	1.3270	1.2521	1.6616	0.9250	53
38	0.6632	0.6157	0.7880	0.7813	1.2799	1.2690	1.6243	0.9076	52
39	0.6807	0.6293	0.7771	0.8098	1.2349	1.2868	1.5890	0.8901	51
40	0.6981	0.6428	0.7660	0.8391	1.1918	1.3054	1.5557	0.8727	50
41	0.7156	0.6561	0.7547	0.8693	1.1504	1.3250	1.5243	0.8552	49
42	0.7330	0.6691	0.7431	0.9004	1.1106	1.3456	1.4945	0.8378	48
43	0.7505	0.6820	0.7314	0.9325	1.0724	1.3673	1.4663	0.8203	47
44	0.7679	0.6947	0.7193	0.9657	1.0355	1.3902	1.4396	0.8029	46
45	0.7854	0.7071	0.7071	1.0000	1.0000	1.4142	1.4142	0.7854	45
		Cos	Sin	Ctn	Tan	Csc	Sec	Rad	Deg.

Table XXVI

NATURAL LOGARITHMS OF NUMBERS—0.00 TO 5.99

(Base e = 2.718 · · ·)

N	0	1	2	3	4	5	6	7	8	9
0.0		5.395	6.088	6.493	6.781	7.004	7.187	7.341	7.474	7.592
0.1	7.697	7.793	7.880	7.960	8.034	8.103	8.167	8.228	8.285	8.339
0.2	8.391	8.439	8.486	8.530	8.573	8.614	8.653	8.691	8.727	8.762
0.3	8.796	8.829	8.861	8.891	8.921	8.950	8.978	9.006	9.032	9.058
0.4	9.084	9.108	9.132	9.156	9.179	9.201	9.223	9.245	9.266	9.287
0.5	9.307	9.327	9.346	9.365	9.384	9.402	9.420	9.438	9.455	9.472
0.6	9.489	9.506	9.522	9.538	9.554	9.569	9.584	9.600	9.614	9.629
0.7	9.643	9.658	9.671	9.685	9.699	9.712	9.726	9.739	9.752	9.764
0.8	9.777	9.789	9.802	9.814	9.826	9.837	9.849	9.861	9.872	9.883
0.9	9.895	9.906	9.917	9.927	9.938	9.949	9.959	9.970	9.980	9.990

(Rows 0.0–0.9: Take tabular value −10)

N	0	1	2	3	4	5	6	7	8	9
1.0	0.0 0000	0995	1980	2956	3922	4879	5827	6766	7696	8618
1.1	9531	*0436	*1333	*2222	*3103	*3976	*4842	*5700	*6551	*7395
1.2	0.1 8232	9062	9885	*0701	*1511	*2314	*3111	*3902	*4686	*5464
1.3	0.2 6236	7003	7763	8518	9267	*0010	*0748	*1481	*2208	*2930
1.4	0.3 3647	4359	5066	5767	6464	7156	7844	8526	9204	9878
1.5	0.4 0547	1211	1871	2527	3178	3825	4469	5108	5742	6373
1.6	7000	7623	8243	8858	9470	*0078	*0682	*1282	*1879	*2473
1.7	0.5 3063	3649	4232	4812	5389	5962	6531	7098	7661	8222
1.8	8779	9333	9884	*0432	*0977	*1519	*2058	*2594	*3127	*3658
1.9	0.6 4185	4710	5233	5752	6269	6783	7294	7803	8310	8813
2.0	9315	9813	*0310	*0804	*1295	*1784	*2271	*2755	*3237	*3716
2.1	0.7 4194	4669	5142	5612	6081	6547	7011	7473	7932	8390
2.2	8846	9299	9751	*0200	*0648	*1093	*1536	*1978	*2418	*2855
2.3	0.8 3291	3725	4157	4587	5015	5442	5866	6289	6710	7129
2.4	7547	7963	8377	8789	9200	9609	*0016	*0422	*0826	*1228
2.5	0.9 1629	2028	2426	2822	3216	3609	4001	4391	4779	5166
2.6	5551	5935	6317	6698	7078	7456	7833	8208	8582	8954
2.7	9325	9695	*0063	*0430	*0796	*1160	*1523	*1885	*2245	*2604
2.8	1.0 2962	3318	3674	4028	4380	4732	5082	5431	5779	6126
2.9	6471	6815	7158	7500	7841	8181	8519	8856	9192	9527
3.0	9861	*0194	*0526	*0856	*1186	*1514	*1841	*2168	*2493	*2817
3.1	1.1 3140	3462	3783	4103	4422	4740	5057	5373	5688	6002
3.2	6315	6627	6938	7248	7557	7865	8173	8479	8784	9089
3.3	9392	9695	9996	*0297	*0597	*0896	*1194	*1491	*1788	*2083
3.4	1.2 2378	2671	2964	3256	3547	3837	4127	4415	4703	4990
3.5	5276	5562	5846	6130	6413	6695	6976	7257	7536	7815
3.6	8093	8371	8647	8923	9198	9473	9746	*0019	*0291	*0563
3.7	1.3 0833	1103	1372	1641	1909	2176	2442	2708	2972	3237
3.8	3500	3763	4025	4286	4547	4807	5067	5325	5584	5841
3.9	6098	6354	6609	6864	7118	7372	7624	7877	8128	8379
4.0	8629	8879	9128	9377	9624	9872	*0118	*0364	*0610	*0854
4.1	1.4 1099	1342	1585	1828	2070	2311	2552	2792	3031	3270
4.2	3508	3746	3984	4220	4456	4692	4927	5161	5395	5629
4.3	5862	6094	6326	6557	6787	7018	7247	7476	7705	7933
4.4	8160	8387	8614	8840	9065	9290	9515	9739	9962	*0185
4.5	1.5 0408	0630	0851	1072	1293	1513	1732	1951	2170	2388
4.6	2606	2823	3039	3256	3471	3687	3902	4116	4330	4543
4.7	4756	4969	5181	5393	5604	5814	6025	6235	6444	6653
4.8	6862	7070	7277	7485	7691	7898	8104	8309	8515	8719
4.9	8924	9127	9331	9534	9737	9939	*0141	*0342	*0543	*0744
5.0	1.6 0944	1144	1343	1542	1741	1939	2137	2334	2531	2728
5.1	2924	3120	3315	3511	3705	3900	4094	4287	4481	4673
5.2	4866	5058	5250	5441	5632	5823	6013	6203	6393	6582
5.3	6771	6959	7147	7335	7523	7710	7896	8083	8269	8455
5.4	8640	8825	9010	9194	9378	9562	9745	9928	*0111	*0293
5.5	1.7 0475	0656	0838	1019	1199	1380	1560	1740	1919	2098
5.6	2277	2455	2633	2811	2988	3166	3342	3519	3695	3871
5.7	4047	4222	4397	4572	4746	4920	5094	5267	5440	5613
5.8	5786	5958	6130	6302	6473	6644	6815	6985	7156	7326
5.9	7495	7665	7834	8002	8171	8339	8507	8675	8842	9009
N	0	1	2	3	4	5	6	7	8	9

$$\log_e 0.10 = 7.69741\ 49070 - 10$$

NATURAL LOGARITHMS OF NUMBERS—6.00 TO 10.09

N	0	1	2	3	4	5	6	7	8	9
6.0	1.7 9176	9342	9509	9675	9840	*0006	*0171	*0336	*0500	*0665
6.1	1.8 0829	0993	1156	1319	1482	1645	1808	1970	2132	2294
6.2	2455	2616	2777	2938	3098	3258	3418	3578	3737	3896
6.3	4055	4214	4372	4530	4688	4845	5003	5160	5317	5473
6.4	5630	5786	5942	6097	6253	6408	6563	6718	6872	7026
6.5	7180	7334	7487	7641	7794	7947	8099	8251	8403	8555
6.6	8707	8858	9010	9160	9311	9462	9612	9762	9912	*0061
6.7	1.9 0211	0360	0509	0658	0806	0954	1102	1250	1398	1545
6.8	1692	1839	1986	2132	2279	2425	2571	2716	2862	3007
6.9	3152	3297	3442	3586	3730	3874	4018	4162	4305	4448
7.0	4591	4734	4876	5019	5161	5303	5445	5586	5727	5869
7.1	6009	6150	6291	6431	6571	6711	6851	6991	7130	7269
7.2	7408	7547	7685	7824	7962	8100	8238	8376	8513	8650
7.3	8787	8924	9061	9198	9334	9470	9606	9742	9877	*0013
7.4	2.0 0148	0283	0418	0553	0687	0821	0956	1089	1223	1357
7.5	1490	1624	1757	1890	2022	2155	2287	2419	2551	2683
7.6	2815	2946	3078	3209	3340	3471	3601	3732	3862	3992
7.7	4122	4252	4381	4511	4640	4769	4898	5027	5156	5284
7.8	5412	5540	5668	5796	5924	6051	6179	6306	6433	6560
7.9	6686	6813	6939	7065	7191	7317	7443	7568	7694	7819
8.0	7944	8069	8194	8318	8443	8567	8691	8815	8939	9063
8.1	9186	9310	9433	9556	9679	9802	9924	*0047	*0169	*0291
8.2	2.1 0413	0535	0657	0779	0900	1021	1142	1263	1384	1505
8.3	1626	1746	1866	1986	2106	2226	2346	2465	2585	2704
8.4	2823	2942	3061	3180	3298	3417	3535	3653	3771	3889
8.5	4007	4124	4242	4359	4476	4593	4710	4827	4943	5060
8.6	5176	5292	5409	5524	5640	5756	5871	5987	6102	6217
8.7	6332	6447	6562	6677	6791	6905	7020	7134	7248	7361
8.8	7475	7589	7702	7816	7929	8042	8155	8267	8380	8493
8.9	8605	8717	8830	8942	9054	9165	9277	9389	9500	9611
9.0	9722	9834	9944	*0055	*0166	*0276	*0387	*0497	*0607	*0717
9.1	2.2 0827	0937	1047	1157	1266	1375	1485	1594	1703	1812
9.2	1920	2029	2138	2246	2354	2462	2570	2678	2786	2894
9.3	3001	3109	3216	3324	3431	3538	3645	3751	3858	3965
9.4	4071	4177	4284	4390	4496	4601	4707	4813	4918	5024
9.5	5129	5234	5339	5444	5549	5654	5759	5863	5968	6072
9.6	6176	6280	6384	6488	6592	6696	6799	6903	7006	7109
9.7	7213	7316	7419	7521	7624	7727	7829	7932	8034	8136
9.8	8238	8340	8442	8544	8646	8747	8849	8950	9051	9152
9.9	9253	9354	9455	9556	9657	9757	9858	9958	*0058	*0158
10.0	2.3 0259	0358	0458	0558	0658	0757	0857	0956	1055	1154
N	0	1	2	3	4	5	6	7	8	9

NATURAL LOGARITHMS OF NUMBERS—10 TO 99

N	0	1	2	3	4	5	6	7	8	9
1	2.30259	39790	48491	56495	63906	70805	77259	83321	89037	94444
2	99573	*04452	*09104	*13549	*17805	*21888	*25810	*29584	*33220	*36730
3	3.40120	43399	46574	49651	52636	55535	58352	61092	63759	66356
4	68888	71357	73767	76120	78419	80666	82864	85015	87120	89182
5	91202	93183	95124	97029	98898	*00733	*02535	*04305	*06044	*07754
6	4.09434	11087	12713	14313	15888	17439	18965	20469	21951	23411
7	24850	26268	27667	29046	30407	31749	33073	34381	35671	36945
8	38203	39445	40672	41884	43082	44265	45435	46591	47734	48864
9	49981	51086	52179	53260	54329	55388	56435	57471	58497	59512

$$\log_e 10 = 2.30258\ 50930$$

Table XXVI continued   423

NATURAL LOGARITHMS OF NUMBERS—100 TO 609

N	0	1	2	3	4	5	6	7	8	9
10	4.6 0517	1512	2497	3473	4439	5396	6344	7283	8213	9135
11	4.7 0048	0953	1850	2739	3620	4493	5359	6217	7068	7912
12	8749	9579	*0402	*1218	*2028	*2831	*3628	*4419	*5203	*5981
13	4.8 6753	7520	8280	9035	9784	*0527	*1265	*1998	*2725	*3447
14	4.9 4164	4876	5583	6284	6981	7673	8361	9043	9721	*0395
15	5.0 1064	1728	2388	3044	3695	4343	4986	5625	6260	6890
16	7517	8140	8760	9375	9987	*0595	*1199	*1799	*2396	*2990
17	5.1 3580	4166	4749	5329	5906	6479	7048	7615	8178	8739
18	9296	9850	*0401	*0949	*1494	*2036	*2575	*3111	*3644	*4175
19	5.2 4702	5227	5750	6269	6786	7300	7811	8320	8827	9330
20	9832	*0330	*0827	*1321	*1812	*2301	*2788	*3272	*3754	*4233
21	5.3 4711	5186	5659	6129	6598	7064	7528	7990	8450	8907
22	9363	9816	*0268	*0717	*1165	*1610	*2053	*2495	*2935	*3372
23	5.4 3808	4242	4674	5104	5532	5959	6383	6806	7227	7646
24	8064	8480	8894	9306	9717	*0126	*0533	*0939	*1343	*1745
25	5.5 2146	2545	2943	3339	3733	4126	4518	4908	5296	5683
26	6068	6452	6834	7215	7595	7973	8350	8725	9099	9471
27	9842	*0212	*0580	*0947	*1313	*1677	*2040	*2402	*2762	*3121
28	5.6 3479	3835	4191	4545	4897	5249	5599	5948	6296	6643
29	6988	7332	7675	8017	8358	8698	9036	9373	9709	*0044
30	5.7 0378	0711	1043	1373	1703	2031	2359	2685	3010	3334
31	3657	3979	4300	4620	4939	5257	5574	5890	6205	6519
32	6832	7144	7455	7765	8074	8383	8690	8996	9301	9606
33	9909	*0212	*0513	*0814	*1114	*1413	*1711	*2008	*2305	*2600
34	5.8 2895	3188	3481	3773	4064	4354	4644	4932	5220	5507
35	5793	6079	6363	6647	6930	7212	7493	7774	8053	8332
36	8610	8888	9164	9440	9715	9990	*0263	*0536	*0808	*1080
37	5.9 1350	1620	1889	2158	2426	2693	2959	3225	3489	3754
38	4017	4280	4542	4803	5064	5324	5584	5842	6101	6358
39	6615	6871	7126	7381	7635	7889	8141	8394	8645	8896
40	9146	9396	9645	9894	*0141	*0389	*0635	*0881	*1127	*1372
41	6.0 1616	1859	2102	2345	2587	2828	3069	3309	3548	3787
42	4025	4263	4501	4737	4973	5209	5444	5678	5912	6146
43	6379	6611	6843	7074	7304	7535	7764	7993	8222	8450
44	8677	8904	9131	9357	9582	9807	*0032	*0256	*0479	*0702
45	6.1 0925	1147	1368	1589	1810	2030	2249	2468	2687	2905
46	3123	3340	3556	3773	3988	4204	4419	4633	4847	5060
47	5273	5486	5698	5910	6121	6331	6542	6752	6961	7170
48	7379	7587	7794	8002	8208	8415	8621	8826	9032	9236
49	9441	9644	9848	*0051	*0254	*0456	*0658	*0859	*1060	*1261
50	6.2 1461	1661	1860	2059	2258	2456	2654	2851	3048	3245
51	3441	3637	3832	4028	4222	4417	4611	4804	4998	5190
52	5383	5575	5767	5958	6149	6340	6530	6720	6910	7099
53	7288	7476	7664	7852	8040	8227	8413	8600	8786	8972
54	9157	9342	9527	9711	9895	*0079	*0262	*0445	*0628	*0810
55	6.3 0992	1173	1355	1536	1716	1897	2077	2257	2436	2615
56	2794	2972	3150	3328	3505	3683	3859	4036	4212	4388
57	4564	4739	4914	5089	5263	5437	5611	5784	5957	6130
58	6303	6475	6647	6819	6990	7161	7332	7502	7673	7843
59	8012	8182	8351	8519	8688	8856	9024	9192	9359	9526
60	9693	9859	*0026	*0192	*0357	*0523	*0688	*0853	*1017	*1182
N	0	1	2	3	4	5	6	7	8	9

$$\log_e 100 = 4.60517\ 01860$$

NATURAL LOGARITHMS OF NUMBERS—600 TO 1109

N	0	1	.2	3	4	5	6	7	8	9
60	6.3 9693	9859	*0026	*0192	*0357	*0523	*0688	*0853	*1017	*1182
61	6.4 1346	1510	1673	1836	1999	2162	2325	2487	2649	2811
62	2972	3133	3294	3455	3615	3775	3935	4095	4254	4413
63	4572	4731	4889	5047	5205	5362	5520	5677	5834	5990
64	6147	6303	6459	6614	6770	6925	7080	7235	7389	7543
65	7697	7851	8004	8158	8311	8464	8616	8768	8920	9072
66	9224	9375	9527	9677	9828	9979	*0129	*0279	*0429	*0578
67	6.5 0728	0877	1026	1175	1323	1471	1619	1767	1915	2062
68	2209	2356	2503	2649	2796	2942	3088	3233	3379	3524
69	3669	3814	3959	4103	4247	4391	4535	4679	4822	4965
70	5108	5251	5393	5536	5678	5820	5962	6103	6244	6386
71	6526	6667	6808	6948	7088	7228	7368	7508	7647	7786
72	7925	8064	8203	8341	8479	8617	8755	8893	9030	9167
73	9304	9441	9578	9715	9851	9987	*0123	*0259	*0394	*0530
74	6.6 0665	0800	0935	1070	1204	1338	1473	1607	1740	1874
75	2007	2141	2274	2407	2539	2672	2804	2936	3068	3200
76	3332	3463	3595	3726	3857	3988	4118	4249	4379	4509
77	4639	4769	4898	5028	5157	5286	5415	5544	5673	5801
78	5929	6058	6185	6313	6441	6568	6696	6823	6950	7077
79	7203	7330	7456	7582	7708	7834	7960	8085	8211	8336
80	8461	8586	8711	8835	8960	9084	9208	9332	9456	9580
81	9703	9827	9950	*0073	*0196	*0319	*0441	*0564	*0686	*0808
82	6.7 0930	1052	1174	1296	1417	1538	1659	1780	1901	2022
83	2143	2263	2383	2503	2623	2743	2863	2982	3102	3221
84	3340	3459	3578	3697	3815	3934	4052	4170	4288	4406
85	4524	4641	4759	4876	4993	5110	5227	5344	5460	5577
86	5693	5809	5926	6041	6157	6273	6388	6504	6619	6734
87	6849	6964	7079	7194	7308	7422	7537	7651	7765	7878
88	7992	8106	8219	8333	8446	8559	8672	8784	8897	9010
89	9122	9234	9347	9459	9571	9682	9794	9906	*0017	*0128
90	6.8 0239	0351	0461	0572	0683	0793	0904	1014	1124	1235
91	1344	1454	1564	1674	1783	1892	2002	2111	2220	2329
92	2437	2546	2655	2763	2871	2979	3087	3195	3303	3411
93	3518	3626	3733	3841	3948	4055	4162	4268	4375	4482
94	4588	4694	4801	4907	5013	5118	5224	5330	5435	5541
95	5646	5751	5857	5961	6066	6171	6276	6380	6485	6589
96	6693	6797	6901	7005	7109	7213	7316	7420	7523	7626
97	7730	7833	7936	8038	8141	8244	8346	8449	8551	8653
98	8755	8857	8959	9061	9163	9264	9366	9467	9568	9669
99	9770	9871	9972	*0073	*0174	*0274	*0375	*0475	*0575	*0675
100	6.9 0776	0875	0975	1075	1175	1274	1374	1473	1572	1672
101	1771	1870	1968	2067	2166	2264	2363	2461	2560	2658
102	2756	2854	2952	3049	3147	3245	3342	3440	3537	3634
103	3731	3828	3925	4022	4119	4216	4312	4409	4505	4601
104	4698	4794	4890	4986	5081	5177	5273	5368	5464	5559
105	5655	5750	5845	5940	6035	6130	6224	6319	6414	6508
106	6602	6697	6791	6885	6979	7073	7167	7261	7354	7448
107	7541	7635	7728	7821	7915	8008	8101	8193	8286	8379
108	8472	8564	8657	8749	8841	8934	9026	9118	9210	9302
109	9393	9485	9577	9668	9760	9851	9942	*0033	*0125	*0216
110	7.0 0307	0397	0488	0579	0670	0760	0851	0941	1031	1121
N	0	1	2	3	4	5	6	7	8	9

$\log_e 1000 = 6.90775\ 52790$
To find the logarithm of a number which is 10 (or 1/10) times a number whose logarithm is given, add to (or subtract from) the given logarithm the logarithm of 10.

Table XXVII

COMMON LOGARITHMS OF NUMBERS

N	0	1	2	3	4	5	6	7	8	9	Proportional parts 1	2	3	4	5
10	0000	0043	0086	0128	0170	0212	0253	0294	0334	0374	4	8	12	17	21
11	0414	0453	0492	0531	0569	0607	0645	0682	0719	0755	4	8	11	15	19
12	0792	0828	0864	0899	0934	0969	1004	1038	1072	1106	3	7	10	14	17
13	1139	1173	1206	1239	1271	1303	1335	1367	1399	1430	3	6	10	13	16
14	1461	1492	1523	1553	1584	1614	1644	1673	1703	1732	3	6	9	12	15
15	1761	1790	1818	1847	1875	1903	1931	1959	1987	2014	3	6	8	11	14
16	2041	2068	2095	2122	2148	2175	2201	2227	2253	2279	3	5	8	11	13
17	2304	2330	2355	2380	2405	2430	2455	2480	2504	2529	2	5	7	10	12
18	2553	2577	2601	2625	2648	2672	2695	2718	2742	2765	2	5	7	9	12
19	2788	2810	2833	2856	2878	2900	2923	2945	2967	2989	2	4	7	9	11
20	3010	3032	3054	3075	3096	3118	3139	3160	3181	3201	2	4	6	8	11
21	3222	3243	3263	3284	3304	3324	3345	3365	3385	3404	2	4	6	8	10
22	3424	3444	3464	3483	3502	3522	3541	3560	3579	3598	2	4	6	8	10
23	3617	3636	3655	3674	3692	3711	3729	3747	3766	3784	2	4	6	7	9
24	3802	3820	3838	3856	3874	3892	3909	3927	3945	3962	2	4	5	7	9
25	3979	3997	4014	4031	4048	4065	4082	4099	4116	4133	2	4	5	7	9
26	4150	4166	4183	4200	4216	4232	4249	4265	4281	4298	2	3	5	7	8
27	4314	4330	4346	4362	4378	4393	4409	4425	4440	4456	2	3	5	6	8
28	4472	4487	4502	4518	4533	4548	4564	4579	4594	4609	2	3	5	6	8
29	4624	4639	4654	4669	4683	4698	4713	4728	4742	4757	1	3	4	6	7
30	4771	4786	4800	4814	4829	4843	4857	4871	4886	4900	1	3	4	6	7
31	4914	4928	4942	4955	4969	4983	4997	5011	5024	5038	1	3	4	5	7
32	5051	5065	5079	5092	5105	5119	5132	5145	5159	5172	1	3	4	5	7
33	5185	5198	5211	5224	5237	5250	5263	5276	5289	5302	1	3	4	5	7
34	5315	5328	5340	5353	5366	5378	5391	5403	5416	5428	1	2	4	5	6
35	5441	5453	5465	5478	5490	5502	5514	5527	5539	5551	1	2	4	5	6
36	5563	5575	5587	5599	5611	5623	5635	5647	5658	5670	1	2	4	5	6
37	5682	5694	5705	5717	5729	5740	5752	5763	5775	5786	1	2	4	5	6
38	5798	5809	5821	5832	5843	5855	5866	5877	5888	5899	1	2	3	5	6
39	5911	5922	5933	5944	5955	5966	5977	5988	5999	6010	1	2	3	4	5
40	6021	6031	6042	6053	6064	6075	6085	6096	6107	6117	1	2	3	4	5
41	6128	6138	6149	6160	6170	6180	6191	6201	6212	6222	1	2	3	4	5
42	6232	6243	6253	6263	6274	6284	6294	6304	6314	6325	1	2	3	4	5
43	6335	6345	6355	6365	6375	6385	6395	6405	6415	6425	1	2	3	4	5
44	6435	6444	6454	6464	6474	6484	6493	6503	6513	6522	1	2	3	4	5
45	6532	6542	6551	6561	6571	6580	6590	6599	6609	6618	1	2	3	4	5
46	6628	6637	6646	6656	6665	6675	6684	6693	6702	6712	1	2	3	4	5
47	6721	6730	6739	6749	6758	6767	6776	6785	6794	6803	1	2	3	4	5
48	6812	6821	6830	6839	6848	6857	6866	6875	6884	6893	1	2	3	4	5
49	6902	6911	6920	6928	6937	6946	6955	6964	6972	6981	1	2	3	4	4
50	6990	6998	7007	7016	7024	7033	7042	7050	7059	7067	1	2	3	3	4
51	7076	7084	7093	7101	7110	7118	7126	7135	7143	7152	1	2	3	3	4
52	7160	7168	7177	7185	7193	7202	7210	7218	7226	7235	1	2	3	3	4
53	7243	7251	7259	7267	7275	7284	7292	7300	7308	7316	1	2	2	3	4
54	7324	7332	7340	7348	7356	7364	7372	7380	7388	7396	1	2	2	3	4
N	0	1	2	3	4	5	6	7	8	9	1	2	3	4	5

COMMON LOGARITHMS OF NUMBERS

N	0	1	2	3	4	5	6	7	8	9	1	2	3	4	5
												Proportional parts			
55	7404	7412	7419	7427	7435	7443	7451	7459	7466	7474	1	2	2	3	4
56	7482	7490	7497	7505	7513	7520	7528	7536	7543	7551	1	2	2	3	4
57	7559	7566	7574	7582	7589	7597	7604	7612	7619	7627	1	1	2	3	4
58	7634	7642	7649	7657	7664	7672	7679	7686	7694	7701	1	1	2	3	4
59	7709	7716	7723	7731	7738	7745	7752	7760	7767	7774	1	1	2	3	4
60	7782	7789	7796	7803	7810	7818	7825	7832	7839	7846	1	1	2	3	4
61	7853	7860	7868	7875	7882	7889	7896	7903	7910	7917	1	1	2	3	3
62	7924	7931	7938	7945	7952	7959	7966	7973	7980	7987	1	1	2	3	3
63	7993	8000	8007	8014	8021	8028	8035	8041	8048	8055	1	1	2	3	3
64	8062	8069	8075	8082	8089	8096	8102	8109	8116	8122	1	1	2	3	3
65	8129	8136	8142	8149	8156	8162	8169	8176	8182	8189	1	1	2	3	3
66	8195	8202	8209	8215	8222	8228	8235	8241	8248	8254	1	1	2	3	3
67	8261	8267	8274	8280	8287	8293	8299	8306	8312	8319	1	1	2	3	3
68	8325	8331	8338	8344	8351	8357	8363	8370	8376	8382	1	1	2	3	3
69	8388	8395	8401	8407	8414	8420	8426	8432	8439	8445	1	1	2	3	3
70	8451	8457	8463	8470	8476	8482	8488	8494	8500	8506	1	1	2	3	3
71	8513	8519	8525	8531	8537	8543	8549	8555	8561	8567	1	1	2	3	3
72	8573	8579	8585	8591	8597	8603	8609	8615	8621	8627	1	1	2	3	3
73	8633	8639	8645	8651	8657	8663	8669	8675	8681	8686	1	1	2	2	3
74	8692	8698	8704	8710	8716	8722	8727	8733	8739	8745	1	1	2	2	3
75	8751	8756	8762	8768	8774	8779	8785	8791	8797	8802	1	1	2	2	3
76	8808	8814	8820	8825	8831	8837	8842	8848	8854	8859	1	1	2	2	3
77	8865	8871	8876	8882	8887	8893	8899	8904	8910	8915	1	1	2	2	3
78	8921	8927	8932	8938	8943	8949	8954	8960	8965	8971	1	1	2	2	3
79	8976	8982	8987	8993	8998	9004	9009	9015	9020	9025	1	1	2	2	3
80	9031	9036	9042	9047	9053	9058	9063	9069	9074	9079	1	1	2	2	3
81	9085	9090	9096	9101	9106	9112	9117	9122	9128	9133	1	1	2	2	3
82	9138	9143	9149	9154	9159	9165	9170	9175	9180	9186	1	1	2	2	3
83	9191	9196	9201	9206	9212	9217	9222	9227	9232	9238	1	1	2	2	3
84	9243	9248	9253	9258	9263	9269	9274	9279	9284	9289	1	1	2	2	3
85	9294	9299	9304	9309	9315	9320	9325	9330	9335	9340	1	1	2	2	3
86	9345	9350	9355	9360	9365	9370	9375	9380	9385	9390	1	1	2	2	3
87	9395	9400	9405	9410	9415	9420	9425	9430	9435	9440	1	1	2	2	3
88	9445	9450	9455	9460	9465	9469	9474	9479	9484	9489	0	1	1	2	2
89	9494	9499	9504	9509	9513	9518	9523	9528	9533	9538	0	1	1	2	2
90	9542	9547	9552	9557	9562	9566	9571	9576	9581	9586	0	1	1	2	2
91	9590	9595	9600	9605	9609	9614	9619	9624	9628	9633	0	1	1	2	2
92	9638	9643	9647	9652	9657	9661	9666	9671	9675	9680	0	1	1	2	2
93	9685	9689	9694	9699	9703	9708	9713	9717	9722	9727	0	1	1	2	2
94	9731	9736	9741	9745	9750	9754	9759	9763	9768	9773	0	1	1	2	2
95	9777	9782	9786	9791	9795	9800	9805	9809	9814	9818	0	1	1	2	2
96	9823	9827	9832	9836	9841	9845	9850	9854	9859	9863	0	1	1	2	2
97	9868	9872	9877	9881	9886	9890	9894	9899	9903	9908	0	1	1	2	2
98	9912	9917	9921	9926	9930	9934	9939	9943	9948	9952	0	1	1	2	2
99	9956	9961	9965	9969	9974	9978	9983	9987	9991	9996	0	1	1	2	2
N	0	1	2	3	4	5	6	7	8	9	1	2	3	4	5

REFERENCES

Treatises and textbooks

Anderson, T. W., *An Introduction to Multivariate Statistical Analysis*, John Wiley & Sons, Inc., New York, 1958.

Arden, B. W., *An Introduction to Digital Computing*, Addison-Wesley Publishing Company, Inc., Reading, Mass., 1963.

Arley, N., and Buch, K. R., *Introduction to the Theory of Probability and Statistics*, John Wiley & Sons, Inc., New York, 1950.

ASTM Manual of Quality Control of Materials, American Society for Testing and Materials, Philadelphia, Pa., 1951.

Barlow, R. E., and Proschan, F., *Mathematical Theory of Reliability*, John Wiley & Sons, Inc., New York, 1965.

Bazovsky, I., *Reliability Theory and Practice*, Prentice-Hall, Inc., Englewood Cliffs, N.J., 1961.

Bowker, A. H., and Lieberman, G. J., *Engineering Statistics*, Prentice-Hall, Inc., Englewood Cliffs, N.J., 1961.

Brownlee, K. A., *Statistical Theory and Methodology in Science and Engineering*, John Wiley & Sons, Inc., New York, 1960.

Burington, R. S., and Torrance, C. C., *Higher Mathematics*, McGraw-Hill Book Company, New York, 1939.

Burr, I. W., *Engineering Statistics and Quality Control*, McGraw-Hill Book Company, New York, 1953.

Calabro, S. R., *Reliability Principles and Practices*, McGraw-Hill Book Company, New York, 1962.

Cochran, W. G., *Sampling Techniques*, 2d ed., John Wiley & Sons, Inc., New York, 1953.

Cochran, W. G., and Cox, G. M., *Experimental Designs*, 2d ed., John Wiley & Sons, Inc., New York, 1958.

Cox, D. R., *Planning of Experiments*, John Wiley & Sons, Inc., New York, 1958.

Cramer, H., *Mathematical Methods of Statistics*, 6th ed., Princeton University Press, Princeton, N.J., 1964.

Croxton, F. E., and Cowden, D. J., *Applied General Statistics*, Prentice-Hall, Inc., Englewood Cliffs, N.J., 1945.

Deming, W. E., *Some Theory of Sampling*, John Wiley & Sons, Inc., New York, 1950.

Dixon, W. J., and Massey, F. J., Jr., *Introduction to Statistical Analysis*, 2d ed., McGraw-Hill Book Company, 1957.

Doob, J. L., *Stochastic Processes*, John Wiley & Sons, Inc., New York, 1953.

Duncan, A. J., *Quality Control and Industrial Statistics*, Richard D. Irwin, Inc., Homewood, Ill., 1959.

Eisenhart, C., Hastay, M. W., and Wallis, W. A., *Selected Techniques of Statistical Analysis*, McGraw-Hill Book Company, New York, 1947.

Epstein, B., "Sampling Procedures and Tables for Life and Reliability Testing," *Quality Control and Reliability Handbook H-108*, Government Printing Office, Washington, D.C., 1960.

Feller, W., *An Introduction to Probability Theory and Its Applications*, Vols. I, II, John Wiley & Sons, Inc., New York, 1957, 1966.

Fisher, R. A., *Statistical Methods for Research Workers*, 10th ed., Oliver & Boyd Ltd., Edinburgh, 1946.

Fisher, R. A., *The Design of Experiments*, 4th ed., Oliver & Boyd Ltd., Edinburgh, 1946.

Freeman, H. A., Friedman, M., Mosteller, F., and Wallis, W. A., *Sampling Inspection*, McGraw-Hill Book Company, New York, 1948.

Fry, T. C., *Probability and Its Engineering Uses*, D. Van Nostrand Company, Inc., New York, 1929.

Goldman, A. S., and Slattery, T. B., *Maintainability*, John Wiley & Sons, Inc., New York, 1964.

Grant, E. L., *Statistical Quality Control*, 3d ed., McGraw-Hill Book Company, New York, 1964.

Guest, P. G., *Numerical Methods of Curve Fitting*, Cambridge University Press, New York, 1961.

Guttman, I., and Wilks, S. S., *Introductory Engineering Statistics*, John Wiley & Sons, Inc., New York, 1965.

Hald, A., *Statistical Theory with Engineering Applications*, John Wiley & Sons, Inc., New York, 1952.

Hammersley, J. M., and Handscomb, D. C., *Monte Carlo Methods*, John Wiley & Sons, Inc., New York, 1964.

Hemmerle, W. J., *Statistical Computations on a Digital Computer*, Blaisdell Publishing Company, Waltham, Mass., 1967.

Hildebrand, F. B., *Introduction to Numerial Analysis*, McGraw-Hill Book Company, New York, 1956.

Hull, T. E., and Dobell, A. R., "Random Number Generators," *Society for Industrial and Applied Mathematics, Review*, Vol. 4, pp. 230–254, 1962.

IBM Reference Manual, *Random Number Generation and Testing*, Manual C 20–8011, White Plains, N.Y., 1959.

Ireson, W. Grant (ed.), *Reliability Handbook*, McGraw-Hill Book Company, New York, 1966.

Johnson, N. L., and Leone, F. C., *Statistics and Experimental Design*, Vols. I, II, John Wiley & Sons, Inc., New York, 1964.

Kempthorne, O., *The Design and Analysis of Experiments*, John Wiley & Sons, Inc., New York, 1952.

Kendall, M. G., *The Advanced Theory of Statistics*, Vols. I, II, McGraw-Hill Book Company, New York, 1950.

Kenney, J. F., *Mathematics of Statistics*, Parts 1, 2, D. Van Nostrand Company, Inc., Princeton, N.J., 1947, 1941.

Lawley, D. N., and Maxwell, A. E., *Factor Analysis as a Statistical Method*, Butterworth & Co. (Publishers), Ltd., London, 1963.

Lehmann, E. L., *Testing Statistical Hypotheses*, John Wiley & Sons, Inc., New York, 1959.

Lloyd, D. K., and Lipow, M., *Reliability: Management Methods, and Mathematics*, Prentice-Hall, Inc., Englewood Cliffs, N.J., 1962.

Loève, M., *Probability Theory*, 2d ed., D. Van Nostrand Company, Inc., Princeton, N.J., 1960.

Mann, H. B., *Analysis and Design of Experiments*, Dover Publications, Inc., New York, 1949.

Middleton, D., *An Introduction to Statistical Communication Theory*, McGraw-Hill Book Company, New York, 1960.

Milne, W. E., *Numerical Calculus*, Princeton University Press, Princeton, N.J., 1949.

Mood, A. M., and Graybill, F. A., *Introduction to the Theory of Statistics*, 2d ed., McGraw-Hill Book Company, New York, 1963.

Natrella, M. G., "Experimental Statistics," *National Bureau of Standards Handbook 91*, Government Printing Office, Washington, D.C., 1963.

Neyman, J., *First Course in Probability and Statistics*, Henry Holt and Company, Inc., New York, 1950.

Ostle, B., *Statistics in Research*, 2d ed., The Iowa State University Press, Ames, Iowa, 1963.

Parzen, E., *Modern Probability Theory and Its Applications*, John Wiley & Sons, Inc., New York, 1960.

Rice, W. B., *Control Charts in Factory Management*, 2d ed., John Wiley & Sons, Inc., New York, 1947.

Riordan, J., *An Introduction to Combinatorial Analysis*, John Wiley & Sons, Inc., New York, 1958.

Ryser, H. J., *Combinatorial Mathematics*, Mathematical Association of America, 1963.

Sheffé, Henry, *The Analysis of Variance*, John Wiley & Sons, Inc., New York, 1959

Shewhart, W. A., *Economic Control of Manufactured Product*, D. Van Nostrand Company, Inc., Princeton, N.J., 1931.

Simon, L. E., *Engineer's Manual of Statistical Methods*, John Wiley & Sons, Inc., New York, 1941.

Snedecor, G. W., *Statistical Methods*, 5th ed., The Iowa State University Press, Ames, Iowa, 1965.

Tippett, L. H. C., *The Methods of Statistics*, Williams & Norgate, Ltd., London, 1945.

Von Alven, William H., *Reliability Engineering*, ARINC Research Corp., Prentice-Hall, Inc., Englewood Cliffs, N.J., 1964.

Wald, A., *Sequential Analysis*, John Wiley & Sons, Inc., New York, 1947.

Wilks, S. S., *Mathematical Statistics*, Princeton University Press, Princeton, N.J., 1944.

Wilks, S. S., *Mathematical Statistics*, John Wiley & Sons, Inc., New York, 1962.

Yule, G. U., and Kendall, M. G., *An Introduction to the Theory of Statistics*, Charles Griffin & Company, Ltd., London, 1949.

Tables

Burington, R. S., *Handbook of Mathematical Tables and Formulas*, 4th ed., McGraw-Hill Book Company, New York, 1965.

Davis, H. T. (ed.), *Tables of the Higher Mathematical Functions*, Parts I and II, Principia Press, Bloomington, Ind., 1933–1935.

Dodge, H. F., and Romig, H. G., *Sampling Inspection Tables*, John Wiley & Sons, Inc., New York, 1959.

Fisher and Yates, *Statistical Tables for Biological, Agricultural and Medical Research*, 6th ed., Hafner Publishing Company, Inc., New York, 1963.

Freeman, H. A., Friedman, M., Mosteller, F., and Wallis, W. A., *Sampling Inspection Tables*, McGraw-Hill Book Company, New York, 1948.

Hald, A., *Statistical Tables and Formulas*, John Wiley & Sons, Inc., New York, 1952.

Kelley, T. L., *The Kelley Statistical Tables*, Harvard University Press, Cambridge, Mass., 1938.

Kendall, M. G., and Smith, B. B., *Tables of Random Sampling Numbers*, Tracts for Computers, No. 24, Cambridge University Press, New York, 1939.

Military Standard 105 D, *Sampling Procedures and Tables for Inspection by Attributes*, Government Printing Office, Washington, D.C., 1963.

Ministry of Defence, Defence Specification DEF-131, "Sampling Procedures and Tables for Inspection by Attributes," H.M. Stationery Office, London, 1961.

National Bureau of Standards, *Tables of the Binomial Probability Distribution*, Applied Mathematics Series 6, Government Printing Office, Washington, D.C., 1949.

Owen, D. B., *Handbook of Statistical Tables*, Addison-Wesley Publishing Company, Inc., Reading, Mass., 1962.

Pearson, E. S., and Hartley, H. O., *Biometrika Tables for Statisticians*, 3d ed., Vol. I, Cambridge University Press, New York, 1966.

Pearson, K., *Tables for Statisticians and Biometricians*, Biometrika Office, University College, London, Vol. I, 1930; Vol. II, 1931.

Pearson, K. (ed.), *Tables of the Incomplete Beta-Function*, Biometrika Office, University College, London, 1934.

Pearson, K. (ed.), *Tables of the Incomplete Gamma-Function*, Biometrika Office, University College, London, 1934.

The Rand Corporation, *A Million Random Digits with 100,000 Normal Deviates*, The Free Press of Glencoe, Illinois, 1955.

Robertson, W. H., *Tables of the Binomial Distribution Function for Small Values of p*, Sandia Corporation Monograph SCR-143, Albuquerque, N.M., January, 1960.

Romig, H. G., *50–100 Binomial Tables*, John Wiley & Sons, Inc., New York, 1953.

Sheppard, W. F., *Tables of the Probability Integral*, British Association Mathematical Tables, Vol. 7, Cambridge University Press, New York, 1939.

Walsh, J. E., *Handbook of Nonparametric Statistics*, Vols. I, II, D. Van Nostrand Company, Inc., Princeton, N.J., 1965.

General guides to tables

Fletcher, A., Miller, J. C. P., Rosenfield, L., and Comrie, L. J., *An Index of Mathematical Tables*, 2d ed. (in 2 vols.), Addison-Wesley Publishing Company, Inc., Reading, Mass., 1962.

Greenwood, J. A., and Hartley, H. O., *Guide to Tables in Mathematical Statistics*, Princeton University Press, Princeton, N.J., 1962.

National Bureau of Standards, *Applied Mathematics Series*, Government Printing Office, Washington, D.C., (There are many comprehensive mathematical and statistical tables in this series.)

GLOSSARY OF SYMBOLS

Symbol	Name or meaning

Algebra

$\|a\|$	absolute value of a
$n!$	n factorial
$0!$	zero factorial, $0! = 1$
$\binom{n}{r} = \dfrac{n!}{(n-r)!r!}$	binomial coefficient; $\binom{n}{r}$ is coefficient of $a^{n-r}b^r$ in expansion of $(a+b)^n$
$\sqrt[n]{a} = a^{1/n}$	the positive nth root of a, $a > 0$
$\sqrt[q]{a^p} = a^{p/q}$	the positive qth root of a^p, $a > 0$
$\sum\limits_{i=1}^{n} a_i$	$\sum\limits_{i=1}^{n} a_i = a_1 + a_2 + \cdots + a_n$

Sets

$\in$	belongs to; is an element of; is a member of
$\notin$	does not belong to; is not an element of; is not a member of
$\supset$	contains
$\subseteq$	is included in; is a subset of
$\subset$	is a proper subset of; is properly included in; *inclusion*
$\not\subset$	is not a proper subset of; is not included in
$\cup$	union; cup; join; logical sum
$\cap$	intersection; cap; logical product
A' or $C_U A$	complement of A relative to universe U
$\emptyset$	the null set; the empty set
$=$	is equal to; equals; is identical with; is equivalent to
$\neq$	is not equal to; not equal
$\approx$ or $\leftrightarrow$	is equivalent to
$\not\approx$	is not equivalent to
$\{\ \}$	used to represent a set
$\{x\}$	the set of objects x
$\mid$ or $:$	such that; for which
$\{x \mid x \text{ has property } p\}$	the set of objects x having the property p
$\{x \mid \ \}$	a set builder, that is, a set consisting of all x such that
$[a,b]$	the set $a \leqq x \leqq b$
$[a,b)$	the set $a \leqq x < b$

$(a,b]$	the set $a < x \leqq b$
(a,b)	the set $a < x < b$
$\langle x, y \rangle$	ordered pair
$A \times B$	cartesian product of A and B, A cross B
$\cup A_r$ or $\displaystyle\bigcup_{r=1}^{n} A_r = A_1 \cup A_2 \cup \cdots \cup A_n$	
$\cap A_r$ or $\displaystyle\bigcap_{r=1}^{n} A_r = A_1 \cap A_2 \cap \cdots \cap A_n$	
$\aleph_0$	*cardinal number* of set of all natural numbers 1, 2, 3, $\ldots$; aleph null
$\aleph$	*cardinal number* of set of all real numbers; aleph

Logic

$\wedge$, $\&$	and; *conjunction*
$\vee$	inclusive or; *inclusive disjunction*
$\underline{\vee}$	exclusive or; *exclusive disjunction*
$\rightarrow$ or $\Rightarrow$	implies; *implication*
$\leftrightarrow$ or $\Leftrightarrow$	is equivalent to; *equivalence*
$\sim$ or $'$	not; *negation*
V	for all, for every; *universal quantifier*
V_x	for all x
$\exists$	there exists at least one; for some; *existential quantifier*
$\exists_x$	there exists an x such that; for some x
$\exists\vert$	there exists uniquely
$\mathrm{E}!v$	there exists exactly one v such that; *uniqueness quantifier*
$\vdash.$	it is asserted that

Relations

R	has the *relation R* to
$\not{R}$ or R'	does not have the relation R to
$\check{R}$	the *converse* of relation R
$\approx$ or $=$	is identical (or equivalent) to; *identity* or *equivalence* relation
$\not\approx$ or $\neq$	is not identical (or not equivalent) to; *diversity relation*
$\subset$	is included in
R^{-1}	the *inverse* of relation R
$x\,R\,y$	x stands in the (binary) relation R to y

$x \not\mathrel{R} y$	x does not stand in the (binary) relation R to y
$\leftrightarrow$	one-to-one reciprocal correspondence
$\rightarrow$ or $\curvearrowright$	corresponds to; gives; is transformed into; has for an image
$\xrightarrow{T}$ or $\underset{\curvearrowright}{T}$	passage by means of T; is transformed by T into
$F: S \rightarrow T$	F maps S into T

Functions

$f(x)$	*value* of function f at x
f^{-1}	*inverse* of function f
$f \circ g$	*composite* of function f and function g
$f^{-1} \circ f = f \circ f^{-1} = E$	
$f: (x, y)$	the function f whose ordered pairs are (x, y)

Probabilities

$P(E)$	probability of E
$P(E \cap F)$ or $P(E \wedge F)$	probability of both E and F
$P(E \mid F)$	probability of E, given F
$P(E \cup F)$ or $P(E \vee F)$	probability of E or F or both
$P(E \veebar F)$	probability of E or F, but not both E and F

Algebraic Structures

$\circ, \odot, \oplus, \otimes, \star,$ $\cup, \cap, +, -$	symbols for *operations*
z or 0	zero
u or 1	unity
a^{-1}	*inverse* of a
$\oslash$	symbol used in ordering
$\oslash$	symbol used in ordering

Vectors

$V_1 \cdot V_2$	scalar product
$V_1 \times V_2$	vector product
∇S, grad S	gradient of S
div V, $\nabla \cdot V$	divergence of V

curl V, rot V,
$\quad \nabla \times V$ curl of V
$\nabla^2 S$, div grad S divergence of grad S

Matrices

$\begin{pmatrix} a_{11} \cdots a_{1n} \\ \cdots \cdots \cdots \\ a_{n1} \cdots a_{nn} \end{pmatrix}$ or $[a_{rs}]$ or (a_{ij}) or A matrix

$(x_1, x_2, \ldots, x_n)$ or $\overrightarrow{x_p}$ or $\|x_1, x_2, \ldots, x_n\|$ row matrix, row vector

$\begin{Bmatrix} y_1 \\ \cdots \\ y_n \end{Bmatrix}$ or $y_p{\uparrow}$ or $\left\| \begin{matrix} y_1 \\ \cdots \\ y_n \end{matrix} \right\|$ column matrix, column vector

$\det A = |A| = \begin{vmatrix} a_{11} \cdots a_{1n} \\ \cdots \cdots \cdots \\ a_{n1} \cdots a_{nn} \end{vmatrix}$ *determinate* of matrix A

INDEX OF SYMBOLS

INDEX OF GREEK SYMBOLS

INDEX OF NUMERICAL TABLES